GEORGE ALBERT BOULENGER

CATALOGUE
OF THE
SNAKES
IN THE
BRITISH MUSEUM
(NATURAL HISTORY)

VOLUME II

Elibron Classics
www.elibron.com

Elibron Classics series.

© 2005 Adamant Media Corporation.

ISBN 1-4021-8518-9 (paperback)
ISBN 1-4021-0858-3 (hardcover)

This Elibron Classics Replica Edition is an unabridged facsimile
of the edition published in 1894 by Order of the Trustees,
London.

Elibron and Elibron Classics are trademarks of
Adamant Media Corporation. All rights reserved.

This book is an accurate reproduction of the original. Any marks, names, colophons, imprints, logos or other symbols or identifiers that appear on or in this book, except for those of Adamant Media Corporation and BookSurge, LLC, are used only for historical reference and accuracy and are not meant to designate origin or imply any sponsorship by or license from any third party.

CATALOGUE

OF THE

SNAKES

IN THE

BRITISH MUSEUM
(NATURAL HISTORY).

VOLUME II.,
CONTAINING THE CONCLUSION OF THE
COLUBRIDÆ AGLYPHÆ.

BY

GEORGE ALBERT BOULENGER, F.R.S.

LONDON:
PRINTED BY ORDER OF THE TRUSTEES.
SOLD BY
LONGMANS & Co., 39 PATERNOSTER ROW;
B. QUARITCH, 15 PICCADILLY; DULAU & Co., 37 SOHO SQUARE, W.;
KEGAN PAUL & Co., PATERNOSTER HOUSE, CHARING CROSS ROAD;
AND AT THE
BRITISH MUSEUM (NATURAL HISTORY), CROMWELL ROAD, S.W.
1894.

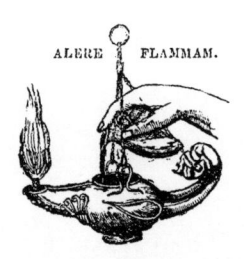

PRINTED BY TAYLOR AND FRANCIS,
RED LION COURT, FLEET STREET.

PREFACE.

The present volume concludes the account of the Aglyphodont Colubrine Snakes.

The collection in the British Museum has received by gifts several important additions since the publication of the first volume. Messrs. Godman and Salvin have deposited a complete series of the species obtained by their collectors and described in their great work on the Fauna of Central America. The contributions to the African Fauna have been scarcely less numerous, the Trustees being indebted to Professor Barboza du Bocage for many specimens of East-African species described by him; to H. H. Johnston, Esq., C.B., F. J. Jackson, Esq., and others for species collected in various parts of British Central Africa; and to Dr. J. Anderson, F.R.S., for a remarkably interesting collection of Egyptian Snakes. The collections made by E. W. Oates, Esq., in Upper Burmah contained also a fair proportion of Snakes which were of value as illustrating the Fauna of this previously little explored region.

ALBERT GÜNTHER,
Keeper of the Department of Zoology.

British Museum (N. H.),
September 23, 1894.

INTRODUCTION.

In dealing with the multitude of forms of Aglyphodont Colubroids, all of comparatively simple structure, the lack of prominent characters on which to establish genera is seriously felt, and, but for the study of the dentition, satisfactory definitions could not be drawn up. I have therefore, as in the previous volume, supplied numerous outline-figures of jaws to facilitate the understanding of the diagnoses.

I have endeavoured to avoid both extremes of excessive fusion or separation in my attempt to render genera approximately equivalent to those of other groups of Reptiles.

Descriptions of 427 species are given. Of these, 347 are represented in the Collection of the Museum, the number of specimens amounting to 2528.

G. A. BOULENGER.

Zoological Department,
July 28th, 1894.

SYSTEMATIC INDEX.

OPHIDIA.

Fam. 7. COLUBRIDÆ.

Subfam. 2. COLUBRINÆ
(*continued*).

	Page
56. Xenelaphis, *Gthr.*	7
1. hexagonotus, *Cant.*	8
57. Drymobius, *Cope*	8
1. bifossatus, *Raddi*	10
2. boddaertii, *Sentz.*	11, 357
3. affinis, *Stdr.*	14
4. rhombifer, *Gthr.*	14, 357
5. bivittatus, *D. & B.*	15
6. dendrophis, *Schleg.*	15, 357
7. chloroticus, *Cope*	16
8. margaritiferus, *Schleg.*	17
58. Phrynonax, *Cope*	18
1. sulphureus, *Wagl.*	19
2. pœcilonotus, *Gthr.*	20
3. guentheri, *Blgr.*	20
4. lunulatus, *Cope*	21
5. fasciatus, *Ptrs*	21
6. eutropis, *Blgr.*	22
7. chrysobronchus, *Cope*	22
59. Spilotes, *Wagl.*	23
1. pullatus, *L.*	23
2. megalolepis, *Gthr.*	24
60. Coluber, *L.*	24
1. dichrous, *Ptrs.*	30
2. corais, *Boie*	31
3. novæ-hispaniæ, *Gm.*	33
4. ? melanotropis, *Cope*	33
5. porphyraceus, *Cant.*	34
6. cantoris, *Blgr.*	35
7. hodgsonii, *Gthr.*	35
8. helena, *Daud.*	36, 357
9. triaspis, *Cope*	37
10. chlorosoma, *Gthr.*	38
11. flavirufus, *Cope*	39
12. guttatus, *L.*	39

	Page
13. leopardinus, *Bp.*	41
14. hohenackeri, *Strauch*	42
15. mandarinus, *Cant.*	42
16. rufodorsatus, *Cant.*	43
17. dione, *Pall.*	44
18. quatuorlineatus, *Lacép.*	45
19. tæniurus, *Cope*	47
20. schrenckii, *Strauch*	48
21. vulpinus, *B. & G.*	49
22. lætus, *B. & G.*	49
23. obsoletus, *Say*	50
24. conspicillatus, *Boie*	51
25. longissimus, *Laur.*	52, 357
26. climacophorus, *Boie*	54
27. phyllophis, *Blgr.*	55
28. davidi, *Sauv.*	56
29. moellendorffii, *Bttgr.*	56
30. oxycephalus, *Boie*	56, 357
31. janseni, *Blkr.*	57, 357
32. frenatus, *Gray*	58
33. prasinus, *Blyth*	59
34. quadrivirgatus, *Boie*	59
35. melanurus, *Schleg.*	60
36. radiatus, *Schleg.*	61
37. erythrurus, *D. & B.*	62, 358
38. enganensis, *Vincig.*	63
39. subradiatus, *Schleg.*	64
40. lineaticollis, *Cope*	64
41. scalaris, *Schinz*	65
42. arizonæ, *Blgr.*	66
43. deppii, *D. & B.*	66
44. catenifer, *Blainv.*	67
45. melanoleucus, *Daud.*	68
61. Synchalinus, *Cope*	70
1. corallioides, *Cope*	70
62. Gonyophis, *Blgr.*	70
1. margaritatus, *Ptrs.*	71
63. Herpetodryas, *Boie*	71
1. sexcarinatus, *Wagl.*	72
2. carinatus, *L.*	73
3. fuscus, *L.*	75

SYSTEMATIC INDEX.

	Page
4. melas, *Cope*	76
5. grandisquamis, *Ptrs*.	76
64. Dendrophis, *Boie*	77
1. pictus, *Boie*	78, 358
2. bifrenalis, *Blgr*.	80, 358
3. calligaster, *Gthr.*	80
4. punctulatus, *Gray*	82
5. grandoculis, *Blgr.*	84
6. formosus, *Boie*	84
7. caudolineolatus, *Gthr.*.	85
8. lineolatus, *Hombr. & Jacq.*	85
9. gastrostictus, *Blgr.*	86
65. Dendrelaphis, *Blgr.*	87
1. tristis, *Daud.*	88, 358
2. subocularis, *Blgr.*	89
3. caudolineatus, *Gray*	89, 358
4. terrificus, *Ptrs.*	90
5. modestus, *Blgr.*	91
66. Chlorophis, *Hallow.*	91
1. emini, *Gthr.*	92
2. ornatus, *Bocage*	93
3. hoplogaster, *Gthr.*	93
4. neglectus, *Ptrs.*	94
5. natalensis, *Smith*	94
6. angolensis, *Bocage*.	95
7. heterolepidotus, *Gthr.*	95, 358
8. irregularis,*Leach*	96
9. heterodermus,*Hallow.*	97, 358
? subcarinatus, *Jan*	91
67. Philothamnus, *Smith*.	98
1. semivariegatus, *Smith*.	99
2. nitidus, *Gthr*.	100
3. dorsalis, *Bocage*	101
4. thomensis, *Bocage*.	101
5. girardi, *Bocage*	102
68. Gastropyxis, *Cope*	102
1. smaragdina, *Schleg.*	103
69. Hapsidophrys, *Fisch.*	103
1. lineata, *Fisch.*	104
70. Thrasops, *Hallow.*	104
1. flavigularis,*Hallow.*105, 358
71. Leptophis, *Bell*	105
1. æruginosus, *Cope*	107
2. depressirostris, *Cope* .	107
3. modestus, *Gthr.*	107
4. mexicanus, *D. & B.*	108
5. cupreus, *Cope*	109
6. saturatus, *Cope*	110
7. diplotropis, *Gthr.*	110
8. bilineatus, *Gthr.*	111, 358
9. occidentalis, *Gthr.*	111
10. nigromarginatus, *Gthr.*	112

	Page
11. liocercus, *Wied*	113
12. ortonii, *Cope*	114
13. urostictus, *Ptrs.*	114
72. Uromacer, *D. & B.*	115
1. catesbyi, *Schl.*	115
2. frenatus, *Gthr.*	116
3. oxyrhynchus, *D. & B.*	116
73. Hypsirhynchus, *Gthr.*	117
1. ferox, *Gthr.*	117
74. Dromicus, *Bibr.*	118
1. chamissonis, *Wgm.*	119
2. angulifer, *Bibr.*	120
3. ater, *Gosse*	121
4. sanctæ-crucis, *Cope*	122
5. antillensis, *Schleg.*	123
6. leucomelas, *D. & B.*	123
7. rufiventris, *D. & B.*	124
8. anomalus, *Ptrs.*	125
9. exiguus, *Cope.*	126
75. Liophis, *Wagl.*	126
1. pygmæus, *Cope*	129
2. triscalis, *L.*	129
3. tæniurus, *Tsch.*	130
4. albiventris, *Jan*	130
5. fraseri, *Blgr.*	131
6. pœcilogyrus, *Wied*	131
7. perfuscus, *Cope*	133
8. melanotus, *Shaw*	134
9. almadensis, *Wagl.*	134
10. viridis, *Gthr.*	135
11. typhlus, *L.*	136
12. epinephelus, *Cope*	137
13. reginæ, *L.*	137
14. juliæ, *Cope*	139
15. cursor, *Lacép.*	139
16. andreæ, *R. & L.*	140
17. parvifrons, *Cope*	141
18. melanostigma, *Wagl.*	142
19. callilænus, *Gosse*	142
20. temporalis, *Cope*	143
21. flavilatus, *Cope*	143
76. Cyclagras, *Cope*	144
1. gigas, *D. & B.*	144
77. Xenodon, *Boie*	144
1. colubrinus, *Gthr.*	146, 359
2. suspectus, *Cope*	147
3. guentheri, *Blgr.*	147
4. neuwiedii, *Gthr.*	148
5. severus, *L.*	149
6. merremii, *Wagl.*	150
78. Lystrophis, *Cope.*	151
1. dorbignyi, *D. & B.*	151
2. histricus, *Jan*	152
3. semicinctus, *D. & B.*	153
79. Heterodon, *Latr.*	153
1. platyrhinus, *Latr.*	154

SYSTEMATIC INDEX.

2. simus, *L*............ 158
3. nasicus, *B. & G.* 156
80. Aporophis, *Cope* 157
   1. lineatus, *L.*........... 158
   2. flavirenatus, *Cope* .. 158
   3. coralliventris, *Blgr.* ... 159
   4. amœnus, *Jan* 160
81. Rhadinæa, *Cope* 160
   1. leucogaster,' *Jan* 163
   2. breviceps, *Cope* 164
   3. calligaster, *Cope*...... 164
   4. mimus, *Cope* 164
   5. anomala, *Gthr.* 165
   6. sagittifera, *Jan* 165
   7. cobella, *L.* 166
   8. purpurans, *D. & B*.... 167
   9. merremii, *Wied* 168
   10. fusca, *Cope* 169
   11. jægeri, *Gthr.* 170
   12. genimaculata, *Bttgr*... 170
   13. obtusa,' *Cope* 171
   14. serperastra, *Cope* 172
   15. affinis, *Gthr.* 172
   16. pœcilopogon, *Cope* .. 173
   17. lachrymans, *Cope* 174
   18. undulata, *Wied* 174
   19. melanauchen, *Jan* 175
   20. occipitalis, *Jan* 175
   21. decorata, *Gthr.* .. 176, 359
   22. vermiculaticeps, *Cope* . 177
   23. clavata, *Ptrs.* 177
   24. vittata, *Jan* 178
   25. laureata, *Gthr.* 179
   26. godmani, *Gthr.* 179
82. Urotheca, *Bibr.* 180
   1. dumerilii, *Bibr.* 181
   2. lateristriga, *Berth.*..... 181
   3. euryzona, *Cope* 182
   4. elapoides, *Cope* .. 182, 359
   5. bicincta, *Herm.*....... 184
83. Trimetopon, *Cope* 184
   1. gracile, *Gthr.*.......... 184
84. Hydromorphus, *Ptrs.* 185
   1. concolor, *Ptrs.* 185
85. Dimades, *Gray*.......... 185
   1. plicatilis, *L.*........... 186
86. Hydrops, *Wagl.* 186
   1. triangularis, *Wagl.* 187, 359
   2. martii, *Wagl.*......... 187
87. Sympholis, *Cope* 188
   1. lippiens, *Cope*........ 188
88. Coronella, *Laur.*......... 188
   1. austriaca, *Laur.* 191
   2. amaliæ, *Bttgr.* .. 193, 359
   3. girondica, *Daud.* 194

4. semiornata, *Ptrs.* 195, 359
5. coronata, *Schleg.*....... 196
6. regularis, *Fisch.*....... 196
7. getula, *L.* 197
8. calligaster, *Harl.* 198
9. leonis, *Gthr.* 199
10. triangulum, *Daud.*..... 200
11. gentilis, *B. & G.*...... 201
12. zonata, *Blainv.* 202
13. micropholis, *Cope* 203
14. doliata, *L.* 205
15. brachyura, *Gthr.* 206
16. punctata, *L.* 206
17. amabilis, *B. & G.* 207
18. regalis, *B. & G.*...... 208
89. Hypsiglena, *Cope*........ 208
   1. ochrorhynchus, *Cope*.. 209
   2. torquata, *Gthr.* ... 210, 359
   3. affinis, *Blgr.* 210
   4. discolor, *Gthr.* 211
   5. latifasciata, *Gthr.* 211
   6. ornata, *Bocourt* 211
90. Rhinochilus, *B. & G.* 212
   1. lecontii, *B. & G.* 212
   2. thominoti, *Bocourt*.... 213
   3. antonii, *Dugès* 213
91. Cemophora, *Cope*........ 213
   1. coccinea, *Blumenb.* .. 214
92. Simotes, *D. & B.*......... 214
   1. splendidus, *Gthr.* 217
   2. purpurascens, *Schleg*... 218
   3. cyclurus, *Cant.* 219
   4. albocinctus, *Cant.* 220
   5. formosanus, *Gthr.* 222, 359
   6. violaceus, *Cant.*...... 222
   7. woodmasoni, *W. Scl.*.. 223
   8. octolineatus, *Schn.*..... 224
   9. phænochalinus, *Cope*.. 225
   10. forbesii, *Blgr.*......... 225
   11. signatus, *Gthr.* 226
   12. subcarinatus, *Gthr.* .. 226
   13. annulifer, *Blgr.*....... 226
   14. tæniatus, *Gthr.* 227
   15. chinensis, *Gthr.* 228
   16. vaillanti, *Sauv.* 228
   17. beddomii, *Blgr.* 229
   18. arnensis, *Shaw* .. 229, 359
   19. theobaldi, *Gthr.*....... 230
   20. cruentatus, *Gthr.* 231
   21. torquatus, *Blgr.*....... 232
   22. planiceps, *Blgr.* 232
93. Oligodon, *Boie* 233
   1. venustus, *Jerd.* 235
   2. travancoricus, *Bedd.*... 236
   3. affinis, *Gthr.* 237

SYSTEMATIC INDEX.

	Page
4. bitorquatus, *Boie*	237
5. trilineatus, *D. & B.*	238
6. modestus, *Gthr.*	238
7. notospilus, *Gthr.*	239
8. everetti, *Blgr.*	239
9. propinquus, *Jan*	240
10. brevicauda, *Gthr.*	240
11. dorsalis, *Gray*	241
12. templetonii, *Gthr.*	241, 359
13. sublineatus, *D. & B.*	242
14. ellioti, *Gthr.*	242
15. subgriseus, *D. & B.*	243
16. vertebralis, *Gthr.*	245
16 *a*. tæniurus, *F. Müll.*	360
17. waandersii, *Blkr.*	245
18. melanocephalus, *Jan*	246
dorsalis, *Berth.*	234
94. Prosymna, *Gray*	246
1. sundevallii, *Smith*	247
2. frontalis, *Ptrs.*	248
3. ambigua, *Bocage*	248
4. meleagris, *Rhdt.*	249
5. jani, *Bianc.*	249
95. Leptocalamus, *Gthr.*	249
1. torquatus, *Gthr.*	250
2. sumichrasti, *Bocourt*	250
3. sclateri, *Blgr.*	251
96. Arrhyton, *Gthr.*	251
1. tæniatum, *Gthr.*	252
2. vittatum, *Gundl. & Ptrs.*	252
3. redimitum, *Cope*	252
97. Simophis, *Ptrs.*	253
1. rhinostoma, *Schleg.*	253
2. rohdii, *Bttgr.*	254
98. Scaphiophis, *Ptrs.*	254
1. albopunctatus, *Ptrs.*	254
99. Contia, *B. & G.*	255
1. æstiva, *L.*	258
2. vernalis, *Harl.*	258
3. agassizii, *Jan*	259
4. decemlineata, *D. & B.*	260
5. collaris, *Mén.*	260
6. fasciata, *Jan*	262
7. angusticeps, *Blgr.*	262
8. rothi, *Jan*	262
9. persica, *And.*	263
10. walteri, *Bttgr.*	263
11. coronella, *Schleg.*	264
12. taylori, *Blgr.*	265
13. episcopa, *Kenn.*	265
14. torquata, *Cope*	266
15. isozona, *Cope*	266
16. occipitalis, *Hallow.*	266
17. pachyura, *Cope*	267
18. mitis, *B. & G.*	267
19. semiannulata, *B. & G.*	268
20. nasus, *Gthr.*	268
lineata, *Kenn.*	269
21. frontalis, *Cope*	270
100. Ficimia, *Gray*	270
1. olivacea, *Gray*	271
2. cana, *Cope*	272
3. quadrangularis, *Gthr.*	272
101. Chilomeniscus, *Cope*	272
1. stramineus, *Cope*	273
2. ephippicus, *Cope*	273
102. Homalosoma, *Wagl.*	273
1. lutrix, *L.*	274
2. shiranum, *Blgr.*	276
3. abyssinicum, *Blgr.*	276
4. variegatum, *Ptrs.*	276
103. Ablabes, *D. & B.*	277
1. semicarinatus, *Hallow.*	278
2. major, *Gthr.*	279
3. doriæ, *Blgr.*	279
4. frenatus, *Gthr.*	280
5. stoliczkæ, *W. Scl.*	281
6. tricolor, *Schleg.*	281
7. calamaria, *Gthr.*	282
8. rappii, *Gthr.*	282
9. baliodirus, *Boie*	283
10. scriptus, *Theob.*	284
11. longicauda, *Ptrs.*	284
bipunctatus, *Jeude*	285
12. nicobariensis, *Stol.*	285
104. Grayia, *Gthr.*	286
1. smythii, *Leach*	286
2. furcata, *Mocq.*	287
3. giardi, *Dollo*	288
105. Xenurophis, *Gthr.*	288
1. cæsar, *Gthr.*	288
106. Virginia, *B. & G.*	288
1. valeriæ, *B. & G.*	289
2. elegans, *Kenn.*	289
107. Abastor, *Gray*	289
1. erythrogrammus, *Daud.*	290
108. Farancia, *Gray*	290
1. abacura, *Holbr.*	291
109. Petalognathus, *D. & B.*	292
1. nebulatus, *L.*	293
110. Tropidodipsas, *Gthr.*	294
1. fasciata, *Gthr.*	295
2. philippii, *Jan*	295
3. fischeri, *Blgr.*	296
4. sartorii, *Cope*	296
5. annulifera, *Blgr.*	297
6. anthracops, *Cope*	297
111. Dirosema, *Blgr.*	298
1. bicolor, *Gthr.*	298

SYSTEMATIC INDEX.

2. omiltemanum, *Gthr.* 299
3. psephotum, *Cope* 299
4. brachycephalum, *Cope* 299
112. Atractus, *Wagl.* 300
 1. elaps, *Gthr.* 302
 2. latifrons, *Gthr.* 303
 3. modestus, *Blgr.* 304
 4. vittatus, *Blgr.* 304
 5. latifrontalis, *Garm...* 304
 6. longiceps, *Cope* 305
 7. peruvianus, *Jan* 305
 8. guentheri, *Wuch.* .. 305
 9. bocourti, *Blgr.* 306
 10. maculatus, *Gthr.* 306
 11. major, *Blgr.* 307
 12. isthmicus, *Blgr.* 307
 13. badius, *Boie* 308
 14. torquatus, *D. & B.* ... 309
 15. crassicaudatus, *D.&B.* 310
 16. duboisi, *Blgr.* 310
 17. occipitoalbus, *Jan* .. 310
 18. reticulatus, *Blgr.* .. 311
 19. emmeli, *Bttgr.* 311
 20. trilineatus, *Wagl.* .. 312
 21. quadrivirgatus, *Jan* 312
 22. favæ, *Filippi* 313
113. Geophis, *Wagl.* 314
 1. poeppigii, *Jan* 316
 2. semidoliatus, *D. & B.* 316
 3. rhodogaster, *Cope* .. 317
 4. dugesii, *Bocourt* 317
 5. sallæi, *Blgr.* 318
 6. chalybæus, *Wagl.* .. 318
 7. hoffmanni, *Ptrs.* 319
 8. dolichocephalus, *Cope* 320
 9. petersii, *Blgr.* 321
 10. championi, *Blgr.* 321
 11. godmani, *Blgr.* 322
 12. dubius, *Ptrs.* 322
 13. rostralis, *Jan* 323
113*a*. Agrophis, *F. Müll.* .. 360
 1. sarasinorum, *F. Müll.* 360
114. Carphophis, *Gerv.* 324
 1. amœnus, *Say* 324
115. Stilosoma, *A. E. Brown* 325
 1. extenuatum,*Brown*.. 325
116. Geagras, *Cope* 326
 1. redimitus, *Cope* 326
117. Macrocalamus, *Gthr.* .. 327
 1. lateralis, *Gthr.* 327
118. Idiopholis, *Mocq.* 327
 1. collaris, *Mocq.* 327

119. Rhabdophidium, *Blgr*... 328
 1. forsteni, *D. & B.* .. 328
120. Pseudorhabdium, *Jan* .. 328
 1. longiceps, *Cant.* 329
 2. oxycephalum, *Gthr.* 329
121. Calamaria, *Boie.* 330
 1. lumbricoidea, *Boie* .. 333
 2. vermiformis, *D. & B.* 333
 3. stahlknechtii, *Stol.* .. 335
 4. baluensis, *Blgr.* 335
 5. grabowskii, *Fisch.* .. 336
 6. albiventer, *Gray* 336
 7. margaritifera, *Blkr...* 336
  hoevenii, *Edel.* 337
 8. prakkii, *Jeude* 337
 9. grayi, *Gthr.* 338
 10. bitorques, *Ptrs.* 338
 11. gervaisii, *D. & B.* .. 338
 12. sumatrana, *Edel.* 339
 13. everetti, *Blgr.* 340
 14. virgulata, *Boie* 340
 15. leucogaster, *Blkr.* .. 341
 16. occipitalis, *Jan* 342
 17. bicolor, *D. & B.* 342
 18. lateralis, *Mocq.* 342
 19. beccarii, *Ptrs.* 343
 20. rebentischii, *Blkr.* .. 343
 21. agamensis, *Blkr.* 343
 22. leucocephala, *D. & B.* 344
 23. schlegelii, *D. & B.* ... 345
 24. linnæi, *Boie* 345
 25. borneensis, *Blkr.* 347
 26. benjaminsii, *Edel.* .. 347
 27. javanica, *Blgr.* 347
 28. brevis, *Blgr.* 348
 29. pavimentata, *D. & B.* 348
 30. septentrionalis, *Blgr.* 349
 31. melanota, *Jan* 349
 32. lovii, *Blgr.* 350
 33. gracillima, *Gthr.* 350
  catenata, *Blyth* 351
122. Typhlogeophis, *Gthr.* .. 351
 1. brevis, *Gthr.* 351
  Amastridium, *Cope*.... 352
  veliferum, *Cope* 352
  Anoplophallus, *Cope* .. 353
  maculatus, *Hallow...* 353

Subfam. 3. RHACHIODONTINÆ.

123. Dasypeltis, *Wagl.* 353
 1. scabra, *L.* 354

CATALOGUE

OF

SNAKES.

Fam. 7. COLUBRIDÆ.

Subfam. 2. *COLUBRINÆ.*
(Continued.)

Synopsis of the Genera.
(Continued from Vol. I. p. 180.)

II. Hypapophyses absent in the posterior dorsal vertebræ, the lower surface of which is smooth or with a low keel.

    A. Posterior maxillary teeth increasing in size; anterior maxillary and mandibular teeth strongly enlarged; pupil vertically elliptic.

Anterior maxillary teeth separated from the rest by an interspace.
48. **Lycodon**, I. p. 348.

Two or three middle maxillary teeth small, separated from the anterior and posterior by distinct interspaces.
49. **Dinodon**, I. p. 360.

    B. Anterior maxillary teeth not enlarged, middle and posterior longest; pupil vertically elliptic.
50. **Stegonotus**, I. p. 364.

    C. Anterior and middle maxillary teeth not enlarged, posterior increasing in size or the last strongly enlarged.

        1. Palatine and pterygoid bones toothed; pupil vertically elliptic.

Maxillary teeth 8 to 10; scales with apical pits.
51. **Dryocalamus**, I. p. 369.

Maxillary teeth 6 to 9; scales without pits; snout cuneiform, projecting.................. 55. **Lytorhynchus**, I. p. 414.

Maxillary teeth 12 to 16; scales with apical pits.
89. **Hypsiglena**, II. p. 208.

2. Palatine and pterygoid bones toothed; pupil round; anterior mandibular teeth much larger than the posterior; scales with apical pits.

    a. Internasal shield entering the nostril; maxillary teeth 12 to 14 .... 52. **Pseudaspis**, I. p. 373.

    b. Internasal not entering the nostril.

        α. Lateral scales disposed obliquely, at least on the anterior part of the body; eyes large.

Maxillary teeth 15 to 20, the posterior slightly enlarged; scales in 21 to 25 rows .... 58. **Phrynonax**, II. p. 18.

Maxillary teeth 20 to 32, the posterior strongly enlarged; scales in 13 or 15 rows .... 71. **Leptophis**, II. p. 105.

        β. Scales forming longitudinal series; maxillary teeth 11 to 20.

Posterior maxillary teeth strongly enlarged and separated from the rest by an interspace; scales narrow, lanceolate.
72. **Uromacer**, II. p. 115.

Posterior maxillary teeth slightly enlarged.
73. **Hypsirhynchus**, II. p. 117.

Posterior maxillary teeth strongly enlarged and separated from the rest by an interspace; scales moderately elongate.
74. **Dromicus**, II. p. 118.

3. Palatine and pterygoid bones toothed; pupil round; mandibular teeth subequal, or anterior a little enlarged.

    a. Longitudinal series of scales in even numbers; maxillary teeth 20 to 33.

Scales forming longitudinal series.
53. **Zaocys**, I. p. 374.

Lateral scales oblique .... 63. **Herpetodryas**, II. p. 71.

    b. Longitudinal series of scales in odd numbers; ventrals rounded or obtusely angulate laterally; subcaudals not angulate laterally.

        α. Maxillary teeth forming a continuous series, or the last two or three separated from the rest by a very short interspace.

* Head distinct from neck; eye large or rather large; body much elongate; scales with apical pits.

† Scales forming longitudinal series.

Maxillary teeth 12 to 20 .... 54. **Zamenis**, I. p. 379.

Maxillary teeth 25 to 30; scales of vertebral row enlarged.
56. **Xenelaphis**, II. p. 7.

Maxillary teeth 22 to 32; no suboculars.
57. **Drymobius**, II. p. 8.

†† Scales disposed obliquely, at least on the anterior part of the body; maxillary teeth 20 to 25.

Lateral scales as long as dorsals. 66. **Chlorophis**, II. p. 91.

Lateral scales much shorter than dorsals.
70. **Thrasops**, II. p. 104.

** Head not or but slightly distinct from neck; eye moderate or small.

† Internasals, nasals, and præfrontals distinct.

‡ Subcaudals in two rows.

Maxillary teeth 14 to 24; scales without pits; tail conical.
81. **Rhadinæa**, II. p. 160.

Maxillary teeth 13 to 16; scales without pits; usually one or more suboculars; tail long, thick throughout, ending obtusely.
82. **Urotheca**, II. p. 180.

Maxillary teeth 12 to 20, posterior not compressed; scales with apical pits; no suboculars .. 88. **Coronella**, II. p. 188.

Maxillary teeth 8 to 11, posterior not compressed; snout strongly projecting .... 91. **Cemophora**, II. p. 213.

Maxillary teeth 8 to 12, posterior compressed; snout more or less projecting .... 92. **Simotes**, II. p. 214.

‡‡ Subcaudals single; snout strongly projecting; maxillary teeth 16 to 19.
90. **Rhinochilus**, II. p. 212.

†† A single præfrontal; two internasals.

Eye small; nostril between two nasals.
83. **Trimetopon**, II. p. 184.

Eye very small; nasal single.. 84. **Hydromorphus**, II. p. 185.

††† A single internasal; two præfrontals; nasal semidivided, in contact with the præocular.

Eye rather small; scales as broad as long.
85. **Dimades**, II. p. 185.

Eye very small .... 86. **Hydrops**, II. p. 186.

†††† Internasals fused with the præfrontals, nasal fused with the first labial.
87. **Sympholis**, II. p. 188.

β. Posterior maxillary teeth strongly enlarged and separated from the rest by a very distinct interspace.

* Scales with apical pits; snout rounded or obtusely conical.

Scales forming longitudinal series; no suboculars.
75. **Liophis**, II. p. 126.

Scales forming longitudinal series; a series of suboculars.
76. **Cyclagras**, II. p. 144.

Scales disposed obliquely; no suboculars.
77. **Xenodon**, II. p. 144.

** Scales with apical pits; snout projecting, cuneiform.

Scales smooth, in 19 or 21 rows.
78. **Lystrophis**, II. p. 151.

Scales keeled, in 23 to 27 rows. 79. **Heterodon**, II. p. 153.

*** Scales without pits; snout obtuse; head narrow.
80. **Aporophis**, II. p. 157.

c. Longitudinal series of scales in odd numbers; ventrals with a suture-like lateral keel, and a notch on each side corresponding to the keel.

a. Scales of vertebral row not enlarged.

Scales smooth; subcaudals keeled and notched like the ventrals.
67. **Philothamnus**, II. p. 98.

Scales keeled; subcaudals keeled and notched like the ventrals.
68. **Gastropyxis**, II. p. 102.

Scales keeled; subcaudals without keel or notch.
69. **Hapsidophrys**, II. p. 103.

β. Scales of vertebral row enlarged.
64. **Dendrophis**, II. p. 77.

4. Palatine and pterygoid teeth absent or few; maxillary bone short, with 5 to 8 teeth.

a. Maxillary teeth forming an uninterrupted series.

Pupil round; snout projecting, obtuse.
93. **Oligodon**, II. p. 233.

Pupil vertically subelliptic; snout with angular horizontal edge.
94. **Prosymna**, II. p. 246.

  b. Posterior maxillary teeth separated from the rest by an interspace.

Scales with apical pits ...... 95. **Leptocalamus**, II. p. 249.

Scales without pits.......... 96. **Arrhyton**, II. p. 251.

 D. Maxillary teeth equal or nearly so, or posterior decreasing in size.

  1. Scales with apical pits.

   a. Longitudinal series of scales in even numbers; eye large; no suboculars .... 59. **Spilotes**, II. p. 23.

   b. Longitudinal series of scales in odd numbers.

    α. Head distinct from neck; eye moderate or large; body usually much elongate.

     * Maxillary teeth nearly equal; scales forming straight longitudinal series.

Nasals distinct, not fused with the loreal; ventrals rounded or more or less distinctly angulate laterally.
       60. **Coluber**, II. p. 24.

Nasal single, fused with the loreal, in contact with the præocular; ventrals sharply angulate laterally.
       61. **Synchalinus**, II. p. 70.

Nasals distinct; ventrals and subcaudals with a suture-like lateral keel and a notch on each side corresponding to the keel.
       62. **Gonyophis**, II. p. 70.

     ** Posterior maxillary teeth decreasing in size; scales narrow and oblique; ventrals and subcaudals with a suture-like lateral keel and a notch on each side corresponding to the keel.
       65. **Dendrelaphis**, II. p. 87.

    β. Head not or but slightly distinct from neck.

     * Maxillary teeth equal or subequal.

Snout cuneiform; rostral shield with obtuse horizontal keel.
       97. **Simophis**, II. p. 253.

Snout beak-shaped, hooked; rostral shield with sharp horizontal keel.................. 98. **Scaphiophis**, II. p. 254.

Snout obtuse or pointed; nasal distinct from the labials.
       99. **Contia**, II. p. 255.

Snout pointed; rostral in contact with or approaching the frontal; nasal fused with the first labial.
       100. **Ficimia**, II. p. 270.

Snout much depressed, strongly projecting; nasal fused with the internasal .... 101. **Chilomeniscus**, II. p. 272.

** Anterior maxillary and mandibular teeth longest.
102. **Homalosoma**, II. p. 273.

2. Scales without pits.

  a. Maxillary and mandibular teeth subequal, 15 to 35; pupil round; anterior temporal present.

  α. Loreal and præfrontal not entering the eye.

Head not or but slightly distinct from neck; eye moderate or rather small; maxillary teeth 15 to 30.
103. **Ablabes**, II. p. 277.

Head distinct from neck; eye moderate; body stout; maxillary teeth 22 to 25............ 104. **Grayia**, II. p. 286.

Head distinct from neck; eye large; maxillary teeth 35.
105. **Xenurophis**, II. p. 288.

  β. Loreal and præfrontal entering the eye, which is small.

Nostril lateral, between two nasals.
106. **Virginia**, II. p. 288.

Nostrils directed upwards, in a semidivided nasal; two internasals.
107. **Abastor**, II. p. 289.

Nostril directed upwards, in a semidivided nasal; a single internasal .... 108. **Farancia**, II. p. 290.

  b. Posterior maxillary and mandibular teeth shortest; anterior temporal usually present.

  α. Head distinct from neck; pupil vertically elliptic.

Vertebral row of scales enlarged; eye large.
109. **Petalognathus**, II. p. 292.

Dorsal scales equal; eye moderate or large.
110. **Tropidodipsas**, II. p. 294.

Dorsal scales equal; eye small. 111. **Dirosema**, II. p. 298.

  β. Head not distinct from neck; pupil round or vertically subelliptic .... 112. **Atractus**, II. p. 299.

  c. Maxillary teeth small, subequal, 8 to 12; anterior temporal present; eye very small.

Nostril in a single nasal; loreal and præfrontal entering the eye.
114. **Carphophis**, II. p. 324.

Nostril between two nasals and the internasal; snout with **angular** horizontal edge .... 116. **Geagras**, II. p. 326.

Nostril between a nasal and the first labial.
                                117. **Macrocalamus**, II. p. 327.

 d. Maxillary teeth small, subequal; parietal in contact with labials; eye small or very small.

  a. Nostril between two nasals.

Loreal present, entering the eye.
                                113. **Geophis**, II. p. 314.
No loreal; an azygous shield between the internasals.
                                118. **Idiopholis**, II. p. 327.

  β. Nostril in a single nasal; no loreal.

   * Nasal small.

Eye moderately small; body extremely slender.
                                115. **Stilosoma**, II. p. 325.
Eye very small .... 119. **Rhabdophidium**, II. p. 328.

   ** Nasal minute.

Internasals present.......... 120. **Pseudorhabdium**, II. p. 328.
No internasals; eye free .... 121. **Calamaria**, II. p. 330.
Eye concealed under the ocular shield.
                                122. **Typhlogeophis**, II. p. 351.

56. XENELAPHIS.

Coryphodon (*non Owen*), part., *Dum. & Bibr. Erp. Gén.* vii. p. 180 (1854); *Günth. Cat. Col. Sn.* p. 107 (1858); *Jan, Elenco sist. Ofid.* p. 63 (1863).
Xenelaphis, *Günth. Rept. Brit. Ind.* p. 250 (1864); *Bouleng. Faun. Ind., Rept.* p. 336 (1890).

Maxillary teeth 25 to 30, gradually increasing in size; anterior mandibular teeth a little enlarged. Head moderately elongate, distinct from neck; eye moderate, with round pupil; a subocular

Fig. 1.

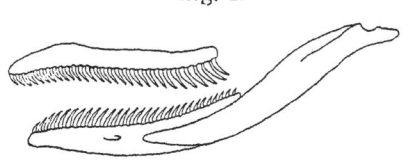

Maxillary and mandible of *Xenelaphis hexagonotus*.

below the præocular. Body elongate, cylindrical; scales smooth, with feebly marked apical pits, in 17 rows, the vertebral row slightly enlarged and six-sided; ventrals rounded. Tail long; subcaudals in two rows.

Burma, Malay Peninsula and Archipelago.

1. Xenelaphis hexagonotus.

Coluber hexahonotus, *Cantor, Cat. Mal. Rept.* p. 74 (1847).
Coryphodon sublutescens, *Dum. & Bibr.* vii. p. 187 (1854).
——— hexanotus, *Günth. Cat.* p. 110 (1858).
Xenelaphis hexahonotus, *Günth. Rept. Brit. Ind.* p. 251, pl. xxi. fig. C (1864).
Ptyas hexagonotus, *Theob. Cat. Rept. Brit. Ind.* p. 168 (1876).
Xenelaphis hexagonotus, *Bouleng. Faun. Ind., Rept.* p. 336 (1890).

Rostral broader than deep, visible from above; internasals as long as or a little longer than the præfrontals; frontal once and one third to once and a half as long as broad, as long as its distance from the end of the snout, a little shorter than the parietals; loreal nearly as long as deep; one præocular, with a rather large subocular below it, wedged in between the third and fourth upper labials; two postoculars, with a large subocular below, separating the eye from the fifth and sixth labials; a third subocular occasionally present, separating the eye altogether from the labials; temporals 2+2; normally eight upper labials, fourth entering the eye; four or five lower labials in contact with the anterior chinshields, which are as long as the posterior. Scales smooth, in 17 rows. Ventrals 185-198; anal divided; subcaudals 140-179. Brown above, with black cross bands, which become indistinct with age, on the anterior half of the body; old specimens showing mere traces of the outer ends of these bands; lower parts uniform yellowish.

Total length 1650 millim.; tail 600.

Burma, Malay Peninsula, Borneo, Sumatra, Java.

a.	Yg. (V. 188; C. 148).	Pinang.	Dr. Cantor. (Type.)
b, c.	♂ (V. 187; C. 140) & ♀ (V. 198; C. 140).	Singapore.	Gen. Hardwicke [C.].
d.	♂ (V. 191; C. 149).	Singapore.	W. T. Blanford, Esq. [P.].
e.	♀ (V. 190; C. 144).	Borneo.	L. L. Dillwyn, Esq. [P.].
f.	♂ (V. 185; C. 179).	Borneo.	Sir J. Brooke [P.].
g.	♂ (V. 190; C. ?).	Rejang R., Sarawak.	Brooke Low, Esq. [P.].
h.	♂ (V. 187; C. 173).	——?	Dr. Bleeker. (*Dendrophis dumerilii*, Blkr.)
i.	Yg. (V. 192; C. 175).	——?	Dr. Bleeker. (*Ablabes polyhemizona*, Blkr.)
k.	Skull.	Sarawak.	

57. DRYMOBIUS.

Coluber, part., *Schleg. Phys. Serp.* ii. p. 125 (1837); *Dum. & Bibr. Mém. Ac. Sc.* xxiii. 1853, p. 455.
Herpetodryas, part., *Schleg. l. c.* p. 173; *Dum. & Bibr. Erp. Gén.* vii

p. 203 (1854); *Günth. Cat. Col. Sn.* p. 113 (1858); *Jan, Elenco sist. Ofid.* p. 80 (1863).
Coryphodon (non Owen), part., *Dum. & Bibr. t. c.* p. 180; *Günth. l. c.* p. 107; *Jan, l. c.* p. 63.
Leptophis, part., *Dum. & Bibr. t. c.* p. 528.
Dromicus, part., *Dum. & Bibr. t. c.* p. 646; *Günth. l. c.* p. 126.
Drymobius, part., *Cope, Proc. Ac. Philad.* 1860, p. 560; *Bocourt, Miss. Sc. Mex., Rept.* p. 715 (1890).
Dendrophidium, *Cope, l. c.* p. 561; *Bocourt, l. c.* p. 730.
Ablabes, part., *Jan, l. c.* p. 51.
Thamnosophis, part., *Jan, l. c.* p. 82.
Tropidonotus, part., *Jan, l. c.* p. 68.
Eudryas, *Bocourt, l. c.* p. 716.
Crossanthera, *Cope, Am. Nat.* 1893, p. 481.

Maxillary teeth 22 to 38, increasing in size posteriorly, the two or three last stouter if not longer than those preceding them; mandibular teeth subequal or anterior somewhat enlarged. Head elongate, distinct from neck; eye large, with round pupil; no suboculars. Body elongate, cylindrical; scales smooth or keeled, with apical pits, in 15 or 17 rows; ventrals rounded or with an obtuse lateral keel. Tail long; subcaudals in two rows.

America, from Texas to Peru and Southern Brazil.

This genus is very closely allied to, and intermediate between, both *Zamenis* and *Coluber*. It is just separable from the former by the greater number of teeth; and it is connected with the latter by *D. bifossatus* and *D. boddaertii*, the dentition of some specimens of which approaches the isodont type. The subocular shield below the præocular, present in most of the species referred to *Zamenis*, is constantly absent in *Drymobius*.

Fig. 2.

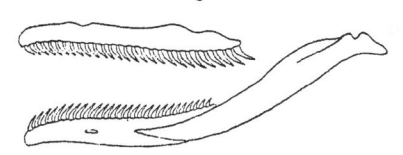

Maxillary and mandible of *Drymobius bifossatus*.

Synopsis of the Species.

I. Scales smooth.

Scales in 15 rows; two labials entering the eye	1. *bifossatus*, p. 10.
Scales in 17 rows; usually three labials entering the eye	2. *boddaertii*, p. 11.

II. Scales keeled.

A. Scales in 15 rows............ ... 3. *affinis*, p. 14.

B. Scales in 17 rows.
   1. Maxillary teeth 30 to 38.
Subcaudals 84–96 .... 4. *rhombifer*, p. 14.
Subcaudals 109–128; loreal as deep as long. 5. *bivittatus*, p. 15.
Subcaudals 113–155; loreal usually longer
   than deep .... 6. *dendrophis*, p. 15.
   2. Maxillary teeth not more than 25.
Frontal hardly once and a half as long as
   broad; loreal not or but slightly longer
   than deep; ventrals 158–169 .... 7. *chloroticus*, p. 16.
Frontal at least once and a half as long as
   broad; loreal longer than deep; ventrals   [p. 17.
   144–159......................... 8. *margaritiferus*,

1. Drymobius bifossatus.

Coluber bifossatus, *Raddi, Mem. Soc. Ital. Modena,* xviii. (Fis.), 1820, p. 333.
—— capistratus, *Lichtenst. Verz. Doubl. Mus. Berl.* p. 104 (1823).
—— lichtensteinii, *Wied, N. Acta Ac. Leop.-Carol.* xii. ii. 1825, p. 493, pl. xlvi., and *Beitr. Nat. Bras.* i. p. 305 (1825); *Wagl. Icon. Amph.* pl. iv. (1828); *Wied, Abbild.* (1831).
—— pantherinus (*non Daud.*), *Schleg. Phys. Serp.* ii. p. 143, pl. v. figs. 13 & 14 (1837).
Coryphodon pantherinus, *Dum. & Bibr.* vii. p. 181 (1854); *Günth. Cat.* p. 107 (1858); *Jan, Icon. Gén.* 24, pl. iii. fig. 1 (1867).
Drymobius pantherinus, *Cope, Proc. Am. Philos. Soc.* xxii. 1885, p. 192.
Ptyas pantherinus, *Boettg. Zeitschr. f. ges. Naturw.* lviii. 1885, p. 233.
Drymobius bifossatus, *Bouleng. Ann. & Mag. N. H.* (6) xiii. 1894, p. 346.

Maxillary teeth 27 to 30, posterior feebly enlarged. Rostral nearly as deep as broad, visible from above; internasals much shorter than the præfrontals; frontal once and two thirds or once and three fourths as long as broad, as long as or longer than its distance from the rostral, a little shorter than the parietals; loreal as long as deep or deeper than long; one præocular, usually not touching the frontal; two postoculars; temporals 2+2; eight upper labials, fourth and fifth entering the eye; four or five lower labials in contact with the anterior chin-shields, which are shorter than the posterior. Scales smooth, in 15 rows. Ventrals 160–183; anal divided; subcaudals 85–102. Above with three parallel series of large, roundish or subquadrate, blackish or dark brown black-edged spots, separated by narrow yellowish or yellowish-brown interspaces; the spots of the lateral series extending to the sides of the belly; a pair of longitudinal dark bands or a U-shaped band on the back of the head and nape, and a dark cross-band between the eyes; labials yellowish, black-edged; lower parts yellowish, with brown or black spots.

Total length 1750 millim.; tail 450.

Brazil, Paraguay.

a, b. ♀ (V. 180; C. 100) & yg. (V. 176; C. 92).		Pernambuco.	J. P. G. Smith, Esq. [P.].
c. ♂ (V. 166; C. 102).		Pernambuco.	
d. ♀ (V. 177; C.?).		Bahia.	Dr. O. Wucherer [C.].
e. ♀ (V. 179; C.?).		Rio Janeiro.	Haslar Collection.
f–h. ♀ (V. 174; C. 95) & yg. (V. 167, 160; C. 98, ?).		Porto Real, Prov. Rio Janeiro.	M. Hardy du Dréneuf [C.].
i. Yg. (V. 176; C. 90).		Corumba, Prov. Matto Grosso.	S. Moore, Esq. [P.].
k. ♂ (V. 171; C. 92).		Porto Alegre.	Dr. H. v. Ihering [C.].
l. ♀ (V. 183; C. 93).		Asuncion, Paraguay.	Dr. J. Bohls [C.].

2. Drymobius boddaertii.

Coluber boddaertii, *Sentzen, Meyer's Zool. Arch.* ii. 1796, p. 59; *Merr. Tent.* p. 110 (1820).
Herpetodryas boddaertii, *Schleg. Phys. Serp.* ii. p. 185 (1837); *Dum. & Bibr.* vii. p. 210 (1854); *Günth. Cat.* p. 115 (1858); *Jan, Icon. Gén.* 49, pl. i. fig. 1 (1879).
Coluber fuscus, *Hallow. Proc. Ac. Philad.* 1845, p. 241.
Dromicus pleii, *Dum. & Bibr. t. c.* p. 661; *Jan, op. cit.* 24, pl. v. fig. 1 (1867).
Herpetodryas rappii, *Günth. l. c.* p. 116, and *Proc. Zool. Soc.* 1859, p. 412.
Ablabes tessellatus, *Berthold, Götting. Anz.* iii. 1859, p. 180; *Jan, Arch. Zool. Anat. Phys.* ii. 1863; p. 281.
Drymobius boddaertii, *Cope, Proc. Ac. Philad.* 1860, p. 561; *Günth. Biol. C.-Am., Rept.* p. 125 (1894).
—— rappii, *Cope, l. c.*
Herpetodryas reticulata, *Peters, Mon. Berl. Ac.* 1863, p. 285; *Garm. Bull. Essex Inst.* xxiv. 1892, p. 91.
? Herpetodryas (Drymobius) reissii, *Peters, Mon. Berl. Ac.* 1868, p. 640.
? Masticophis pulchriceps, *Cope, Proc. Ac. Philad.* 1868, p. 105.
? Masticophis melanolomus, *Cope, t. c.* p. 134.
Herpetodryas bilineatus, *Jan, op. cit.* p. 31, pl. iv. fig. 3 (1869).
—— quinquelineatus, *Steind. Sitzb. Ak. Wien,* lxii. 1870, p. 346.
Drymobius heathii, *Cope, Journ. Acad. Philad.* (2) viii. 1876, p. 179.
Dromicus (Alsophis) maculivittis, *Peters, Mon. Berl. Ac.* 1877, p. 458.
Drymobius bilineatus, *Cope, Proc. Amer. Philos. Soc.* xvii. 1877, p. 35.
Herpetodryas lævis, *Fischer, Arch. f. Nat.* 1881, p. 227, pl. xi. figs. 4–6.
Coryphodon alternatus, *Bocourt, Bull. Soc. Philom.* (7) viii. 1884, p. 133, *and Miss. Sc. Mex., Rept.* pl. xlv. fig. 3 (1888).
Dromicus cæruleus, *Fischer, Jahrb. Hamb. Wiss. Anst.* ii. 1885, p. 103, pl. iv. fig. 7.
Alsophis pulcher, *Garm. Proc. Amer. Philos. Soc.* xxiv. 1887, p. 283.
Drymobius alternatus, *Cope, Bull. U.S. Nat. Mus.* no. 32, 1887, p. 70.
? Drymobius melanolomus, *Cope, l. c.*
Drymobius lævis, *Cope, l. c.; Bocourt, Miss. Sc. Mex., Rept.* p. 722, pl. li. fig. 6 (1890).
Herpetodryas (Drymobius) boddaerti, var. heathii, *Boettg. Ber. Senck. Ges.* 1889, p. 313.

Drymobius (Eudryas) boddaertii, *Bocourt, l. c.* p. 720, pl. li. figs. 1 & 2.
—— dorsalis, *Bocourt, l. c.* p. 722, fig. 2.
—— cæruleus, *Bocourt, l. c.* p. 727, fig. 4.
—— pleii, *Bocourt, l. c.* p. 728, fig. 3.
Zamenis melanolomus, *Günth. Biol. C.-Am., Rept.* p. 122 (1894).

Maxillary teeth 22 to 25, posterior moderately or feebly enlarged. Rostral broader than deep, just visible from above; internasals shorter than the præfrontals; frontal once and a half to twice as long as broad, as long as or longer than its distance from the end of the snout, as long as or a little shorter than the parietals; loreal longer than deep; one præocular, rarely touching the frontal; two (rarely three) postoculars; temporals 2+2 or 2+3 (rarely 1+2); nine (rarely eight) upper labials, fourth, fifth, and sixth (or fourth and fifth, or third, fourth, and fifth) entering the eye; five lower labials in contact with the anterior chin-shields, which are shorter than the posterior. Scales smooth, in 17 rows. Ventrals obtusely angulate, 163–204; anal divided; subcaudals 84–136. Coloration very variable.

Total length 1090 millim.; tail 310.

Tropical America.

A. Brownish, greyish, or olive above, uniform or with a pair of light streaks along part or the whole of the body; the scales sometimes black-edged; a more or less distinct dark streak on each side of the head, through the eye; belly yellowish, uniform or dotted with olive; throat sometimes spotted or marbled with brown. (*C. boddaertii*, Sentz.)

a. ♂ (V. 170; C. 111).	Mexico.	M. Sallé [C.].
b. ♂ (V. 186; C. 112).	Zacuapan, Mexico.	Mr. H. Fink [C.].
c–d. ♂ (V. 185; C. 109) & ♀ (V. 190; C. ?).	Tres Marias Ids.	Hr. A. Forrer [C.].
e–f. ♂ (V. 179; C. 125) & ♀ (V. 185; C. ?).	Yucatan.	
g–h. ♂ (V. 176; C. 120) & yg. (V. 174; C. ?).	Teapa, Tabasco.	F. D. Godman, Esq. [P.].
i. Yg. (V. 172; C. 106).	Hacienda del Hobo, S. Mexico.	P. Geddes, Esq. [P.].
k–l. ♂ (V. 176, 174; C. 111, 116).	Huatuzco, Vera Cruz.	F. D. Godman & O. Salvin, Esqs. [P.].
m. ♂ (V. 178; C. ?).	Vera Paz, low forest.	O. Salvin, Esq. [C.].
n. Hgr. (V. 180; C. 120).	Peten, Guatemala.	O. Salvin, Esq. [C.].
o. ♀ (V. 184; C. 109).	W. Coast of Central America.	H.M.S. 'Beagle.'
p. ♀ (V. 196; C. 113).	Venezuela.	Mr. Dyson [C.].
q. ♀ (V. 204; C. 117).	St. Vincent, W.I.	Mus. Comp. Zoology [E.].
r–s. ♂ (V. 192; C. 92) & yg. (V. 200; C. 120).	St. Vincent.	Mr. H. H. Smith [C.]; F. D. Godman, Esq. [P.].
t. ♂ (V. 196; C. 122).	Id. of Grenada, W.I.	Mus. Comp. Zoology [E.].

u. ♂ (V. 198; C. 117).	Moustiques Id., Grenadines.	Mr. H. H. Smith [C.]; F. D. Godman, Esq. [P.].
v. ♂ (V. 185; C. ?).	Brit. Guiana.	G. A. Boulenger, Esq. [P.].
w. ♂ (V. 182; C. 101).	Demerara.	
x, y. ♂ (V. 176, 186; C. 108, 112).	Vryheids Lust, Demerara.	Rev. W. T. Turner [C.].
z–α. ♂ (V. 179; C. 107) & ♀ (V. 186; C. 106).	Surinam.	
β. ♀ (V. 193; C. 115).	Maracaibo.	Mr. Lowe [C.].
γ. ♀ (V. 190; C. 99).	Para.	
δ. ♀ (V. 196; C. 101).	Chyavetas, N.E. Peru.	

B. Pale olive-brown above, with two whitish longitudinal lines and narrow black cross-lines between them.

a. ♀ (V. 176; C. 101).	Pallatanga, Ecuador.	Mr. Buckley [C.].
b. Hgr. (V. 197; C.?).	Yurimaguas, N.E. Peru.	

C. Pale olive-brown above, with a dorsal series of large transverse quadrangular blackish spots alternating with another series on each side; a black streak through the eye; labials black-edged; throat and anterior ventrals spotted or marbled with black, or blackish spotted with yellowish. (*H. rappii*, Gthr.; *C. alternatus*, Bocourt.)

a. Yg. (V. 186; C. 108).	Tres Marias Ids.	Hr. A. Forrer [C.].
b. Yg. (V. 177; C. 108).	Trinidad.	F. W. Urich, Esq. [P.].
c, d–e. Yg. (V. 177, 186, 183; C. 126, 120, 123).	Venezuela.	Mr. Dyson [C.]. (Types of *H. rappii*.)
f. Yg. (V. 178; C. 96).	British Guiana.	G. A. Boulenger, Esq. [P.].
g. Yg. (V. 187; C. 101).	Berbice.	
h. Yg. (V. 178; C. 102).	Berbice.	Lady Essex [P.].
i. Yg. (V. 173; C. ?).	Brazil.	
		(Types of *H. rappii*.)
k, l, m. ♀ (V. 185, 176; C. 100, 93) & yg. (V. 178; C. 96).	W. Ecuador.	Mr. Fraser [C.].

D. Markings as in the preceding, but restricted to the anterior part of the body; upper lip and lower parts uniform white.

a–b. Yg. (V. 183, 176; C. 130, 136).	Dueñas, Guatemala.	O. Salvin, Esq. [C.].

E. Brown above, with two or four black longitudinal streaks connected, at least on the anterior part of the body, by black cross-lines or bars; labials dark-edged; throat spotted or marbled with brown, or dark, spotted with white; belly yellowish, with a series of black spots along the outer ends of the ventrals. (*D. pleii*, D. & B.)

a. ♀ (V. 196; C. 113).	Venezuela.	Mr. Dyson [C.]. (One of the types of *H. rappii*.)

F. Pale olive-brown above, with five dark longitudinal streaks; a dark streak through the eye; labials black-edged; throat dark olive, spotted with white. (*H. quinquelineatus*, Steind.)

a. ♀ (V. 198; C. 96).   Rosario de Cucuta, Colombia.   Mr. C. Webber [C.].

G. Olive above, whitish or pale olive beneath; a dark brown vertebral stripe, and a less distinct dark stripe along each side. (*D. dorsalis*, Bocourt.)

a, b. ♂ (V. 178; C. 120) & ♀ (V. 193; C. 112).   Dueñas, Guatemala.   O. Salvin, Esq. [C.].

3. Drymobius affinis.

Herpetodryas affinis, *Steindachn. Sitzb. Ak. Wien,* lxii. 1870, p. 348, pl. vii. figs. 4 & 5.

Rostral just visible from above; internasals a little shorter than the præfrontals; frontal very broad anteriorly, very narrow posteriorly, the lateral border very concave; loreal twice as long as deep; one præocular, in contact with the frontal; two postoculars; temporals 1 + 2; nine upper labials, fifth and sixth entering the eye. Scales in 15 rows, all except the outer row keeled. Olive above, yellowish brown below; upper lip yellow, black-edged below; a black stripe on each side of the head; upper surface of head dotted with black; anterior part of body with five narrow black stripes; gular region blackish, spotted with yellow.

Brazil.

4. Drymobius rhombifer.

Coryphodon rhombifer, *Günth. Proc. Zool. Soc.* 1860, p. 236.
Herpetodryas dendrophis, part., *Jan, Icon. Gén.* 31, pl. iii. fig. 1 (1869).
Spilotes rhombifer, *Peters, Mon. Berl. Ac.* 1879, p. 777.
Drymobius rhombifer, *Bocourt, Miss. Sc. Mex., Rept.* pl. xliii. fig. 1 (1888).
Zamenis rhombifer, *Günth. Biol. C.-Am., Rept.* p. 120, pl. xlv. (1894).

Maxillary teeth 30 to 32, posterior moderately enlarged. Rostral broader than deep, just visible from above; internasals much shorter than the præfrontals; frontal once and a half to once and two thirds as long as broad, as long as its distance from the rostral or the end of the snout, a little shorter than the parietals; loreal as long as deep or a little longer; one præocular, not touching the frontal; two postoculars; temporals 2 + 2; nine upper labials, fourth, fifth, and sixth entering the eye; five lower labials in contact with the anterior chin-shields, which are shorter than the posterior. Scales strongly keeled, in 17 rows. Ventrals 148–163; anal divided; subcaudals 84–96. Pale brown above, with a more or less distinct series of large rhomboidal dark spots edged with blackish; lower parts yellowish, with a lateral series of black spots.

Total length 950 millim.; tail 300.

Colombia, Ecuador, Northern Peru.

a. Hgr. (V. 153; C. 85).   Chiriqui.   F. D. Godman, Esq. [P.].

b. Hgr.(V. 160; C. 94).	Colombia.	
c. ♀ (V. 163 ; C. ?).	Esmeraldas, Ecuador.	Mr. Fraser [C.]. (Type).
d, e. ♂ (V. 148; C. 93) & ♀ (V. 158; C. 84).	Moyobamba, Peru.	Mr. A. H. Roff [C.].
f. Yg. (V. 150; C. 96).	——— ?	v. Lidth de Jeude Collection.

5. Drymobius bivittatus.

Leptophis bivittatus, *Dum. & Bibr.* vii. p. 540 (1854).
Thamnosophis bivittatus, *Jan, Elenco*, p. 82 (1863).
Tropidonotus subradiatus, *Jan, Arch. Zool. Anat. Phys.* iii. 1865, p. 227, *and Icon. Gén.* 27, pl. iii. figs. 1 & 2 (1868).
Herpetodryas tetratænia, *Günth. Ann. & Mag. N. H.* (4) ix. 1872, p. 23.

Maxillary teeth 32, posterior very feebly enlarged. Rostral broader than deep, just visible from above ; internasals shorter than the præfrontals; frontal once and a half as long as broad, a little longer than its distance from the end of the snout, a little shorter than the parietals; loreal as long as deep; one præocular, not touching the frontal; two postoculars; temporals 2 + 2; nine upper labials, fourth, fifth, and sixth entering the eye ; five lower labials in contact with the anterior chin-shields, which are much shorter than the posterior. Scales strongly keeled, in 17 rows. Ventrals 147-152; anal divided; subcaudals 109-128. Olive-grey above, neck with black spots; body and tail with two broad black dorsal stripes separated by the width of one scale, and with a black lateral line running along the second series of scales ; a black streak on each side of the head, above the lip, which is white; lower parts white, the shields olive-grey at the outer ends and black-edged in front.

Total length 730 millim.; tail 290.
Colombia.

a. ♂ (V. 151; C. 128).   Bogota.   (Type of *H. tetratænia*).

6. Drymobius dendrophis.

Herpetodryas dendrophis, *Schleg. Phys. Serp.* ii. p. 196 (1837), *and Abbild.* p. 132, pl. xliv. figs. 25-28 (1844) ; *Günth. Cat.* p. 118 (1858), *and Ann. & Mag. N. H.* (3) xii. 1863, p. 358, pl. vi. fig. B; *Jan, Icon. Gén.* 31, pl. iii. fig. 2 (1869).
——— æstivus, *Berthold, Abh. Ges. Götting.* iii. 1847, p. 11.
——— poitei, *Dum. & Bibr.* vii. p. 208 (1854).
——— brunneus, *Günth. Cat.* p. 116, *and Ann. & Mag. N. H.* (3) xii. 1863, p. 358, pl. vi. fig. A ; *Salvin, Proc. Zool. Soc.* 1860, p. 456 ; *Garm. Bull. Essex Inst.* xxiv. 1892, p. 91.
Drymobius dendrophis, *Cope, Proc. Ac. Philad.* 1860, p. 561; *Günth. Biol. C.-Am., Rept.* p. 127 (1894).
Herpetodryas nuchalis, *Peters, Mon. Berl. Ac.* 1863, p. 285.
Masticophis brunneus, *Cope, Proc. Ac. Philad.* 1868, p. 105.
Dendrophidium dendrophis, *Cope, Proc. Am. Philos. Soc.* xxiii. 1886, p. 278 ; *Bocourt, Le Natur.* 1889, p. 46, figs., *and Miss. Sc. Mex., Rept.* p. 730, pl. xlix. fig. 4 (1890) ; *Dugès, La Naturaleza*, (2) ii. 1892, p. 100, pl. v.

Drymobius percarinatus, *Cope, Proc. Amer. Philos. Soc.* xxxi. 1893, p. 344.

Maxillary teeth 32 to 38, posterior moderately enlarged. Rostral broader than deep, visible from above; internasals as long as or a little shorter than the præfrontals; frontal once and a half to once and three fourths as long as broad, as long as its distance from the end of the snout (longer in the young), as long as or a little shorter than the parietals; loreal usually longer than deep; one præocular, not touching the frontal; two or three postoculars; temporals 2 + 2; nine (exceptionally ten) upper labials, fourth, fifth, and sixth (or fifth, sixth, and seventh) entering the eye; five (rarely four) lower labials in contact with the anterior chin-shields, which are shorter than the posterior. Scales more or less strongly keeled, in 17 rows. Ventrals 150–167; anal entire or divided; subcaudals 113–155. Olive-brown above, uniform or with blackish cross-bands enclosing round whitish spots, or with whitish black-edged cross-bands; upper lip whitish; ventrals and subcaudals olive on the sides, whitish, uniform or dotted or edged with olive, in the middle.

Total length 1020 millim.; tail 420.

From Mexico to Ecuador and the Guianas.

a. Hgr. ♂ (V. 160; C. 123).	Mexico.	M. Sallé [C.].
b–e. ♀ (V. 160, 167, 160; C. 113, 139, 116) & yg. (V. 166; C. 145).	Vera Paz, low forest.	O. Salvin, Esq. [C.].
f. ♂ (V. 152; C. 119).	Guatemala.	O. Salvin, Esq. [C.].
g–h. ♀ (V. 160; C. 128) & yg. (V. 150; C. 150).	Chiriqui.	F. D. Godman, Esq. [P.].
i. ♀ (V. 156; C. ?).	Guayaquil.	(Type of *H. brunneus*.)
k–m. ♂ (V. 162, 151, 156; C. 117, 125, ?) & ♀ (V. 157; C. ?).	W. Ecuador.	Mr. Fraser [C.].
n. Yg. (V. 159; C. 140).	Quito.	
o. Yg. (V. 150; C. 155).	Zaruma, Ecuador.	J. F. Gunter, Esq. [C.].
p. Yg. (V. 162; C. 142).	Venezuela.	Mr. Dyson [C.].
q. Skull.	Cayenne.	

7. Drymobius chloroticus.

Dendrophidium chloroticum, *Cope, Proc. Amer. Philos. Soc.* xxiii. 1886, p. 278.
Drymobius chloroticus, *Cope, Bull. U.S. Nat. Mus.* no. 32, 1887, p. 69; *Bocourt, Miss. Sc. Mex., Rept.* p. 718, pl. l. fig. 7 (1890).
—— brunneus (*non Günth.*), *Bocourt, l. c.* pl. l. fig. 6.

Maxillary teeth not more than 25, posterior rather strongly enlarged. Rostral broader than deep, visible from above; internasals shorter than the præfrontals; frontal hardly once and a half as long as broad, as long as its distance from the end of the snout, a little shorter than the parietals; loreal not or but slightly longer than deep; one præ- and two postoculars; temporals 2+2; nine upper labials, fourth, fifth, and sixth entering the eye; five lower labials in contact with the anterior chin-shields, which are shorter than the posterior. Scales in 17 rows, dorsals keeled. Ventrals 158-169; anal divided; subcaudals 114-117. Green above; ventrals yellow in the middle, green on the sides; young sometimes with black cross bars.

Total length 1280 millim.; tail 400.

Guatemala.

8. Drymobius margaritiferus.

Herpetodryas margaritiferus, *Schleg. Phys. Serp.* ii. p. 184 (1837), *and Abbild.* pl. xliv. figs. 19 & 20 (1844).
Leptophis margaritiferus, *Dum. & Bibr.* vii. p. 539 (1854).
Zamenis tricolor, *Hallow. Journ. Ac. Philad.* (2) iii. 1855, p. 34, pl. iii.
Dromicus margaritiferus, *Günth. Cat.* p. 126 (1858); *Salvin, Proc. Zool. Soc.* 1860, p. 456.
Drymobius margaritiferus, *Cope, Proc. Ac. Philad.* 1860, p. 561; *Bocourt, Miss. Sc. Mex., Rept.* p. 716, pl. xlix. fig. 2 (1890); *Stejneger, Proc. U.S. Nat. Mus.* xiv. 1891, p. 504.
Thamnosophis margaritiferus, *Jan, Elenco,* p. 82 (1863), *and Icon. Gén.* 31, pl. vi. fig. 3 (1869); *Dugès, La Naturaleza,* (2) i. 1888, p. 132.
Drymobius margaritiferus, var. occidentalis, *Bocourt, l. c.* p. 718.

Maxillary teeth 24 or 25, posterior strongly enlarged. Rostral broader than deep, visible from above; internasals shorter than the præfrontals; frontal once and a half to once and three fourths as long as broad, as long as its distance from the rostral or the end of the snout, as long as or slightly shorter than the parietals; loreal longer than deep; one præocular, not touching the frontal; two or three postoculars; temporals 2+2; nine upper labials, fourth, fifth, and sixth entering the eye; five (rarely four) lower labials in contact with the anterior chin-shields, which are shorter than the posterior. Scales more or less distinctly keeled, in 17 rows. Ventrals 144-159; anal divided; subcaudals 101-126. Black above, each scale green at the base and with a yellow elongate spot or longitudinal streak; head pale brown or olive above, black on the temples; labials yellow, with black sutures; ventrals and subcaudals yellow, black-edged.

Total length 1260 millim.; tail 430.

From South-western Texas and Mexico to Colombia and Venezuela.

a-h. ♂ (V. 153, 154, 154, 152; C. ?, 121, ?, ?) & ♀ (V. 152, 154, 151, 151; C. ?, ?, 107, 120).	Mexico.	Hr. Hugo Finck [C.].
i. ♂ (V. 148; C. 113).	Mexico.	M. Sallé [C.].
k. ♀ (V. 145; C. ?).	Presidio, nr. Mazatlan.	Hr. A. Forrer [C.].
l. Yg. (V. 151; C. 122).	Jalapa.	Hr. Hoege [C.].

m-p. ♂ (V. 154; C. ?) & ♀ (V. 149, 150, 154; C. ?, ?, 114).	Teapa, Tabasco.	F. D. Godman, Esq. [P.].
q-s. ♂ (V. 147, 152, 151; C. ?, 119, 123).	Atoyac, Guerrero.	F. D. Godman, Esq. [P.].
t. ♀ (V. 152; C. 126).	Omilteme, Guerrero.	F. D. Godman, Esq. [P.].
u-v. ♂ (V. 151, 153; C. 101, 125).	Sto. Domingo de Guzman.	Dr. A. C. Buller [C.].
w. ♂ (V. 156; C. 117).	Belize.	Dr. Günther [P.].
x, y. ♀ (V. 150, 148; C. ?, ?).	Honduras.	
z. ♀ (V. 154; C. ?).	Dueñas, Guatemala.	O. Salvin, Esq. [C.].
a. Hgr. (V. 154; C. 120).	Caracas.	
β. Skull.	Vera Cruz.	

58. PHRYNONAX.

Spilotes, part., *Dum. & Bibr. Erp. Gén.* vii. p. 218 (1854); *Günth. Cat. Col. Sn.* p. 96 (1858); *Jan, Elenco sist. Ofid.* p. 62 (1863); *Bocourt, Miss. Sc. Mex., Rept.* p. 685 (1888).
Phrynonax, *Cope, Proc. Ac. Philad.* 1862, p. 348.

Maxillary teeth 15 to 20, the posterior slightly enlarged; anterior mandibular teeth longest. Head distinct from neck; eye large, with round pupil; no suboculars. Body elongate, compressed; scales more or less distinctly keeled, disposed obliquely on the sides, in 21 to 25 rows; ventrals obtusely angulate laterally. Tail long; subcaudals in two rows.

Tropical America.

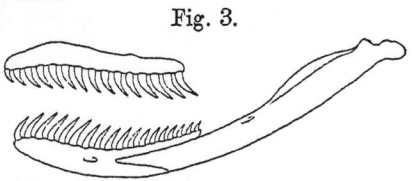

Fig. 3.

Maxillary and mandible of *Phrynonax sulphureus*.

Synopsis of the Species.

I. Scales in 21 or 23 rows.

A. Scales strongly keeled on the middle of the back.

Three postoculars; temporals 1+1.... 1. *sulphureus*, p. 19.
Two postoculars; temporals 2+2 .... 2. *pœcilonotus*, p. 20.

B. Scales feebly keeled.

Præocular separated from the frontal; ten upper labials .... 3. *guentheri*, p. 20.
Præocular separated from the frontal; nine upper labials .... 4. *lunulatus*, p. 21.

Præocular in contact with or narrowly
  separated from the frontal; eight or
  nine upper labials................ 5. *fasciatus*, p. 21.

II. Scales in 25 rows.

Dorsal scales very strongly keeled .... 6. *eutropis*, p. 22.
Scales smooth, except the row next to
  the vertebral, which is feebly keeled . 7. *chrysobronchus*, p. 22.

1. Phrynonax sulphureus.

Natrix sulphurea, *Wagl. in Spix, Serp. Bras.* p. 26, pl. ix. (1824).
Coluber pœcilostoma, *Wied, Beitr. Nat. Bras.* i. p. 250 (1825), *and Abbild.* (1831); *Schleg. Phys. Serp.* ii. p. 153, pl. vi. figs. 5 & 6 (1837).
Spilotes pœcilostoma, *Dum. & Bibr.* vii. p. 221 (1854); *Günth. Cat.* p. 100 (1858); *Jan, Icon. Gén.* 48, pl. v. figs. 3 & 4 (1876).

Rostral broader than deep, visible from above; internasals as long as or a little shorter than the præfrontals; frontal as long as broad or slightly longer, as long as its distance from the rostral or the end of the snout, shorter than the parietals; loreal longer than deep; one præocular, not reaching the frontal; three postoculars; temporals 1+1; eight upper labials, fourth and fifth entering the eye; five lower labials in contact with the anterior chin-shields, which are as long as or a little shorter than the posterior. Scales in 21 rows, all except the outer row strongly keeled. Ventrals distinctly angulate, 208–226; anal entire; subcaudals 126–144. Yellowish, olive, or brown above, with or without black spots or irregular cross bars; the keels of the scales black; belly yellow or olive, sometimes turning to black posteriorly.

Total length 2700 millim.; tail 720.

Guianas, Brazil, North-eastern Peru, Ecuador.

A. Labials without black margins; tail olive or brown.

a. ♂ (V. 214; C. 129).	Cabaros, Upper Amazon.	Mr. E. Bartlett [C.].
b. Hgr. (V. 217; C. 139).	Moyobamba, N.E. Peru.	Mr. A. H. Roff [C.].
c. ♀ (V. 216; C. 134).	Guayaquil.	Mr. Fraser [C.].
d. Yg. (V. 216; C. 136).	S. America.	Zoological Society.
e. Yg. (V. 208; C. 140).	S. America.	Dr. Pye-Smith [P.].

B. Sutures between some of the labials black; tail black.

f. ♂ (V. 217; C. 144).	Demerara Falls.	
g. ♂ (V. 215; C. 139).	S. America.	Zoological Society.

---

h. Skull.   Brazil.

2. Phrynonax pœcilonotus.

Spilotes pœcilonotus, part., *Günth. Cat.* p. 100 (1858), *and Biol. C.-Am., Rept* p. 117, pl. xliii. (1894).
? Spilotes pœcilonotus, *Bocourt, Miss. Sc. Mex., Rept.* p. 691, pl. xliii. fig. 4 (1888).
Spilotes argus, *Bocourt, l. c.* p. 692, pl. xlviii. fig. 10.*

Rostral broader than deep, just visible from above; internasals shorter than the præfrontals; frontal a little longer than broad, as long as its distance from the rostral, nearly as long as the parietals; loreal longer than deep; one præocular, not in contact with the frontal; two postoculars; temporals 2+2; eight or nine upper labials, fourth and fifth, or fourth, fifth, and sixth, entering the eye; six or seven lower labials in contact with the anterior chin-shields, which are as long as or a little shorter than the posterior. Scales in 23 rows, the dorsals very strongly keeled, the keels forming continuous raised lines in the middle of the body. Ventrals obtusely angulate laterally, 201-208; anal entire; subcaudals 126-133. Blackish brown above, with yellow spots which may form a vertebral stripe; upper lip yellow, with black edges to the shields; ventrals yellow with blackish edge; posterior portion of body and tail dark brown or black above and below.

Total length 1800 millim.; tail 500.

Central America.

a. ♀ (V. 208; C. 133).	Honduras.	(Type.)
b. ♀ (V. 202; C. 132).	Vera Paz.	O. Salvin, Esq. [C.].
c. ♀ (V. 201; C. 126).	Guatemala.	L. Greening, Esq. [P.].

3. Phrynonax guentheri.

Spilotes argus (*non Bocourt*), *Günth. Biol. C.-Am., Rept.* p. 118, pl. xliv. (1894).

Rostral a little broader than deep, just visible from above; internasals a little shorter than the præfrontals; frontal a little longer than broad, as long as its distance from the rostral, a little shorter than the parietals; loreal longer than deep; one præocular, not in contact with the frontal; two postoculars; temporals 2+2; ten upper labials, sixth and seventh, or fifth, sixth, and seventh, entering the eye; seven lower labials in contact with the anterior chin-shields, which are much longer than the posterior. Scales in 23 rows, median dorsals faintly keeled. Ventrals obtusely angulate laterally, 213; anal entire. Black above, each scale with a yellowish spot; the spots on the three vertebral series larger, forming a light stripe bordered with black; lateral scales greyish, with black border; head black above, with small yellowish spots; tail black, with yellowish spots forming four stripes, one above, one beneath, and one on each side; anterior ventrals yellowish, black-edged, posterior dark grey to black.

---

* I have examined the type specimen in the Brussels Museum.

Total length 1360 millim.
Mexico.

a. ♀ (V. 213; C. ?).   Atoyac, Vera Cruz.   Mr. H. H. Smith [C.];
F. D. Godman, Esq. [P.].

4. Phrynonax lunulatus.

Spilotes pœcilonotus, part., *Günth. Cat.* p. 100 (1858), *and Biol. C.-Am., Rept.* p. 117 (1894).
Tropidodipsas lunulata, *Cope, Proc. Ac. Philad.* 1860, p. 517, and 1862, p. 348.
Spilotes lunulatus, *Cope, Bull. U.S. Nat. Mus.* no. 32, p. 71 (1887).
—— lunulatus, part., *Bocourt, Miss. Sc. Mex., Rept.* p. 694, pl. xlii. fig. 1 (1888).

Rostral a little broader than deep, just visible from above; internasals shorter than the præfrontals; frontal once and one third as long as broad, as long as its distance from the rostral, slightly shorter than the parietals; loreal a little longer than deep; one præocular, widely separated from the frontal; two postoculars; temporals 2+3; nine upper labials, fourth, fifth, and sixth entering the eye; six or seven lower labials in contact with the anterior chin-shields, which are a little longer than the posterior. Scales in 21 or 23 rows, dorsals feebly keeled. Ventrals very obtusely angulate, 202-204; anal entire; subcaudals 126-136. Olive-brown above; young with crescent-shaped brown, black-edged cross bands on the back, of which mere traces are left in the adult; belly dotted with brown.

Total length 1380 millim.; tail 380.
Central America.

a. ♂ (V. 202; C. 126).   Mexico.   Hr. Hugo Finck [C.].
(One of the types of *S. pœcilonotus.*)

5. Phrynonax fasciatus.

Spilotes fasciatus, *Peters, Mon. Berl. Ac.* 1869, p. 443.
—— fasciatus, part., *Günth. Ann. & Mag. N. H.* (4) ix. 1872, p. 20.
Ahætulla polylepis, *Peters, Mon. Berl. Ac.* 1867, p. 709.
Spilotes lunulatus, part., *Bocourt, Miss. Sc. Mex., Rept.* p. 694 (1888).

Rostral broader than deep, just visible from above; internasals as long as or a little shorter than the præfrontals; frontal a little longer than broad, as long as its distance from the end of the snout, nearly as long as the parietals; loreal as long as deep or a little longer; one præocular, in contact with or very narrowly separated from the frontal; two postoculars; temporals 2+2; eight (or nine) upper labials, fourth, fifth, and sixth (or fifth, sixth, and seventh) entering the eye, last very long; seven or eight lower labials in contact with the anterior chin-shields, which are shorter than the posterior. Scales in 23 rows, dorsals faintly keeled. Ventrals very obtusely angulate, 191-207; anal entire; subcaudals 120-132. Young pale brown above, with crescentic brown, black-edged cross bands; head pale brown, speckled with darker; belly closely speckled or powdered with brown. Adult uniform dark olive-brown.

Total length 1500 millim.; tail 400.

Guianas and Amazons.

a. ♀ (V. 201; C. 129).   British Guiana.   Demerara Museum.
b. Yg. (V. 200; C. 125). Surinam.
c. ♀ (V. 207; C. 120).   Peruvian Amazon. Mr. E. Bartlett [C.].

6. Phrynonax eutropis. (PLATE I. fig. 1.)

Spilotes fasciatus, part., *Günth. Ann. & Mag. N. H.* (4) ix. 1872, p. 20.

Rostral broader than deep, just visible from above; internasals as long as the præfrontals; frontal slightly longer than broad, as long as its distance from the end of the snout, as long as the parietals; loreal longer than deep; one præocular, forming a suture with the frontal; two postoculars; temporals 2+2; eight upper labials, fourth, fifth, and sixth entering the eye, eighth very long; seven or eight lower labials in contact with the anterior chinshields, which are much shorter than the posterior. Scales in 25 rows, the dorsals very strongly keeled, the keels forming continuous raised lines in the middle of the body. Ventrals very obtusely angulate, 191; anal entire; subcaudals 126. Uniform brown above; upper lip and anterior lower surfaces yellow; belly turning to olive-brown towards the tail.

Total length 1450 millim.; tail 430.

Habitat unknown.

a. ♂ (V. 191; C. 126).   ——?

7. Phrynonax chrysobronchus.

Spilotes chrysobronchus, *Cope, Journ. Ac. Philad.* (2) viii. 1876, p. 136, pl. xxviii. fig. 11; *Bocourt, Miss. Sc. Mex., Rept.* p. 695 (1888).

Rostral as deep as broad; internasals nearly as long as the præfrontals; frontal about once and two fifths as long as broad, as long as its distance from the end of the snout, as long as the parietals; loreal longer than deep; one præocular, nearly touching the frontal; two postoculars; temporals 2+2; seven upper labials, third, fourth, and fifth entering the eye, last very long; seven lower labials in contact with the anterior chin-shields, which are nearly as long as the posterior. Scales in 25 rows, all smooth except the row next to the vertebral, which is weakly keeled. Ventrals 220; anal entire; subcaudals 117. Brown, the scales dotted with lighter; head darker; one or more borders of the scales black; upper lip, throat, and anterior part of belly yellow; remainder of lower surfaces passing from brown to black below the tail.

Total length 1670 millim.; tail 422.

Coast of Costa Rica.

59. SPILOTES.

Tyria, part., *Fitzing. N. Class. Rept.* p. 29 (1826).
Spilotes, *Wagler, Syst. Amph.* p. 179 (1830).
Spilotes, part., *Dum. & Bibr. Erp. Gén.* vii. p. 218 (1854); *Günth. Cat. Col. Sn.* p. 96 (1858); *Jan, Elenco sist. Ofid.* p. 62 (1863); *Bocourt, Miss. Sc. Mex., Rept.* p. 685 (1888).

Maxillary teeth equal, 19 to 22; anterior mandibular teeth longest. Head slightly distinct from neck; eye moderate, with round pupil; no suboculars, loreal often absent. Body elongate, slightly compressed; scales keeled, very strongly imbricate, acutely pointed, with apical pits, in 14 or 16 rows; ventrals rounded. Tail rather long; subcaudals in two rows.

South America.

1. Spilotes pullatus.

Coluber pullatus, *Linn. Mus. Ad. Frid.* p. 35, pl. xx. fig. 3 (1754), *and S. N.* i. p. 388 (1766).
Cerastes coronatus, *Laur. Syn. Rept.* p. 83 (1768).
Coluber coronatus, *Gmel. S. N.* i. p. 1088 (1788).
—— variabilis, *Merr. Beitr.* ii. p. 40, pl. xii. (1790); *Wied, Beitr. Nat. Bras.* i. p. 271 (1825), *and Abbild.* (1831); *Schleg. Phys. Serp.* ii. p. 149, pl. vi. figs. 1 & 2 (1837).
—— plutonius, *Daud. Rept.* vi. p. 324 (1803).
Natrix caninana, *Merr. Tent.* p. 121 (1820).
Tyria pullata, *Fitz. N. Class. Rept.* p. 60 (1826).
Spilotes pullatus, *Wagl. Syst. Amph.* p. 179 (1830); *Bocourt, Miss. Sc. Mex., Rept.* pl. xliv. fig. 2 (1888).
—— variabilis, *Dum. & Bibr.* vii. p. 220 (1854); *Wucherer, Proc. Zool. Soc.* 1861, p. 324.
—— variabilis, part., *Günth. Cat.* p. 99 (1858).
—— pullatus, var. anomalolepis, *Bocourt, l. c.* p. 685, pl. xliv. fig. 3.

Rostral slightly broader than deep, visible from above; internasals broader than long, much shorter than the præfrontals; frontal once and one third to once and a half as long as broad, as long as or a little shorter than its distance from the end of the snout, a little shorter than the parietals; loreal very small or absent, the posterior nasal often in contact with the præocular, which is single; two postoculars; temporals 1+1; six or seven upper labials, third and fourth entering the eye, fifth or sixth very large; four lower labials in contact with the anterior chin-shields, which are as long as or a little shorter than the posterior. Scales in 16 rows. Ventrals 198–232; anal entire; subcaudals 90–120. Black above, uniform or with yellow spots which may form cross bands; snout and sides of head yellow, the sutures between the shields black; belly yellow, with irregular black cross bands.

Total length 2100 millim.; tail 490.

South America east of the Andes.

A. Black and yellow.

a. ♀ (V. 232; C. 103).	Berbice.	
b. ♀ (V. 217; C. 98).	Brit. Guiana.	Demerara Museum.

$c, d.$ ♀ (V. 222; C. 105) & yg. (V. 218; C. 111).		Upper Amazon.	Mr. E. Bartlett [C.].
$e.$ Yg. (V. 227; C. 108).		Moyobamba,N.E.Peru.	Mr. A. H. Roff [C.].
$f.$ ♂ (V. 222; C. 117).		Pernambuco.	W.A.Forbes,Esq.[P.].
$g.$ Hgr. ♀ (V. 211; C. 120).		Minas Geraes.	
$h.$ ♂ (V. 198; C. 111).		Rio Grande do Sul.	Dr. H. v. Ihering [C.].
$i.$ Skeleton.		Surinam.	
$k.$ Skull.		Brazil.	

B. Nearly entirely black, only the anterior ventrals being partly yellow.

$l.$ ♂ (V. 219; C. 111).   ——?

2. Spilotes megalolepis. (Plate II.)

Spilotes megalolepis, *Günth. Ann. & Mag. N. H.* (3) xv. 1865, p. 93.

Closely allied to the preceding, but internasals not broader than long; frontal scarcely longer than broad; eight upper labials, fourth and fifth entering the eye; and scales much larger, the largest quite as large as the parietal shields, and in 14 rows only. Coloration as in *S. pullatus*, but darker, the posterior half of the body and the tail being entirely black, above and below.

Total length 2300 millim.

South America (?).

$a.$ ♀ (V. 220; C. ?).   ——?   (Type.)

60. COLUBER.

Coluber, part., *Linn. S. N.* i. p. 375 (1766); *Wagl. Syst. Amph.* p. 179 (1830); *Boie, Isis,* 1827, p. 518; *Schleg. Phys. Serp.* ii. p. 125 (1837); *Günth. Cat.* p. 87 (1858), *and Rept. Brit. Ind.* p. 237 (1864).

Natrix, part., *Laur. Syn. Rept.* p. 73 (1768).

Chironius, part., *Fitzing. N. Class. Rept.* p. 29 (1826).

Tyria, part., *Fitzing. l. c.*

Gonyosoma, *Wagler, Icon. Amph.* (1828), *and Syst. Amph.* p. 184; *Dum. & Bibr. Erp. Gén.* vii. p. 213 (1854); *Günth. Cat.* p. 122, *and Rept. Brit. Ind.* p. 293.

Zamenis, part., *Wagl. Syst. Amph.* p. 188.

Rhinechis, *Michahelles, in Wagl. Icon. Amph.* (1833); *Dum. & Bibr. t. c.* p. 225; *Günth. Cat.* p. 85; *Cope, Proc. U.S. Nat. Mus.* xiv. 1892, p. 637.

Xenodon, part., *Schleg. l. c.* p. 80.

Herpetodryas, part., *Schleg. l. c.* p. 173.

Callopeltis, *Bonap. Mem. Acc. Torin.* (2) ii. 1839, p. 101.

Elaphis, *Bonap. l. c.* p. 102; *Dum. & Bibr. t. c.* p. 241; *Günth. Cat.* p. 92, *and Rept. Brit. Ind.* p. 240; *Bocourt, Miss. Sc. Mex., Rept.* p. 683 (1888).

Pituophis, *Holbr. N. Am. Herp.* iv. p. 7 (1842); *Dum. & Bibr. t. c.* p. 232; *Günth. Cat.* p. 85; *Bocourt, l. c.* p. 665; *Cope, Proc. U.S. Nat. Mus.* xiv. 1892, p. 638.

Cynophis, *Gray, Ann. & Mag. N. H.* (2) iv. 1849, p. 246; *Günth. Cat.* p. 95, *and Rept. Brit. Ind.* p. 246.

Alopecophis, *Gray, l. c.* p. 247.
Churchillia, *Baird & Gir. in Stansbury's Explor. Great Salt Lake,* p. 350 (1852).
Plagiodon, *Dum. & Bibr. Mém. Ac. Sc.* xxiii. 1853, p. 447, *and Erp. Gén.* vii. p. 170.
Compsosoma (*non Serv.*), *Dum. & Bibr. ll. cc.* pp. 453, 291; *Günth. Rept. Brit. Ind.* p. 243.
Scotophis, *Baird & Gir. Cat. N. Amer. Rept.* p. 73 (1853); *Bocourt, l. c.* p. 678.
Georgia, *Baird & Gir. l. c.* p. 92.
Spilotes, part., *Dum. & Bibr. Erp. Gén.* vii. p. 218; *Günth. Cat.* p. 96; *Bocourt, l. c.* p. 685; *Cope, Proc. U.S. Nat. Mus.* xiv. 1892, p. 636.
Ablabes, part., *Dum. & Bibr. t. c.* p. 304.
Arizona, *Kennicott, Rep. U.S. Mex. Bound. Surv.* ii. *Rept.* p. 18 (1859); *Bocourt, l. c.* p. 676.
Drymobius, part., *Cope, Proc. Ac. Philad.* 1860, p. 560; *Bocourt, l. c.* p. 715.
Cœlognathus, *Cope, l. c.* p. 563.
Natrix, *Cope, Proc. Ac. Philad.* 1862, p. 338, *and Bull. U.S. Nat. Mus.* no. 32, 1887, p. 71.
Coronella, part., *Jan, Arch. Zool. Anat. Phys.* ii. 1863, p. 236.
Phyllophis, *Günth. Rept. Brit. Ind.* p. 295.
Geoptyas, *Steindachn. Sitzb. Ak. Wien,* lv. 1867, p. 269.
Allophis, *Peters, Mon. Berl. Ac.* 1872, p. 686.
Coluber, *Cope, Proc. U.S. Nat. Mus.* xi. 1888, p. 389; *Bouleng. Faun. Ind., Rept.* p. 330 (1890); *Cope, Proc. U.S. Nat. Mus.* xiv. 1892, p. 630.
Pantherophis, *Garman, Bull. Essex Inst.* xxiv. 1892, p. 109.

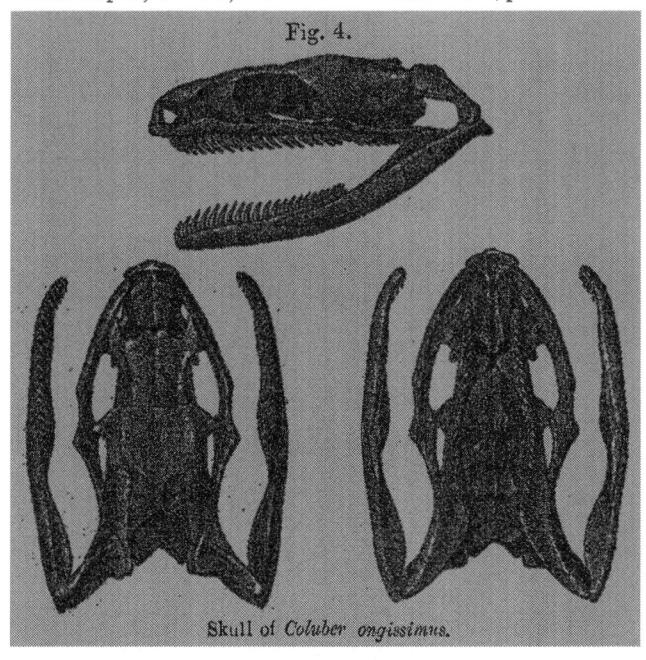

Fig. 4.

Skull of *Coluber ongissimus*.

Maxillary teeth 12 to 22, subequal in size; anterior mandibular teeth longest. Head distinct from neck, elongate; eye moderate or rather large, with round pupil; loreal sometimes absent. Body elongate, cylindrical or feebly compressed; scales smooth or keeled, with apical pits, in 15 to 35 rows; ventrals rounded or angulate laterally. Tail moderate or long; subcaudals in two rows.

Europe, Asia, North and Tropical America.

The species comprised under this genus form a series nearly parallel to that obtained in *Zamenis*, the extreme forms of both these genera showing much the same amount of differentiation.

Synopsis of the Species.

I. Rostral shield not deeper than broad, the portion visible from above not half as long as its distance from the frontal.

A. Ventrals without or with a rather indistinct lateral keel*.

1. Scales in 15 to 19 rows.

   a. Anal entire.

Scales in 15 rows, smooth..............	1. *dichrous*, p. 30.
Scales in 17 rows, smooth or feebly keeled.	2. *corais*, p. 31.
Scales in 19 rows; dorsals more or less distinctly keeled......................	[p. 33. 3. *novæ-hispaniæ*,

   b. Anal divided.

Scales in 17 rows, keeled	4. *melanotropis*, p. 33.
Scales in 19 rows, smooth..............	5. *porphyraceus*, p. 34.

2. Scales in 21 rows or more (rarely 19 in *C. cantoris*), smooth or feebly keeled.

   a. Anal usually entire.

Scales in 19 or 21 rows; subcaudals 65-76.	6. *cantoris*, p. 35.
Scales in 23 to 27 rows; subcaudals 75-94.	8. *helena*, p. 36.

   b. Anal usually divided.

      α. Scales smooth or faintly keeled.

         * Rostral as broad as deep.

Scales in 23 rows; subcaudals 79-90	7. *hodgsonii*, p. 35.
Scales in 31 to 35 rows; subcaudals 91-119.	9. *triaspis*, p. 37.

---

* In some doubtful cases it will be necessary to try, also, division B of this key.

60. COLUBER.

** Rostral broader than deep.

† No subocular below the præocular.

‡ Scales in 23 rows or more; third labial not entering the eye.

Scales in 33 rows, smooth or faintly keeled; ventrals 247–260; subcaudals 112 .... 10. *chlorosoma*, p. 38.
Scales in 27 to 33 rows, faintly keeled; ventrals 252–266; subcaudals 102–119. 11. *flavirufus*, p. 39.
Scales in 25 to 31 rows, faintly keeled on the posterior part of the body; ventrals 200–239; subcaudals 60–88 .... 12. *guttatus*, p. 39.
Scales in 25 or 27 rows, all perfectly smooth; ventrals 222–260; subcaudals 68–89 .. 13. *leopardinus*, p. 41.
Scales in 23 or 25 rows; ventrals 201–228; subcaudals 60–66 .... 14. *hohenackeri*, p. 42.

‡‡ Scales in 23 rows; third and fourth labials entering the eye; ventrals 208–222; subcaudals 62–74...... 15. *mandarinus*, p. 42.

‡‡‡ Scales in 21 rows; ventrals 162–190; subcaudals 47–68...... 16. *rufodorsatus*, p. 43.

†† A subocular below the præocular; scales in 23 to 27 rows; ventrals 172–214; subcaudals 50–80 .... 17. *dione*, p. 44.

β. Scales feebly but distinctly keeled, in 23 or 25 rows.

* Frontal considerably longer than broad. [p. 45.

Ventrals 195–234; subcaudals 63–90 .... 18. *quatuorlineatus*,
Ventrals 230–284; subcaudals 90–107 .. 19. *tæniurus*, p. 47.

** Frontal not or but slightly longer than broad.

Scales in 23 rows; ventrals 208–221; subcaudals 61–76; loreal considerably longer than deep .... 20. *schrenckii*, p. 48.
Scales in 25 rows (rarely 23 or 27); ventrals 200–234; subcaudals 67–85 .... 21. *vulpinus*, p. 49.

3. Scales strongly keeled, in 27 rows . 40. *lineaticollis*, p. 64.

B. Ventrals with a distinct, though often very obtuse, lateral keel.

1. Anal usually divided; scales in 21 rows or more.

   a. Subcaudals less than 125.

   α. No subocular below the præocular.

   * Scales in 25 to 29 rows.

Scales smooth or very faintly keeled...... 22. *lætus*, p. 49.
Scales very distinctly keeled .... 23. *obsoletus*, p. 50.

** Scales in 21 or 23 rows. [p. 51.

Third and fourth labials entering the eye .. 24. *conspicillatus*,
Fourth and fifth labials entering the eye .. 25. *longissimus*, p. 52.

β. A subocular below the præocular.

* Scales in 23 or 25 rows.

Scales feebly keeled; ventrals 226–244; [p. 54.
  subcaudals 97–122 .... 26. *climacophorus*,
Scales strongly keeled; ventrals 209–220;
  subcaudals 80–96 .... 27. *phyllophis*, p. 55.
Scales strongly keeled; ventrals 173; sub-
  caudals 70 .... 28. *davidi*, p. 56.

** Scales in 27 rows; ventrals 268–274; subcaudals
97–99 .... 29. *moellendorffii*, p. 56.

b. Subcaudals 122–149; scales smooth or faintly keeled, in 23 to 27 rows; ventrals 236–263.

Two labials entering the eye; green above. 30. *oxycephalus*, p. 56
Three labials entering the eye; tail black .. 31. *janseni*, p. 57.

2. Anal usually divided; scales in 17 or 19 rows; ventrals 193–210.

a. Scales in 19 rows.

α. Three labials entering the eye.

No loreal; subcaudals 120–121 .... 32. *frenatus*, p. 58.
A loreal; subcaudals 91–107 .... 33. *prasinus*, p. 59.

β. Two labials entering the eye; subcaudals 70–96.
3. Anal entire; scales keeled. 34. *quadrivirgatus*, p. 59.

a. Scales in 19 rows, 21 on the neck.

All caudal scales keeled .... 35. *melanurus*, p. 60.
Outer caudal scales smooth .... 36. *radiatus*, p. 61.

b. Scales in 21 rows, 23 on the neck. 37. *erythrurus*, p. 62.

c. Scales in 23 or 25 rows, 25 or 27 on the neck.

Rostral a little broader than deep; three
  labials entering the eye .... 38. *enganensis*, p. 63.
Rostral much broader than deep; two la-
  bials entering the eye .... 39. *subradiatus*, p. 64.

II. The portion of the rostral shield visible from above measuring at least half its distance from the rostral.

A. Scales smooth.

Rostral deeper than broad; anal usually
  divided .... 41. *scalaris*, p. 65.
Rostral as deep as broad; anal entire .... 42. *arizonæ*, p. 66.

B. Scales keeled; anal entire.

A pair of præfrontals; two labials entering the eye; rostral not or but slightly deeper than broad; scales in 27 or 29 rows .... 43. *deppii*, p. 66.

Præfrontals usually broken up into two pairs or more; not more than one labial entering the eye; rostral not or but slightly deeper than broad; scales in 29 to 35 rows .... 44. *catenifer*, p. 67.

Præfrontals broken up into two pairs or more; rostral much deeper than broad, the portion visible from above measuring more than half its distance from the frontal; scales in 29 to 35 rows ...... 45. *melanoleucus*, p. 68.

TABLE SHOWING NUMBERS OF SCALES AND SHIELDS.

I. Old-World species.

	Sc.	V.	A.	C.	Labials entering eye
porphyraceus	19	190–215	2	52–76	4th, 5th.
cantoris	19–21	213–232	1–2	65–76	4th, 5th.
hodgsonii	23	233–246	2	79–90	4th, 5th.
helena	23–27	220–265	1	75–94	5th, 6th, or 4th, 5th, 6th.
leopardinus	25–27	222–260	2	68–89	4th, 5th.
hohenackeri	23–25	201–228	2	60–66	4th, 5th.
mandarinus	23	208–222	2	62–74	3rd, 4th.
rufodorsatus	21	162–190	2	47–68	3rd, 4th, or 4th, 5th.
dione	23–27	172–214	2	50–80	4th, 5th, or 5th, 6th.
quadrilineatus	23–25	195–234	2	63–90	4th, 5th, or 5th, 6th.
tæniurus	23–25	230–284	1–2	90–107	4th, 5th, or 5th, 6th.
schrenckii	23	208–221	2	61–76	4th, 5th.
conspicillatus	21	200–224	2	60–76	3rd, 4th.
longissimus	21–23	212–248	2	60–91	4th, 5th, or 5th, 6th.
climacophorus	23–25	226–244	1–2	97–122	4th, 5th.
phyllophis	23	209–220	1–2	80–96	4th, 5th, or 5th, 6th.
davidi	23	173	2	70	4th, 5th.
moellendorffii	27	268–274	2	97–99	6th, 7th.
oxycephalus	23–27	233–263	2	122–149	5th, 6th, or 6th, 7th.
jansenii	23–25	247–255	2	133–140	5th, 6th, 7th.
frenatus	19	203–204	2	120–121	4th, 5th, 6th.
prasinus	19	198–206	1–2	91–107	4th, 5th, 6th.
quadrivirgatus	19	193–210	1–2	70–96	4th, 5th.
melanurus	19	193–234	1	89–109	4th, 5th, 6th.
radiatus	19	224–242	1	85–100	3rd, 4th, 5th, or 4th, 5th, or 4th, 5th, 6th.
erythrurus	21	211–233	1	86–100	4th, 5th, 6th.
enganensis	23	239–243	1	107–108	5th, 6th.
subradiatus	23–25	228–242	1	80–100	5th, 6th.
scalaris	25–29	201–220	1–2	48–68	4th or 4th, 5th.

II. American species.

	Sc.	V.	A.	C.	Labials entering eye.
dichrous	15	161–176	1	87–101	4th, 5th, or 3rd, 4th, 5th.
corais	17	184–215	1	53–83	3rd, 4th, or 4th, 5th.
novæ-hispaniæ	19	204–222	1	115–138	4th, 5th.
melanotropis	17	152	2	94	4th, 5th, 6th.
triaspis	31–35	243–282	2	91–119	4th, 5th, or 5th, 6th.
chlorosoma	33	247–260	2	107–112	4th, 5th, or 5th, 6th.
flavirufus	27–33	252–266	2	102–119	4th, 5th, or 4th, 5th, 6th.
guttatus	25–31	200–239	2	60–88	4th, 5th.
vulpinus	23–27	200–234	2	67–85	4th, 5th.
lætus	25–29	220–240	2	72–81	4th, 5th.
obsoletus	25–29	217–245	2	72–93	4th, 5th.
lineaticollis	27	244	1	69	4th, 5th, or 4th, 5th, 6th.
arizonæ	27–31	212–227	1	46–59	4th, 5th.
deppii	27–29	209–233	1	51–67	3rd, 4th, or 4th, 5th.
catenifer	29–35	204–245	1	66–72	4th or 0.
melanoleucus	29–35	209–239	1	45–65	4th or 5th.

1. Coluber dichrous.

Herpetodryas dichroa, *Peters, Mon. Berl. Ac.* 1863, p. 284.
—— occipitalis, *Günth. Ann. & Mag. N. H.* (4) i. 1868, p. 420, and ix. 1872, p. 23.
Spilotes piceus, *Cope, Proc. Ac. Philad.* 1868, p. 105, *and Journ. Ac. Philad.* (2) viii. 1876, p. 180.

Rostral broader than deep, visible from above; internasals broader than long, shorter than the præfrontals; frontal nearly once and a half as long as broad, a little longer than its distance from the end of the snout, shorter than the parietals; loreal nearly as long as deep; one præ- and two postoculars; temporals 2+2, rarely 1+2; eight upper labials, fourth and fifth, and usually also the third, entering the eye; four or five lower labials in contact with the anterior chin-shields, which are much shorter than the posterior. Scales smooth, in 15 rows. Ventrals 161–176; anal entire; subcaudals 87–101. Adult uniform olive-brown or blackish olive above and on the sides of the ventrals, yellowish white inferiorly. Young black above, with narrow yellowish cross bands; head yellow or pale olive, the shields on the snout edged with black, with a large black patch covering the forehead, a black spot below the eye, and a blackish bar across the occipital region, or a pair of large black blotches on the parietal shields and on the occipit.

Total length 1310 millim.; tail 360.

Brazil and North-eastern Peru.

a, b. ♂ (V. 166, 169; C. 98, 97).	Bahia.	Dr. O. Wucherer [C.].
c. Yg. (V. 175; C. 96).	Pebas.	Mr. Hauxwell [C.].
		(Type of *H. occipitalis*.)
d–e. ♂ (V. 161, 169; C. 94, 87).	Pebas.	Mr. E. Bartlett [C.].

f. ♀ (V. 168; C.?).	Yurimaguas, Huallaga River, N.E. Peru.	Dr. Hahnel [C.].
g. Yg. (V. 170; C. 101).	Moyobamba, N.E. Peru.	Hr. A. H. Roff [C.].
h. Yg. (V. 176; C. 97).	Sarayacu, N.E. Peru.	Mr. W. Davis [C.]; Messrs. Veitch [P.].

2. Coluber corais.

? Coluber reticularis, *Daud. Rept.* vi. p. 281 (1803).
Coluber corais, *Boie, Isis,* 1827, p. 537; *Schleg. Phys. Serp.* ii. p. 139, pl. v. figs. 9 & 10 (1837), *and Abbild.* p. 102, pl. xxviii. figs. 9–11 (1844).
—— obsoletus (*non Say*), *Holbr. N. Am. Herp.* iii. p. 61, pl. xii. (1842).
—— couperi, *Holbr. l. c.* p. 75, pl. xvi.
Georgia couperi, *Baird & Gir. Cat. N. Am. Rept.* p. 92 (1853).
—— obsoleta, *Baird & Gir. l. c.* p. 158, *and U.S. Mex. Bound. Surv.* ii., *Rept.* pl. xv. (1859).
Spilotes corais, *Dum. & Bibr.* vii. p. 223 (1853); *Günth. Cat.* p. 98 (1858); *Wucherer, Proc. Zool. Soc.* 1861, p. 324; *Günth. Zool. Rec.* 1867, p. 140; *Jan, Icon. Gén.* 48, pl. iv. fig. 6 (1876); *Garm. N. Am. Rept.* p. 48 (1883); *Cope, Proc. U.S. Nat. Mus.* xiv. 1892, p. 636; *Günth. Biol. C.-Am., Rept.* p. 116 (1894).
—— melanurus, *Dum. & Bibr. t. c.* p. 224; *Cope, Proc. Ac. Philad.* 1860, p. 564; *Jan, l. c.* pl. v. fig. 2.
—— erebennus, *Cope, l. c.* pp. 342 & 364, *and Proc. Ac. Philad.* 1864, p. 167; *Sumichrast, Arch. Sc. Phys. Nat.* (2) xlvi. 1873, p. 259.
—— couperi, *Cope, Proc. Ac. Philad.* 1860, pp. 342 & 364; *Garm. l. c.* p. 48.
Geoptyas collaris, *Steindachn. Sitzb. Ak. Wien,* lv. 1867, p. 271, pl. iii. figs. 4–7.
—— flaviventris, *Steindachn. l. c.* p. 269, pl. iv. figs. 4–7.
Spilotes corais, var. melanurus, *Cope, Journ. Ac. Philad.* (2) viii. 1876, p. 135; *Bocourt, Miss. Sc. Mex., Rept.* p. 687, pl. xliv. fig. 1 (1888).
—— ——, var. erebennus, *Cope, l. c.*
Coryphodon constrictor, *Jan, l. c.* pl. vi. fig. 1.
Spilotes obsoletus, *Garm. l. c.* p. 48.
—— corais, var. obsoletus, *Boccourt, l. c.* p. 689.
—— —— xanthurus, *A. E. Brown, Proc. Ac. Philad.* 1893, p. 433.

Rostral broader than deep, visible from above; internasals as long as broad or a little broader, shorter than the præfrontals; frontal as long as broad or a little longer, as long as its distance from the rostral or the end of the snout, much shorter than the parietals; loreal nearly as long as deep; one præ- or two postoculars; rarely a small subocular below the præocular; temporals 2+2; eight (rarely seven) upper labials, fourth and fifth (or third and fourth) entering the eye; four lower labials in contact with the anterior chin-shields, which are as long as or a little longer than the posterior. Scales in 17 rows (19 or 21 on the neck), smooth, or with a short and feeble keel, disposed somewhat obliquely in the

32 COLUBRIDÆ.

young. Ventrals without keel, 184–215; anal entire; subcaudals 53–83. Brown to black above, yellow to black inferiorly.
Total length 1880 millim.; tail 400.
From the Southern United States to Brazil and Bolivia.

A. Belly and lower surface of tail uniform yellow; no black on the labials. (*C. corais*, Boie; *C. flaviventris*, Steind.)

a. ♀ (V. 215; C. 72).	Trinidad.	
b. ♀ (V. 210; C. 71).	Demerara.	
c. ♂ (V. 203; C. 79).	Berbice	
d. ♀ (V. 206; C. 78).	Caripe, Para.	J. P. G. Smith, Esq. [P.].
e. ♂ (V. 178; C. 81).	Para.	
f, g. ♀ (V. 211; C. 77) & yg. (V. 202; C. 77).	Pernambuco.	J. P. G. Smith, Esq. [P.].
h. ♂ (V. 212; C. ?).	Pernambuco.	W. A. Forbes, Esq. [P.].
i. ♀ (V. 202; C. 72).	Bahia.	Dr. O. Wucherer [C.].
k. ♂ (V. 203; C. 73).	Bahia.	
l. ♂ (V. 204; C. 76).	Brazil.	Zoological Society.
m. ♀ (V. 215; C. 76).	Bolivia.	M. Suarez [P.].
n. Skull.	Cayenne.	

B. Lower surface of tail black; belly with the shields often black-edged or entirely black towards the tail; at least the labials below the eye with deep black bars on the sutures; usually an oblique black streak on each side of the neck. (*C. melanurus*, D. & B.; *C. collaris*, Steind.)

a. Yg. (V. 184; C. 61).	Duval Co., Texas.	W. Taylor, Esq. [C.].
b. ♀ (V. 190; C. 69).	Presidio, near Mazatlan.	Hr. A. Forrer [C.]; F. D. Godman, Esq. [P.].
c–d. ♀ (V. 203, 200; C. 77, 81).	Tres Marias Ids.	Hr. A. Forrer [C.].
e. ♀ (V. 189; C. 60).	Tampico.	F. D. Godman, Esq. [P.].
f. Hgr. (V. 197; C. 71).	Teapa, Tabasco.	F. D. Godman, Esq. [P.].
g. Yg. (V. 193; C. 77).	Oaxaca.	M. Sallé [C.].
h, i. ♂ (V. 196; C. 77) & yg. (V. 195; C. 76).	Yucatan.	
k. Hgr. (V. 199; C. 77).	Stann Creek, Brit. Honduras.	Rev. J. Robertson [C.].
l. Yg. (V. 209; C. 73).	San Gerónimo.	O. Salvin, Esq. [C.].
m, n. ♂ (V. 198; C. 75) & ♀ (V. 198; C. 71).	Dueñas, Guatemala.	O. Salvin, Esq. [C.].
o. ♂ (V. 198; C. 76).	Lanquin, Guatemala.	O. Salvin, Esq. [C.].
p. ♂ (V. 202; C. 77).	Guatemala.	O. Salvin, Esq. [C.].
q–r. Yg. (V. 192, 203; C. 78, 76).	Cartago, Costa Rica.	

C. Nearly entirely black. (*C. obsoletus*, Holbr.; *C. couperi*, Holbr.; *C. erebennus*, Cope.)

a, b. ♂ (V. 189; C. 68) & hgr. (V. 189; C. 73).	Mexico.	Mr. H. Finck [C.].
c. ♀ (V. 198; C. 65).	Atoyac, Mexico.	F. D. Godman, Esq. [P.].

3. Coluber novæ-hispaniæ.

Seba, Thes. ii. pl. xx. fig. 1 (1735).
Cerastes mexicanus, *Laur. Syn. Rept.* p. 83 (1768).
Coluber novæ-hispaniæ, *Gmel. S. N.* i. p. 1088 (1788).
Spilotes variabilis, part., *Günth. Cat.* p. 99 (1858).
—— pullatus auribundus, *Cope, Proc. Ac. Philad.* 1861, p. 300.
—— salvinii, *Günth. Ann. & Mag. N. H.* (3) ix. 1862, p. 125, pl. ix. fig. 5, and *Biol. C.-Am., Rept.* p. 116, pl. xlii. (1894).
—— variabilis, *Sumichr. Arch. Sc. Phys. Nat.* (2) xlvi. 1873, p. 259.
—— variabilis, var. auribundus, *Garm. N. Am. Rept.* p. 50 (1883).
—— auribundus, *Cope, Bull. U.S. Nat. Mus.* no. 32, 1887, p. 71; *Bocourt, Miss. Sc. Mex., Rept.* p. 689, pl. xliv. fig. 5 (1888).

Rostral broader than deep, visible from above; internasals as long as broad or a little broader than long, shorter than the præfrontals; frontal as long as or a little longer than broad, as long as its distance from the rostral or the end of the snout, as long as or a little shorter than the parietals; loreal small, as long as deep or a little longer; one præ- and two postoculars; temporals 1+1 or 1+2; eight upper labials, fourth and fifth entering the eye, seventh and eighth very large; three to five lower labials in contact with the anterior chin-shields, which are as long as or a little shorter than the posterior. Scales in 19 rows (17 and 18 in spec. *h*), dorsals more or less distinctly keeled. Ventrals 204–222; anal entire; subcaudals 115–138. Black above, varied with yellow, the yellow forming regular cross bands on the posterior part of the body and on the tail; belly yellow, with irregular black spots.

Total length 2120 millim.; tail 550.

Mexico and Central America.

a. ♀ (V. 213; C. 119).	Mexico.	Hr. Hugo Finck [C.].
b. Yg. (V. 210; C. 120).	Mexico.	M. Sallé [C.].
c. ♀ (V. 212; C. 116).	Mexico.	
d. Yg. (V. 204; C. 127).	Belize.	
e, f. Hgr. (V. 212; C. 132) & yg. (V. 217; C. 130.	StannCreek,Brit.Honduras.	Rev. J. Robertson [C.].
g. ♂ (V. 212; C. ?).	Atoyac, Guerrero.	F. D. Godman, Esq. [P.].
h–i. ♂ (V. 208; C. 134) & vg. (V. 212; C. 123).	Huatuzco, Vera Cruz.	F. D. Godman, Esq. [P.].
k. Hgr. (V. 215; C. 138).	Yzabal, Guatemala.	O. Salvin, Esq. [C.]. (Type of *S. salvini*.)
l. ♀ (V. 220; C. 127).	Rio Motagua.	F. D. Godman, Esq. [P.].

4. Coluber? melanotropis.

Dendrophidium melanotropis, *Cope, Journ. Ac. Philad.* (2) viii. 1876, p. 134, pl. xxvi. fig. 1.
Elaphis melanotropis, *Cope, Bull. U.S. Nat. Mus.* no. 32, 1887, p. 71.

Rostral broader than deep; frontal bell-shaped, wide in front, contracted behind, as long as the parietals; loreal much longer than deep; one præocular, with a subocular below it; two postoculars; temporals 2+2; nine upper labials, fourth, fifth, and sixth entering the eye. Scales in 17 rows, all keeled except the two outer on each side; the keels on the row on each side of the vertebral strongest. Ventrals not angulate, 152; anal divided; subcaudals 94. Green above and on the outer fourth of the ventral shields, yellow beneath; the keels of the median three dorsal rows black.

Total length 1240 millim.; tail 365.

Costa Rica.

5. Coluber porphyraceus.

Coluber porphyraceus, *Cantor, Proc. Zool. Soc.* 1839, p. 51; *Günth. Rept. Brit. Ind.* p. 239, pl. xx. fig. J (1864); *Anders. Proc. Zool. Soc.* 1871, p. 172, *and An. Zool. Res. Yunnan*, p. 812 (1879); *Hubrecht, in Snelleman, Nat. Hist. Midden-Sumatra, Rept.* p. 5, pl. —. fig. 1 (1887).

Psammophis nigrofasciatus, *Cantor, l. c.* p. 53.

Coronella callicephalus, *Gray, Ann. & Mag. N. H.* (2) xii. 1853, p. 390; *Blyth, Journ. As. Soc. Beng.* xxiii. 1855, p. 289.

Coluber callicephalus, *Günth. Cat.* p. 92 (1858).

Ablabes porphyraceus, *Bouleng. Faun. Ind., Rept.* p. 308 (1890).

Rostral nearly twice as broad as deep, visible from above; internasals shorter than the præfrontals; frontal broad, as long as or a little longer than its distance from the end of the snout, shorter than the parietals; loreal rather small, longer than deep; a large præocular; two postoculars; temporals 1+2; eight upper labials, fourth and fifth entering the eye; four or five lower labials in contact with the anterior chin-shields, which are longer than the posterior. Scales smooth, in 19 rows. Ventrals without keel, 190–215; anal divided; subcaudals 52–76. Pale reddish brown above, with dark brown black-edged cross bands; a black streak along the middle of the head, and another on each side, from the eye to the first transverse band; the whole or posterior part of body and tail with two longitudinal black lines, in addition to the cross bands; lower parts uniform yellow.

Total length 760 millim.; tail 140.

Eastern Himalayas, hills of Assam, Burma, Yunnan, Malay Peninsula, Sumatra.

a. ♂ (V. 197; C. 68).	Khasi Hills.	Sir J. Hooker [P.]. (Type of *C. callicephalus*.)
b-c. ♂ (V. 208; C. 76) & ♀ (V. 206; C. 73).	Khasi Hills?	Dr. Griffith.
d-f. ♀ (V. 213, 213; C. ?, 72) & yg. (V. 199; C. 66).	Khasi Hills.	T. C. Jerdon, Esq. [P.].
g. Yg. (V. 195; C. 72).	Himalayas.	Col. Beddome [C.].
h-l. ♀ (V. 212; C. 60), hgr. (V. 202; C. 63), & yg. (V. 205, 190; C. 59, 65).	Darjeeling.	W. T. Blanford, Esq. [P.].
m-o. ♀ (V. 208; C. 57) & yg. (V. 204, 190; C. 60, 64).	Toungyi, Shan States, 5000 ft.	Lieut. Blakeway [C.].
p. Yg. (V. 215; C. 76).	Singapore.	Dr. Cantor. (Type of *P. nigrofasciatus*.)

6. Coluber cantoris.

Coluber reticularis (*non Daud.*), *Cantor, Proc. Zool. Soc.* 1839, p. 51; *Bouleng. Faun. Ind., Rept.* p. 332 (1890); *W. Sclater, Journ. As. Soc. Beng.* lx. 1891, p. 239.
Spilotes reticularis, *Günth. Cat.* p. 98 (1858).
Compsosoma reticulare, *Günth. Rept. Brit. Ind.* p. 245, pl. xxi. fig. D (1864); *Anders. Proc. Zool. Soc.* 1871, p. 172; *Theob. Cat. Rept. Brit. Ind.* p. 166 (1876).

Rostral as deep as broad, or a little broader than deep, visible from above; internasals broader than long, shorter than the præfrontals; frontal once and one fourth to once and a half as long as broad, as long as or a little shorter than its distance from the end of the snout, as long as or a little shorter than the parietals; loreal longer than deep; one large præocular, often with a small subocular below; two (exceptionally three) postoculars; temporals 2+2 or 2+3; eight upper labials, fourth and fifth entering the eye; five lower labials in contact with the anterior chin-shields, which are as long as or a little longer than the posterior. Scales in 19 or 21 rows, feebly keeled on the posterior part of the body. Ventrals 213–232; anal usually entire; subcaudals 65–76. Brown above, darker behind, anteriorly with squarish dark brown spots, posteriorly with more or less distinct light cross bands; head uniform pale brown; lower parts yellowish, spotted with brown or black, or nearly uniform dark brown.

Total length 1200 millim.; tail 225.

Himalayas, Khasi and Garo Hills, Burma.

a, b. ♂ (Sc. 19; V. 220; A. 2; C. 72) & yg. (Sc. 19; V. 213; A. 1; C. 65).	Nepal.	B. H. Hodgson, Esq. [P.].
c. ♀ (Sc. 21; V. 230; A. 1; C. 73).	Sikkim.	Messrs. v. Schlagintweit [C.].
d, e, f. ♀ (Sc. 21; V. 229; A. 1; C. 73) & yg. (Sc. 21, 21; V. 221, 227; A. 1, 2; C. 75, 70).	Sikkim.	Sir J. Hooker [P.].
g–h. ♀ (Sc. 21; V. 225; A. 1; C. 74) & yg. (Sc. 21; V. 221; A. 1; C. 73).	Darjeeling.	W. T. Blanford, Esq. [P.].
i, k, l. ♂ (Sc. 21; V. 222; A. 1; C. 75) & ♀ (Sc. 21, 21; V. 232, 232; A. 1, 1; C. 70, 76).	Khasi Hills.	Sir J. Hooker [P.].

7. Coluber hodgsonii.

Spilotes melanurus, part., *Günth. Cat.* p. 97 (1858).
—— hodgsonii, *Günth. Proc. Zool. Soc.* 1860, p. 156, pl. xxvii.
Compsosoma hodgsonii, *Günth. Rept. Brit. Ind.* p. 246 (1864); *Stoliczka, Journ. As. Soc. Beng.* xxxix. 1870, p. 189, and xl. 1871, p. 430; *Theob. Cat. Rep. Brit. Ind.* p. 166 (1876).
Coluber hodgsonii, *Bouleng. Faun. Ind., Rept.* p. 332 (1890).

Rostral as deep as broad, visible from above; internasals broader than long, much shorter than the præfrontals; frontal once and one third to once and a half as long as broad, as long as or a little shorter than its distance from the end of the snout, shorter than the parietals; loreal longer than deep, often united with the præfrontal; one large præocular, rarely with a small subocular below it; two postoculars; temporals 2+2 or 2+3; normally eight upper labials, fourth and fifth entering the eye; five lower labials in contact with the anterior chin-shields, which are as long as the posterior or a little longer. Scales in 23 rows, feebly keeled on the posterior part of the body. Ventrals 233–246; anal divided; subcaudals 79–90. Brownish olive above, most of the scales black-edged; young with blackish cross bands; lower parts yellowish, the outer part of the margin of each ventral shield blackish.

Total length 600 millim.; tail 230.

Himalayas.

a. ♀ (V. 245; C. 90).	Tsomoriri, Ladak.	Messrs. v. Schagintweit [C.].	(Types.)
b-c. ♂ (V. 246; C. 79) & ♀ (V. 233; C. 88).	Nepal.	B. H. Hodgson, Esq. [P.].	
d. ♀ (V. 234; C. 83).	Darjeeling.	W. T. Blanford, Esq. [P.].	

8. Coluber helena.

Russell, Ind. Serp. i. pl. xxxii. (1796).
Coluber helena, *Daud. Rept.* vi. p. 277 (1803); *Boulen*g. *Faun. Ind., Rept.* p. 331 (1890).
Herpetodryas helena, *Schleg. Phys. Serp.* ii. p. 192 (1837).
Cynophis bistrigatus, *Gray, Ann. & Mag. N. H.* (2) iv. 1849, p. 246.
Plagiodon helena, *Dum. & Bibr.* vii. p. 170 (1854); *Jan, Icon. Gén.* 20, pl. iv. fig. 1 (1867).
Herpetodryas malabaricus, *Jerdon, Journ. As. Soc. Beng.* xxii. 1854, p. 530.
Cynophis helena, *Günth. Cat.* p. 95 (1858); *Cope, Proc. Ac. Philad.* 1860, p. 565; *Günth. Rept. Brit. Ind.* p. 247 (1864); *Theob. Cat. Rept. Brit. Ind.* p. 167 (1876); *Blanf. Journ. As. Soc. Beng.* xlviii. 1879, p. 125; *Murray, Zool. Sind*, p. 376 (1883).
—— malabaricus, *Günth. Rept. Brit. Ind.* p. 248, pl. xxi. fig. A; *Theob. l. c.* p. 167.
—— malabaricus, var. carinata, *F. Müll. Verh. nat. Ges. Basel*, vi. 1878, p. 671.

Rostral a little broader than deep, visible from above; internasals broader than long, much shorter than the præfrontals; frontal once and one third to once and three fourths as long as broad, as long as its distance from the end of the snout, shorter than the parietals; loreal somewhat longer than deep; one large præocular, sometimes in contact with the frontal; two postoculars; temporals 2+2 or 2+3; nine (exceptionally ten or eleven) upper labials, fifth and sixth (or fourth, fifth, and sixth) entering the eye; five or six lower labials in contact with the anterior chin-shields, which are as long as or a little longer than the posterior. Scales in 23 to 27 rows, smooth, or feebly keeled on the posterior part of the body and on the tail. Ventrals 220-265; anal entire; subcaudals 75-94. Young pale brown above, with black cross bands, each enclosing four to six white ocelli; adult darker brown, with transverse series of black squarish spots, or with more or less distinct traces of the livery of the young; a vertical black streak below the eye and an oblique one behind the eye; some specimens have a white black-edged collar, others two black longitudinal streaks on the head. Lower parts yellowish, with or without a few small black spots; sometimes with a more or less distinct festooned marking on each side.

Total length 1260 millim.; tail 250.

India and Ceylon.

a. ♂ (Sc. 27; V. 228; C. 94).		Madras.	Sir W. Elliot [P.].
b-d. ♀ (Sc. 25; V. 242; C. 85) & yg. (Sc. 25, 25; V. 232, 228; C. 90, 78).		Matheran.	Dr. Leith [P.].
e-f. ♀ (Sc. 25, 25; V. 245, 265; C. 85, ?).		Malabar.	Col. Beddome [C.].
g-h. Yg. (Sc. 25, 25; V. 236, 247; C. 94, 75).		Anamallays.	Col. Beddome [C.]. (Types of *C. malabaricus*.)
i. Yg. (Sc. 27; V. 226; C. 80).		Anamallays.	Col. Beddome [C.].
k. ♀ (Sc. 25; V. 261; C. 80).		Peermad, Travancore.	H. S. Ferguson, Esq. [P.].
l. Hgr. (Sc. 27; V. 231; C. 81).		Ceylon.	R. Templeton, Esq. [P.]. (Type of *C. bistrigatus*.)
m-r. ♂ (Sc. 27; V. 235; C. 88) & yg. (Sc. 27, 27, 27, 27, 27; V. 224, 225, 231, 233, 220; C. 84, 94, 80, 82, ?).		Ceylon.	

9. Coluber triaspis.

Coluber triaspis, *Cope, Proc. Ac. Philad.* 1866, p. 128, *and Proc. Am. Philos. Soc.* xxii. 1885, p. 175; *Günth. Biol. C.-Am., Rept.* p. 115 (1894).

Pityophis intermedius, *Boettg. Ber. Offenb. Ver. Nat.* 23-24, p. 148 (1883).

Coluber mutabilis, *Cope, Proc. Am. Philos. Soc.* xxii. 1885, p. 175.

Natrix mutabilis, *Cope, Bull. U.S. Nat. Mus.* no. 32, 1887, p. 71.

—— triaspis, *Cope, l. c.*

Scotophis mutabilis, *Bocourt, Miss. Sc. Mex., Rept.* p. 680, pl. xlvi. fig. 2 (1888).

Rostral nearly as deep as broad, visible from above; internasals broader than long, much shorter than the præfrontals; frontal once and a half as long as broad, as long as its distance from the end of the snout, shorter than the parietals; loreal longer than deep; one præocular, sometimes in contact with the frontal; two postoculars; temporals 2 or 3+3 or 4; eight (rarely nine) upper labials, fourth and fifth (or fifth and sixth) entering the eye; four or five lower labials in contact with the anterior chin-shields, which are as long as the posterior. Scales in 31 to 35 rows, posterior dorsals faintly keeled. Ventrals 243–282; anal divided; subcaudals 91–119. Young pale brown above, with one or two dorsal series of large dark brown black-edged spots, and an alternating lateral series of smaller spots; two or three dark brown stripes on the occiput and nape; two dark brown curved bands across the snout; adult uniform brown; lower parts uniform yellowish.

Total length 610 millim.; tail 115. Reaches a length of over 1 metre.

Mexico and Central America.

a.	Hgr. (Sc. 35; V. 264; C. 95).	Mexico.	M. Sallé [C.].
b.	Yg. (Sc. 35; V. 254; C. 91).	City of Mexico.	Mr. Doorman [C.].
c.	Yg. (Sc. 35; V. 260; C. 102).	Mexico.	
d.	Hgr. (Sc. 35; V. 267; C. 100).	Mezquital del Oro, Zacatecas.	Dr. A. C. Buller [C.].

10. Coluber chlorosoma.

Coluber chlorosoma, *Günth. Biol. C.-Am., Rept.* p. 115, pl. xli. (1894).

Rostral considerably broader than deep, just visible from above; internasals nearly as long as broad, or broader than long, much shorter than the præfrontals; frontal once and one fourth or once and one third as long as broad, as long as its distance from the rostral or the end of the snout, shorter than the parietals; loreal longer than deep; one præocular; two postoculars; temporals 3+3 or 3+4; eight or nine upper labials, fourth and fifth or fifth and sixth entering the eye; four lower labials in contact with the anterior chin-shields, which are not longer than the posterior. Scales in 33 rows, smooth or posterior dorsals faintly keeled. Ventrals 247–260; anal divided; subcaudals 107–112. Uniform pale greyish above and on the outer ends of the ventrals, white beneath.

Total length 1050 millim.

Mexico.

a.	♂ (V. 260; C. ?).	Atoyak, Guerrero.	Mr. H. H. Smith [C.]; F. D. Godman, Esq. [P.].
b.	♂ (V. 247; C. 112).	Amula, Guerrero.	Mr. H. H. Smith [C.]; F. D. Godman, Esq. [P.]. (Types.)
c.	Head and neck.	S. Ramon, Jalisco, 1500 ft.	Dr. A. C. Buller [C.].
d.	♂ (V. 247; C. 107).	S. Mexico.	F. D. Godman, Esq. [P.].

Should perhaps rank as a variety of the following species.

11. Coluber flavirufus.

Coluber flavirufus, *Cope, Proc. Ac. Philad.* 1866, p. 319, *and Proc. Am. Philos. Soc.* xxii. 1885, p. 175; *Günth. Biol. C.-Am., Rept.* p. 115 (1894).
Elaphis pardalinus, *Peters, Mon. Berl. Ac.* 1868, p. 642.
—— rodriguezii, *Bocourt, Le Natur.* 1887, p. 168, *and Miss. Sc. Mex., Rept.* p. 683, pl. xlvi. fig. 1 (1888).
Natrix flavirufus, *Cope, Bull. U.S. Nat. Mus.* no. 32, 1887, p. 71.

Rostral considerably broader than deep, just visible from above; internasals much broader than long, much shorter than the præfrontals; frontal once and one third to once and a half as long as broad, nearly as long as its distance from the end of the snout, shorter than the parietals; loreal longer than deep; one large præocular (which may be divided into two); two postoculars; temporals 2 or 3+3 or 4; eight or nine upper labials, fourth and fifth or fourth, fifth, and sixth entering the eye; four or five lower labials in contact with the anterior chin-shields, which are as long as or a little longer than the posterior. Scales in 27 to 33 rows, dorsals faintly keeled. Ventrals 252-266; anal divided; subcaudals 102-119. Yellowish or pale brown above, with a dorsal series of reddish or chestnut-brown, black-edged, transverse spots, which may alternate or be confluent into a zigzag band; a lateral series of spots, alternating with the larger dorsal; two curved dark bands across the snout, some symmetrical markings on the crown and occiput, and two longitudinal bands on the occiput and nape, sometimes confluent in front and behind; lower parts yellowish, with or without small brown spots.

Total length 1220 millim.; tail 260.

Mexico, Guatemala, Honduras.

a. ♀ (Sc. 31; V. 252; C. 102). Mexico. Hr. H. Finck [C.].
b. Yg. (Sc. 31; V. 263; C. 116). Belize.
c. Yg. (Sc. 33; V. 262; C. 117). Ruatan Id. F. D. Godman, Esq. [P.].

12. Coluber guttatus.

Catesby, *Nat. Hist. Carol.* ii. pl. lv. (1771); *Merrem, Beitr.* ii. p. 39, pl. xi. (1790).
Coluber guttatus, *Linn. S. N.* i. p. 385 (1766); *Harl. Journ. Ac. Philad.* v. 1827, p. 363, *and Med. Phys. Res.* p. 126 (1835); *Holbr. N. Am. Herp.* iii. p. 65, pl. xiv. (1842); *Cope, Proc. U.S. Nat. Mus.* xiv. 1892, p. 632; *Hay, Batr. & Rept. Indiana,* p. 92 (1893); *Günth. Biol. C.-Am., Rept.* p. 114 (1894).
—— maculatus, *Bonnat. Encycl. Méth., Ophiol.* p. 9 (1789); *Harl. ll. cc.* pp. 362, 125.
—— compressus, *Donnd. Zool. Beitr.* p. 206 (198).
—— carolinianus, *Shaw, Zool.* iii. p. 460, pl. cxix. (1802).
—— molossus, *Daud. Rept.* i.p 269 (1803); *Harl. ll. cc.* pp. 363, 126.

Coluber pantherinus, *Daud. l. c.* p. 318.
—— floridanus, *Harl. ll. cc.* pp. 360, 124.
Scotophis guttatus, *Baird & Gir. Cat. N. Am. Rept.* p. 78 (1853); *Bocourt, Miss. Sc. Mex., Rept.* p. 678, pl. xlvi. fig. 4 (1888).
—— emoryi, *Baird & Gir. l. c.* p. 157; *Kennicott, Rep. U.S. Mex. Bound. Surv.* ii., *Rept.* p. 19, pl. xii. (1859).
Elaphis guttatus, *Dum. & Bibr.* vii. p. 273 (1854); *Jan, Icon. Gén.* 21, pl. vi. (1867); *Garm. N. Am. Rept.* p. 55, pl. iv. fig. 1 (1883); *H. Garm. Bull. Illin. Lab.* iii. 1892, p. 292.
—— rubriceps, *Dum. & Bibr. t. c.* p. 270.
Coluber guttatus, part., *Günth. Cat.* p. 89 (1858).
Scotophis calligaster, *Kennicott, Proc. Ac. Philad.* 1859, p. 98.
Coluber rhinomegas, *Cope, Proc. Ac. Philad.* 1860, p. 255.
Elaphis alleghaniensis, part., *Jan, l. c.* 24, pl. ii. fig. 2.
Coluber emoryi, *Cope, Check-List N. Am. Rept.* p. 39 (1875).
? Coluber bairdi, *Yarrow, in Cope, Bull. U.S. Nat. Mus.* no. 17, 1880, p. 41.
Natrix emoryi, *Cope, Bull. U.S. Nat. Mus.* no. 32, 1887, p. 71.
Coluber guttatus sellatus, *Cope, Proc. U.S. Nat. Mus.* xi. 1888, p. 387.

Rostral broader than deep, just visible from above; internasals broader than long, much shorter than the præfrontals; frontal once and a half to once and two thirds as long as broad, as long as its distance from the rostral or the end of the snout, shorter than the parietals; loreal longer than deep; one præ- and two or three postoculars; temporals 2 or 3+3 or 4; eight upper labials, fourth and fifth entering the eye; four or five lower labials in contact with the anterior chin-shields, which are as long as or a little longer than the posterior. Scales in 25 to 29 (rarely 31) rows; dorsals very feebly keeled. Ventrals not or but very obtusely angulate laterally, 200–239; anal divided; subcaudals 60–88. Yellowish or pale brown above, with a dorsal series of large, brown or red, black-edged spots, and an alternating lateral series of smaller spots; a curved dark band from eye to eye across the præfrontals, continued behind the eye to the angle of the mouth; a ∩ or ⋂-shaped marking from the frontal shield to the nape; labials usually with black sutures or spots; belly yellow, with large squarish black blotches.
Total length 1040 millim.; tail 160.
United States east of the Rocky Mountains; North Mexico.

A. 25 to 40 dorsal spots from the nape to the base of the tail; labials with black sutures. (*C. guttatus*, L.)

a. ♀ (Sc. 25; V. 234; N. America. E. Doubleday, Esq. [P.].
 C. 67).
b. Yg. (Sc. 27; V. 231; New York.
 C. 67).
c. ♂ (Sc. 27; V. 207; Tennessee. Christiania Museum.
 C. 65).
d. ♂ (Sc. 27; V. 220; Marion Co., Flo- A. Erwin Brown, Esq.
 C. 71). rida. [P.].
e. Skull. Charleston.

B. 35 to 50 dorsal spots from the nape to the base of the tail; labials without black sutures. (*C. emoryi*, B. & G.)

a. Yg. (Sc. 27; V. 207; Kansas. Smithsonian Instit. [P.]. C. 67).
b. ♀ (Sc. 31; V. 225; Duval Co., Texas. W. Taylor, Esq. [C.]. C. 65).

13. Coluber leopardinus.

Coluber quadrilineatus (*non Lacép.*), *Pall. Zoogr. Ross.-As.* iii. p. 40 (1811); *Günth. Cat.* p. 88 (1858); *Strauch, Schl. Russ. R.* p. 73 (1873); *Bedriaga, Bull. Soc. Nat. Mosc.* lvi. 1882, p. 298.
—— leopardinus, *Bonap. Icon. Faun. Ital.* (1834); *Schleg. Phys. Serp.* ii. p. 169 (1837).
—— cruentatus, *Steven, Bull. Soc. Nat. Mosc.* viii. 1835, p. 317.
Callopeltis leopardinus, *Bonap. Mem. Acc. Tor.* (2) ii. 1839, p. 432; *Nordm. in Demidoff, Voy. Russ. Mér.* iii. p. 348, *Rept.* pls. vi., viii., ix. (1840); *De Betta, Atti Ist. Ven.* (3) xiii. 1868, p. 934, *and Faun. Ital., Rett. Anf.* p. 38 (1874); *Tomasini, Wiss. Mitth. aus Bosn. u. Herzeg.* ii. p. 618 (1894).
Ablabes quadrilineatus, *Dum. & Bibr.* vii. p. 319 (1854).
Natrix leopardina, *Cope, Proc. Ac. Philad.* 1862, p. 338.
Coronella quadrilineata, *Jan, Arch. Zool. Anat. Phys.* ii. 1863, p. 247, *and Icon. Gén.* 13, pl. v. (1865).
Callopeltis quadrilineatus, *Schreib. Herp. Eur.* p. 277 (1875); *Camerano, Mon. Ofid. Ital., Colubr.* p. 52 (1891); *Minà-Palumbo, Nat. Sicil.* xii. 1893, p. 127.

Rostral broader than deep, just visible from above; internasals broader than long, shorter than the præfrontals; frontal once and one third to once and a half as long as broad, as long as its distance from the end of the snout, shorter than the parietals; loreal longer than deep; one præ- and two postoculars; temporals 1 or 2 + 2 or 3; eight upper labials, fourth and fifth entering the eye; four or five lower labials in contact with the anterior chin-shields, which are longer than the posterior. Scales smooth, in 25 or 27 rows. Ventrals 222–260, rounded; anal divided; subcaudals 68–89. Greyish or pale brown above, with a dorsal series of dark brown or reddish black-edged transverse spots and a lateral alternating series of smaller black spots, or with two dark brown black-edged stripes bordering a yellowish vertebral stripe; usually a Λ-shaped dark marking on the occiput and nape; a crescentic black band from eye to eye, an oblique black band from the postoculars to the angle of the mouth, and a black spot below the eye; lower parts white, checkered with black, or nearly entirely black.

Total length 900 millim.; tail 160.

Southern Italy, Malta, Dalmatia, Balkan Peninsula, Crimea, Asia Minor.

A. Spotted form. (*C. leopardinus*, Bp.)

a, b. ♂ (Sc. 25; V. 232; Malta. Miss Attersoll [P.]. C. 85) & yg. (Sc. 27; V. 241; C. 79).

c. Hgr. (Sc. 27; V. 245; Malta. C. A. Wright, Esq. [P.].
   C. 81).
d. ♀ (Sc. 27; V. 247; Zara, Dalmatia. Count M. G. Peracca [P.].
   C. 75).
e–g. ♂ (Sc. 27, 25; V. 241, Zara. Dr. F. Werner [E.].
   238; C. 84, 84) & ♀
   (Sc. 27; V. 245; C. ?).
h–i. ♂ (Sc. 25; V. 234; Zara. Hr. Spada-Novak [C.].
   C. 83) & ♀ (Sc. 25;
   V. 238; C. 68).

B. Striped form. (*C. quadrilineatus*, Pall.)

a. ♀ (Sc. 27; V. 247; Trieste. Dr. Rüppell [P.].
   C. 77).
b. Yg. (Sc. 25; V. 226; Ionian Ids. Dr. Mann [P.].
   C. 79).
c. ♂ (Sc. 25; V. 245; ——?
   C. 81).

14. Coluber hohenackeri.

Coluber rubriventris, *Dwigubsky, Essay Nat. Hist. Russ. Emp.* p. 67 [Russian] (1832).
—— hohenackeri, *Strauch, Schl. Russ. R.* p. 70 (1873); *Boettg. Ber. Senck. Ges.* 1890, p. 294.

Rostral broader than deep, just visible from above; internasals broader than long, shorter than the præfrontals; frontal a little longer than broad, as long as its distance from the end of the snout, shorter than the parietals; loreal a little longer than deep; one præ- and two postoculars; temporals 2+3; eight upper labials, fourth and fifth entering the eye; five lower labials in contact with the anterior chin-shields, which are longer than the posterior. Scales smooth or faintly keeled, in 23 (rarely 25) rows. Ventrals 201–228, rounded; anal divided; subcaudals 60–66. Grey above, with four alternating series of dark spots; a ∧-shaped black marking on the nape, a black streak from the eye to the angle of the mouth, and a black line below the eye; belly yellowish or reddish, closely spotted or marbled with blackish grey.

Total length 650 millim.; tail 110.

Transcaucasia and Asia Minor.

a–b. ♂ (V. 211; C. 63) & yg. Amasia, Asia Minor.
   (V. 221; C. 62).

15. Coluber mandarinus.

Coluber mandarinus, *Cantor, Zool. Chusan*, pl. xii. (1840), *and Ann. & Mag. N. H.* ix. 1842, p. 488; *Günth. Cat.* p. 91 (1858), *and Rept. Brit. Ind.* p. 238, pl. xx. fig. H (1864)*.

Rostral twice as broad as deep, just visible from above; internasals shorter than the præfrontals; frontal a little longer than

---

* Moellendorff's *Coluber mandarinus* (Journ. N. China Br. As. Soc. [2] xi. 1877, p. 104), "a black water-snake with orange-red marks," from the province of Chihli, Northern China, must be a different snake.

broad, longer than its distance from the end of the snout, as long as or a trifle shorter than the parietals; loreal small, as long as deep; one præ- and two postoculars; temporals 1+2 or 3; seven upper labials, third and fourth entering the eye; four lower labials in contact with the anterior chin-shields, which are longer than the posterior. Scales in 23 rows. Ventrals 208-222, rounded; anal divided: subcaudals 62-74. Scarlet above, with black spots and a dorsal series of large rhomboidal black spots with a yellow centre; head with black symmetrical markings as in *Simotes*, viz. a cross band on the end of the snout, a crescentic band across the forehead, passing through the eyes, and a ∧-shaped one on the occiput; lower parts yellow, with large black transverse spots or cross bands.

Total length 700 millim.; tail 120.

Chusan.

a. ♀ (V. 222; C. 62).   Chusan.   Dr. Cantor. (Type.)
b. Hgr. (V. 208; C. 74).   Chusan.

16. Coluber rufodorsatus.

Tropidonotus rufodorsatus, *Cantor, Zool. Chusan*, pl. xiii. (1840), and *Ann. & Mag. N. H.* ix. 1842, p. 483.
Ablabes sexlineatus, *Dum. & Bibr.* vii. p. 324 (1854).
Coluber rufodorsatus, *Günth. Cat.* p. 89 (1858), *and Rept. Brit. Ind.* p. 238, pl. xx. fig. G (1864); *Strauch, Schl. Russ. R.* p. 79 (1873); *Boettg. Ber. Offenb. Ver. Nat.* 1888, p. 70.
Coronella sexlineata, *Jan, Arch. Zool. Anat. Phys.* ii. 1863, p. 249, and *Icon. Gén.* 14, pl. vi. figs. 2 & 3 (1865).
Simotes herzi, *Boettg. Zool. Anz.* 1886, p. 519.
Ablabes rufodorsatus, *Bouleng. Ann. & Mag. N. H.* (6) v. 1890, p. 138.

Rostral broader than deep, scarcely visible from above; internasals as broad as long or a little longer than broad, as long as or a little shorter than the præfrontals; frontal once and a half to twice as long as broad, as long as its distance from the end of the snout, shorter than the parietals; loreal as long as deep or a little longer; one (rarely two) præ- and two postoculars; temporals 2+3; seven (rarely eight) upper labials, third and fourth (or fourth and fifth) entering the eye; four or five lower labials in contact with the anterior chin-shields, which are as long as the posterior. Scales smooth, in 21 rows. Ventrals 162-190, rounded; anal divided; subcaudals 47-68. Yellowish olive above, with an orange vertebral line and four series of olive, black-edged, elongate spots, which are posteriorly, or throughout, confluent into longitudinal bands; the two median of these bands meet on the vertex; a ∧-shaped black band on the snout, extending from eye to eye, and a black band from the eye to the angle of the mouth; belly bright yellow, checkered with black; tail with four or five black longitudinal streaks.

Total length 650 millim.; tail 95.

Eastern Siberia and China.

a. ♀ (V. 190; C. 48).		Western hills of Pekin.	S. W. Bushell, Esq. [P.].
b-c. ♂ (V. 167; C. 57) & ♀ (V. 174; C. 50).		Shanghai.	R. Swinhoe, Esq. [C.].
d-e. ♂ (V. 165; C. 58) & ♀ (V. 170; C. 47).		Mountains north of Kiu Kiang.	A. E. Pratt, Esq. [C.].
f. ♀ (V. 174; C. 49).		Chikiang.	Mr. Fortune [C.].
g. ♂ (V. 172; C. 59).		Hang-Chau.	J. J. Walker, Esq. [P.].
h. ♀ (V. 177; C. 50).		Ningpo.	
i-k. ♀ (V.178,178; C.51, 50).		Chusan.	Dr. Cantor. (Types.)
l. ♀ (V. 182; C. 50).		Chusan.	J. J. Walker, Esq. [P.].
m-n. ♀ (V. 183, 178; C. 54, 49).		Formosa.	R. Swinhoe, Esq. [C.].
o-r. ♂ (V.166,162; C.60, 53) & hgr. (V. 186,184; C. 53, 51).		Hoi-How, Hainan.	J. Neumann, Esq. [P.].
s. Yg. (V. 167; C. 68).		China.	R. Swinhoe, Esq. [C.].
t-u. ♀ (V. 179; C. 51) & yg. (V. 169; C. 59).		China.	

17. Coluber dione.

Coluber dione, *Pall. Reise*, ii. p. 717 (1773), *and Zoogr. Ross.-As.* iii. p. 39 (1811); *Ménétr. Cat. Rais.* p. 68 (1832).
Chironius dione, *Fitzing. N. Class. Rept.* p. 60 (1826).
Coluber eremita, *Eichw. Zool. Spec.* iii. p. 174 (1831).
—— trabalis, *Schleg. Phys. Serp.* ii. p. 167 (1837).
—— mæoticus, *Rathke, Mém. Sav. Etr. Ac. St. Pétersb.* iii. 1837, p. 433, pl. i. figs. 9–12.
Cœlopeltis dione, *Eichw. Faun. Casp.-Cauc.* p. 120, pl. xxviii. figs. 1–3 (1841).
Elaphis dione, *Dum. & Bibr.* vii. p. 218 (1854); *Günth. Cat.* p. 92 (1858), *and Rept. Brit. Ind.* p. 240 (1864); *Jan, Icon. Gén.* 21, pl. iii. fig. A (1867); *De Betta, Atti Ist. Ven.* (3) xiii. 1868, p. 931; *Strauch, Schl. Russ. R.* p. 83 (1873); *Schreib. Herp. Eu.* . p. 246 (1875).

Rostral broader than deep, visible from above; internasals broader than long, shorter than the præfrontals; frontal once and one fourth to once and a half as long as broad, as long as its distance from the end of the snout, shorter than the parietals; loreal as long as deep, or a little longer; a large præocular, with a subocular below it; two or three postoculars; temporals 2+3 or 3+3; eight or nine upper labials, fourth and fifth or fifth and sixth entering the eye; four or five lower labials in contact with the anterior chin-shields, which are nearly as long as the posterior. Scales in 25 or 27 (rarely 23) rows, smooth or faintly keeled. Ventrals not or but very obtusely angulate laterally, 172–214; anal divided; subcaudals 50–80. Pale brown or greyish olive above, with blackish cross lines or dark brown or reddish black-edged spots and usually three more or less distinct pale stripes; two dark longitudinal bands on the nape, usually united on the head and terminating on the frontal shield; a curved dark cross band from eye to eye, and continued from the eye to the angle of the mouth; belly yellowish, usually dotted or spotted with blackish.

Total length 900 millim.; tail 170.

Southern Russia, Transcaucasia, and Temperate Asia and Japan.

a. ♀ (Sc. 25; V. 193; C. 50).		Caucasus.	
b. Yg. (Sc. 27; V. 180; C. 66).		Mangyschlak.	St. Petersburg Mus. [E.].
c. ♀ (Sc. 25; V. 198; C. 60).		Wernoje.	St. Petersburg Mus. [E.].
d. ♀ (Sc. 25; V. 198; C. 58).		Semipolatinsk.	St. Petersburg Mus. [E.].
e. ♀ (Sc. 25; V. 203; C. 61).		R. Ili.	Rev. H. Lansdell [C.].
f, g. ♂ (Sc. 27; V. 198; C. 75) & ♀ (Sc. 25; V. 206; C. 60).		R. Emba.	St. Petersburg Mus. [E.].
h. ♂ (Sc. 23; V. 179; C. 60).		Ussuri R.	Warsaw Mus. [E.].
i. ♀ (Sc. 25; V. 200; C. 64).		Pekin.	R. Swinhoe, Esq. [C.].
k–l. Hgr. (Sc. 27, 25; V. 196, 187; C. 59, 64).		Western hills of Pekin.	S. W. Bushell, Esq. [P.].
m–n, o–p. ♂ (Sc. 25, 27; V. 188, 186; C. 69, 71) & yg. (Sc. 25, 25; V. 198, 198; C. 65, ?).		Chefoo.	R. Swinhoe, Esq. [C.].
q. Yg. (Sc. 25; V. 209; C. 64).		Chen Lang Kuan, Gan King.	Mr. John Brock [P.].
r–s. ♀ (Sc. 27; V. 202; C. 68) & yg. (Sc. 23; V. 187; C. 76).		Mountains north of Kiu Kiang.	A. E. Pratt, Esq. [C.].
t. Yg. (Sc. 25; V. 204; C. 80).		Hoi-How, Hainan.	J. Neumann, Esq. [P.].
u. ♂ (Sc. 27; V. 200; C. 79).		China.	Sir E. Belcher [P.].
v. ♂ (Sc. 25; V. 180; C. 68).		China.	J. Brenchley, Esq. [P.].

18. Coluber quatuorlineatus.

Coluber quatuorlineatus, *Lacép. Serp.* pp. 82, 163, pl. vii. fig. 1 (1789); *Bonnat. Encycl. Méth., Ophiol.* p. 44, pl. xxxix. fig. 1 (1790).
—— nauii, *Donnd. Zool. Beitr.* iii. p. 206 (1798).
—— quadristriatus, *Donnd. l. c.* p. 207.
—— quaterradiatus, *Gmel. Der Naturf.* xxviii. 1799, p. 158, pl. iii. fig. 1; *Desmarest, Faune Franç., Rept.* pl. xiv. (1826); *Schleg. Phys. Serp.* ii. p. 159, pl. vi. figs. 9 & 10 (1837); *Werner, Verh. zool.-bot. Ges. Wien,* xli. 1891, p. 759.
—— elaphis, *Shaw, Zool.* iii. p. 450 (1802); *Metaxa, Mon. Serp. Rom.* p. 37 (1823); *Frivaldsky, Mon. Serp. Hung.* p. 44 (1823).
—— quadrilineatus, *Daud. Rept.* vi. p. 266 (1803); *Bouleng. Ann. & Mag. N. H.* (6) vii. 1891, p. 280.
—— sauromates, *Pall. Zoogr. Ross.-As.* iii. p. 42 (1811); *Eichw. Zool. Spec.* iii. p. 174 (1831); *Nordm. in Demidoff, Voy. Russ. Mér.* iii. p. 345, *Rept.* pl. iii. (1840).

Coluber pictus, *Pall. l. c.* p. 45.
—— xanthogaster, *Andrzejowsky, Nouv. Mém. Soc. Nat. Mosc.* ii. 1832, p. 333, pl. xxiii.
—— alpestris, *Ménétr. Cat. Rais.* p. 68 (1832).
—— pœcilocephalus, *Fisch. de Waldh. Bull. Soc. Nat. Mosc.* iv. 1832, p. 575.
Elaphe parreysii, *Wagl. Icon. Amph.* pl. xxvii. (1833).
Natrix elaphis, *Bonap. Icon. Faun. Ital.* (1834).
Elaphis quadrilineatus, *Bonap. Mem. Acc. Tor.* (2) ii. 1839, p. 433; *De Betta, Atti Ist. Ven.* (3) xiii. 1868, p. 929, *and Faun. Ital., Rett. Anf.* p. 43 (1874).
—— parreysii, *Bonap. l. c.* p. 434.
Tropidonotus sauromates, *Eichw. Faun. Casp.-Cauc.* p. 111, pl. xxv. (1841).
Elaphis quaterradiatus, *Dum. & Bibr.* vii. p. 254 (1854); *Günth. Cat.* p. 93 (1858); *Bedriaga, Bull. Soc. Nat. Mosc.* lvi. 1882, p. 302; *Camerano, Mon. Ofid. Ital., Colubr.* p. 37 (1891); *Minà-Palumbo, Nat. Sicil.* xii. 1892, p. 52.
—— sauromates, *Dum. & Bibr. t. c.* p. 288; *Jan, Icon. Gén.* 21, pl. ii. (1867); *Strauch, Schl. Russ. R.* p. 92 (1873); *Schreib. Herp. Eur.* p. 250 (1875); *Bedriaga, l. c.* p. 306.
—— dione, var. græca, *Jan, Elenco,* p. 61 (1863), *and l. c.* pl. iii. fig. B.
—— cervone, *Schreib. l. c.* p. 254; *Tomasini, Wiss. Mitth. aus Bosn. u. Herzeg.* ii. p. 608 (1894).

Rostral broader than deep, just visible from above; internasals broader than long, shorter than the præfrontals; frontal once and one fourth to once and a half as long as broad, as long as its distance from the rostral, shorter than the parietals; loreal nearly as long as deep; one præocular, with a subocular below it; two or three postoculars; temporals 2+3 or 3+4; eight (exceptionally nine) upper labials, fourth and fifth (or fifth and sixth) entering the eye; four or five lower labials in contact with the anterior chin-shields, which are longer than the posterior. Scales in 25 (rarely 23) rows, feebly but distinctly keeled, outer rows smooth. Ventrals 195–234, rounded; anal divided; subcaudals 63–90. Young with three or five alternating series of dark brown black-edged spots on a pale brown ground, the dorsal series largest, transversely oval or rhomboidal; in the specimens from Italy and Dalmatia (*C. quatuorlineatus*) these spots gradually disappear with age, and are replaced by a pair of black streaks along each side of the back; a black oblique streak from the eye to the angle of the mouth; belly yellow, closely spotted or marbled with brown, the spots usually disappearing with age.

Total length 1350 millim.; tail 240.

Mediterranean coast of France (?), Italy, Dalmatia, Greece, Hungary, Southern Russia, Cis- and Trancaucasia.

A. Spots disappearing in the adult, replaced by black longitudinal lines. (*C. quatuorlineatus*, Lacép.)

a. ♂ (V. 213; C. 81).   Bologna.   Prof. J. J. Bianconi [P.].

b–c, d. ♀ (V. 224; C. ?), hgr. (V. 210; C. 77), & yg. (V. 223; C. 70).	Naples.	Count M. G. Peracca [P.].
e–f. ♂ (V. 208; C. 79) & hgr. (V. 219; C. 71).	Naples.	Dr. F. S. Monticelli [P.].
g, h. ♂ (V. 210; C. 79) & yg. (V. 218; C. 66).	Zara, Dalmatia.	Dr. F. Werner [E.].
i–k. ♂ (V. 205; C. 79) & ♀ (V. 218; C. 68).	Zara, Dalmatia.	
l. ♀ (V. 217; C. 70).	S. Europe.	Dr. Günther [P.].

B. Spots persistent throughout life. (*C. sauromates*, Pall.)

a. ♀ (V. 210; C. 66).	Malo, Derbentsky Uluss, Gov. Astrakhan.	St. Petersburg Mus. [E.].
b. ♀ (V. 206; C. 64).	Shirvan.	Berlin Museum [E.].
c. ♀ (V. 211; C. 71).	—— ?	Sir A. Smith [P.].

19. Coluber tæniurus.

Elaphis virgatus, part., *Günth. Cat.* p. 79 (1858).
—— tæniurus, *Cope, Proc. Ac. Philad.* 1860, p. 565; *Günth. Rept. Brit. Ind.* p. 242 (1864); *Strauch, Schl. Russ. R.* p. 103 (1873); *Bouleng. Ann. & Mag. N. H.* (5) xix. 1887, p. 170.
Coluber nuthalli, *Theob. Cat. Rept. As. Soc. Mus.* p. 51 (1868).
Elaphis yunnanensis, *Anders. An. Zool. Res. Yunnan*, p. 813 (1879).
—— grabowskyi, *Fischer, Arch. f. Nat.* 1885, p. 59, pl. iv. fig. 3.
Coluber tæniurus, *Bouleng. Ann. & Mag. N. H.* (6) v. 1890, p. 139, and *Faun. Ind., Rept.* p. 333 (1890); *W. Sclater, Journ. As. Soc. Beng.* lx. 1891, p. 239.

Rostral broader than deep, visible from above; internasals a little broader than long, much shorter than the præfrontals; frontal once and one third to once and two thirds as long as broad, as long as or shorter than its distance from the end of the snout, as long as or a little shorter than the parietals; loreal longer than deep; one large præocular, sometimes in contact with the frontal, usually with a small subocular below it; two postoculars; temporals $2+2$ or $2+3$; usually nine (sometimes eight) upper labials, fifth and sixth (or fourth and fifth) entering the eye; five or six lower labials in contact with the anterior chin-shields, which are as long as or a little longer than the posterior. Scales in 23 or 25 rows, dorsals feebly but distinctly keeled, outer rows smooth. Ventrals distinctly angulate laterally, 230–284; anal divided, rarely entire; subcaudals 90–107. Grey-brown or olive above, head and nape uniform; a black stripe on each side of the head, passing through the eye; anterior part of back with black transverse lines or network, posterior part with a pale vertebral stripe between two broad black ones; belly yellowish anteriorly, greyish posteriorly; a black stripe along each side of the posterior part of the belly and along each side of the tail, separated from the upper lateral stripe by a whitish streak.

Total length 1500 millim.; tail 300.

Manchuria, China, Sikkim, Cochinchina, Siam, Borneo, Sumatra.

a–c. ♂ (Sc. 23; V. 255; A. 2; C. 107), hgr. (Sc. 25; V. 249; A. 2; C. 107), & yg. (Sc. 23; V. 240; A. 1; C. 90). Western hills of Pekin. W. Bushell, Esq. [P.].

d, e, f. ♂ (Sc. 25; V. 232; A. 2; C. 95), ♀ (Sc. 25; V. 233; A. 2; C. 99), & yg. (Sc. 23; V. 230; A. 2; C. 101). Shanghai. R. Swinhoe, Esq. [C.].

g. Yg. (Sc. 25; V. 240; A. 2; C. 94). Chikiang. Mr. Fortune [C.].

h. Yg. (Sc. 10; V. 231; A. 1; C. 92). Mountains of Kiu Kiang. C. Maries, Esq. [C.].

i. ♂ (Sc. 25; V. 239; A. 2; C. ?). Mountains of Kiu Kiang. A. E. Pratt, Esq. [C.].

k. Hgr. (Sc. 23; V. 255; A. 2; C. 99). Darjeeling. T. C. Jerdon, Esq. [P.].

l. ♀ (Sc. 23; V. 255; A. 2; C. 95). Darjeeling. W. T. Blanford, Esq. [P.].

m–o. ♂ (Sc. 25; V. 282; A. 2; C. 94), hgr. (Sc. 25; V. 278; A. 2; C. 98), & head. S.E. Borneo. Hr. Grabowski [C.]. (Types of *E. grabowskyi.*)

p. ♂ (Sc. 25; V. 282; A. 2; C. 97). Pajo, Sumatra. Hr. C. Bock [C.].

20. Coluber schrenckii.

Elaphis schrenckii, *Strauch, Schl. Russ. R.* p. 100 (1873).
Coluber schrenckii, *Bouleng. Ann. & Mag. N. H.* (6) v. 1890, p. 139.

Eye rather small. Rostral broader than deep, just visible from above; internasals broader than long, shorter than the præfrontals; frontal slightly longer than broad, as long as or shorter than its distance from the end of the snout, shorter than the parietals; loreal considerably longer than deep; one præocular, with (or without) a small subocular below it; two postoculars; temporals 2+3; eight upper labials, fourth and fifth entering the eye; five lower labials in contact with the anterior chin-shields, which are a little longer than the posterior. Scales in 23 rows, feebly but distinctly keeled, outer rows smooth. Ventrals not distinctly angulate laterally, 208–221; anal divided; subcaudals 61–76. Brown or black above, uniform or with more or less regular pale brown cross bands, disposed obliquely, and bifurcating on the sides; lips black and yellow; belly yellowish, uniform or closely spotted or checkered with black.

Total length 1430 millim.; tail 230.

Amoorland, Corea, and Northern Japan.

A. Brown (young) or black above, with paler oblique cross bands; belly closely spotted with black.

a. ♀ (V. 220; C. 68). Ussuri R. Warsaw Museum [E.].
b. Yg. (V. 210; C. 68). Seoul, Corea. C. W. Campbell, Esq. [P.].

B. Uniform brown above, yellowish beneath, the caudal shields edged with brown.

c. ♂ (V. 216; C. 76).   Seoul, Corea.   C. W. Campbell, Esq. [P.].

21. Coluber vulpinus.

Scotophis vulpinus, *Baird & Gir. Cat. N. Am. Rept.* p. 75 (1853), and *Rep. Explor. Surv. R. Miss.* R. x. pl. xxix. fig. 51 (1859); *Hay, Batr. & Rept. Indiana,* p. 90 (1893).
Coluber vulpinus, *Cope, Check-List N. Am. Rept.* p. 39 (1875).
Elaphis guttatus, var. vulpinus, *Garm. N. Am. Rept.* p. 56 (1883).

Eye rather small. Rostral broader than deep, the portion visible from above measuring one third its distance from the frontal; internasals broader than long, much shorter than the præfrontals; frontal nearly as long as broad, as long as its distance from the rostral, shorter than the parietals; loreal as long as deep; one præ- and two postoculars; temporals 2+3; eight upper labials, fourth and fifth entering the eye; five lower labials in contact with the anterior chin-shields, which are a little longer than the posterior. Scales in 25 (rarely 23 or 27) rows, dorsals feebly but very distinctly keeled. Ventrals 200-234; anal divided; subcaudals 67-85. Yellowish brown above, with a dorsal series of large chestnut-brown black-edged spots and an alternating lateral series of smaller spots; head with indistinct dark markings; two dark stripes along the occiput and nape; lower parts yellowish, checkered with black.

Total length 960 millim.

Northern United States east of the Rocky Mountains (Minnesota, Michigan, Wisconsin, Indiana).

d. ♀ (V. 207; C. ?).   Wisconsin.   Smithsonian Institution [P.].

22. Coluber lætus.

Scotophis lætus, *Baird & Gir. Cat. N. Am. Rept.* p. 77, *and in Marcy's Rep. Red Riv.* p. 227, pl. vi. (1853).
—— confinis, *Baird & Gir. Cat.* p. 76.
? Coluber rosaceus, *Cope, Proc. U.S. Nat. Mus.* xi. 1888, p. 388, pl. xxxvi. fig. 3.

Rostral broader than deep, just visible from above; internasals much shorter than the præfrontals; frontal a little longer than broad, as long as its distance from the end of the snout, shorter than the parietals; loreal a little longer than deep; one præ- and two postoculars; temporals 2+3; eight upper labials, fourth and fifth entering the eye; five lower labials in contact with the anterior chin-shields, which are as long as the posterior. Scales smooth or faintly keeled, in 25 to 29 rows. Ventrals obtusely angulate laterally, 220-240; anal divided; subcaudals 72-81. Above with a series of large squarish brown black-edged blotches, separated by brownish-white or grey interspaces; two alternating lateral series of small spots on each side; a brown band across the end of the snout, another from one angle of the mouth to the other, passing through the eyes and across the forehead; a brown spot below the eye; lower parts yellow, with square brown spots, which are few anteriorly and numerous posteriorly.

Total length 700 millim.; tail 125.

South-eastern United States and Mexico.

a. ♂ (Sc. 29; V. 220; C. 72). Mexico.

C. rosaceus, Cope, is probably a colour-variety of this species.

23. Coluber obsoletus.

Coluber obsoletus, *Say, in Long's Exped. Rocky M.* i. p. 140 (1823); *Harl. Journ. Ac. Philad.* v. 1827, p. 347, *and Med. Phys. Res.* p. 112 (1835); *Cope, Proc. U.S. Nat. Mus.* xiv. 1892, p. 634; *Hay, Batr. & Rept. Indiana*, p. 93 (1893).
—— alleghaniensis, *Holbr. N. Am. Herp.* p. 111, pl. xx. (1836), *and* 2nd ed. iii. p. 85, pl. xix. (1842); *Dekay, N. York Faun., Rept.* p. 36, pl. xii. fig. 26 (1842).
—— quadrivittatus, *Holbr. op. cit.* 2nd ed. iii. p. 89, pl. xx.; *Günth. Cat.* p. 88 (1858).
Scotophis alleghaniensis, *Baird & Gir. Cat. N. Am. Rept.* p. 73 (1853).
—— lindheimeri, *Baird & Gir. l. c.* p. 74.
Elaphis quadrivittatus, *Dum. & Bibr.* vii. p. 265 (1854); *Garm. N. Am. Rept.* p. 56 (1883).
—— spiloides, *Dum. & Bibr. t. c.* p. 269.
—— holbrookii, *Dum. & Bibr. t. c.* p. 272.
—— alleghaniensis, *Hallow. Proc. Ac. Philad.* 1856, p. 243; *Jan, Icon. Gén.* 24, pl. ii. fig. 1 (1867).
Coluber spiloides, *Günth. l. c.* p. 90.
Scotophis obsoletus, *Kennicott, Proc. Acad. Philad.* 1860, p. 330.
Coluber lindheimerii, *Cope, Check-List N. Am. Rept.* p. 39 (1875).
Elaphis obsoletus, *Garm. l. c.* p. 54 (1883); *H. Garm. Bull. Illin. Lab.* iii. 1892, p. 290.
Pantherophis lindheimeri, *Garm. Bull. Essex Inst.* xxiv. 1892, p. 108.
—— alleghaniensis, *Garm. l. c.*

Rostral broader than deep, just visible from above; internasals broader than long, much shorter than the præfrontals; frontal a little longer than broad, as long as its distance from the rostral, shorter than the parietals; loreal longer than deep; one præ- and two postoculars; temporals 2+3; eight upper labials, fourth and fifth entering the eye; four or five lower labials in contact with the anterior chin-shields, which are as long as the posterior. Scales very distinctly keeled, in 25 to 29 rows. Ventrals obtusely angulate laterally, 217–245; anal divided; subcaudals 72–93. Brown to black above; young with a series of large dorsal dark spots and two lateral series of smaller alternating spots, which may disappear or be replaced by four dark stripes.

Total length 1360 millim.; tail 310.

United States east of the Rocky Mountains.

A. Nearly uniform black above and on the posterior two thirds of the lower parts. (*C. óbsoletus*, Say.)

a. ♂ (Sc. 27; V. 233; C. 80). N. America. Lord Ampthill [P.].

b-c. ♂ (Sc. 27; V. 231; Raleigh, N. Carolina.   Messrs. Brimley [C.].
C. 88) & hgr. (Sc. 27;
V. 236; C. 78).

B. Dark brown above, but with distinct traces of the large spots; belly largely blotched with blackish. (*C. alleghaniensis*, Holbr., *C. lindheimeri*, B. & G.)

a. ♂ (Sc. 25; V. 232; Bloomington, Indiana. C. H. Bollman, Esq.
C. 80).   [C.].
b. ♀ (Sc. 27; V. 233; Illinois.   Smithsonian Instit.
C. 77).   [P.].

C. Pale brown above, with large dark brown spots; yellowish inferiorly. (*C. spiloides*, D. & B.)

a. ♂ (Sc. 29; V. 232; New York.
C. ?).
b. ♀ (Sc. 29; V. 231; Duval Co., Texas. W. Taylor, Esq. [C.].
C. 78).

D. Spots as in C, combined with stripes as in E; belly with large brown spots.

a. ♀ (Sc. 27; V. 241; N. America.
C. 77).

E. Pale brown above, with four dark brown or blackish stripes; belly yellowish, without or with a few brown spots. (*C. quadrivittatus*, Holbr.)

a, b. ♂ (Sc. 27, 27; V.   N. America.
237, 241; C. 84, 89).

24. Coluber conspicillatus.

Coluber conspicillatus, *Boie, Isis*, 1826, p. 211; *Schleg. Phys. Serp.* ii. p. 171 (1837), *and Faun. Japon., Rept.* p. 85, pl. iii. (1838); *Günth. Cat.* p. 91 (1858).
Elaphis conspicillatus, *Dum. & Bibr.* vii. p. 285 (1854); *Hilgendorf, Sitzb. ges. Naturf. Freunde*, 1880, p. 114.
Coronella conspicillata, *Jan, Arch. Zool. Anat. Phys.* ii. 1863, p. 249, *and Icon. Gén.* 14, pl. vi. fig. 1 (1865).
Callopeltis conspicillatus, *Giglioli, Proc. Zool. Soc.* 1887, p. 595.

Rostral nearly twice as broad as deep, visible from above; internasals much broader than long, shorter than the præfrontals; frontal a little longer than broad, as long as its distance from the end of the snout, shorter than the parietals; loreal longer than deep; one præ- and two postoculars; temporals 1+2 or 2+2; seven upper labials, third and fourth entering the eye; four lower labials in contact with the anterior chin-shields, which are a little longer than the posterior. Scales smooth, or faintly keeled on the posterior part of the body, in 21 rows. Ventrals distinctly angulate laterally, 200–224; anal divided; subcaudals 60–76. Pale brown above, with small black spots or a dorsal series of black cross bars; some adult specimens almost uniform olive-brown above; a black streak across the end of the snout, passing through the nostrils, a curved one from eye to eye, another from the eye to the angle of the mouth, a Λ-shaped one on the back of the head, with the point on the frontal, and a black spot or vertical streak below the eye; lower parts whitish, checkered with black. Young (*e*) vermilion above, with small black spots edged with yellowish white.

Total length 830 millim.; tail 150.

Japan; Corea.

a, b. ♂ (V. 219; C. 67) & yg. (V. 217; C. 66).	Japan.	Leyden Museum.
c. ♂ (V. 215; C. 67).	Yokohama.	H.M.S. 'Challenger.'
d. Yg. (V. 224; C. 63).	Nikko.	H.M.S. 'Challenger.'
e. Yg. (V. 206; C. 68).	Haruna Hills, Japan, 2500 ft.	Lord Dormer [P.].

25. Coluber longissimus.

Natrix longissima, *Laur. Syn. Rept.* p. 74 (1768).
Coluber flavescens, *Gmel. S. N.* i. p. 1115 (1788); *Daud. Rept.* vi. p. 272 (1803); *Frivaldsky, Mon. Serp. Hung.* p. 40 (1823); *Lenz, Schlangenk.* p. 509, pl. vi. (1832); *Bonap. Icon. Faun. Ital.* (1833); *De Betta, Erp. Veron., Mem. Acc. Verona,* xxxv. 1857, p. 197.
—— æsculapii (*non L.*), *Lacép. Serp.* pp. 98, 165, pl. vii. fig. 2 (1789); *Bonnat. Encycl. Méth., Ophiol.* p. 43, pl. xxxix. fig. 2 (1790); *Host, in Jacq. Coll. Botan.* iv. p. 356, pl. xxvii. (1790); *Wolf, in Sturm, Deutschl. Faun.* iii. Heft 2 (1799); *Daud. Rept.* vii. p. 30 (1803); *Metaxa, Mon. Serp. Rom.* p. 37 (1823); *Frivaldsky, l. c.* p. 42; *Desmarest, Faune Franç., Rept.* pl. xv. (1826); *Millet, Faune de Maine-et-Loire,* p. 632, pl. iv. (1828); *Schleg. Phys. Serp.* ii. p. 130, pl. v. figs. 1 & 2 (1837); *Günth. Cat.* p. 88 (1858); *Strauch, Schl. Russ. R.* p. 57 (1873); *Boettg. Ber. Senckenb. Ges.* 1889, p. 271; *Méhely, Beitr. Mon. Kronstadt, Herp.* p. 29 (1892); *Sarauw, Nat. og Mennesk. Copenh.* xi. 1894, p. 258; *Dürigen, Deutschl. Amph. u. Rept.* p. 308, pl. viii. fig. 1 (1894).
—— longissimus, *Bonnat. l. c.* p. 59; *Blanf. Zool. E. Pers.* p. 420 (1876).
—— asclepiadeus, *Donnd. Zool. Beytr.* iii. p. 205 (1798).
—— sellmanni, *Donnd. l. c.* p. 207.
—— pannonicus, *Donnd. l. c.* p. 208.
—— leprosus, *Donnd. l. c.*
—— romanus, *Suckow, Anfangsgr. d. Naturg.* iii. p. 198 (1798).
—— scopolii, *Merr. Tent.* p. 104 (1820).
—— fugax, *Eichw. Zool. Spec.* iii. p. 174 (1831).
Zamenis æsculapii, *Fitzing. Beitr. Landesk. Oesterr.* i. p. 326 (1832); *Eichw. Faun. Casp.-Cauc.* p. 150 (1841).
Callopeltis flavescens, *Bonap. Mem. Acc. Torin.* (2) ii. 1839, p. 432.
Coluber sauromates, juv., *Nordm. in Demidoff, Voy. Russ. Mér.* iii. p. 346, *Rept.* pl. vi. fig. 2 (1840).
Elaphis æsculapii, *Dum. & Bibr.* vii. p. 278 (1854); *Jan, Icon. Gén.* 24, pl. i. fig. 4 (1867); *Viaud-Grandm. Et. Serp. Vendée,* p. 10 (1868); *Fatio, Vert. Suisse,* iii. p. 136 (1872); *De Betta, Faun. Ital., Rett. Anf.* p. 43 (1874).

Callopeltis æsculapii, *Schreib. Herp. Eur.* p. 281 (1875); *Tomasini, Wiss. Mitth. aus Bosn. u. Herzeg.* ii. p. 620 (1894).
Elaphis flavescens, *Leydig, Abh. Senck. Ges.* xiii. 1883, p. 176.
Callopeltis longissimus, *Camerano, Mon. Ofid. Ital., Colubr.* p. 54 (1891); *Minà-Palumbo, Nat. Sicil.* xii. 1893, p. 129.
Coronella austriaca, part., *Sarauw, Nat. og Mennesk. Copenh.* x. 1893, p. 216.

Rostral broader than deep, just visible from above; internasals broader than long, shorter than the præfrontals; frontal once and one fourth to once and one third as long as broad, as long as its distance from the rostral or the end of the snout, shorter than the parietals; loreal as long as deep or longer; one præ- and two postoculars; temporals 2+3; eight or nine upper labials, fourth and fifth or fifth and sixth entering the eye; four or five lower labials in contact with the anterior chin-shields, which are as long as or a little longer than the posterior. Scales smooth, or faintly keeled on the posterior part of the body, in 21 or 23 rows. Ventrals distinctly angulate laterally, 212-248; anal divided; subcaudals 60-91. Grey or olive-brown above, some of the scales with whitish lines on the margins; sometimes with four darker stripes along the body*; a dark streak behind the eye; upper lip and a triangular patch on each side behind the temple yellow; belly uniform pale yellow. Young with dark brown dorsal spots, forming four or five longitudinal series, a Λ-shaped black marking on the nape, behind the yellow temporal blotches, a dark brown bar across the forehead, and a black vertical line below the eye; belly greyish or yellowish olive.

Total length 110 millim.; tail 240.

Central Europe, Denmark †, Italy, Dalmatia, Balkan Peninsula, Cis- and Transcaucasia.

a-b. ♂ (Sc. 23, 23; V. 225, 222; C. 78, 82).	France.	
c. ♂ (Sc. 21; V. 223; C. 82.)	Nantes.	
d-f. ♂ (Sc. 23; V. 220; C. 83), ♀ (Sc. 23; V. 231; C. 77), & hgr. (Sc. 23; V. 226; C. 86).	Schlangenbad.	Dr. Günther [P.].

---

* The report of the occurrence of *C. quatuorlineatus* at Saumur, Maine-et-Loire (Millet, Faune de Maine-et-Loire, p. 629), is evidently based on such a striped specimen, as pointed out by Viaud-Grandmarais.

† I have expressed doubts (Zool. 1894, p. 14) as to whether a large specimen (1280 millim. long) with 218 ventral shields, from the island of Seeland, had been correctly referred by Sarauw to *Coronella austriaca*. Further notes on that specimen, captured at Peterswarft in 1863, and now preserved in the Seminary at Skaarup, which Mr. Sarauw subsequently sent me at my request, accompanied by a sketch of the head, changed my doubts into certainty: the Peterswarft snake belongs to a distinct species, which turns out to be *Coluber longissimus*. Three specimens have been examined by Mr. Sarauw, which are perhaps the last survivors of a snake formerly inhabiting the forests of Southern Seeland. It appears that, so far, no authentic specimen of *Coronella austriaca* is known from Denmark.

g. ♀ (Sc. 23; V. 224; C. 75).		Schlangenbad.	Dr. Günther [P.].
h. Yg. (Sc. 21; V. 229; C. 76).		Weidling am Bach, near Vienna.	Dr. F. Werner [E.].
i–k. ♂ (Sc. 23, 23; V. 226, 219; C. 84, 80).		Near Vienna.	Dr. F. Werner [E.].
l. ♀ (Sc. 23; V. 226; C. 75).		Bologna.	Prof. J. J. Bianconi [P.].
m. ♂ (Sc. 23; V. 224; C. 83).		Travnik, Bosnia.	Dr. F. Werner [E.].
n, o. Skulls.		France.	

26. Coluber climacophorus.

Coluber climacophorus, *Boie, Isis,* 1826, p. 210.
—— virgatus, *Schleg. Phys. Serp.* ii. p. 145 (1837), *and Faun. Japon., Rept.* p. 83, pl. ii. (1838).
Elaphis virgatus, *Dum. & Bibr.* vii. p. 261 (1854); *Jan, Icon. Gén.* 21, pl. i. (1867).
—— virgatus, part., *Günth. Cat.* p. 94 (1858); *Hilgendorf, Sitzb. Ges. naturf. Freunde,* 1880, p. 113.

Rostral broader than deep, just visible from above; internasals broader than long, shorter than the præfrontals; frontal about once and a half as long as broad, as long as its distance from the end of the snout, shorter than the parietals; loreal longer than deep; one præocular, with a subocular below it; two (rarely three) postoculars; temporals 2+3; eight upper labials, fourth and fifth entering the eye; four or five lower labials in contact with the anterior chin-shields, which are as long as or a little longer than the posterior. Scales in 23 (rarely 25) rows, feebly but distinctly keeled, outer rows smooth. Ventrals distinctly angulate laterally, 226–244; anal divided (rarely entire); subcaudals 97–122. Young grey-brown or pale brown above, with a dorsal series of large transverse, brown, dark-edged spots, and a lateral series of smaller spots; these spots usually disappear in the adult and are replaced by a pair of dark stripes along each side of the back; a blackish streak from the eye to the angle of the mouth; lower parts yellowish, uniform or dotted or clouded with olive, or with small blackish spots. Old specimens nearly uniform olive-brown above and below, with the lateral ventral keel yellowish.

Total length 1680 millim.; tail 350.
Japan.

a, b. ♀ (Sc. 23; V. 231; C. 100) & hgr. (Sc. 23; V. 234; C. 112).		Japan.	Leyden Museum.
c. Yg. (Sc. 23; V. 222; C. 108).		Japan.	Sir R. Owen [P.].
d. Yg. (Sc. 23; V. 232; A. 1; C. ?)		Japan.	Mr. Adams [C.].
e–f. ♀ (Sc. 23; V. 234; C. 103) & yg. (Sc. 23; V. 235; C. 116).		Nagasaki.	Mr. Whitely [C.].

g. Yg. (Sc. 23; V. 229; Nikko. Mr. Whitely [C.].
C. 103).
h–i. Yg. (Sc. 25, 23; V. Myanoshita. J. H. Leech, Esq.
232, 226; C. 100, 97). [P.].
k. ♂ (Sc. 25; V. 234; Onsen Mt., Shimabara. Mr. Holst [C.].
C. 99).

27. Coluber phyllophis.

Elaphis sauromates (*non Pall.*), *Günth. Cat.* p. 93 (1858).
—— sauromates, part., *Günth. Rept. Brit. Ind.* p. 241, pl. xxi.
  fig. E (1864).
Phyllophis carinata, *Günth. l. c.* p. 295, pl. xxi. fig. B, *and Ann. &
  Mag. N. H.* (6) i. 1888, p. 170.
Elaphis, sp., *Strauch, Schl. Russ. R.* p. 99 (1873); *Bedriaga, Bull.
  Soc. Nat. Mosc.* lvi. 1882, p. 309.
Coluber phyllophis, *Bouleng. Ann. & Mag. N. H.* (6) vii. 1891,
  p. 280.

Snout projecting; eye rather large. Rostral much broader than deep, visible from above; nasal sometimes entire or semi-divided; internasals as long as broad, or a little longer, at least as long as the præfrontals; frontal once and one third to once and two thirds as long as broad, as long as its distance from the rostral or the end of the snout, a little shorter than the parietals; loreal longer than deep; one præocular, with a subocular below it; two postoculars; temporals 2+3 or 3+3; eight (rarely nine) upper labials, fourth and fifth (or fifth and sixth) entering the eye; four or five lower labials in contact with the anterior chin-shields, which are nearly as long as or a little shorter than the posterior. Scales in 23 rows very strongly keeled, outer row smooth. Ventrals obtusely angulate laterally, 209–220; anal divided (rarely entire); subcaudals 80–96. Young pale olive above, with traces of a few black transverse bands on the anterior part of the body, and a brown lateral line on the posterior part of the body and along the tail; labials yellowish, with brown sutures; belly yellowish, with a series of black dots on each side. The adult of a darker coloration, most of the scales and shields having black borders; anterior part of back usually with more or less distinct black cross bands; belly more or less dotted or spotted with black, the posterior ventrals and the subcaudals usually edged with black.

Total length 1800 millim.; tail 380.

China.

a–c, d. ♂ (V. 213, 215; Kiu Kiang. A. E. Pratt, Esq.
  A. 2, 1; C. 93, 87), ♀ [C.].
  (V. 215; A. 2; C. 83);
  & yg. (V. 210; A. 1;
  C. 86).
e. ♂ (V. 209; A. 2; C. 90). Wu-lee Lake, 25 miles P. W. Bassett-Smith,
  S. of Ningpo. Esq. [P.].
f. ♀ (V. 213; A. 2; C. Near Ningpo.
  80).

g. Yg. (V. 220; A. 2; C. 95).		China.	Dr. A. Günther [P.]. (Type of *Phyllophis carinata*.)
h, i. ♂ (V. 215; A. 2; C. 96) & ♀ (V. 209; A. 2; C. 90).		China.	R. Swinhoe, Esq. [C.].

28. Coluber davidi.

Tropidonotus davidi, *Sauvage, Bull. Soc. Philom.* (7) viii. 1884, p. 144 *.

Snout slightly projecting; eye rather large. Rostral visible from above; internasals as long as broad, two thirds the length of the præfrontals; frontal twice as long as broad, as long as its distance from the end of the snout, not longer than the parietals; loreal longer than deep; two præoculars, with a subocular below them; two postoculars; temporals 2+3 or 3+4; eight upper labials, fourth and fifth entering the eye. Scales in 23 rows, strongly keeled, outer row smooth. Ventrals 173; anal divided; subcaudals 70. Yellowish brown, with large irregular brown spots; head without any markings.

Total length 750 millim.; tail 160.

China.

29. Coluber moellendorffii.

Cynophis moellendorffii, *Boettg. Zool. Anz.* 1886, p. 520, *and Ber. Offenb. Ver. Naturk.* 1888, p. 72, pl. i. fig. 1.

Rostral broader than deep, visible from above; internasals broader than long, hardly half as long as the præfrontals; frontal a little longer than broad, not quite as long as its distance from the rostral, a little shorter than the parietals; loreal twice as long as deep; a large præocular, in contact with the frontal, with a subocular below it; two postoculars; temporals 2+3 or 2+2; ten upper labials, sixth and seventh entering the eye; five lower labials in contact with the anterior chin-shields, which are as long as the posterior. Scales in 27 rows, feebly but distinctly keeled, outer rows smooth. Ventrals obtusely keeled laterally, 268–274; anal divided; subcaudals 97–99. Grey above, with a dorsal series of large, dark grey, black-edged hexagonal spots, and a lateral series of alternating smaller spots; yellowish-white inferiorly, largely checkered with black.

Total length 1655 millim.; tail 310.

Nanning and Canton, China.

30. Coluber oxycephalus.

Coluber oxycephalus, *Boie, Isis,* 1827, p. 537; *Bouleng. Faun. Ind., Rept.* p. 335 (1890); *W. Sclater, Journ. As. Soc. Beng.* lx. 1891, p. 239.

---

* I am indebted to Dr. Mocquard for notes on, and a sketch of, the head of the type specimen preserved in the Paris Museum.

Gonyosoma viride, *Wagl.* Icon. Amph. pl. ix. (1828).
Herpetodryas oxycephalus, *Schleg.* Phys. Serp. ii. p. 189, pl. vii.
  figs. 8 & 9 (1837), *and Abbild.* pl. xliv. figs. 1-9 (1844); *Cantor*,
  Cat. Mal. Rept. p. 80 (1847).
Alopecophis chalybeus, *Gray*, Ann. & Mag. N. H. (2) iv. 1849,
  p. 247.
Gonyosoma oxycephalum, *Dum. & Bibr.* vii. p. 213 (1854); *Günth.*
  Cat. p. 122 (1858), *and Rept. Brit. Ind.* p. 294 (1864); *Jan*, Icon.
  Gén. 31, pl. i. (1869); *Stoliczka*, Journ. As. Soc. Beng. xxxix. 1870,
  p. 193, & xlii. 1873, p. 123; *Theob.* Cat. Rept. Brit. Ind. p. 189
  (1876).

Body rather compressed. Snout subacuminate, elongate, obliquely truncate and projecting; rostral nearly as deep as broad, visible from above; internasals one half to two thirds as long as the præfrontals, which are more than half as long as the parietals; frontal as long as broad, or a little longer, as long as its distance from the rostral or the end of the snout, a little shorter than the parietals; loreal very elongate; one large præocular, in contact with the frontal; two postoculars; temporals 2+3; nine to eleven upper labials, two (rarely three) of which (fifth and sixth or sixth and seventh) enter the eye; six lower labials in contact with the anterior chin-shields, which are much longer than the posterior. Scales in 23 to 27 rows, smooth or faintly keeled. Ventrals with a lateral keel, 233 to 263; anal divided; subcaudals 122-149. Bright green above, the scales usually finely edged with black; yellowish or pale green below; a blackish streak along each side of the head, passing through the eye; tail usually yellowish brown.

Total length 2300 millim.; tail 480.

Eastern Himalayas, Malay Peninsula and Archipelago.

a. ♀ (Sc. 25; V. 258; C. 123).		Tenasserim.	Prof. Oldham [P.].
b. Hgr. (Sc. 27; V. 248; C. 142).		Pinang.	Dr. Cantor.
c, d. ♀ (Sc. 25, 23; V. 252, 238; C. 136, 139).		Singapore.	Gen. Hardwicke [P.].
e. ♂ (Sc. 25; V. 233; C. 133).		Borneo.	Sir J. Brooke [P.].
f. ♂ (Sc. 25; V. 241; C. 136).		Borneo.	L. L. Dillwyn, Esq. [P.].
g, h. ♀ (Sc. 27; V. 251; C. 122) & yg. (Sc. 25; V. 245; C. 122).		Philippines.	H. Cuming, Esq. [C.].
i-k, l. ♂ (Sc. 25; V. 240; C. 143), hgr. (Sc. 25; V. 240; C. 139), & yg. (Sc. 25; V. 239; C. 128).		Java.	
m. Yg. (Sc. 25; V. 236; C. 135).		[Mauritius.]	Zoological Society. (Type of *Alopecophis chalybeus*.)

31. **Coluber janseni.** (Plate I. fig. 2.)

Gonyosoma jansenii, *Bleek.* Nat. Tijdschr. Nederl. Ind. xvi. 1858, p. 242.

Allophis (Elaphis) nigricaudus, *Peters. Mon. Berl. Ac.* 1872, p. 686.
Elaphis nigricaudus, *Peters & Doria, Ann. Mus. Genova,* xiii. 1878, p. 387.

Very much like *C. oxycephalus*, but snout more obtuse and rather shorter. Rostral broader than deep, just visible from above; internasals as long as broad, a little shorter than the præfrontals; frontal once and one fourth to once and a half as long as broad, as long as or a little shorter than its distance from the end of the snout, a little shorter than the parietals; loreal at least twice as long as deep; one large præocular, in contact with the frontal; two postoculars; temporals 1+2 or 2+3; nine or ten upper labials, fifth to seventh entering the eye; five or six lower labials in contact with the anterior chin-shields, which are much longer than the posterior. Scales smooth or faintly keeled, in 23 or 25 rows. Ventrals angulate laterally, 247–255; anal divided; subcaudals 133–140. Olive or yellowish brown above, some or all of the scales black-edged; entirely black posteriorly and on the tail; a black lateral stripe may be present.

Total length 1990 millim.; tail 450.

North Celebes.

a. ♂ (Sc. 23; V. 247; C. 140).		Manado.	Dr. Bleeker. (Type.)
b, c. ♂ (Sc. 23, 25; V. 253, 255; C. 140, 133).		Manado.	Dr. A. B. Meyer [C.].

32. Coluber frenatus. (PLATE III. fig. 1.)

Herpetodryas frenatus, *Gray, Ann. & Mag. N. H.* (2) xii. 1853, p. 390.
Gonyosoma frenatum, *Günth. Cat.* p. 123 (1858), *and Rept. Brit. Ind.* p. 295 (1864); *Theob. Cat. Rept. Brit. Ind.* p. 190 (1876).
Coluber frenatus, *Bouleng. Faun. Ind., Rept.* p. 335 (1890).

Snout subacuminate, obliquely truncate and projecting. Rostral a little broader than deep, hardly visible from above; internasals much shorter than the præfrontals; frontal once and one third to once and a half as long as broad, as long as its distance from the end of the snout, shorter than the parietals; no loreal, præfrontal in contact with the labials; one large præocular; two postoculars; temporals 2+2 or 2+3; nine (or eight) upper labials, fourth, fifth, and sixth entering the eye; five lower labials in contact with the anterior chin-shields, which are as long as the posterior. Scales in 19 rows, dorsals faintly keeled. Ventrals with a lateral keel, 203–204; anal divided; subcaudals 120–121. Uniform bright green above; a black streak along each side of the head, passing through the eye; upper lip and lower parts pale green, ventral keel whitish.

Total length 830 millim.; tail 240.

Khasi Hills.

a. ♀ (V. 203; C. 120).	Khasi Hills.	Sir J. Hooker [P.].	(Type.)
b. ♀ (V. 204; C. 121).	Khasi Hills.	T. C. Jerdon, Esq. [P.].	

33. Coluber prasinus.

Coluber prasinus, *Blyth, Journ. As. Soc. Beng.* xxiii. 1854, p. 291; *Bouleng. Faun. Ind., Rept.* p. 334 (1890); *W. Sclater, Journ. As. Soc. Beng.* lx. 1891, p. 239.
Gonyosoma gramineum, *Günth. Rept. Brit. Ind.* p. 294, pl. xxiii. fig. D (1864); *Theob. Cat. Rept. Brit. Ind.* p. 190 (1876); *Anders. An. Zool. Res. Yunnan,* p. 824 (1879).

Rostral a little broader than deep, just visible from above; internasals a little broader than long, shorter than the præfrontals; frontal once and one fourth to once and one third as long as broad, as long as its distance from the end of the snout, shorter than the parietals; loreal square or longer than deep; one præocular (sometimes divided), usually in contact with the frontal; two postoculars; temporals 1+2 or 2+2; nine upper labials, fourth, fifth, and sixth entering the eye; five lower labials in contact with the anterior chin-shields, which are as long as or a little longer than the posterior. Scales in 19 rows; dorsals feebly keeled in the adult, smooth in the young. Ventrals with a lateral keel, 198-206; anal divided (rarely entire); subcaudals 91-107. Uniform bright green above; upper lip and lower surfaces yellowish or greenish white.

Total length 900 millim.; tail 230.

Eastern Himalayas, Khasi Hills, Assam, Burma.

a. Yg. (V. 203; C. 100).	Khasi Hills (?).	Dr. Griffith.
		(Type of *G. gramineum*.)
b, c. ♂ (V. 198; C. 107) & hgr. (V. 200; C. ?).	Khasi Hills.	T. C. Jerdon, Esq. [P.].
d. ♀ (V. 198; C. 91).	Toungyi, Shan States, 5000 ft.	Lieut. Blakeway [C.].

34. Coluber quadrivirgatus.

Coluber quadrivirgatus, *Boie, Isis,* 1826, p. 209; *Schleg. Phys. Serp.* ii. p. 147, pl. v. figs. 15 & 16 (1837), *and Faun. Japon.* p. 84, pl. i. (1838).
—— vulneratus, *Boie, l. c.* p. 212.
Compsosoma quadrivirgatum, *Dum. & Bibr.* vii. p. 301 (1854).
Elaphis quadrivirgatus, *Günth. Cat.* p. 94 (1858); *Jan, Icon. Gén.* 24, pl. i. figs. 1-3 (1867); *Hilgendorf, Sitz. Ges. naturf. Freunde,* 1880, p. 114.
—— bilineatus, *Hallow. Proc. Ac. Philad.* 1860, p. 497.

Eye rather large; rostral broader than deep, just visible from above; internasals broader than long, shorter than the præfrontals; frontal once and three fifths to twice as long as broad, longer than its distance from the end of the snout, as long as or a little shorter than the parietals; loreal usually as long as deep, or a little deeper; one præocular, usually with a subocular below it; two postoculars; temporals 2+2 or 2+3; eight upper labials, fourth and fifth entering the eye; four or five lower labials in contact with the anterior chin-shields, which are as long as or longer than the posterior. Scales in 19 rows, feebly but distinctly keeled (sometimes nearly smooth in the young), outer rows smooth. Ventrals distinctly angulate laterally, 193–210; anal divided, rarely entire; subcaudals 70–96. Greyish olive or pale brown above, with darker spots or cross bars, or with a pair of dark stripes running along each side of the back; a dark streak from the eye to the angle of the mouth; lower parts yellowish clouded with olive-grey, or olive-grey to blackish, the lateral ventral keels always lighter. Some specimens entirely black, with the exception of the lips and throat, which are yellowish.

Total length 1030 millim.; tail 240.

Japan, Corea.

A. Olive or brown above, with more or less distinct spots or stripes.

a. ♂ (V. 208; A. 2; C. 88).		Japan.	Leyden Museum.
b–d. ♂ (V. 204, 204; A. 2, 2; C. 81, 84) & ♀ (V. 200; A. 2; C. 70).		Yokohama.	H.M.S. 'Challenger.'
e–f. Yg. (V. 209, 200; A. 2, 1; C. 80, 87).		Nikko.	C. Maries, Esq. [C.].
g. Yg. (V. 195; A. 2; C. 80).		Myanoshita, Japan.	J. H. Leech, Esq. [P.].
h. Hgr. (V. 202; A. 2; C. 80).		Prov. Satsuma, Japan.	J. H. Leech, Esq. [P.].
i. ♂ (V. 199; A. 2; C. 90).		Gensan, Corea.	J. H. Leech, Esq. [P.].

B. Uniform black above and below, or blackish above with more or less distinct black stripes.

k, l. ♀ (V. 200, 202; A. 1, 2; C. 84, 83).		Japan.	Leyden Museum.
m. ♂ (V. 203; A. 2; C. 85).		Japan.	Zoological Society.
n. Yg. (V. 203; A. 2; C. 83).		Japan.	Sir R. Owen [P.].
o–p. Yg. (V. 203, 205; A. 2, 2; C. 83, 89).		Kumamoto, Japan.	J. H. Leech, Esq. [P.].

35. Coluber melanurus.

Coluber melanurus, *Schleg. Phys. Serp.* ii. p. 141, pl. v. figs. 11 & 12 (1837), *and Abbild.* pl. v. (1837); *Bouleng. Faun. Ind., Rept.* p. 334 (1890).

Compsosoma melanurum, *Dum. & Bibr.* vii. p. 299 (1854); *Günth. Rept. Brit. Ind.* p. 244 (1864); *Stoliczka, Journ. As. Soc. Beng.* xxxix. 1870, p. 188; *Theob. Cat. Rept. Brit. Ind.* p. 165 (1876).

Spilotes melanurus, *Günth. Cat.* p. 97 (1858).

Elaphis melanurus, *Jan, Icon. Gén.* 21, pl. iv. fig. 1, and pl. v. fig. 1 (1867).

Rostral broader than deep, just visible from above; internasals as long as broad, or broader, shorter than the præfrontals; frontal once and one sixth to once and one third as long as broad, as long as or a little longer than its distance from the end of the snout or from the rostral, shorter than the parietals; loreal not or but slightly longer than deep; one large præocular; two postoculars; temporals 2+2 or 2+3; nine upper labials, fourth to sixth entering the eye; five or six lower labials in contact with the anterior chin-shields, which are as long as the posterior. Scales in 19 rows (21 on the neck), rather strongly keeled; all the caudal scales keeled. Ventrals with an obtuse lateral keel, 193-234; anal entire; subcaudals 89-109. Pale brown anteriorly, with a yellow, black-edged vertebral stripe, which becomes gradually more and more indistinct towards the hinder part of the body, which, like the tail, is darker brown or black; a series of black spots on each side of the anterior part of the body, of ocelli with bright yellow centres in the young; some adult nearly uniform blackish brown; a black streak below the eye; an oblique black streak from the eye to the angle of the mouth, and another on the temple and neck.

Total length 1500 millim.; tail 380.

Southern China, Burma, Malay Peninsula, Sumatra, Borneo, Java.

a. ♂ (V. 214; C. ?).	China.	— Lindsay, Esq. [P.].
b. Yg. (V. 222; C. 100).	Pinang.	Gen. Hardwicke [P.].
c. ♂ (V. 217; C. 90).	Singapore.	Dr. Dennys [P.].
d. ♀ (V. 219; C. 101).	Sumatra.	Leyden Museum.
e-f. ♂ (V. 217; C. 99) & ♀ (V. 222; C. 100).	Deli, Sumatra.	Mr. Iversen [C.].
g. ♀ (V. 221; C. ?).	Nias.	Hr. Sundermann [C.].
h. Yg. (V. 216; C. 103).	Borneo.	Sir E. Belcher [P.].
i-k. Yg. (V. 203, 193; C. 89, 90).	Borneo.	L. L. Dillwyn, Esq [P.].
l. Hgr. (V. 212; C. 106).	Borneo.	Sir J. Brooke [P.].
m. ♀ (V. 230; C. 102).	Borneo.	Sir Hugh Low [P.].
n-p. ♂ (V. 206; C. 107) & yg. (V. 207, 230; C. 101, 99).	Rejang R., Sarawak.	Brooke Low, Esq. [P.]
q. ♀ (V. 234; C. 107).	Sarawak.	A. Everett, Esq. [C.].
r. ♂ (V. 208; C. 98).	Baram, Borneo.	C. Hose, Esq. [C.].
s. Hgr. (V. 221; C. 91).	Java.	
t. Skull.	Sarawak.	

36. Coluber radiatus.

Russell, Ind. Serp. ii. pl. xlii. (1801).
Coluber radiatus, *Schleg. Phys. Serp.* ii. p. 135, pl. v. figs. 5 & 6 (1837); *Cantor, Cat. Mal. Rept.* p. 73 (1847); *Bouleng. Faun. Ind., Rept.* p. 333 (1890); *W. Sclater, Journ. As. Soc. Beng.* lx. 1891, p. 239.
—— quadrifasciatus, *Cantor, Proc. Zool. Soc.* 1839, p. 51.
Tropidonotus quinque, *Cantor, l. c.* p. 54.
Compsosoma radiatum, *Dum. & Bibr.* vii. p. 292 (1854); *Günth. Rept. Brit. Ind.* p. 243 (1864); *Theob. Cat. Rept. Brit. Ind.* p. 165 (1876); *Anders. An. Zool. Res. Yunnan*, p. 815 (1879).

Spilotes radiatus, *Günth. Cat.* p. 96 (1858).
Elaphis radiatus, *Jan, Elenco*, p. 61 (1863).

Rostral a little broader than deep, visible from above; internasals as long as broad or a little broader, shorter than the præfrontals; frontal once and one sixth to once and a half as long as broad, as long as or a little shorter than its distance from the end of the snout, shorter than the parietals; loreal usually longer than deep; one large præocular; two postoculars; temporals usually 2+2; eight or nine upper labials, third to fifth, or fourth and fifth, or fourth to sixth entering the eye; four or five lower labials in contact with the anterior chin-shields, which are as long as the posterior. Scales in 19 rows (21 on the neck), more or less strongly keeled on the posterior half of the body, the outer series without trace of a keel. Ventrals with an obtuse lateral keel, 224-242; anal entire; subcaudals 85-100. Yellowish brown above, with one or two black stripes on each side of the anterior half of the back, the lower usually broken up; a black line across the occiput; three black lines radiating from the eye; lower parts uniform yellow or speckled with olive.

Total length 1610 millim.; tail 330.

Southern China, Eastern Himalayas, Bengal, Assam, Burma, Cochinchina, Malay Peninsula, Sumatra, Java.

a.	♀ (V. 242; C. 86).	India.	Capt. Stafford [P.].
b.	♀ (V. 235; C. 95).	Sikkim.	Sir J. Hooker [P.].
c.	Yg. (V. 239; C. 85).	Khasi Hills.	Sir J. Hooker [P.].
d, e, f, g.	♂ (V. 231, 235; C. 95, ?), ♀ (V. 225; C. ?), & hgr. (V. 234; C. 94).	Toungoo.	E. W. Oates, Esq. [P.].
h.	Yg. (V. 222; C. 85).	Fort Stedman, Shan States, 3000 ft.	E. W. Oates, Esq. [P.].
i.	Yg. (V. 239; C. 92).	Mergui.	Dr. Cantor. (Type of *T. quinque*.)
k.	Yg. (V. 235; C. 87).	Mergui.	Prof. Oldham [P.].
l-m.	♂ (V. 228; C. 86) & yg. (V. 224; C. 94).	Pinang.	Dr. Cantor.
n.	Yg. (V. 234; C. 96).	Pinang.	Gen. Hardwicke [P.].

37. Coluber erythrurus.

Plagiodon erythrurus, *Dum. & Bibr.* vii. p. 175 (1854); *Peters, Mon. Berl. Ac.* 1861, p. 684; *Jan, Icon. Gén.* 20, pl. iv. fig. 2 (1867); *Steind. Sitzb. Ak. Wien*, c. i. 1891, p. 141.
Compsosoma melanurum, var., *Dum. & Bibr. t. c.* p. 301; *Günth. Proc. Zool. Soc.* 1873, p. 169.
Elaphis subradiatus, part., *Günth. Cat.* p. 95 (1858).
Spilotes melanurus, part., *Günth. l. c.* p. 97.
Elaphis melanurus, var. manillensis, *Jan, Icon. Gén.* 21, pl. iv. fig. 2 (1867).
—— melanurus, var. celebensis, *Jan, l. c.* pl. v. fig. 2.
Compsosoma melanurum, var. erythrurum, *Fischer, Jahrb. Wiss. Anst. Hamb.* ii. 1885, p. 101.

Very closely allied to *C. radiatus* and *C. melanurus*, but differing from both in the more numerous scales, viz., in 21 rows on the body and 23 on the neck. Scales less strongly keeled than in *C. melanurus*, more as in *C. radiatus*; the outer row of caudal scales smooth or very faintly keeled. Loreal usually not or but slightly longer than deep; fourth to sixth upper labials entering the eye. Ventrals 211–233; anal entire; subcaudals 86–100. No dorsal stripes.

Total length 1670 millim.; tail 370.

Philippines, Sooloo Islands, Celebes.

A. Pale brown or olive above, the scales edged with blackish; a more or less distinct oblique dark streak behind the eye, and another (often absent) on each side of the neck. (*C. erythrurus*, D. & B.; *manillensis*, Jan.)

   a. Lower parts yellow, usually turning to red posteriorly and on the tail.

a, b, c. ♂ (V. 217, 228; C. 87, ?) & yg. (V. 225; C. 95).	Philippines.	H. Cuming, Esq. [C.].
d. ♀ (V. 233; C. ?).	Luzon.	Dr. A. B. Meyer [C.].
e. ♂ (V. 221; C. ?).	N. Leyte.	A. Everett, Esq. [C.].

   b. Upper and lower parts turning to blackish posteriorly; tail black.

f. ♀ (V. 226; C. 97).	Negros.	Dr. A. B. Meyer [C.].

B. Brown above, reddish anteriorly; no lines or spots on the head; a large V- or ⋏-shaped black marking on the neck; young with black vertical bars on the sides. (*C. celebensis*, Jan.)

   a. Lower parts bright yellow; posterior part of belly and tail speckled with brown.

a–b. ♂ (V. 222; C. 96) & yg. (V. 225; C. 100).	Macassar.	Dr. A. B. Meyer [C.].

   b. Upper and lower parts turning to black posteriorly.

c. ♂ (V. 216; C. 96).	Celebes.	
d, e. ♀ (V. 228, 232; C. ?, 90).	N. Celebes.	Dr. A. B. Meyer [C.].
f. ♂ (V. 223; C. 98).	Manado.	Dr. A. B. Meyer [C.].

38. Coluber enganensis.

Coluber enganensis, *Vinciguerra, Ann. Mus. Genova,* (2) xii. 1892, p. 524.

Intermediate between *C. erythrurus* and *C. subradiatus*. Differing from the former in the more numerous scales, viz. 23 rows on the body and 27 on the neck; from the latter in the absence of a subocular separating the fourth labial from the eye, the deeper rostral, which is but little broader than deep, and the shorter posterior chin-shields, which are not or but slightly longer than the anterior. Ventrals 239–243; anal entire; subcaudals 107–108. Brown above, darker along the back, uniform or with black spots disposed in longitudinal series on the anterior part of the body; a short dark streak behind the eye; lower parts yellowish, spotted with pale brown posteriorly.

Total length 1420 millim.; tail 310.

Engano Island, S.W. Sumatra.

39. Coluber subradiatus.

Coluber subradiatus, *Schleg. Phys. Serp.* ii. p. 136 (1837), *and Abbild.* p. 101, pls. xxviii. figs. 7 & 8, and pl. xxix. (1840).
Compsosoma subradiatum, *Dum. & Bibr.* vii. p. 297 (1854).
Elaphis subradiatus, part., *Günth. Cat.* p. 95 (1858).
—— nyctenurus, *Jan, Elenco*, p. 61 (1863).
—— subradiatus, *Van Lidth de Jeude, in M. Weber, Zool. Ergebn.* p. 184 (1890).

Nearly allied to *C. radiatus*, but differing in the following points:—Rostral much broader than deep; a subocular below the ocular; fifth and sixth labials entering the eye; anterior chin-shields much shorter than the posterior; scales in 23 or 25 rows (25 or 27 on the neck). Ventrals 228–242; anal entire; subcaudals 80–100. Pale reddish brown above, with a pair of black stripes along each side of the anterior half of the body; these stripes may be broken up into series of spots; a short black streak behind the eye; lower parts uniform yellowish.

Total length 1250 millim.; tail 260.

Timor, Flores.

a. ♂ (Sc. 23; V. 233; C. 93).		Timor.	Leyden Museum. (One of the types.)
b. Yg. (Sc. 23; V. 230; C. 96).		——?	Dr. Bleeker.

40. Coluber lineaticollis.

Arizona lineaticollis, *Cope, Proc. Ac. Philad.* 1861, p. 300.
Spilotes lineaticollis, *Cope, Bull. U.S. Nat. Mus.* no. 32, 1887, p. 72.
Pituophis lineaticollis, *Günth. Biol. C.-Am., Rept.* p. 124, pl. xlvii. (1894).

Rostral broader than deep, visible from above; internasals broader than long, shorter than the præfrontals; frontal once and a half as long as broad, as long as its distance from the rostral, a little shorter than the parietals; loreal a little longer than deep; one (or two) præ- and three postoculars; temporals 2+3 or 4; eight or nine upper labials, fourth and fifth, or fourth, fifth, and sixth, entering the eye; four or five lower labials in contact with the anterior chin-shields. Scales in 27 rows, dorsals strongly keeled. Ventrals 244; anal entire; subcaudals 69. Pale brown above; head uniform;

neck with four interrupted black stripes; a dorsal series of ∞-shaped black markings, and a series of smaller markings on each side; lower parts yellow, posteriorly with squarish black blotches.

Total length 1500 millim.; tail 230.

Mexico and Guatemala.

a. ♂ (V. 244; C. 69).   Dueñas, Guatemala.   O. Salvin, Esq. [C.].

41. Coluber scalaris.

Coluber scalaris, *Schinz, Cuv. Thierr.* ii. p. 123 (1822); *Boie, Isis*, 1827, p. 536.
—— hermanni, *Desmarest, Faune Franç., Rept.* pl. xix. (1826); *Lesson, Act. Soc. Linn. Bord.* xii. 1841, p. 58.
Rhinechis agassizii, *Michahelles, in Wagl. Icon. Amph.* pl. xxv. (1833).
Coluber agassizii, *Dugès, Ann. Sc. Nat.* (2) iii. 1835, p. 139.
Xenodon michahelles, *Schleg. Phys. Serp.* ii. p. 92 (1837).
Rhinechis scalaris, *Bonap. Icon. Faun. Ital.* (1838); *Dum. & Bibr.* vii. p. 227 (1854); *Günth. Cat.* p. 85 (1858); *Jan, Icon. Gén.* 20, pl. i. (1867); *De Betta, Faun. Ital., Rett. Anf.* p. 39 (1874) *Schreib. Herp. Eur.* p. 290 (1875); *Bedriaga, Amph. & Rept. Portug.* p. 73 (1889).

Snout strongly projecting; rostral deeper than broad, forming an acute angle posteriorly, wedged in between the internasals, the portion visible from above nearly as long as its distance from the frontal; internasals shorter than the præfrontals; frontal about once and one third as long as broad, as long as or shorter than its distance from the end of the snout, nearly as long as the parietals; loreal longer than deep; one præ- and two or three postoculars; temporals 2+3 or 2+4; seven or eight upper labials, fourth or fourth and fifth entering the eye; four or five lower labials in contact with the anterior chin-shields, which may be either longer or shorter than the posterior. Scales smooth, in 27 (rarely 25 or 29) rows. Ventrals 201–220; anal divided (rarely entire); subcaudals 48–68. Young yellowish or pale brown above, with a series of regular H-shaped black or blackish-brown markings along the back, and small black spots on the sides; a V-shaped black marking on the snout, a black streak from the eye to the angle of the mouth, and a black spot below the eye; belly yellow, spotted or checkered with black, or nearly entirely black. These dorsal markings disappearing in the adult, and replaced by a pair of brown stripes running along the back; the belly uniform yellow.

Total length 925 millim.; tail 160.

South of France, Pyrenean Peninsula.

a. ♂ (Sc. 27; V. 207; C. 62).   Near Nice.   Dr. F. Werner [E.]
b. Yg. (Sc. 27; V. 210; C. 58).   Montpellier.
c. ♀ (Sc. 27; V. 213; C. ?).   Spain.   Lord Lilford [P.].

d. Yg. (Sc. 27; V. 201; C. 60).		Coimbra.	Dr. Lopes Vieira [P.].
e. Hgr. (Sc. 27; V. 204; A. 1; C. 58).		Coimbra.	Prof. Barboza du Bocage [P.].
f. ♂ (Sc. 27; V. 205; C. 63).		Alfeite, Portugal.	Prof. Barboza du Bocage [P.].
g. Yg. (Sc. 27; V. 213; C. 60).		Aldegallega, Portugal.	Prof. Barboza du Bocage [P.].
h. Skull.		Montpellier.	

42. Coluber arizonæ.

Arizona elegans, *Kennicott, Rep. U.S. Mex. Bound. Surv.* ii. *Rept.* p. 18, pl. xiii. (1859); *Bocourt, Miss. Sc. Mex., Rept.* p. 676 (1888).
Pityophis elegans, *Cope, Check-list N. Am. Rept.* p. 39 (1875).
Rhinechis elegans, *Cope, Proc. Am. Philos. Soc.* xxiii. 1886, p. 284, and *Proc. U.S. Nat. Mus.* xiv. 1892, p. 638.

Snout strongly projecting; rostral as deep as broad, forming an acute angle posteriorly, the portion visible from above measuring nearly two thirds its distance from the frontal; internasals nearly as long as the præfrontals; frontal a little longer than broad, a little shorter than its distance from the end of the snout or than the parietals; loreal longer than deep; one præ- and two postoculars; temporals 2 + 3 or 4; eight upper labials, fourth and fifth entering the eye; five lower labials in contact with the anterior chin-shields, which are as long as or longer than the posterior. Scales smooth, in 27 to 31 rows. Ventrals 212–227; anal entire; subcaudals 46–59. Pale brown above; the vertebral line yellowish, but interrupted by a series of larger, oval, transverse, dark-edged brown spots; a brown streak from the eye to the angle of the mouth; lower parts white.

Total length 1020 millim.; tail 160.

Southern United States, from California to the Mississippi; North Mexico.

a. ♂ (Sc. 29; V. 212; C. 59).		Duval Co., Texas.	W. Taylor, Esq. [C.].
b. ♀ (Sc. 27; V. 227; C. 46).		Warner's Ranch, San Diego Co., California.	Prof. C. Eigenmann [C.].

43. Coluber deppii.

Elaphis pleurostictus, *Dum. & Bibr.* vii. p. 244 (1854).
—— deppei, *Dum. & Bibr. t. c.* p. 268.
Pituophis vertebralis, *Günth. Cat.* p. 86 (1858).
Arizona jani, *Cope, Proc. Ac. Philad.* 1860, p. 369.
Pituophis mexicanus (*non D. & B.*), *Jan, Icon. Gén.* 22, pl. ii. fig. 1 (1867).
—— deppei, *Jan, l. c.* fig. 2; *Dugès, La Naturaleza,* (2) i. 1888, p. 123, pl. xlii. fig. 15; *Günth. Biol. C.-Am., Rept.* p. 124 (1894).

Spilotes deppei, *Cope, Bull. U.S. Nat. Mus.* no. 32, 1887, p. 72.
Pituophis pleurostictus, *Bocourt, Miss. Sc. Mex., Rept.* p. 666, pl. xlii. figs. 2 & 3 (1888).

Snout projecting; rostral as broad as deep or slightly deeper, the portion visible from above measuring one half or three fifths its distance from the frontal; internasals much shorter than the præfrontals; frontal once and a half as long as broad, a little longer than its distance from the rostral, as long or a little shorter than the parietals; loreal a little longer than deep; one præ- and two or three postoculars; temporals 3+4; seven or eight upper labials, fourth and fifth or third and fourth entering the eye; four or five lower labials in contact with the anterior chin-shields, which are longer than the posterior. Scales in 27 or 29 rows, dorsals feebly but distinctly keeled. Ventrals 209–233; anal entire; subcaudals 51–67. Yellowish brown above, with a dorsal series of large quadrangular spots; sides with smaller spots; young with a dark band across the forehead, from eye to eye; labials with black sutures; lower parts yellowish, with square brown blotches.

Total length 1690 millim.; tail 180.

Mexico.

a. ♂ (Sc. 29; V. 220; C. 55). City of Mexico. Mr. Doorman [C.].
b. ♀ (Sc. 29; V. 229; C. 61). Hacienda del Castillo, Guadalajara. Dr. A. C. Buller [C.].
c. ♀ (Sc. 27; V. 221; C. 59). Tehuantepec. M. F. Sumichrast [C.].
d–e. ♂ (Sc. 27; V. 228; C. 56) & hgr. (Sc. 29; V. 219; C. 58). Mexico. M. Sallé [C.].
f, g. ♀ (Sc. 27, 27; V. 212, 221; C. 55, 63). Mexico.
h. ♂ (Sc. 27; V. 218; C. 66). S. Mexico. F. D. Godman, Esq. [P.].

44. Coluber catenifer.

Coluber catenifer, *Blainv. Nouv. Ann. Mus.* iv. 1835, p. 290, pl. xxvi. fig. 2.
—— vertebralis, *Blainv. l. c.* p. 293, pl. xxvii. fig. 2.
Pituophis catenifer, *Baird & Gir. Cat. N. Am. Rept.* p. 69 (1853); *Günth. Cat.* p. 87 (1858); *Girard, U.S. Explor. Exped., Herp.* p. 135, pl. viii. figs. 1–7 (1858); *Garm. N. Am. Rept.* p. 52 (1883); *Bocourt, Miss. Sc. Mex., Rept.* p. 670, pl. xlvii. fig. 4 (1888); *Cope, Proc. U.S. Nat. Mus.* xiv. 1892, p. 641; *Stejneger, N. Am. Faun.* no. 7, pt. ii. p. 206 (1893).
—— wilkesii, *Baird & Gir. l. c.* p. 71; *Gir. l. c.* p. 137, pl. ix.
—— annectens, *Baird & Gir. l. c.* p. 72, and *Rep. Explor. Surv. R. Miss. R.* x., *Rept.* pl. xxix. fig. 48 (1859).
—— vertebralis, *Dum. & Bibr.* vii. p. 238 (1854); *Bocourt, l. c.* p. 672, pl. xlii. fig. 1.
Elaphis reticulatus, *Dum. & Bibr. t. c.* p. 246.
Pityophis hæmatois, *Cope, Proc. Ac. Philad.* 1860, p. 342.

Pituophis melanoleucus, vars. catenifer *and* vertebralis, *Jan, Icon. Gén.* 22, pl. i. figs. 1 & 3 (1867).
—— reticulatus, *Bocourt, l. c.* pl. xlvii. fig. 3.
—— catenifer deserticola, *Stejneger, l. c.*

Snout projecting; rostral as broad as deep or slightly deeper, the portion visible from above measuring one half its distance from the frontal; internasals much shorter than the præfrontals, which are usually broken up into four shields; frontal once and one fourth to once and a half as long as broad, as long as its distance from the rostral or the end of the snout, as long as the parietals; loreal longer than deep; one præocular, usually with a subocular below it; two or three postoculars and one or two suboculars; temporals 3 or 4+4 or 5; eight upper labials, fourth usually entering the eye; four or five lower labials in contact with the anterior chin-shields, which are longer than the posterior. Scales in 29 to 35 rows, dorsals more or less strongly keeled. Ventrals 204-245'; anal entire; subcaudals 66-72. Yellowish or pale brown above; a dorsal series of large dark brown or black spots; sides with small dark spots; young with a dark band across the forehead, from eye to eye, an oblique streak behind and another below the eye; lower parts yellowish, uniform or with brown blotches.

Total length 1140 millim.; tail 190.

United States west of the Rocky Mountains; Western Mexico.

A. Fourth labial entering the eye.

a. ♀ (Sc. 31; V. 218; Monterey, California. J. H. Gurney, Esq. C. 66). [P.].
b. Yg. (Sc. 31; V. 204; San Francisco. Christiania Museum. C. 68).
c. ♂ (Sc. 29; V. 212; California. Smithsonian Instit. C. 70). [P.].
d. ♂ (Sc. 33; V. 218; California. C. 70).
e. ♂ (Sc. 31; V. 241; Great Salt Lake. Christiania Museum. C. 70).
f. ♂ (Sc. 29; V. 223; W. Mexico. Haslar Coll. C. 70).
g. Yg. (Sc. 29; V. 209; ——? C. 71).

B. Eye separated from the labials by a series of suboculars.

h. ♂ (Sc. 33; V. 219; San Diego, California. Christiania Museum. C. ?).

45. Coluber melanoleucus.

Coluber melanoleucus, *Daud. Rept.* vi. p. 409 (1803); *Harl. Journ. Ac. Philad.* v. 1827, p. 359, *and Med. Phys. Res.* p. 122 (1835).
—— sayi, *Schleg. Phys. Serp.* ii. p. 157 (1837).
Pituophis melanoleucus, *Holbr. N. Am. Herp.* iv. p. 7, pl. i. (1842); *Baird & Gir. Cat. N. Am. Rept.* p. 65 (1853); *Dum. & Bibr.* vii. p. 233 (1854); *Günth. Cat.* p. 87 (1858); *Baird, Rep. Explor.*

Surv. R. Miss. R. x. pl. xxix. fig. 44 (1859); *Jan, Icon. Gén.* 22, pl. i. fig. 2 (1867); *Garm. N. Am. Rept.* p. 51 (1883); *Bocourt, Miss. Sc. Mex., Rept.* pl. xlvii. fig. 5 (1888); *Günth. Biol. C.-Am., Rept.* p. 125 (1894).

Churchillia bellona, *Baird & Gir. in Stansbury's Expl. Great Salt Lake*, p. 350 (1852).

Pituophis affinis, *Hallow. Proc. Ac. Philad.* 1852, p. 181.

—— bellona, *Baird & Gir. l. c.* p. 66, *and Rep. Explor. Surv. R. Miss. R.* x. pl. xxix. fig. 46; *Bocourt, l. c.* fig. 2.

—— macclellanii, *Baird & Gir. ll. cc.* p. 68, fig. 47.

—— mexicanus, *Dum. & Bibr.* vii. p. 236, pl. lxii.; *Günth. Cat.* p. 88; *Bocourt, l. c.* p. 674, pls. xlii. fig. 4 & xlvii. fig. 6.

—— sayi, *Baird, Explor. Surv. R. Miss. R.* x. pl. xxix. fig. 45; *Cope, Proc. U.S. Nat. Mus.* xiv. 1892, p. 640.

Pityophis catenifer, var. sayi, *H. Garm. Bull. Illin. Lab.* iii. 1892, p. 288.

Snout projecting; rostral much deeper than broad, the portion visible from above measuring one half to two thirds its distance from the frontal; præfrontals broken up into four shields or more; frontal once and one third to once and two thirds as long as broad, as long as or a little longer than its distance from the rostral; parietals usually broken up behind; loreal longer than deep; one large præocular, sometimes with a small subocular below it; two or three postoculars and one or two posterior suboculars; temporals 3 or 4 + 4 or 5; eight or nine upper labials, fourth or fifth entering the eye; five or six lower labials in contact with the anterior chin-shields, which are much longer than the posterior. Scales in 29 to 35 rows, dorsals strongly keeled. Ventrals 209–239; anal entire; subcaudals 45–65. Yellowish or pale brown above; a dorsal series of large squarish, brown, black-edged spots; sides with small spots; young with a dark band across the forehead, from eye to eye, and another from the eye to the angle of the mouth; some or all of the labials with black sutures; yellowish white inferiorly, usually with brown spots.

Total length 1480 millim.; tail 190.

North America and Mexico.

a, b, c. ♂ (Sc. 31, 29, 29; V. 218, 216, 215; C. 61, 62, 62).	Brit. Columbia.	J. K. Lord, Esq. [P.].
d. ♀ (Sc. 31; V. 220; C. 53).	Medicine Hat, Assiniboia.	Prof. C. Eigenmann [C.].
e. ♀ (Sc. 31; V. 221; C. 52).	Nebraska.	Smithsonian Instit. [P.].
f. ♀ (Sc. 31; V. 229; C. 56).	New Mexico.	Smithsonian Instit. [P.].
g, h. ♀ (Sc. 35, 35; V. 230, 237; C. 47, 47).	Duval Co., Texas.	W. Taylor, Esq. [C.].
i. ♂ (Sc. 33; V. 230; C. 45).	Nuevo Leon.	W. Taylor, Esq. [C.].
k, l. ♂ (Sc. 31; V. 215; C. 57) & ♀ (Sc. 33; V. 224; C. 52).	N. America.	

m, n, o. ♂ (Sc. 29; V. 211; C. 57), ♀ (Sc. 33; V. 238; C. 53), & hgr. (Sc. 35; V. 223; C. 55).   Mexico.

61. SYNCHALINUS.

Synchalinus, *Cope, Proc. Amer. Philos. Soc.* xxxi. 1893, p. 345.

Teeth equal. Head distinct from neck; eye large, with round pupil; nasals and loreal fused into a single elongate shield. Body elongate, compressed; scales smooth or faintly keeled, with apical pits, in 23 rows; ventrals sharply angulate laterally. Tail long; subcaudals in two rows.

Central America.

1. Synchalinus corallioides.

Synchalinus corallioides, *Cope, l. c.*

Similar in general appearance to *Corallus hortulanus*. Snout subtruncate; rostral slightly visible from above; internasals subquadrate; præfrontals wider than long; frontal shorter than supraocular, with concave lateral borders; parietals as broad as long; nasal in contact with the single præocular; two postoculars; temporals 2+2; eight upper labials, fourth, fifth, and sixth entering the eye, last very long; posterior chin-shields longer than the anterior. Scales in 23 rows, three or four median rows faintly keeled. Ventrals 209; anal entire; subcaudals 134. Yellowish brown above, with a dorsal series of parallelogrammic spots of an iron-rust colour, each of which has a small blackish spot at its anterior extremity; sides with wide vertical spots of the colour of the dorsals, with which some may be confluent, forming with them broad cross bands; a dark brown band from the eye to the side of the neck; three dark bars below the eye; a dark marking on the vertex; belly dark mahogany on the middle line; a dark spot on the outer ends of the ventrals, opposite to the lateral vertical bars.

Total length 450 millim.; tail 125.

Buenos Ayres, Costa Rica.

62. GONYOPHIS.

Gonyophis, *Bouleng. Ann. & Mag. N. H.* (6) viii. 1891, p. 290.

Maxillary teeth 23, equal; mandibular teeth subequal. Head distinct from neck, elongate; eye moderate, with round pupil. Body elongate, compressed; scales feebly keeled, with apical pits, in 19 rows; ventrals with a suture-like lateral keel, and a notch on each side corresponding to the keel. Tail long; subcaudals in two rows, keeled and notched like the ventrals.

Malay Peninsula and Borneo.

1. Gonyophis margaritatus.

Gonyosoma margaritatum, *Peters, Mon. Berl. Ac.* 1871, p. 578, *a Ann. Mus. Genova,* iii. 1872, p. 39, pl. v. fig. 3.
Gonyophis margaritatus, *Bouleng. Ann. & Mag. N. H.* (6) viii. 1891, p. 290; *Werner, Verh. zool.-bot. Ges. Wien,* 1893, p. 357.

Rostral broader than deep, just visible from above; internasals as long as broad, shorter than the præfrontals; frontal a little longer than broad, as long as its distance from the end of the snout, shorter than the parietals; loreal longer than deep; one præocular, not in contact with the frontal, or just touching; two postoculars; temporals 2+2; nine upper labials, fourth, fifth, and sixth, or fifth and sixth, entering the eye; four or five lower labials in contact with the anterior chin-shields, which are as long as the posterior. Scales in 19 rows, feebly keeled. Ventrals 230–241; anal divided; subcaudals 108–120. Black above, each scale with a yellowish-green spot, or green with black borders to the scales; hinder part of body and tail with bright orange rings; lower parts yellowish, the shields with or without a black edge; a black streak on each side of the head behind the eye, and another along the suture between the parietals.

Total length 1530 millim.; tail 370.

Borneo, Singapore.

a. Yg. (V. 233; C. 108).   Mt. Dulit, Borneo.   C. Hose, Esq. [C.].

63. HERPETODRYAS.

Chironius, part., *Fitz. N. Class. Rept.* p. 31 (1826).
Erpetodryas, *Boie, in Férussac, Bull. Sc. Nat.* ix. 1826, p. 235, *and Isis,* 1827, p. 521.
Herpetodryas, *Wagl. Syst. Amph.* p. 180 (1830); *Cope, Proc. Ac. Philad.* 1860, p. 562; *Bocourt, Miss. Sc. Mex., Rept.* p. 732 (1890).
Macrops, *Wagl. l. c.* p. 182.
Herpetodryas, part., *Schleg. Phys. Serp.* ii. p. 173 (1837); *Dum. & Bibr. Erp. Gén.* vii. p. 203 (1854); *Günth. Cat. Col. Sn.* p. 113 (1858).
Dendrophis, part., *Dum. & Bibr. t. c.* p. 193.
Phyllosira, *Cope, Proc. Ac. Philad.* 1862, p. 349.

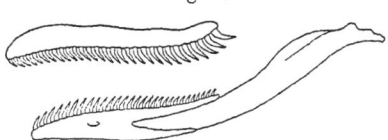

Fig. 5.

Maxillary and mandible of *Herpetodryas fuscus.*

Maxillary teeth 28 to 32, posterior largest, stouter if not longer than the rest; mandibular teeth subequal or decreasing in size

posteriorly. Head distinct from neck; eye large or rather large, with round pupil. Body elongate, more or less compressed; scales smooth or keeled, the laterals smaller and oblique, with apical pits, in 10 or 12 rows; ventrals rounded or obtusely angulate laterally. Tail long; subcaudals in two rows.
Central and South America.

Synopsis of the Species.

I. Scales in 12 rows; anal usually divided.

Four or more rows of scales keeled .... 1. *sexcarinatus*, p. 72.
Scales smooth, or only the two middle rows keeled .... 2. *carinatus*, p. 73.

II. Scales in 10 rows.

Anal entire; scales smooth, or only the two middle rows keeled .... 3. *fuscus*, p. 75.
Anal divided; scales smooth .... 4. *melas*, p. 76.
Anal divided; scales keeled .... 5. *grandisquamis*, p. 76.

1. Herpetodryas sexcarinatus.

Natrix sexcarinata, *Wagl. in Spix, Serp. Bras.* p. 35, pl. xii. (1824), and *Syst. Amph.* p. 180 (1830).
? Natrix cinnamomea, *Wagl. l. c.* p. 20, pl. vi. fig. 1
Erpetodryas quadricarinatus, *Boie, Isis*, 1827, p. 548.
Herpetodryas carinatus, part., *Schleg. Phys. Serp.* ii. p. 175, pl. vii. figs. 5-7 (1837); *Dum. & Bibr.* vii. p. 207 (1854); *Jan, Icon. Gén.* 31, pl. ii. fig. 1 (1869).
—— carinatus, *Boettg. Zeitschr. f. ges. Naturw.* lviii. 1885, p. 233.
—— sexcarinatus, *Bouleng. Proc. Zool. Soc.* 1891, p. 355.

Eye smaller than in *H. carinatus*. Rostral broader than deep, just visible from above; internasals as long as or a little shorter than the præfrontals; frontal once and one third to once and two thirds as long as broad, as long as or a little longer than its distance from the end of the snout, a little shorter than the parietals; loreal longer than deep; one præ- and two postoculars; temporals 1+2; eight or nine upper labials, fourth and fifth or fifth and sixth entering the eye; five lower labials in contact with the anterior chin-shields, which are a little shorter than the posterior. Scales in 12 rows, four or more of which are distinctly keeled. Ventrals indistinctly angulate laterally, 142-159; anal divided (rarely entire); subcaudals 100-128. Uniform brown or olive above and on the outer ends of the ventrals; head often reddish; upper lip and lower parts yellowish.

Total length 1350 millim.; tail 450.

Venezuela, Brazil, Paraguay.

A. Uniform pale brown above.

a–c, d–f. ♂ (V. 143, 143, 142; C. 118, 122, 125), hgr. (V. 145, 145; C. 113, ?), & yg. (V. 144, C. 120).	Porto Real, Prov. Rio Janeiro.	M. F. Hardy du Dréneuf [C.].
g. ♂ (V. 145; C. 128).	S. José dos Campos, Prov. S. Paulo.	Mr. A. Thomson [P.].
h–i. ♂ (V. 146; C. ?) & ♀ (V. 150; C. 119).	Asuncion, Paraguay.	Dr. J. Bohls [C.].

B. Olive above, tail with two black stripes.

k–l. ♂ (V. 155; C. 116) & ♀ (V. 151; C. 119).	Venezuela.	Mr. Dyson [C.].

2. Herpetodryas carinatus.

Coluber carinatus, *Linn. Mus. Ad. Frid.* p. 31 (1754), *and S. N.* i. p. 384 (1766); *Daud. Rept.* vii. p. 115 (1803).
—— bicarinatus, *Wied, Reis. Bras.* i. p. 181 (1820), *and Beitr.* i. p. 284 (1825), *and Abbild.* (1831).
Natrix bicarinata, *Wagl. in Spix, Serp. Bras.* p. 23, pl. vii. (1824).
Coluber pyrrhopogon, *Wied, Beitr.* i. p. 291, *and Abbild.*
—— lævicollis, *Wied, Beitr.* i. p. 296.
Chironius carinatus, *Fitz. N. Class. Rept.* p. 60 (1826).
Erpetodryas carinatus, *Boie, Isis,* 1827, p. 548.
Herpetodryas carinatus, part., *Schleg. Phys. Serp.* ii. p. 175, pl. vii. figs. 3 & 4 (1837); *Dum. & Bibr.* vii. p. 207 (1854); *Günth. Cat.* p. 115 (1858); *Cope, Proc. Ac. Philad.* 1860, p. 562; *Jan, Icon. Gén.* 31, pl. ii. figs. 2–4 (1869); *Garm. Proc. Am. Philos. Soc.* xxiv. 1887, p. 284; *Günth. Biol. C.-Am., Rept.* p. 128 (1894).
—— fuscus, part., *Dum. & Bibr. t. c.* p. 209; *Günth. l. c.* p. 114.
—— carinatus, *Wucherer, Proc. Zool. Soc.* 1861, p. 324.
—— carinatus, vars. flavolineata *and* macrophthalma, *Jan, Elenco,* p. 80 (1863).
Dendrophis, sp., *F. Müller, Verh. nat. Ges. Basel,* vi. 1878, p. 678.
Herpetodryas flavolineatus, *Boettg. Zeitschr. f. ges. Naturw.* lviii. 1885, p. 234.
—— carinatus, var. vincenti, *Bouleng. Proc. Zool. Soc.* 1891, p. 355.

Eye very large. Rostral broader than deep, visible from above; internasals as long as or slightly shorter than the præfrontals; frontal once and one fourth to once and two thirds as long as broad, as long as or a little shorter than its distance from the end of the snout or the parietals; loreal longer than deep; one præ- and two (rarely three) postoculars; temporals $1+1$ or $1+2$; eight or nine upper labials, fourth and fifth, fifth and sixth, or fourth, fifth, and sixth, entering the eye; five or six lower labials in contact with the anterior chin-shields, which are as long as or shorter than the posterior. Scales in 12 rows, all smooth, or the two middle rows faintly keeled (females), or the two middle rows more or less strongly keeled (males). Ventrals feebly angulate laterally, 145–173; anal divided (very rarely entire); subcaudals 110–171.

Total length 1530 millim.; tail 500.

Central America; South America east of the Andes and north of the Rio de la Plata; Lesser West Indies (Trinidad, Guadeloupe, St. Vincent).

A. Dark brown or blackish above; the vertebral line pale brown; a series of large yellowish spots along each side of the body on the outer series of scales; ventral and subcaudal shields yellow, with a fine black edge. Temporals 1+2; two labials entering the eye. (*C. carinatus*, L.)

a.	♂ (V. 156; C. ?).	Demerara.	
b.	♂ (V. 156; C. 124).	Berbice.	Lady Essex [P.].
c.	♀ (V. 165; C. 125).	Trinidad.	
d.	♀ (V. 153; C. 110).	Guadeloupe.	

B. Olive above, with whitish spots, most of the scales black-edged; belly greyish; a black line along the middle of the lower surface of the tail. Temporals 1+2; two labials entering the eye.

a.	Hgr. (V. 154; C. 132).	Berbice.	Lady Essex [P.].
b.	Yg. (V. 155; C. 114).	Ecuador.	Mr. Fraser [C.].

C. Olive or dark green above, usually with a more or less distinct pale green or yellowish black-edged vertebral stripe; caudal scales black-edged; lower parts pale olive or greenish yellow. Temporals usually 1+2; usually three labials entering the eye. (*C. bicarinatus*, Wied.)

a.	♀ (V. 147; C. ?).	Chiriqui.	F. D. Godman, Esq. [P.].
b.	♀ (V. 164; C. 171).	Venezuela.	Mr. Dyson [C.].
c.	♀ (V. 145; C. 134).	Pampa del Sacramento, N.E. Peru.	Mr. W. Davis [C.]; Messrs. Veitch [P.].
d.	Hgr. (V. 146; C. 126).	Caballo Cocha.	Mr. W. Davis [C.]; Messrs. Veitch [P.].
e.	♀ (V. 145; C. ?).	Pebas.	H. W. Bates, Esq. [C.].
f.	♀ (V. 150; C. 137).	Para.	R. Graham, Esq. [P.].
g.	♀ (V. 157; C. ?).	Bahia.	Dr. O. Wucherer [C.].
h, i–k.	♂ (V. 155; C. 141) & ♀ (V. 154, 163; C. 139, 141).	Rio Janeiro.	A. Fry, Esq. [P.].
l.	♀ (V. 162; C. 140).	Porto Real, Prov. Rio Janeiro.	M. Hardy du Dréneuf [C.].
m.	♂ (V. 160; C. 139).	Rio Grande do Sul.	Dr. H. von Ihering [C.].

D. Blackish brown or black above; upper lip and gular region yellowish; belly plumbeous or blackish. Temporals 1 + 2; three labials entering the eye; three postoculars. (Var. *vincenti*, Blgr.)

a–b.	♀ (V. 168, 166; C. 155, 148).	St. Vincent.	F. D. Godman, Esq. [P.].
c.	♀ (V. 164; C. 156).	——?	

E. Pale brown or buff above, turning to blackish on the neck; a bright yellow vertebral stripe on the anterior half of the body, gradually vanishing behind; uniform yellowish white inferiorly. Temporals 1+1; usually three labials entering the eye. (*H. flavolineatus*, Jan.)

a. ♀ (V. 148; C. 131).	Bahia.	
b. Hgr. (V. 159; C. 125).	Paraguay.	
c. Hgr. (V. 159; C. 147).	——?	

3. Herpetodryas fuscus.

Coluber fuscus, *Linn. Mus. Ad. Frid.* p. 32, pl. xvii. fig. 1 (1754), and *S. N.* i. p. 383 (1766); *Daud. Rept.* vii. p. 112 (1803).
—— saturninus, *Linn. ll. cc.* p. 32, pl. ix. fig. 1, & p. 384.
Natrix scurrula, *Wagl. in Spix, Serp. Bras.* p. 24, pl. viii. (1824).
Herpetodryas carinatus, part., *Schleg. Phys. Serp.* ii. p. 175, pl. vii. figs. 5-7 (1837); *Cope, Proc. Ac. Philad.* 1860, p. 562.
Coluber spixii, *Hallow. Proc. Ac. Philad.* 1845, p. 241.
—— pickeringii, *Hallow. l. c.* p. 242.
Herpetodryas fuscus, part., *Dum. & Bibr.* vii. p. 209 (1854); *Günth. Cat.* p. 114 (1858).
Dendrophis viridis, *Dum. & Bibr. t. c.* p. 202, pl. lxxix. fig. 1.
Herpetodryas sebastus, *Cope, l. c.*
Phyllosira flavescens, *Cope, Proc. Ac. Philad.* 1862, p. 349.
Herpetodryas carinatus, vars. glabra, decalepis, scurrula, *Jan, Elenco*, p. 80 (1863).
—— holochlorus, *Cope, Journ. Ac. Philad.* viii. 1876, p. 178.

Eye large. Rostral broader than deep, just visible from above; internasals as long as or a little shorter than the præfrontals; frontal once and a half to once and two thirds as long as broad, as long as or longer than its distance from the end of the snout, as long as or a little shorter than the parietals; loreal as long as deep or longer; one præ- and two postoculars; temporals 1+1; nine upper labials, fourth, fifth, and sixth, rarely fifth and sixth only, entering the eye; five lower labials in contact with the anterior chin-shields, which are as long as or a little shorter than the posterior. Scales in 10 rows, all smooth, or the vertebral pair faintly keeled. Ventrals not angulate laterally, 144–168; anal entire; subcaudals 111–133.

Total length 2300 millim.; tail 760.

Tropical South America.

A. Olive-brown above, yellowish or whitish inferiorly.
(*C. fuscus*, L.)

a. ♀ (V. 158; C. 114).	Brit. Guiana.	Sir R. Schomburgk [P.].
b, c. ♂ (V. 146; C. 124) & ♀ (V. 151; C. 130).	Moyobamba, N.E. Peru.	Mr. A. H. Roff [C.].
d. ♂ (V. 147; C. 122).	Cayaria, N.E. Peru.	Mr. W. Davis [C.]; Messrs. Veitch [P.].
e. ♂ (V. 156; C.?).	Guayaquil.	Mr. Fraser [C.].
f. Skull.	Cayenne.	
g. Skull.	Brazil.	

B. Dark green above, pale green below. (*D. viridis*, D. & B.)

a, b. ♀ (V. 162; C. 117) & hgr. (V. 155; C. 112).	Berbice.	Lady Essex [P.].
c. ♀ (V. 154; C. 123).	Chyavetas, E. Peru.	
d. ♀ (V. 158; C. ?).	Guayaquil.	Mr. Fraser [C.].

C. Black above, blackish olive below.

a. ♂ (V. 153; C. 111).	Moyobamba, N.E. Peru.	Mr. A. H. Roff [C.].

D. Brown above, with more or less regular light, black-edged cross bands, which are most distinct in the young; belly yellow or olive. (*C. saturninus*, L.)

a. Yg. (V. 145; C. 133).	Caracas.	
b. ♀ (V. 148; C. 127).	Demerara Falls.	
c. ♂ (V. 144; C. 115).	Brazil.	Lord Stuart [P.].
d, e. ♀ (V. 168; C. ?) & yg. (V. 144; C. 122).	Bahia.	Dr. O. Wucherer [C.].

E. Brown above, blotched with black. (*N. scurrula*, Wagl.)

a. ♂ (V. 157; C. ?).	Xeberos.	
b. ♂ (V. 153; C. ?).	Moyobamba.	Mr. A. H. Roff [C.].

4. Herpetodryas melas.

Herpetodryas melas, *Cope, Proc. Amer. Philos. Soc.* xxiii. 1886, p. 278.

Eye large. One præ- and two postoculars; temporals 1+1; nine upper labials, fourth, fifth, and sixth entering the eye; anterior chin-shields shorter than the posterior. Scales in 10 rows, all smooth, those of the median rows rather smaller than the parietal shields. Ventrals 158; anal divided; subcaudals 139. Black, except on the upper lip and the anterior half of the belly, which are cream-coloured.

Total length 1210 millim.; tail 470.

Nicaragua.

5. Herpetodryas grandisquamis.

Spilotes grandisquamis, *Peters, Mon. Berl. Ac.* 1868, p. 451.
Herpetodryas grandisquamis, *Cope, Journ. Ac. Philad.* viii. 1876, p. 135; *Bocourt, Miss. Sc. Mex., Rept.* p. 732, pl. xliii. fig. 5 (1890).
—— carinatus, *Cope, l. c.*

Eye rather large. Rostral broader than deep, just visible from above; internasals as long as the præfrontals; frontal once and a half as long as broad, as long as its distance from the end of the snout, a little shorter than the parietals; loreal as long as deep or a little longer; one præ- and two postoculars; temporals 2+2; nine upper labials, fourth, fifth, and sixth entering the eye; five lower labials in contact with the anterior chin-shields, which are a little shorter than the posterior. Scales in 10 longitudinal rows, about as large as the parietal shields, all except the outer row keeled; the keels on the two median rows stronger. Ventrals 157–162; anal divided; subcaudals 118–135. Black, the throat and anterior part of the belly yellow.

Total length 2200 millim.; tail 500.

Costa Rica.

64. DENDROPHIS.

Ahætulla, part., *Gray, Ann. Phil.* x. 1825, p. 208.
Leptophis, part., *Bell, Zool. Journ.* ii. 1825, p. 328; *Jan, Elenco sist. Ofid.* p. 84 (1863).
Dendrophis, *Boie, Isis,* 1827, p. 520; *Bouleng. Faun. Ind., Rept.* p. 337 (1890).
Dendrophis, part., *Wagl. Syst. Amph.* p. 182 (1830); *Schleg. Phys. Serp.* ii. p. 220 (1837); *Dum. & Bibr. Erp. Gén.* vii. p. 193 (1854); *Günth. Cat. Col. Sn.* p. 148 (1858); *Jan, l. c.* p. 85; *Günth. Rept. Brit. Ind.* p. 296 (1864).

Maxillary teeth 20 to 33, the posterior more or less enlarged, stouter if not longer than the rest; anterior mandibular teeth longest. Head distinct from neck, more or less elongate; eye large, with round pupil. Body elongate, more or less compressed; scales smooth, in 13 or 15 rows, narrow, disposed obliquely, with apical pits, those of the vertebral row more or less enlarged; ventrals with a suture-like lateral keel and a notch on each side, corresponding to the keel. Tail long; subcaudals in two rows, keeled and notched like the ventrals.

South-eastern Asia and Australia.

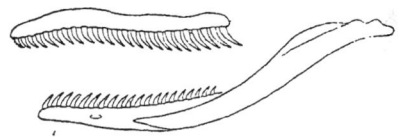

Fig. 6.

Maxillary and mandible of *Dendrophis pictus*.

Synopsis of the Species.

I. Maxillary teeth 20 to 26; eye not longer than its distance from the nostril.

A. Scales in 15 rows.

A single loreal; ventrals 165–190	1. *pictus*, p. 78.
Two loreals; ventrals 154–171	2. *bifrenalis*, p. 80.

B. Scales in 13 rows.

Usually two superposed anterior temporals; ventrals 176–211	3. *calligaster*, p. 80.
Usually a single, much elongate anterior temporal; ventrals 191–220; no dark streak on the side of the head	4. *punctulatus*, p. 82.

II. Maxillary teeth 27 to 33; eye usually longer than its distance from the nostril.

A. Scales in 15 rows.

Two postoculars; vertebral scales scarcely enlarged on the anterior part of the body; ventrals 174–188 .... 5. *grandoculis*, p. 84.

Usually three postoculars; vertebral scales very strongly enlarged; ventrals 179–205 .... 6. *formosus*, p. 84.

B. Scales in 13 rows.

Two postoculars; ventrals 149–161; subcaudals 124–128 .... 7. *caudolineolatus*, p. 85.

Two postoculars; ventrals 173–203; subcaudals 131–151 .... 8. *lineolatus*, p. 85.

Three postoculars; ventrals 162; subcaudals 148 .... 9. *gastrostictus*, p. 86.

1. Dendrophis pictus.

? Coluber ahætulla, *Linn. Mus. Ad. Frid.* p. 35, pl. xxii. fig. 3 (1754).
? Coluber ahætulla, part., *Linn. S. N.* i. p. 387 (1766).
? Natrix ahætulla, *Laur. Syn. Rept.* p. 79 (1768).
Coluber pictus, *Gmel. S. N.* i. p. 1116 (1788).
―― decorus, *Shaw, Zool.* iii. p. 538 (1802).
Dipsas schokari, part., *Kuhl, Beitr. Zool. vergl. Anat.* p. 80 (1820).
Ahætulla decorus, *Gray, Ann. Phil.* x. 1825, p. 208.
Leptophis ahætulla, part., *Bell, Zool. Journ.* ii. 1825, p. 328.
Dendrophis picta, *Boie, Isis,* 1827, p. 530; *Stoliczka, Journ. As. Soc. Beng.* xxxix. 1870, p. 193.
? Dendrophis polychroa, *Boie, l. c.*
Ahætulla bellii, *Gray, Ill. Ind. Zool.* ii. pl. lxxx. (1834).
Dendrophis pictus, part., *Schleg. Phys. Serp.* ii. p. 228, pl. ix. figs. 5–7 (1837); *Dum. & Bibr.* vii. p. 197 (1854); *Girard, U.S. Explor. Exped., Herp.* p. 129 (1858); *Günth. Cat.* p. 148 (1858), *and Rept. Brit. Ind.* p. 297 (1864); *Jan, Icon. Gén.* 32, pl. i. fig. 3 (1869); *Theob. Cat. Rept. Brit. Ind.* p. 190 (1876); *Bouleng. Faun. Ind., Rept.* p. 337 (1890).
Leptophis pictus, *Cantor, Cat. Mal. Rept.* p. 82 (1847).
Ahætulla picta, part., *Cope, Proc. Ac. Philad.* 1860, p. 556.
Dendrophis picta, var. andamanensis, *Anders. Proc. Zool. Soc.* 1871, p. 184.
Leptophis formosus, *Jan, Icon. Gén.* 49, pl. vi. fig. 2 (1879).

Maxillary teeth 23 to 26. Eye as long as its distance from the nostril. Rostral much broader than deep, visible from above; internasals as long as or a little shorter than the præfrontals; frontal once and one third to once and two thirds as long as broad, as long as its distance from the rostral or the end of the snout, shorter than the parietals; loreal elongate (in one specimen divided into two, in another fused with the præfrontal); one præ- and two postoculars; temporals 2+2, rarely 1+1 or 1+2; nine upper labials, fifth and sixth, or fourth, fifth, and sixth, entering the eye, rarely seven or eight; five, rarely four, lower labials in contact with the anterior chin-shields, which are shorter than the posterior. Scales in 15 rows, vertebrals nearly as large as outer; the enlarged vertebral scales originating abruptly a short distance behind the head, by fusion of two nuchal scales. Ventrals 165–190; anal divided; subcaudals 122–164. Olive or bronze-brown above; a black stripe on each side of the head, passing through the eye, usually very strongly marked on the temple and extending to the nape, where it may widen or break up into spots; a yellow lateral stripe, bordered below by a dark, usually black line or stripe along the limit between the lower row of scales and the ventral shields; upper lip yellow; lower parts uniform pale yellow or greenish.

Total length 1180 millim.; tail 440.

Eastern Himalayas, Bengal, hills of Southern India, Burma, Indo-China, Malay Peninsula and Archipelago.

a. Hgr. ♂ (V. 184; C. 122).	Sikkim.	Sir J. Hooker [P.].
b–c. Hgr. ♀ (V. 170, 171; C. ?, ?).	Bengal.	Gen. Hardwicke [P.].
d. Hgr. ♂ (V. 165; C. 164).	Anamallays.	Col. Beddome [C.].
e. Hgr. ♂ (V. 177; C. 149).	Lao Mountains.	M. Mouhot [C.].
f, g. Hgr. ♂ (V. 167, 165; C. 130, 127)	Camboja.	M. Mouhot [C.].
h, i. ♀ (V. 174, 174; C.?, 132).	Pinang.	Dr. Cantor.
k–l. ♀ (V. 167, 170; C. 133, 145).	Sumatra.	Leyden Museum.
m. Hgr. ♀ (V. 165; C. 137).	Nias.	Hr. Sundermann [C.].
n–o, p. ♂ (V. 169; C. 140) & ♀ (V. 179, 174; C. 123, 146).	Borneo.	Sir E. Belcher [P.].
q, r. ♀ (V. 178; C. ?) & yg. (V. 178; C. 139).	Borneo.	Sir Hugh Low [P.].
s. ♀ (V. 171; C. 131).	Borneo.	L. L. Dillwyn, Esq. [P.].
t–v. ♀ (V. 167, 168, 169; C. 123, 139, 130).	Java.	G. Lyon, Esq. [P.].
w. ♀ (V. 166; C. 130).	Java.	A. Scott, Esq. [P.].
x. ♀ (V. 172; C. 141).	Philippines.	H. Cuming, Esq. [C.].
y. Yg. (V. 174; C. 153).	Philippines.	H. J. Veitch, Esq. [P.].
z–a. ♂ (V. 169; C. 130) & hgr. (V. 168; C. 145).	Manado, Celebes.	Dr. A. B. Meyer [P.].
β. ♂ (V. 190; C. 158).	N. Ceram.	
γ. ♀ (V. 185; C. 160).	Misol.	
δ. Yg. (V. 166; C. 145).	Ternate.	H.M.S. 'Challenger.' (Type of *C. decorus*.)
ε. Hgr. (V. 170; C. 136).	—— ?	
ζ. Skeleton.	Borneo.	

2. Dendrophis bifrenalis. (Plate IV. fig. 1.)

Dendrophis picta, var. C, *Günth. Cat.* p. 149 (1858).
—— bifrenalis, *Bouleng. Faun. Ind., Rept.* p. 338 (1890).

Maxillary teeth 24 or 25. Head very narrow and elongate, shorter than its distance from the nostril. Rostral much broader than deep, visible from above; internasals as long as or a little shorter than the præfrontals; frontal once and a half to once and two thirds as long as broad, as long as its distance from the rostral or the end of the snout, shorter than the parietals; two loreals; one præ- and two postoculars; temporals 2+2; nine upper labials, fifth and sixth entering the eye; five lower labials in contact with the anterior chin-shields, which are much shorter than the posterior. Scales in 15 rows, vertebrals larger than outer. Ventrals 154-171; anal divided; subcaudals 144-155. Dark olive above; a yellowish lateral stripe running along the outer row of scales; upper lip yellow; a black streak on each side of the head, passing through the eye; some oblique black bars may be present on each side of the neck; ventrals and subcaudals pale greenish yellow between the keels, dark olive between the keels and the light lateral stripe.

Total length 1030 millim.; tail 380.

Ceylon.

a, b. ♀ (V. 171; C. 144) & yg. Ceylon.
(V. 154; C. 155).   } (Types.)
c. Hgr. ♂ (V. 155; C. ?).   Ceylon.   A. Paul, Esq. [P.].

3. Dendrophis calligaster.

Dendrophis lineolata, part., *Dum. & Bibr.* vii. p. 200 (1854).
—— calligaster, *Günth. Ann. & Mag. N. H.* (3) xx. 1867, p. 53.
—— lineolatus (*non Hombr. & Jacq.*), *Jan, Icon. Gén.* 32, pl. iii. fig. 2 (1869).
—— salomonis, *Günth. Ann. & Mag. N. H.* (4) ix. 1872, p. 25.
—— aruensis, *Doria, Ann. Mus. Genova*, vi. 1874, p. 349, pl. xii. fig. *g*; *Sauvage, Bull. Soc. Philom.* (7) ii. 1878, p. 41.
—— katowensis, *Macleay, Proc. Linn. Soc. N. S. W.* ii. 1877, p. 37.
—— darnleyensis, *Macleay, l. c.* p. 38.
—— punctulatus, part., *Peters & Doria, Ann. Mus. Genova*, xiii. 1878, p. 390.

Maxillary teeth 20 to 26. Eye not longer than its distance from the posterior border of the nostril. Rostral once and two thirds to twice as broad as deep, visible from above; internasals as long as or a little longer than the præfrontals; frontal once and one third to once and three fourths as long as broad, as long as its distance from the rostral or the tip of the snout, usually shorter than the parietals; loreal once and a half to thrice as long as deep (rarely absent, fused with the præfrontal); one præ- and two (rarely three) postoculars; temporals usually 2+2, sometimes 1+2; eight or nine upper labials, fourth and fifth or fifth and sixth entering the eye; five (rarely four) lower labials in contact with the anterior chin-shields, which are much shorter than the posterior. Scales in 13 rows, the vertebrals about as large as the outer. Ventrals 176–211; anal divided; subcaudals 125–151. Coloration very variable, but a more or less distinct black or blackish streak on each side of the head, passing through the eye, is constant, and usually connected with its fellow across the rostral shield; upper lip and throat yellow; some of the scales with more or less distinct yellowish outer border.

Total length 1270 millim.; tail 430.

Moluccas, New Guinea and neighbouring Islands, Solomon Islands, Cape York.

A. Bronzy or olive above; the dark streak on the side of the head usually prolonged to the neck, where it may break up into spots; pale olive beneath, more or less closely speckled with darker. (*D. calligaster*, Gthr.; *D. aruensis*, Doria; *D. darnleyensis*, Macleay.)

a–b. ♀ (V. 194, 191; C. 141, 150).	Mansinam, New Guinea.	M. L. Laglaize [C.].
c–e. ♂ (V. 178; C. 128) & ♀ (V. 184, 185; C. 132, 135).	Islands of Torres Straits.	Rev. S. Macfarlane [C.].
f–g, h. ♀ (V. 184, 190; C. 149, 143) & hgr. (V. 191; C. 136).	Murray Id., Torres Straits.	Rev. S. Macfarlane [C.].
i–l. ♂ (V. 179, 178; C. 143, ?) & ♀ (V. 180; C. ?).	Cornwallis Id., Torres Straits.	Rev. S. Macfarlane [C.].
m. ♂ (V. 176; C. 136).	Cape York.	(Type.)
n. ♀ (V. 178; C. 145).	St. Aignan, Louisiade Archipelago.	B. H. Thomson, Esq. [P.].
o. ♂ (V. 177; C. 142).	Shortland Ids., Solomon Group.	H. B. Guppy, Esq. [P.].
p–s. ♂ (V. 182; C. ?), ♀ (V. 183; C. 139), & hgr. (V. 181, 182; C. 133, 142).	Gela, Solomon Ids.	C. M. Woodford, Esq. [C.].
t. ♀ (V. 184; C. 138).	Guadalcanar, Solomon Ids.	C. M. Woodford, Esq. [C.].

B. Like the preceding, but anterior dorsal region tinged with vermilion.

a–b. ♀ (V. 185, 178; C. 140, 133).	Guadalcanar, Solomon Ids.	C. M. Woodford, Esq. [C.].

C. Like A, but anterior dorsal region dark brown or black, with the scales of the vertebral series yellow.

a. ♀ (V. 183; C. 127).	Aleya, S.E. New Guinea.	
b. ♀ (V. 183; C. 127).	Rossel Id., Louisiade Archipelago.	B. H. Thomson, Esq. [P.].

D. Bronzy above; the black streak on the side of the head prolonged as a broad band along each side of the anterior part of the body; belly pale olive, immaculate. (*D. katowensis*, Macleay.)

a–b. ♂ (V. 189; C. 144) & ♀ (V. 190; C. 137).   N.W. New Guinea.   M. A. Linden [C.].

E. Bronzy or olive above, the scales usually black-edged; the black streak on the side of the head not prolonged on the body, or but feebly marked; belly yellowish or pale olive, immaculate. (*D. salomonis*, Gthr.)

a. ♀ (V. 211; C. ?).   Timor Laut.   H. O. Forbes, Esq. [C.].
b. ♂ (V. 201; C. ?).   Kei Ids.   Capt. Langen [P.].
c–e, f. ♂ (V. 202, 197; C. 148, ?) & ♀ (V. 207, 192; C. 151, 148).   Duke of York Id.   Rev. G. Brown [C.].
g, h. ♂ (V. 193, 194; C. ?, 148).   Solomon Ids.   G. Krefft, Esq. [P.]. (Types of *D. salomonis*.)
i. Yg. (V. 180; C. 126).   Alu, Shortland Ids., Solomon Ids.   C. M. Woodford, Esq. [C.].
k. ♀ (V. 200; C. ?).   San Christoval, Solomon Ids.   H. B. Guppy, Esq. [P.].
l, m. ♀ (V. 202; C. ?) & yg. (V. 203; C. 143).   Santa Anna Id., Solomon Ids.   H. B. Guppy, Esq. [P.].

F. Olive-green to bright green above, uniform bright greenish yellow beneath; black streak on the side of the head well-marked but not continued on the body.

a–e. ♀ (V. 180, 182, 188, 186, 187; C. 137, 126, 136, 131, 137).   Rubiana, New Georgia, Solomon Ids.   C. M. Woodford, Esq. [C.].

G. Black above, dark olive-grey or blackish beneath.

a–b. ♀ (V. 191, 190; C. 125, ?).   Rossel Id., Louisiade Archipelago.   B. H. Thomson, Esq. [P.].
c. ♂ (V. 183; C. 136).   Normanby, Louisiade Archipelago.   B. H. Thomson, Esq. [P.].
d. ♀ (V. 179; C. 126).   Guadalcanar, Solomon Ids.   C. M. Woodford, Esq. [C.].

4. Dendrophis punctulatus.

Leptophis punctulatus, *Gray, in King's Voy. Austral.* ii. p. 432 (1827).
Dendrophis (Ahætula) olivacea, *Gray, Zool. Misc.* p. 54 (1842).
—— (Ahætula) fusca, *Gray, l. c.*
—— prasinus, *Girard, Proc. Ac. Philad.* 1857, p. 181, *and U.S. Explor. Exped., Herp.* p. 131, pl. xii. figs. 7–10 (1858).
—— punctulata, *Günth. Cat.* p. 149 (1858); *Peters & Doria, Ann. Mus. Genova,* xiii. 1878, p. 390.
—— fuscus, *Jan, Elenco,* p. 86 (1863), *and Icon. Gén.* 32, pl. iii. fig. 1 (1869).
—— punctulata, *Krefft, Snakes of Austral.* p. 23, pls. iv. & v. fig. 6 (1869).

64. DENDROPHIS.

Dendrophis gracilis, *Macleay, Proc. Linn. Soc. N. S. W.* i. 1877, p. 15.
—— olivacea, *Macleay, Proc. Linn. Soc. N. S. W.* ii. 1878, p. 220.

Maxillary teeth 24 to 26. Eye not longer than its distance from the posterior border of the nostril. Rostral once and a half to once and two thirds as broad as deep, visible from above; internasals nearly as long as or a little longer than the præfrontals; frontal once and one fourth to once and a half as long as broad, as long as or a little longer than its distance from the end of the snout, shorter than the parietals; loreal once and a half to twice and a half as long as deep; one præ- and two postoculars; temporals usually 1+2, rarely 2+2, or 1+1, anterior very long; seven or eight (rarely nine) upper labials, fourth and fifth (rarely fifth and sixth) entering the eye; five lower labials in contact with the anterior chin-shields, which are much shorter than the posterior. Scales in 13 rows, the vertebrals about as large as the outer. Ventrals 191–220; anal divided; subcaudals 120–144. Brown or olive above, uniform or with black edges to the scales; usually some of the scales with a yellowish outer border; no black stripe on the side of the head; upper lip, throat, and anterior ventrals yellowish.

Total length 1640 millim.; tail 440.

Northern and Eastern Australia.

A. Ventrals and subcaudals pale olive or brownish, with the keels yellowish, but unspotted.

a. ♀, bad state (V. 220; C. 137). Careening Bay, N. coast of Australia. J. Hunter, Esq. [P.]. (Type.)
b, c, d–e, f–g, h. ♂ (V. 207, 211, 209, 210, 208; C. 132, 140, ?, 144, 120), ♀ (V. 215; C. 131), & yg. (V. 211; C. 136). Port Essington. Lord Derby [P.]. (Types of *D. olivacea* and *D. fusca*.)
i. ♀ (V. 204; C. 132). N.W. Australia.
k. ♀ (V. 212; C. 137). N. Australia. J. Stokes, Esq. [C.].
l. Yg. (V. 218; C. 135). N.E. Australia. Capt. Grey [P.].
m. ♂ (V. 196; C. 127). Cooktown, N. Queensland. Capt. Drevar [C.].
n. ♀ (V. 196; C. ?). Cape York.
o. ♂ (V. 193; C. 136). Albany Island. J. Gould, Esq. [C.].
p. Yg. (V. 204; C. 136). Queensland.
q. ♂ (V. 196; C. ?). Moreton Bay. Mr. Strange [P.].
r. ♀ (V. 199; C. ?). North Shore, Sydney. F. M. Rayner, Esq. [C.].
s–t, u. ♀ (V. 194; C. ?), hgr. (V. 204; C. 132), & yg. (V. 191; C. 126). New South Wales. G. Krefft, Esq. [P.].

B. Dark olive above, each scale partly black; posterior ventrals marbled with black, or black in the middle with the posterior border and the sides above the keels yellowish white; subcaudals black, with two yellowish lines following the keels. (*D. gracilis*, Macleay.)

a. ♀ (V. 202; C. 127). Queensland.

5. Dendrophis grandoculis. (Plate IV. fig. 2.)

Dendrophis grandoculis, *Bouleng. Faun. Ind., Rept.* p. 337 (1890).

Maxillary teeth 30 to 32. Eye extremely large, as long as its distance from the rostral or the anterior border of the nostril. Rostral much broader than deep, just visible from above; internasals as long as or a little longer than the præfrontals; frontal once and a half to once and two thirds as long as broad, as long as its distance from the end of the snout, as long as the parietals; loreal elongate; one præ- and two postoculars; temporals 2+2 or 1+2; nine upper labials, fourth, fifth and sixth entering the eye; five lower labials in contact with the anterior chin-shields, which are shorter than the posterior. Scales in 15 rows, vertebrals not quite as large as the outer, scarcely enlarged on the anterior part of the body. Ventrals 174–188; anal divided; subcaudals 117–124. Brown or olive-brown above, with irregular small black blotches; eye bordered with whitish; no lateral stripes; sides of neck sometimes tinged with vermilion; lower parts pale olive, with or without small black spots on the sides; three black lines along the tail, one on each side and one below.

Total length 1200 millim.; tail 370.

Hills of Southern India.

a. ♀ (V. 176; C. 117).	Tinnevelly hills, 2000–3000 feet.		
b. Head only.	Coonoor Ghat, Wynad, 2500 feet.	}	Col. Beddome [C.]. (Types.)
c. ♂ (V. 174; C. 117).	Travancore.		
d. ♀ (V. 188; C. 124).	Peermad, Travancore, 3300 feet.		H. S. Ferguson, Esq. [P.].

6. Dendrophis formosus.

Dendrophis formosa, *Boie, Isis,* 1827, p. 542; *Schleg. Phys. Serp.* ii. p. 232, pl. ix. figs. 3 & 4 (1837); *Dum. & Bibr.* vii. p. 199 (1854); *Günth. Cat.* p. 150 (1858).

Maxillary teeth 28 to 31. Eye very large, as long as its distance from the rostral or the anterior border of the nostril. Rostral twice as broad as deep, just visible from above; internasals a little longer than the præfrontals; frontal once and one third to once and a half as long as broad, as long as its distance from the end of the snout, as long as the parietals; loreal elongate; one præ- and two to four (usually three) postoculars; temporals 2+2; nine upper labials (rarely eight), fifth and sixth or fourth, fifth, and sixth (or third, fourth, and fifth) entering the eye; five lower labials in contact with the anterior chin-shields, which are much shorter than the posterior. Scales in 15 rows, vertebrals strongly enlarged, considerably larger than the outer, originating abruptly on the nape by fusion of the two median scales. Ventrals 179–205; anal divided; subcaudals 132–158. Olive above, scales black-edged; a black stripe on each side of the head, passing through the eye and extending on to the nape, where it considerably widens; no light lateral stripe; two black lines may be present along each side of the posterior part of the body; upper lip greenish yellow; lower parts pale green.

Total length 1310 millim.; tail 450.

Mountains of the Malay Peninsula, Borneo, Java.

a. ♀ (V. 196; C. 158).	Malacca.	Governor of Singapore [P.].
b-c. ♀ (V. 191, 190; C. 148, ?).	Borneo.	L. L. Dillwyn, Esq. [P.].
d. ♂ (V. 182; C. 146).	Java.	Leyden Museum.
e. ♀ (V. 205; C. 147).	Willis Mts., Kediri, Java, 5000 feet.	Baron v. Huegel [C.].
f. Hgr. ♀ (V. 195; C. 132).	Afghanistan (??).	

7. Dendrophis caudolineolatus.

Dendrophis caudolineolatus, *Günth. Proc. Zool. Soc.* 1869, p. 506, pl. xl. fig. 1; *Bouleng. Faun. Ind., Rept.* p. 339 (1890).
? Dendrophis gregorii, *Haly, Taprobanian*, iii. 1888, p. 51.

Maxillary teeth 30 to 32. Eye large, as long as or slightly longer than its distance from the nostril. Rostral once and two thirds as broad as long, visible from above, upper border forming an obtuse angle; internasals as long as or a little shorter than the præfrontals; frontal once and a half to once and two thirds as long as broad, longer than its distance from the end of the snout, as long as the parietals; loreal elongate; one præocular; two postoculars; temporals 1+2 or 2+1; eight or nine upper labials, fourth and fifth or fifth and sixth entering the eye; four or five lower labials in contact with the anterior chin-shields, which are much shorter than the posterior. Scales in 13 rows, vertebrals moderately enlarged, as large as outer. Ventrals 149–161; anal divided; subcaudals 124–128. Bronzy olive above, anteriorly with oblique narrow black streaks; tail with four more or less distinct black lines, one above, one along the meeting-edges of the subcaudals, and one on each side; a black temporal streak; lower surfaces greenish white.

Total length 650 millim.; tail 235.

Ceylon.

a. Hgr. ♂ (V. 149; C. 124).	Ceylon.	B. H. Barnes, Esq. [P.]. (Types.)
b. Yg. (V. 161; C. 126).	Ceylon.	
c. Hgr. ♂ (V. 149; C. 128).	Ceylon.	

8. Dendrophis lineolatus.

Dendrophis lineolata, *Hombr. & Jacq. Voy. Pôle Sud, Zool.* iii., *Rept.* p. 20, pl. ii. fig. 1 (1842); *Sauvage, Bull. Soc. Philom.* (7) ii. 1878, p. 40.
—— lineolata, part., *Dum. & Bibr.* vii. p. 200 (1854).
—— striolatus, *Peters, Mon. Berl. Ac.* 1867, p. 25.
—— punctulatus, var. atrostriata, *Meyer, Mon. Berl. Ac.* 1874, p. 136.
—— punctulatus, var. fasciata, *Meyer, l. c.*
—— macrops, *Günth. Proc. Zool. Soc.* 1877, p. 131, fig.

? Dendrophis breviceps, *Macleay, Proc. Linn. Soc. N. S. W.* ii. 1877, p. 37.

Dendrophis punctulatus, part., *Peters & Doria, Ann. Mus. Genova,* xiii. 1878, p. 390.

? Dendrophis papuæ, *Douglas-Ogilby, Rec. Austral. Mus.* i. 1891, p. 198.

Dendrophis elegans, *Douglas-Ogilby, l. c.* p. 199.

Maxillary teeth 27 to 32. Eye very large, as long as its distance from the centre or the anterior border of the nostril. Rostral once and two thirds to once and three fourths as broad as deep, just visible from above; internasals as long as or a little longer than the præfrontals; frontal once and one fourth to once and a half as long as broad, as long as its distance from the end of the snout, as long as or a little shorter than the parietals; loreal nearly twice as long as deep; one præ- and two postoculars; temporals 2+2, 1+2, or 1+1; eight or nine upper labials, fourth and fifth or fifth and sixth entering the eye; five lower labials in contact with the anterior chin-shields, which are much shorter than the posterior. Scales in 13 rows, vertebrals about as large as outer. Ventrals 173-203; anal divided; subcaudals 131-151. Olive above, some or all of the scales black-edged; some of the lateral scales with more or less distinct yellowish or whitish outer border; head dark olive-brown or blackish above, upper lip yellow; anterior part of body often with black oblique cross bars; throat and lower surface of neck yellow; belly and lower surface of tail greenish or greyish olive, the keels yellowish; a blackish streak along the middle of the lower surface of the tail.

Total length 1820 millim.; tail 570.

Pelew Islands, New Guinea, New Ireland.

a–b. ♂ (V. 173; C. 131) & hgr. (V. 177; C. 139).		Pelew Islands.	Museum Godeffroy.
c–d. ♀ (V. 187; C. ?) & hgr. (V. 191; C. 147).		N.W. New Guinea.	M. A. Linden [C.].
e–g. ♀ (V. 191, 193, 193; C. 139, 136, 143).		Fly River.	Rev. S. Macfarlane [C.].
h. ♀ (V. 203; C. 151).		Aleya, S.E. New Guinea.	
i. ♂ (V. 192; C. 142).		Duke of York Id.	Rev. G. Brown [C.]. (Type of *D. macrops.*)

9. Dendrophis gastrostictus. (PLATE IV. fig. 3.)

Maxillary teeth 33. Eye very large, as long as its distance from the anterior border of the nostril. Rostral nearly twice as broad as deep, just visible from above; internasals as long as the præfrontals; frontal once and one third as long as broad, as long as its distance from the end of the snout, much shorter than the parietals; loreal once and two thirds as long as deep; one præ- and three postoculars; temporals 2+2; eight upper labials, fourth and fifth entering the eye; four or five lower labials in contact with the anterior chin-shields, which are much shorter than the posterior. Scales in 13 rows, vertebrals nearly as large as outer. Ventrals 162; anal divided; subcaudals 148. Bronzy above, head, nape and tail blackish; upper lip and throat yellow; ventrals yellowish, with crowded black dots and small spots.

Total length 1120 millim.; tail 420.

New Guinea.

a. ♂ (V. 162; C. 148).   N.W. New Guinea.   M. A. Linden [C.].

65. DENDRELAPHIS.

Leptophis, part., *Bell, Zool. Journ.* ii. 1825, p. 328.
Dendrophis, part., *Wagl. Syst. Amph.* p. 182 (1830); *Schleg. Phys. Serp.* ii. p. 220 (1837); *Dum. & Bibr. Erp. Gén.* vii. p. 193 (1854); *Günth. Cat. Col. Sn.* p. 148 (1858); *Jan, Elenco sist. Ofid.* p. 85 (1863); *Günth. Rept. Brit. Ind.* p. 296 (1864).
Dendrelaphis, *Bouleng. Faun. Ind., Rept.* p. 339 (1890).

Maxillary teeth 18 to 23; anterior maxillary and mandibular teeth longest. Head elongate, distinct from neck; eye large, with round pupil. Body much elongate, feebly compressed; scales smooth, in 13 or 15 rows, narrow, disposed obliquely, with apical pits, those of the vertebral row not or but very slightly enlarged; ventrals with a suture-like lateral keel and a notch on each side, corresponding to the keel. Tail long; subcaudals in two rows, keeled and notched like the ventrals.

India, Ceylon, Burma, Malay Peninsula and Archipelago.

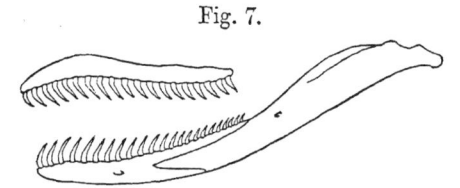

Fig. 7.

Maxillary and mandible of *Dendrelaphis tristis*.

Synopsis of the Species.

I. Scales in 15 rows.

Two labials entering the eye	1. *tristis*, p. 88.
A single labial entering the eye	2. *subocularis*, p. 89.

II. Scales in 13 rows.

Eye longer than its distance from the nostril; ventrals 171–188	3. *caudolineatus*, p. 89.
Eye as long as its distance from the nostril; ventrals 167–182	4. *terrificus*, p. 90.
Eye as long as its distance from the nostril; ventrals 191–193: no black lines along the body and tail	5. *modestus*, p. 91.

1. Dendrelaphis tristis.

Russell, Ind. Serp. i. pl. xxxi. (1796) & ii. pls. xxv. & xxvi. (1801).
Coluber tristis, *Daud. Rept.* vi. p. 430 (1803).
Dipsas schokari, part., *Kuhl, Beitr. Zool. vergl. Anat.* p. 80 (1820).
Leptophis mancas, *Bell, Zool. Journ.* ii. 1825, p. 208.
Dendrophis chairecacos, *Boie, Isis,* 1827, p. 541.
—— maniar, *Boie, l. c.* p. 542.
—— boii, *Cantor, Proc. Zool. Soc.* 1839, p. 53.
—— pictus, part., *Schleg. Phys. Serp.* ii. p. 228 (1837); *Dum. & Bibr.* vii. p. 197 (1853); *Günth. Cat.* p. 148 (1858), *and Rept. Brit. Ind.* p. 297 (1864); *Jan, Icon. Gén.* 32, pl. i. figs. 1 & 2 (1869); *Theob. Cat. Rept. Brit. Ind.* p. 190 (1876); *Bouleng. Faun. Ind., Rept.* p. 337 (1890).
Leptophis pictus, part., *Cantor, Cat. Mal. Rept.* p. 83 (1847).
Dendrophis pictus, var. vertebralis, *Jan, Elenco,* p. 85 (1863).
—— pictus, *F. Müll. Verh. nat. Ges. Basel,* viii. 1878, p. 268.
—— helena, *Werner, Zool. Anz.* 1893, p. 81.

Eye as long as its distance from the nostril. Rostral about once and one third as broad as deep, visible from above; internasals as long as or a little shorter than the præfrontals; frontal once and a half to once and two thirds as long as broad, as long as or a little longer than its distance from the end of the snout, as long as or slightly shorter than the parietals; loreal elongate; one præ- and two post-oculars; temporals 2+2; nine upper labials (rarely eight), fifth and sixth (or fourth and fifth) entering the eye; five lower labials in contact with the anterior chin-shields, which are much shorter than the posterior. Scales in 15 rows. Ventrals 163–192; anal divided; subcaudals 113–140. Olive above, with or without black margins to the scales; scales of outer row, and of middle row on anterior part of body, usually yellowish; usually a round yellow spot in the middle between the parietals; upper lip yellow, bordered by a black streak along the temple; black spots or oblique bars may be present on each side of the neck, and one or two black lateral lines along the posterior part of the body; lower parts yellowish or greenish.

Total length 1310 millim.; tail 420.
Eastern Himalayas, Assam, India, Ceylon.

a. Hgr. ♂ (V. 192; C. 133).	Darjeeling.	T. C. Jerdon, Esq. [P.].
b. ♂ (V. 187; C. ?).	Khasi Hills.	Sir J. Hooker [P.].
c–d. Hgr. ♂ (V. 189; C. 120) & ♀ (V. 185; C. ?).	Khasi Hills (?).	Dr. Griffith.
e. Hgr. ♀ (V. 185; C. 127).	Sind.	Messrs. v. Schlagintweit [C.].
f. ♀ (V. 181; C. ?).	Matheran, Bombay.	Dr. Leith [P.].
g. ♂ (V. 184; C. 140).	India.	Dr. P. Russell.
h. ♀ (V. 166; C. 113).	Ceylon.	Dr. Kelaart.
i. ♂ (V. 184; C. 132).	Ceylon.	E. W. H. Holdsworth, Esq. [C.].
k, l, m. ♀ (V. 163, 164; C. ?, 121) & yg. (V. 165; C. 127).	Ceylon.	

n. Hgr. ♂ (V. 177; C. 128).	Trincomalee.	Major Barrett [P.].
o. Skull.	India.	Haslar Collection.
p. Skull.	——?	

2. Dendrelaphis subocularis.

Dendrophis picta, part., *Anders. An. Zool. Res. Yunnan*, p. 824 (1879).
—— subocularis, *Bouleng. Ann. Mus. Genova*, (2) vi. 1888, p. 600, pl. vi. fig. 2, and *Faun. Ind., Rept.* p. 338 (1890).

Eye as long as its distance from the anterior border of the nostril. Rostral a little broader than deep, visible from above; internasals nearly as long as the præfrontals; frontal once and two thirds as long as broad, longer than its distance from the end of the snout, a little shorter than the parietals; loreal elongate; one præ- and two postoculars; temporals 2+2; eight upper labials, fifth largest and bordering the eye; five lower labials in contact with the anterior chin-shields, which are much longer than the posterior. Scales in 15 rows. Ventrals 167–172; anal divided; subcaudals 74–105. Olive-brown above, upper lip and a lateral streak yellowish white; an olive-brown stripe between the lateral yellowish streak and the yellowish ventrals, occupying the extremities of the latter shields and the lower half of the outer scales; a black line behind the eye, above the labials; anterior labials with black lines on the sutures; a rather indistinct pale greenish line along the nape, bordered on each side, in the young, by a series of large round black spots.

Total length 820 millim.; tail 220.

Upper Burma.

| a. Hgr. (V. 167; C. 105). | Bhamo. | M. L. Fea [C.]. |
| | | (One of the types.) |

3. Dendrelaphis caudolineatus.

Ahætula caudolineata, *Gray, Ill. Ind. Zool.* ii. pl. lxxxi. (1834).
Leptophis caudalineatus, *Cantor, Cat. Mal. Rept.* p. 85 (1847).
Dendrophis octolineata, *Dum. & Bibr.* vii. p. 201 (1835); *Jan, Icon. Gén.* 32, pl. ii. fig. 1 (1869).
—— caudolineata, *Günth. Cat.* p. 150 (1858), *and Rept. Brit. Ind.* p. 297 (1864); *Stoliczka, Journ. As. Soc. Beng.* xxxix. 1870, p. 194; *Günth. Zool. Rec.* 1870, p. 75; *Stoliczka, Journ. As. Soc. Beng.* xlii. 1873, p. 123.
Dendrelaphis caudolineatus, *Bouleng. Faun. Ind., Rept.* p. 339 (1890).

Snout little longer than the eye, which exceeds its distance from the nostril. Rostral a little broader than deep, visible from above; internasals as long as or a little shorter than the præfrontals; frontal once and one third to once and two thirds as long as broad, longer than its distance from the end of the snout, a little shorter than the parietals; loreal elongate; one præ- and two postoculars; temporals 2+2; nine upper labials, fifth and sixth (rarely fourth, fifth, and sixth) entering the eye; five lower labials in contact with the anterior chin-shields, which are shorter than the posterior. Scales in 13 rows. Ventrals 171-188; anal divided; subcaudals 100-112. Light brownish bronze or greenish yellow above, the scales with black edges forming more or less regular longitudinal lines; a yellowish lateral streak edged above and below by a black band, the lower of which extends on to the outer edge of the ventrals; lips and lower surfaces pale metallic citrine, the tail with a black median line.

Total length 1520 millim.; tail 380.

Southern India, Malay Peninsula, Mergui, Borneo, Sumatra, Philippines.

a-b. ♂ (V. 177; C. 112) & ♀ (V. 184; C. 100).	Wynad, Malabar.	Col. Beddome [C.].
c. Hgr. (V. 176; C. 109).	Perak.	J. H. Leech, Esq. [P.].
d-e. ♀ (V. 181, 187; C. 104, 110).	Pinang.	Dr. Cantor.
f-g. ♀ (V. 183; C. 109) & yg. (V. 180; C. 104).	Singapore.	Dr. Dennys [P.].
h. ♀ (V. 188; C. 110).	Malay Peninsula (?).	East India Company [P.].
i. ♂ (V. 174; C. 100).	E. coast of Sumatra.	Mrs. Findlay [P.].
k. ♀ (V. 171; C. ?).	Nias.	Hr. Sundermann [C.].
l. ♀ (V. 173; C. 100).	Bunguran Id., Natuna Ids.	A. Everett, Esq. [C.].
m. ♀ (V. 183; C. 111).	Borneo.	Sir J. Brooke [P.].
n. ♂ (V. 175; C. 108).	Borneo.	Sir H. Low [P.].
o. ♀ (V. 186; C. ?).	Borneo.	L. L. Dillwyn, Esq. [P.].
p. Yg. (V. 176; C. 109).	Matang, Borneo.	
q. Yg. (V. 183; C. ?).	Philippines.	H. Cuming, Esq. [C.].
r. Yg. (V. 172; C. 106).	Puerta Princesa.	A. Everett, Esq. [C.].

4. Dendrelaphis terrificus.

Dendrophis picta, var. B, *Günth. Cat.* p. 149 (1858).
—— punctulata, spec. *o, Günth. l. c.* p. 150.
—— terrificus, *Peters, Mon. Berl. Ac.* 1872, p. 583.
—— philippinensis, *Günth. Proc. Zool. Soc.* 1879, p. 78, pl. iv.
Dendrelaphis terrificus, *Bouleng. Faun. Ind., Rept.* p. 339 (1890).

Eye as long as its distance from the nostril. Rostral a little broader than deep, just visible from above; internasals as long as or a little shorter than the præfrontals; frontal once and a half as long as broad, as long as its distance from the rostral, shorter than the parietals; loreal elongate; one præ- and two postoculars; temporals 2+2; nine upper labials, fifth and sixth or fourth, fifth, and sixth entering the eye; five lower labials in contact with the anterior chin-shields, which are shorter than the posterior. Scales in 13 rows. Ventrals 167-182; anal divided; subcaudals 92-113. Olive above, some of the scales black-edged or with white outer border; a black stripe on each side of the head and neck, passing through the eye, sometimes continued along the body and tail; a black line along the outer edge of the ventrals and subcaudals, which are of a pale yellowish green; a black median line along the lower surface of the tail.

Total length 1050 millim.; tail 300.
Philippines, Celebes.

a. ♀ (V. 167; C. 100).	N. Mindanao.	A. Everett, Esq. [C.].
		(Type of *D. philippinensis*.)
b, c. ♀ (V. 172; C. ?) & yg. (V. 182; C. 113).	Philippines.	H. Cuming, Esq. [C.].
d. ♀ (V. 179; C. 102).	Manado.	Dr. A. B. Meyer [C.].

5. Dendrelaphis modestus. (PLATE IV. fig. 4.)

Eye as long as its distance from the nostril. Rostral broader than deep, just visible from above; internasals as long as the præfrontals; frontal once and one third as long as broad, as long as its distance from the rostral or the end of the snout, shorter than the parietals; loreal elongate; one præ- and two postoculars; temporals 2 + 2; nine upper labials, fifth and sixth entering the eye; five lower labials in contact with the anterior chin-shields, which are shorter than the posterior. Scales in 13 rows. Ventrals 191-193; anal divided; subcaudals 114. Olive above, scales black-edged; a rather indistinct dark streak on each side of the head, passing through the eye; upper lip and lower surface of head yellow; pale green beneath, the posterior ventrals and the subcaudals margined with blackish.

Total length 1120 millim.; tail 315.
Ternate.

a–b. ♀ (V. 193, 191; C. 114, ?).   Ternate.   H.M.S. 'Challenger.'

66. CHLOROPHIS *.

Philothamnus, part., *Smith, Ill. Zool. S. Afr., Rept.* (1840); *Peters, Reise n. Mossamb.* iii. p. 128 (1882); *Bocage, Jorn. Sc. Lisb.* ix. 1882, p. 1.
Leptophis, part., *Dum. & Bibr. Erp. Gén.* vii. p. 528 (1854); *Jan, Elenco sist. Ofid.* p. 84 (1863).
Chlorophis, *Hallow. Proc. Ac. Philad.* 1857, p. 52; *Cope, Proc. Ac. Philad.* 1860, p. 559; *Theob. Cat. Rept. As. Soc. Mus.* p. 49 (1868).
Ahætulla, part., *Günth. Cat. Col. Sn.* p. 151 (1858), *and Ann. & Mag. N. H.* (3) xi. 1863, p. 283.
Herpetæthiops, *Günth. Ann. & Mag. N. H.* (3) xviii. 1866, p. 27.

Fig. 8.

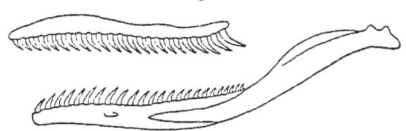

Maxillary and mandible of *Chlorophis natalensis*.

---

* A snake figured, but not described, by Jan, as from Mozambique, *Dendrophis subcarinatus*, Icon. Gén. 32, pl. ii. fig. 2 (1869), perhaps belongs to this genus, but has feebly-keeled dorsal scales and the dentition is represented as isodont.

Maxillary teeth 20 to 25, posterior longest; mandibular teeth subequal. Head more or less elongate, distinct from neck; eye large or rather large, with round pupil. Body cylindrical; scales smooth, with apical pits, in 15 rows, disposed obliquely, at least on the anterior part of the body; ventrals rounded or more or less distinctly keeled on each side, the keel not extending to the subcaudals. Tail long or rather long; subcaudals in two rows.

Tropical and South Africa.

Synopsis of the Species.

I. No trace of ventral keels.

Three labials (fourth, fifth, and sixth) entering the eye; subcaudals 111–122.   1. *emini*, p. 92.
Three labials (third, fourth, and fifth) entering the eye; subcaudals 85–99 ..   2. *ornatus*, p. 93.
Two labials entering the eye; subcaudals 85–105 ....   3. *hoplogaster*, p. 93.

II. Ventrals more or less distinctly keeled on the sides.

  A. Anal divided (rarely entire).

    1. Two labials entering the eye.

Fourth and fifth labials entering the eye; a single anterior temporal; subcaudals 77–114 ....   4. *neglectus*, p. 94.
Fourth and fifth labials entering the eye; usually two superposed anterior temporals; subcaudals 114–124 ....   5. *natalensis*, p. 94.
Fifth and sixth labials entering the eye; subcaudals 90–100 ....   6. *angolensis*, p. 95.

    2. Three labials entering the eye.

Ventrals 175–190; subcaudals 115–135; temporals 1+1; body very slender anteriorly ....   7. *heterolepidotus*, p. 95.
Ventrals 150–182; subcaudals 94–133; temporals usually 1+2 ....   8. *irregularis*, p. 96.

  B. Anal entire; subcaudals 78–92; three labials entering the eye ....   9. *heterodermus*, p. 97.

1. Chlorophis emini. (Plate V. fig. 1.)

Ahætulla irregularis, part., *Günth. Ann. & Mag. N. H.* (3) xi. 1863, p. 285.
—— emini, *Günth. Ann. & Mag. N. H.* (6) i. 1888, p. 325.

Rostral a little broader than deep, the portion visible from above measuring one third its distance from the frontal; internasals a little shorter than the præfrontals; frontal once and a half to once and two thirds as long as broad, a little longer than its distance from the end of the snout, a little shorter than the parietals; loreal nearly twice as long as deep; one præocular, usually not touching the frontal; two postoculars; temporals 1+2 or 1+1; nine upper labials, fourth, fifth, and sixth entering the eye; five lower labials in contact with the anterior chin-shields, which are a little shorter than the posterior. Scales in 15 rows. Ventrals rounded, 155-190; anal divided; subcaudals 111-122. Green above, the interstitial skin black; some of the scales with a white spot on the basal half of the outer margin.

Total length 720 millim.; tail 250.

Eastern Central Africa.

a. ♂ (V. 155; C. 111). Monbuttu. Dr. Emin Pasha [P.].
(Type.)
b-c. Yg. (V. 190, 183; C. 121, 122). 500 miles south of Chartoum. Consul J. Petherick [C.].

2. Chlorophis ornatus.

Philothamnus ornatus, *Bocage, Jorn. Sc. Lisb.* ii. 1872, p. 80, & ix. 1882, p. 15, fig.

Internasals shorter than the præfrontals; frontal with straight lateral borders, once and a half as long as broad, shorter than the parietals; loreal longer than deep; one præocular, in contact with the frontal; two postoculars; temporals 1+1; eight upper labials, third, fourth, and fifth entering the eye. Scales in 15 rows. Ventrals rounded, 152-166; anal divided; subcaudals 85-99. Bronzy green above, with a brown vertebral stripe edged with yellowish; yellowish or pale green beneath.

Total length 710 millim.; tail 200.

Guinea; Angola.

3. Chlorophis hoplogaster. (PLATE V. fig. 2.)

Ahætulla hoplogaster, *Günth. Ann. & Mag. N. H.* (3) xi. 1863, p. 284; *Bouleng. Faun. Ind., Rept.* p. 305 (1890).
Chlorophis oldhami, *Theob. Cat. Rept. As. Soc. Mus.* p. 49 (1868).
Cyclophis oldhami, *Theob. Cat. Rept. Brit. Ind.* p. 159 (1876).
Philothamnus hoplogaster, *Bocage, Jorn. Sc. Lisb.* ix. 1882, p. 17.

Rostral broader than deep, the portion visible from above measuring one third to two fifths its distance from the frontal; internasals as long as or a little shorter than the præfrontals; frontal once and a half as long as broad, as long as its distance from the end of the snout, a little shorter than the parietals; loreal once and two thirds to twice as long as deep; one præocular, not touching the frontal; two postoculars; temporals 1+1 or 1+2; eight upper labials, fourth and fifth entering the eye; five lower labials in contact with the anterior chin-shields, which are shorter than the posterior. Scales in 15 rows. Ventrals rounded, without trace of a keel, 150-169; anal divided; subcaudals 85-105. Uniform green or olive above, greenish yellow inferiorly.

Total length 700 millim.; tail 220.

Central and South Africa.

a, b. ♂ (V. 150; C. 105)   Port Natal.   Mr. T. Ayres [C.].
  & ♀ (V. 156; C. 89).   } (Types.)
c. ♀ (V. 154; C. 93).   Port Natal.   Rev. H. Calloway [P.].
d–e. ♀ (V. 154, 159; C.   Natal.   E. Howlett, Esq. [P.].
  85, 92).
f. ♀ (V. 160; C. 94).   Damaraland.
g. ♀ (V. 169; C. 95).   Victoria Nyanza.

4. Chlorophis neglectus.

Philothamnus neglectus, *Peters, Mon. Berl. Ac.* 1866, p. 890, *and Reise n. Mossamb.* iii. p. 130, pl. xix. A. fig. 2 (1882); *Pfeffer, Jahrb. Hamb. Wiss. Anst.* x. 1893, p. 84.
Ahætulla neglecta, *Günth. Proc. Zool. Soc.* 1893, p. 620.

Snout rather pointed. Rostral broader than deep, the portion visible from above measuring one third or two fifths its distance from the frontal; internasals as long as or a little shorter than the præfrontals; frontal once and a half to once and two thirds as long as broad, as long as its distance from the end of the snout, a little shorter than the parietals; loreal twice as long as deep; one præocular, not touching the frontal; two postoculars; temporals 1+1 or 1+2; eight (exceptionally seven) upper labials, fourth and fifth (or third and fourth) entering the eye; five (rarely four or six) lower labials in contact with the anterior chin-shields, which are as long as or shorter than the posterior. Scales in 15 rows. Ventrals with a slight lateral keel, 149–166; anal divided; subcaudals 77–114. Green above, greenish yellow beneath; some purplish-brown blotches may be present on the anterior part of the body.

Total length 800 millim.; tail 230.

East Africa.

a. ♀ (V. 153; C. 100).   Taita.   Mr. Wray [C.].
b. ♂ (V. 155; C. 100).   Mkonumbi.   Dr. J. W. Gregory [P.].
c. ♀ (V. 159; C. 92).   Lamu.   F. J. Jackson, Esq. [P.].
d. Yg. (V. 149; C. 114).   Zanzibar Coast.
e. ♀ (V. 152; C. 91).   Shiré Highlands.   H. H. Johnston, Esq. [P.].
f–k. ♂ (V. 153, 157; C.   Zomba, Brit. C.   H. H. Johnston, Esq. [P.].
  93, 97), ♀ (V. 149, 150;   Africa.
  C. 86, 77), and hgr. (V.
  155; C. 89).
l. Hgr. (V. 152; C. 84).   Milanji, Brit. C.   H. H. Johnston, Esq. [P.].
                               Africa.

5. Chlorophis natalensis.

Dendrophis (Philothamnus) natalensis, *Smith, Ill. Zool. S. Afr., Rept.* pl. lxiv. (1840).
Ahætulla irregularis, var. natalensis, *Günth. Cat.* p. 152 (1858).
—— natalensis, *Günth. Ann. & Mag. N. H.* (3) xi. 1863, p. 285.
Philothamnus natalensis, *Bocage, Jorn. Sc. Lisb.* ix. 1882, p. 18.

Rostral broader than deep, just visible from above; internasals as long as or a little shorter than the præfrontals; frontal with slightly concave lateral borders, once and a half to once and two thirds as long as broad, longer than its distance from the end of the snout, as long as the parietals; loreal once and a half to twice as long as deep (rarely absent); one præocular, not touching the frontal; two postoculars; temporals 2+2 (rarely 1+2); eight upper labials, fourth and fifth entering the eye; five lower labials in contact with the anterior chin-shields, which are shorter than the posterior. Scales in 15 rows. Ventrals with distinct lateral keel, 151-169; anal divided; subcaudals 114-124. Olive-green above, usually some of the scales with a white spot on the outer border, and the skin between the scales black; greenish-white inferiorly.

Total length 1000 millim.; tail 330.

South Africa.

a. ♂ (V. 167; C. ?).	Damara-land.	
b. ♀ (V. 164; C. ?).	Cape of Good Hope.	Dr. Statham [P.].
c. Hgr. (V. 166; C. 118).	Brit. Caffraria.	F. P. M. Weale, Esq. [P.].
d, e. ♀ (V. 169, 169; C. 114, ?).	King Williamstown.	H. Trevelyan, Esq. [P.].
f. ♂ (V. 156; C. 120).	Port Elizabeth.	J. M. Leslie, Esq. [P.].
g. ♂ (V. 164; C. 124).	Port Natal.	Rev. H. Calloway [P.].
h. ♀ (V. 151; C. 119).	Orange River.	Dr. Kannemeyer [P.].
i. Ad., skel.	S. Africa.	

6. Chlorophis angolensis.

Philothamnus angolensis, *Bocage, Jorn. Sc. Lisb.* ix. 1882, p. 7.

Frontal rather short, with concave sides; loreal scarcely longer than deep; temporals 1+2; nine upper labials, fifth and sixth entering the eye. Ventrals feebly keeled, 150-160; anal divided; subcaudals 90-100. Olive-green above, yellowish green beneath.

Total length 980 millim.; tail 320.

Angola.

7. Chlorophis heterolepidotus. (PLATE V. fig. 3.)

Ahætulla heterolepidota, *Günth. Ann. & Mag. N. H.* (3) xi. 1863, p. 286, & (4) ix. 1872, p. 26.
Leptophis heterolepidota, *Bocage, Jorn. Sc. Lisb.* i. 1867, p. 69.
Philothamnus heterolepidotus, *Bocage, Jorn. Sc. Lisb.* ix. 1882, p. 8, fig.; *Boettg. Ber. Senck. Ges.* 1888, p. 60; *Pfeffer, Jahrb. Hamb. Wiss. Anst.* x. 1893, p. 82.
Ahætulla gracillima, *Günth. Ann. & Mag. N. H.* (6) i. 1888, p. 326.

Rostral broader than deep, the portion visible from above measuring about one third its distance from the frontal; internasals as long as or a little shorter than the præfrontals; frontal with slightly concave lateral borders, once and a half to once and two thirds as long as broad, as long as or a little longer than its distance from the end of the snout, shorter than the parietals; loreal once and two thirds to twice as long as deep; one præocular, not in contact with the frontal; two postoculars; temporals 1+1; eight or nine upper labials, fourth, fifth, and sixth (exceptionally third, fourth, and fifth) entering the eye; five or six lower labials in contact with the anterior chin-shields, which are shorter than the posterior. Scales in 15 rows. Ventrals with feeble lateral keel, 175–190; anal divided; subcaudals 115–135. Green above, yellowish beneath.

Total length 650 millim.; tail 210.

Tropical Africa.

a.	♀ (V. 182; C. 117).	W. Africa?	(Type.)
b.	♂ (V. 178; C. 128).	Lagos.	Dr. A. Günther [P.].
c.	♂ (V. 186; C. 134).	Lagos.	
d.	♀ (V. 180; C. 124).	Lower Congo.	M. A. Linden [C.]. (Type of *A. gracillima*.)
e.	♂ (V. 181; C. ?).	Angola.	Prof. B. du Bocage [P.]
f.	♀ (V. 188; C. 123).	Coast of Zanzibar.	

8. Chlorophis irregularis.

? Coluber cærulescens, *Daud. Rept.* vii. p. 54 (1803).
Coluber irregularis, *Leach, in Bowdich, Miss. Ashantee*, p. 494 (1819).
Dendrophis (Philothamnus) albo-variata, *Smith, Ill. Zool. S. Afr., Rept.* pl. lxv. & pl. lxiv. fig. 3 (1840).
—— chenonii, *Reinh. Vid. Selsk. Skrift.* x. 1843, p. 246, pl. i. figs. 13 & 14.
Leptophis chenonii, *Dum. & Bibr.* vii. p. 545 (1854); *A. Dum. Arch. du Mus.* x. 1860, p. 199; *Jan, Icon. Gén.* 50, pl. i. fig. 2 (1881).
Ahætulla irregularis, part., *Günth. Cat.* p. 152 (1858), *and Ann. & Mag. N. H.* (3) xi. 1863, p. 285.
Philothamnus irregularis, var. longifrenatus, *Buchh. & Peters, Mon. Berl. Ac.* 1875, p. 199.
—— irregularis, *Peters, Mon. Berl. Ac.* 1877, p. 615; *Bocage, Jorn. Sc. Lisb.* ix. 1882, p. 4, fig.; *Boettg. Ber. Senck. Ges.* 1888, p. 61.
Ahætulla shirana, *Günth. Ann. & Mag. N. H.* (6) i. 1888, p. 326.
Chlorophis irregularis, *Bouleng. Proc. Zool. Soc.* 1891, p. 306.
Philothamnus guentheri, *Pfeffer, Jahrb. Hamb. Wiss. Anst.* x. 1893, p. 85, pl. i. figs. 3–5.

Rostral broader than deep, the portion visible from above measuring one fourth to one third its distance from the frontal; internasals as long as or a little shorter than the præfrontals; frontal with concave lateral borders, once and one third to once and a half as long as broad, as long as or a little longer than its distance from the end of the snout, slightly shorter than the parietals; loreal once and a half to twice and a half as long as deep (rarely absent); one præocular, in contact with or narrowly separated from the frontal; two (rarely three) postoculars; temporals usually 1+2, sometimes 1+1 or 2+2; nine upper labials, fourth, fifth, and sixth entering the eye; five lower labials in contact with the anterior chin-shields, which are shorter than the posterior. Scales in 15 rows. Ventrals with moderately marked lateral keel, 150–182; anal divided (rarely entire); subcaudals 94–133. Green or olive above, scales often with a white spot at the base, with or without a black upper border; interstitial skin black; sometimes with black spots or irregular cross bands on the anterior part of the body; greenish yellow inferiorly.

Total length 820 millim.; tail 240.

Tropical Africa.

a. ♂ (V. 173; C. 127).	Gambia.	J. Mitchell, Esq. [P.].
b. ♀ (V. 182; C. 106).	McCarthie Id.	Officers of the Chatham Museum [P.].
c–d. ♂ (V. 175; C. 125) & ♀ (V. 180; C. 116).	Bissao.	V. H. Cornish, Esq. [C.].
e–f. ♂ (V. 165; C. 133) & ♀ (V. 162; C. 111).	Sierra Leone.	Sir A. Kennedy [P.].
g. ♂ (V. 166; C. 122).	Sierra Leone.	H. C. Hart, Esq. [P.].
h–i, k–l. ♂ (V. 168; C. 130) & ♀ (V. 168, 172, 171; C. 111, 108, 103).	Sierra Leone.	
m–n. ♀ (V. 171; C. 122) & head.	Ashantee.	T. E. Bowdich, Esq. [P.]. (Types.)
o. ♀ (V. 176; C. 119).	Guinea?	Leyden Museum.
p. ♀ (V. 157; C. 94).	Gaboon.	
q. ♀ (V. 154; C. ?).	Angola.	Lieut. Cameron [P.].
r. ♀ (V. 158; C. 103).	Caconda, Angola.	Prof. B. du Bocage [P.].
s–t. ♀ (V. 162, 163; C. 105, 103).	Zambesi Expedition.	Sir J. Kirk [C.].
u. Hgr. ♂ (V. 157; C. 114).	Blantyre Mission Station.	H. A. Simons, Esq. [C.]. (Type of *A. shirana*.)
v–w. ♀ (V. 154, 162; C. 98, 108).	Shiré Valley.	
x. ♂ (V. 170; C. 114).	Ugogo, E. Africa.	Capt. Speke [P.].

9. Chlorophis heterodermus.

Chlorophis heterodermus, *Hallow. Proc. Ac. Philad.* 1857, p. 54.
Ahætulla heteroderma, *Günth. Ann. & Mag. N. H.* (3) xi. 1863, p. 285.
Herpetæthiops bellii, *Günth. Ann. & Mag. N. H.* (3) xviii. 1866, p. 27, pl. vii. fig. B.
Philothamnus heterodermus, *Buchh. & Peters, Mon. Berl. Ac.* 1875, p. 199; *Bocage, Jorn. Sc. Lisb.* ix. 1882, p. 18; *Boettg. Ber. Senck. Ges.* 1888, p. 59.

Rostral broader than deep, the portion visible from above measuring hardly one fourth its distance from the frontal; internasals as long as or a little shorter than the præfrontals; frontal with slightly concave lateral borders, as long as or a little longer than its distance from the end of the snout, as long as the parietals; loreal once and a half to twice as long as deep; one præocular, usually not in contact with the frontal; two postoculars; temporals 2 + 2; nine upper labials, fourth, fifth, and sixth entering the eye; five lower labials in contact with the anterior chin-shields, which are as long as the posterior. Scales in 15 rows. Ventrals with moderately strong lateral keel, 148–162; anal entire; subcaudals 78–92. Green or olive above, the skin between the scales black; young with black cross bars; lower parts greenish yellow. Specimens *a* and *c* black above, blackish olive beneath.

Total length 740 millim.; tail 200.

West Africa, from Sierra Leone to the Congo.

a. ♀ (V. 162; C. 90).	Sierra Leone.	Sir A. Kennedy [P.].
b. ♂ (V. 150; C. 78).	Freetown, Sierra Leone.	R. Dinzey, Esq. [C.].
c. ♀ (V. 159; C. 85).	Victoria Sherborough Id., Sierra Leone.	Lieut. Bell [C.]. (Type of *Herpetæthiops bellii*.)
d. Yg. (V. 148; C. 90).	Gold Coast.	W. F. Evans, Esq. [P.].
e. Yg. (V. 156; C. 83).	Lagos.	
f. ♀ (V. 158; C. 79).	Oil River.	H. H. Johnston, Esq. [P.].
g. ♀ (V. 157; C. 83).	W. Africa.	Mr. Rich [C.].

67. PHILOTHAMNUS.

Philothamnus, part., *Smith, Ill. Zool. S. Afr., Rept.* (1840); *Peters, Reise n. Mossamb.* iii. p. 128 (1882); *Bocage, Jorn. Sc. Lisb.* ix. 1882, p. 1.

Ahætulla, part., *Günth. Cat. Col. Sn.* p. 151 (1858), *and Ann. & Mag. N. H.* (3) xi. 1863, p. 283.

Dendrophis, part., *Jan, Elenco sist. Ofid.* p. 85 (1863).

Maxillary teeth 25, posterior longest; mandibular teeth subequal. Head elongate, distinct from neck; eye large, with round pupil. Body cylindrical; scales smooth, with apical pits, in 13 or 15 rows, disposed obliquely, at least on the anterior part of the body; ventrals with a suture-like lateral keel, and a notch on each side corresponding to the keel. Tail long; subcaudals in two rows, keeled and notched like the ventrals.

Tropical and South Africa.

Synopsis of the Species.

I. Scales in 15 rows.

   A. Subcaudals 110–155.

Temporals usually 2+2 .... 1. *semivariegatus*, p. 99.
Temporals 1+2 .... 2. *nitidus*, p. 100.
Temporals 1+1 .... 3. *dorsalis*, p. 101.

   B. Subcaudals 163–175 .... 4. *thomensis*, p. 101.

II. Scales in 13 rows .... 5. *girardi*, p. 102.

1. Philothamnus semivariegatus.

? Coluber albiventris, *Reuss, Mus. Senckenb.* i. 1834, p. 144.
Dendrophis (Philothamnus) semivariegata, *Smith, Ill. Zool. S. Afr., Rept.* pls. lix. & lx., & lxiv. fig. 1 (1840).
Ahætulla semivariegata, *Günth. Ann. & Mag. N. H.* (3) xi. 1863, p. 285.
—— kirkii, *Günth. Ann. & Mag. N. H.* (4) i. 1868, p. 424, *and Zool. Rec.* 1869, p. 116.
Philothamnus punctatus, *Peters, Mon. Berl. Ac.* 1868, p. 889, *and Decken's Reise,* iii. *Amph.* p. 16, pl. i. fig. 2 (1869), *and Reise n. Mossamb.* iii. p. 129, pl. xix. A. fig. 1 (1882); *Bocage, Jorn. Sc. Lisb.* ix. 1882, p. 14.
Dendrophis melanostigma, *Jan, Icon. Gén.* 32, pl. ii. fig. 3 (1869).
? Philothamnus nigrofasciatus, *Buchh. & Peters, Mon. Berl. Ac.* 1875, p. 199.
—— irregularis, var., *Fischer, Abh. nat. Ver. Brem.* vii. 1881, p. 229, pl. xiv. figs. 5–7.
—— smithii, *Bocage, l. c.* p. 12, fig.
Ahætulla bocagii, *Günth. Ann. & Mag. N. H.* (6) i. 1888, p. 326.
Leptophis punctatus, *Mocquard, Mém. Cent. Soc. Philom.* 1888, p. 128.
Philothamnus semivariegatus, *Bouleng. Proc. Zool. Soc.* 1891, p. 307.
—— punctatus, var. sansibaricus, *Pfeffer, Jahrb. Hamb. Wiss. Anst.* x. 1893, p. 83.

Rostral broader than deep, the portion visible from above measuring one fourth or one fifth its distance from the frontal; internasals as long as or a little shorter than the præfrontals; frontal with concave lateral borders, once and a half to once and two thirds as long as broad, as long as or a little longer than its distance from the end of the snout, as long as or a little shorter than the parietals; loreal at least twice as long as deep; one præocular, in contact with or narrowly separated from the frontal; two postoculars; temporals 2+2 (rarely 2+1, 1+2, or 1+1); nine upper labials, fifth and sixth, or fourth, fifth, and sixth, entering the eye; five lower labials in contact with the anterior chin-shields, which are shorter than the posterior. Scales in 15 rows. Ventrals 169–207; anal divided; subcaudals 112–155. Green or olive above, with or without black spots or cross bars; greenish yellow inferiorly.

Total length 1210 millim.; tail 450.

Tropical and South Africa.

A. Irregular black cross bars; some of the scales with a whitish spot.

a. Hgr. (V. 199; C. 138).   W. Africa.

B. No black spots or bars; some of the scales with a whitish spot.

b. ♂ (V. 196; C. 134).   Gambia.   Dr. P. Rendall [C.].

C. With black spots, at least on the anterior part of the back.

c. ♂ (V. 181; C. 122).   Palapye, Kalahari Desert.   R. J. Cuninghame, Esq. [P.].
d. Hgr. (V. 187; C. 138).   E. Africa.   Dr. Livingstone [C.].
e. ♀ (V. 193; C. 137).   Shiré Valley.   Sir J. Kirk [C.].

f. Hgr. (V. 200; C. 144). Blantyre Mission Station. J. Grant, Esq. [P.].
g. ♀ (V. 188; C. 144). Lake Nyassa. Universities Mission [C.].
h. ♂ (V. 181; C. 133). Mozambique. Prof. B. du Bocage [P.].
i. ♀ (V. 175; C. 153). Lake Tanganyika. Sir J. Kirk [C.].
k. ♂ (V. 189; C. 147). Zanzibar. Sir J. Kirk [C.].
l–m. Yg. (V. 170, 175; C. 147, 136). Kilimandjaro. F. J. Jackson, Esq. [P.].
n–o. ♂ (V. 171; C. 141) & ♀ (V. 177; C. ?). Kurawa. Dr. J. W. Gregory [P.].
p. ♂ (V. 166; C. 142). Mkonumbi. Dr. J. W. Gregory [P.].

D. Spots and bars absent or very indistinct.

q. ♀ (V. 179; C. 126). Near Grahamstown. Rev. G. H. R. Fisk [P.].
r. ♀ (V. 185; C. 111). Humbo, Angola. Prof. B. du Bocage [P.]. (One of the types of *P. smithii*.)
s. ♀ (V. 196; C. 126). Angola. Lieut. Cameron [P.]. (Type of *A. bocagii*.)
t. ♂ (V. 199; C. ?). Bissao. V. H. Cornish, Esq. [C.].
u–w. ♀ (V. 171, 173; C. 135, 149) & yg. (V. 171; C. ?). Zanzibar. Sir J. Kirk [C.].
x. ♀ (V. 169; C. 130). E. Africa. Sir J. Kirk [C.].
y. ♀ (V. 173; C. 138). E. Africa. Capt. Speke [C.].
z. ♀ (V. 171; C. ?). Melindi. Dr. J. W. Gregory [P.].

2. Philothamnus nitidus. (PLATE V. fig. 4.)

Ahætulla nitida, part., *Günth. Ann. & Mag. N. H.* (3) xi. 1863, p. 286.
—— lagoensis, *Günth. Ann. & Mag. N. H.* (4) ix. 1872, p. 26.
Philothamnus lagoensis, *Bocage, Jorn. Sc. Lisb.* ix. 1882, p. 6.

Head narrow. Rostral broader than deep, the portion visible from above measuring about one fifth its distance from the frontal; internasals as long as the præfrontals; frontal with concave lateral borders, once and three fourths as long as broad, a little longer than its distance from the end of the snout, a little shorter than the parietals; loreal twice to twice and a half as long as broad; one præocular, in contact with or narrowly separated from the frontal; two postoculars; temporals 1+2; nine upper labials, fourth, fifth, and sixth entering the eye; five lower labials in contact with the anterior chin-shields, which are shorter than the posterior. Scales in 15 rows. Ventrals 163–165; anal divided; subcaudals 150–153. Dark green above, pale green beneath.

Total length 880 millim.; tail 330.

Guinea.

a. ♀ (V. 163; C. ca. 150).   Lagos.   (Type of *A. lagoensis.*)
b. ♂ (V. 165; C. 153).   ———?   (One of the types of *A. nitida.*)

Possibly not specifically distinct from *P. semivariegatus.*

3. Philothamnus dorsalis.

Leptophis dorsalis, *Bocage, Jorn. Sc. Lisb.* i. 1866, p. 69.
Ahætulla dorsalis, *Günth. Ann. & Mag. N. H.* (4) i. 1868, p. 424.
Philothamnus dorsalis, *Peters, Mon. Berl. Ac.* 1876, p. 119; *Bocage, Jorn. Sc. Lisb.* ix. 1882, p. 9, fig.; *Boettg. Ber. Senck. Ges.* 1888, p. 58.

Rostral broader than deep, the portion visible from above measuring about one fourth its distance from the frontal; internasals as long as or a little shorter than the præfrontals; frontal with concave lateral borders, once and a half to once and two thirds as long as broad, a little longer than its distance from the end of the snout, a little shorter than the parietals; loreal at least twice as long as deep; one præocular, in contact with or narrowly separated from the frontal; two postoculars; temporals 1+1; nine upper labials, fourth, fifth, and sixth entering the eye; five lower labials in contact with the anterior chin-shields, which are shorter than the posterior. Scales in 15 rows. Ventrals 170–190; anal divided; subcaudals 118–145. Olive above, with a brown vertebral stripe; snout yellowish brown; lower parts yellowish.

Total length 840 millim.; tail 290.

West Africa, from the Ogowé to Benguela.

a. ♂ (V. 172; C. ?).   Lower Congo.   M. A. Linden [C.].
b. Hgr. ♂ (V. 173; C. 141). Angola.   Prof. B. du Bocage [P.].
c. ♀ (V. 179; C. 131).   Angola.   J. J. Monteiro, Esq. [P.].
d. ♂ (V. 179; C. 145).   Carangigo.   Dr. Welwitsch [P.].
e. ♂ (V. 184; C. 133).   Benguela.   Prof. B. du Bocage [P.].

4. Philothamnus thomensis.

Ahætulla nitida, part., *Günth. Ann. & Mag. N. H.* (3) xi. 1863, p. 286.
Philothamnus thomensis, *Bocage, Jorn. Sc. Lisb.* ix. 1882, p. 11, fig.; *Bedriaga, Amph. et Rept. I. de la Guinée* (*Istituto*, xxxix.), p. 29 (1892).

Rostral broader than deep, the portion visible from above measuring not more than one fourth its distance from the frontal; internasals as long as the præfrontals; frontal with concave lateral borders, once and two thirds as long as broad, longer than its distance from the end of the snout, as long as the parietals; loreal at least twice as long as deep; one præocular, in contact with or narrowly separated from the frontal; two postoculars; temporals 1+1 or 1+2; nine upper labials, fourth, fifth, and sixth entering the eye; five lower labials in contact with the anterior chin-shields, which are shorter than the posterior. Scales in 15 rows. Ventrals

207-220; anal divided; subcaudals 163-175. Olive-green above, some of the scales edged with black; pale green beneath.
Total length 900 millim.; tail 320.
S. Thomé Island, W. Africa.

a. ♀ (V. 207; C. 169). S. Thomé. Prof. B. du Bocage [P.]. (One of the types.)
b. ♂ (V. 207; C. ?). S. Thomé. Prof. B. du Bocage [P.].
c. ♂ (V. 219; C. 167). [Demerara.] Col. Sabine [P.]. (One of the types of *A. nitida*.)

5. Philothamnus girardi.

Philothamnus girardi, *Bocage, Jorn. Sc. Lisb.* (2) iii. 1893, p. 47.

Rostral broader than deep, the portion visible from above measuring about one fourth its distance from the frontal; internasals shorter than the præfrontals; frontal nearly twice as long as broad, with concave lateral borders; loreal much elongate, sometimes divided into two; one præ- and two postoculars; temporals 1+1, rarely 1+2; nine upper labials, fourth, fifth, and sixth entering the eye; five lower labials in contact with the anterior chin-shields, which are shorter than the posterior. Scales in 13 rows. Ventrals 189-194; anal divided; subcaudals 145-153. Olive-green above, with black edges to the scales, forming cross bars on the anterior part of the body; yellowish white beneath, with a black line along the ventral keel.
Total length 780 millim.; tail 280.
Anno-Bom Island, Gulf of Guinea.

a. ♂ (V. 186; C. 145). Anno-Bom Id. Prof. B. du Bocage [P.]. (One of the types.)

68. GASTROPYXIS.

Dendrophis, part., *Schleg. Phys. Serp.* ii. p. 220 (1837).
Leptophis, part., *Dum. & Bibr. Erp. Gén.* vii. p. 528 (1854); *Jan, Elenco sist. Ofid.* p. 84 (1863).
Ahætulla, part., *Günth. Cat. Col. Sn.* p. 151 (1858).
Gastropyxis, *Cope, Proc. Ac. Philad.* 1860, p. 556.
Hapsidophrys, part., *Fischer, Abh. naturw. Ver. Hamb.* iii. 1856, p. 110; *Günth. l. c.* p. 144; *Jan, l. c.* p. 86.

Maxillary teeth 20 to 22, posterior longest; mandibular teeth subequal. Head rather elongate, distinct from neck; eye large, with round pupil. Body cylindrical; scales keeled, without or with very indistinct apical pits, in 15 rows. Ventrals with a suture-like lateral keel, and a notch on each side corresponding to the keel. Tail very long; subcaudals in two rows, strongly keeled and notched like the ventrals.
West Africa.

1. Gastropyxis smaragdina.

Dendrophis smaragdina, *Schleg. Phys. Serp.* ii. p. 237 (1837).
Leptophis gracilis, *Hallow. Proc. Ac. Philad.* 1844, p. 60.
—— smaragdinus, *Dum. & Bibr.* vii. p. 537 (1854); *Hallow. Proc. Ac. Philad.* 1854, p. 100; *A. Dum: Arch. du Mus.* x. 1860, p. 199, pl. xvii. fig. 6; *Jan, Icon. Gén.* 49, pl. vi. fig. 4 (1879).
Hapsidophrys cœruleus, *Fischer, Abh. nat. Ver. Hamb.* iii. 1856, p. 111, pl. ii. fig. 6; *Günth. Cat.* p. 145 (1858).
Ahætulla smaragdina, *Günth. Cat.* p. 151.
Gastropyxis smaragdina, *Cope, Proc. Ac. Philad.* 1860, p. 558.
Hapsidophrys smaragdina, *Reichen. Arch. f. Nat.* 1874, p. 292; *Boettg. Ber. Senck. Ges.* 1888, p. 62, and 1889, p. 279.

Rostral broader than deep, visible from above; internasals longer than the præfrontals; frontal once and a half as long as broad, as long as or a little longer than its distance from the end of the snout, as long as or slightly shorter than the parietals; loreal twice to thrice as long as deep, sometimes confluent with the præfrontal; one præocular, in contact with or narrowly separated from the frontal; two postoculars; temporals 1+2; nine (rarely eight) upper labials, fifth and sixth (or fourth and fifth) entering the eye; five or six lower labials in contact with the anterior chin-shields, which are shorter than the posterior. Scales in 15 rows, all strongly keeled and distinctly striated. Ventrals 150–174; anal divided; subcaudals 140–172. Dark bluish green above, pale green beneath and on the upper lip; a blackish streak on each side of the head, passing through the eye.

Total length 980 millim.; tail 390.

West Africa, from Sierra Leone to the Gaboon.

a. ♂ (V. 157; C. 140).	Sierra Leone.	H. C. Hart, Esq. [P.].
b. ♂ (V. 155; C. 152).	Sierra Leone.	
c, d. ♀ (V. 156; C. 144) & yg. (V. 155; C. ?).	Ashantee.	
e–f. ♀ (V. 154; C. 147) & yg. (V. 157; C. ?).	Ancobra River, Gold Coast.	Major Burton and Capt. Cameron [P.].
g. ♀ (V. 162; C. ?).	Niger.	Mr. Fraser [C.].
h–i. ♀ (V. 169; C. 161) & hgr. (V. 171; C.172).	Akassa, mouth of Niger.	Dr. J. W. Crosse [P.].
k–l. ♀ (V. 171, 172; C. ?, 145).	Oil River.	H. H. Johnston, Esq. [P.].
m. ♂ (V. 164; C. 166).	Fernando Po.	
n. ♀ (V. 153; C. 146).	Guinea.	Leyden Museum.
o–p. ♀ (V.157,161; C.148, 147).	Eloby district, Gaboon.	H. Ansell, Esq. [P.].

69. HAPSIDOPHRYS.

Hapsidophrys, part., *Fischer, Abh. naturw. Ver. Hamb.* iii. 1856, p. 110; *Günth. Cat. Col. Sn.* p. 144 (1858); *Jan, Elenco sist. Ofid.* p. 86 (1863).

Maxillary teeth 30 to 32, posterior longest; mandibular teeth subequal. Head distinct from neck; eye very large, with round pupil. Body cylindrical; scales keeled, with apical pits, in 15 rows;

ventrals with a suture-like lateral keel, and a notch on each side corresponding to the keel. Tail long; subcaudals in two rows.
West Africa.

1. Hapsidophrys lineata.

Hapsidophrys lineatus, *Fischer, Abh. naturw. Ver. Hamb.* iii. 1856, p. 111, pl. ii. fig. 5; *Günth. Cat.* p. 144 (1858); *Jan, Icon. Gén.* 33, pl. i. fig. 2 (1869); *Mocquard, Bull. Soc. Philom.* (7) xi. 1887, p. 76.

Rostral broader than deep, just visible from above; internasals as long as the præfrontals; frontal once and a half to once and two thirds as long as broad, as long as or a little longer than its distance from the end of the snout, as long as the parietals; loreal at least twice as long as deep; one præocular (rarely two), not touching the frontal; two postoculars; temporals 2+2 (rarely 1+2); eight or nine upper labials, fourth and fifth or fifth and sixth entering the eye; four to six lower labials in contact with the anterior chin-shields, which are as long as or a little shorter than the posterior. Scales in 15 rows, striated and keeled, the outer row without or with a very faint keel. Ventrals 158–170; anal entire; subcaudals 95–158. Upper parts striated black and green, each scale green in the middle and black on the sides; upper lip and lower parts pale green or yellowish green.

Total length 1100 millim.; tail 340.
West Africa.

a. ♀ (V. 170; C. 95).	Coast of Guinea.	
b–c. ♀ (V. 167; C. 115) & yg. (V. 167; C. 121).	Ancobra River, Gold Coast.	Major Burton and Capt. Cameron [P.].
d. ♂ (V. 167; C. 117).	Ancobra River.	Capt. Torry [P.].
e. ♀ (V. 165; C. ?).	Oil River.	H. H. Johnston, Esq. [P.].
f. Skeleton.	W. Africa.	

70. THRASOPS.

Thrasops, *Hallow. Proc. Ac. Philad.* 1857, p. 67.
Thrasops, part., *Cope, Proc. Ac. Philad.* 1860, p. 556.

Maxillary teeth 20 to 22, the three or four last longest and

Fig. 9.

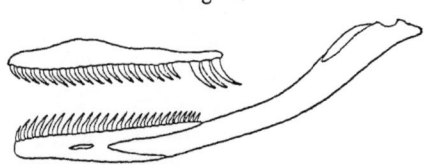

Maxillary and mandible of *Thrasops flavigularis*.

separated from the rest by an interspace; anterior mandibular teeth feebly enlarged. Head rather short, distinct from neck; eye large, with round pupil. Body more or less compressed; scales very strongly overlapping, disposed obliquely, in 13 or 15 rows, the laterals much shorter than the dorsals, which are keeled; apical pits present; ventrals rounded, or with an interrupted lateral keel posteriorly. Tail long; subcaudals in two rows.

West Africa.

1. Thrasops flavigularis.

Dendrophis flavigularis, *Hallow. Proc. Ac. Philad.* 1852, p. 205.
Thrasops flavigularis, *Hallow. Proc. Ac. Philad.* 1857, p. 67; *Peters, Mon. Berl. Ac.* 1876, p. 119.
Hapsidophrys niger, *Günth. Ann. & Mag. N. H.* (4) ix. 1872, p. 25.
Thrasops pustulatus, *Buchh. & Peters, Mon. Berl. Ac.* 1875, p. 199.
—— flavigularis, var. pustulata, *Boettg. Ber. Senck. Ges.* 1889, p. 279.

Rostral with straight vertical sides, a little broader than deep, visible from above; internasals as long as the præfrontals; frontal a little longer than broad, scarcely broader than the supraocular, as long as its distance from the rostral or the end of the snout and as long as the parietals; loreal as deep as long; one præocular, not in contact with the frontal; three postoculars; temporals 1+1; eight upper labials, fourth and fifth entering the eye; chin-shields short, the anterior in contact with four lower labials. Scales in 13 or 15 rows, dorsals keeled. Ventrals 179–206; anal divided; subcaudals 130–140. Black, with or without yellowish spots on the back; lips and throat yellowish or brownish white.

Total length 1700 millim.; tail 500.

West Africa, from Sierra Leone to the Gaboon.

a. ♂ (V. 179; C. 130).	Sierra Leone.	
b. ♂ (V. 187; C. 132).	Monrovia, Liberia.	Christiania Museum.
c. ♀ (V. 203; C. 140).	Gaboon.	(Type of *H. niger*.)
d. Skull.	Cameroons.	

71. LEPTOPHIS.

Leptophis, part., *Bell, Zool. Journ.* ii. 1825, p. 328; *Dum. & Bibr. Erp. Gén.* vii. p. 528 (1854); *Jan, Elenco sist. Ofid.* p. 84 (1863); *Cope, Journ. Ac. Philad.* (2) viii. 1876, p. 132.
Leptophis, *Wagl. Syst. Amph.* p. 183 (1830).
Ahætula, part., *Gray, Griff. A. K.* ix. *Syn.* p. 93 (1831); *Günth. Cat. Col. Sn.* p. 151 (1858).
Dendrophis, part., *Schleg. Phys. Serp.* ii. p. 220 (1837).
Thrasops, part., *Cope, Proc. Ac. Philad.* 1860, p. 556.
Diplotropis, *Günth. Ann. & Mag. N. H.* (4) ix. 1872, p. 24.

Maxillary teeth 20 to 32, posterior longest; anterior mandibular teeth longest. Head elongate, distinct from neck; eye large, with round pupil. Body slender, cylindrical or slightly compressed; scales keeled or smooth, with apical pits, in 13 or 15 rows, disposed obliquely, at least on the anterior portion of the body; ventrals rounded or angulate laterally. Tail long; subcaudals in two rows. Central and South America.

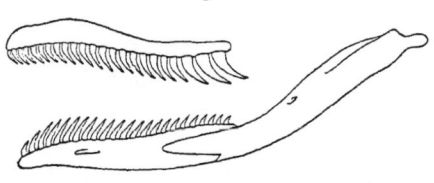

Fig. 10.

Maxillary and mandible of *Leptophis liocercus*.

Synopsis of the Species.

I. Loreal present (rarely absent).

  A. Scales smooth.

Nasals not elongate; loreal subquadrate ..   1. *æruginosus*, p. 107.
Nasals elongate; loreal thrice as long as   [p. 107.
    deep ....   2. *depressirostris*,

  B. Scales keeled, except the outer row.

Loreal once and a half as long as deep;
    ventrals without trace of lateral keel ..   3. *modestus*, p. 107.
Loreal twice to thrice as long as deep;
    ventrals feebly angulate laterally ......   4. *mexicanus*, p. 108.

  C. Five median rows of scales feebly keeled.
                                     5. *cupreus*, p. 109.

  D. Only the scales of the series on each side of the vertebral keeled.

Ventrals not angulate, 160 ....   6. *saturatus*, p. 110.
Ventrals feebly angulate laterally, 175–186   7. *diplotropis*, p. 110.
Ventrals feebly angulate laterally, 144....   8. *bilineatus*, p. 111.

II. Loreal absent.

  A. Scales in 15 rows, keeled; ventrals more or less distinctly angulate laterally.

Scales of vertebral row more feebly keeled
    than the one or two on each side of it;
    frontal not more than once and a half as
    long as broad ....   9. *occidentalis*, p. 111.
Scales more or less distinctly keeled, the
    keel on the middle row more feeble, or
    obsolete; frontal at least once and a half   [p. 112.
    as long as broad.................... 10. *nigromarginatus*,

Scales all except the outer row strongly
keeled............................ 11. *liocercus*, p. 113.

B. Scales in 15 rows, smooth; ventrals not angulate laterally.
12. *ortonii*, p. 114.

C. Scales in 13 rows, all except the outer row keeled.
13. *urostictus*, p. 114.

1. Leptophis æruginosus.

Leptophis æruginosus, *Cope, Journ. Ac. Philad.* (2) viii. 1876, p. 132; *Günth. Biol. C.-Am., Rept.* p. 130 (1894).
Philothamnus æruginosus, *Cope, Proc. Amer. Philos. Soc.* xxiii. 1886, p. 279.

Nasal not elongate; loreal present, subquadrate; præocular scarcely reaching frontal; two postoculars; temporals 1+2; nine upper labials, fifth and sixth entering the eye; posterior chin-shields longer than the anterior. Scales in 15 rows, not keeled, but finely striate. Ventrals faintly angulate laterally, 146; anal divided; subcaudals 142. Golden brown above; vertebral line (one row of scales) yellow on the anterior half of the body; a black streak on each side of the head behind the eye; beneath blue, fading to yellowish on the gular region.

Total length 405 millim.; tail 155.

Costa Rica.

2. Leptophis depressirostris.

Philothamnus depressirostris, *Cope, Proc. Ac. Philad.* 1860, p. 557.
Leptophis depressirostris, *Günth. Biol. C.-Am., Rept.* p. 130 (1894).

Rostral twice as broad as deep; frontal as long as the parietals; nasals elongate; loreal present, thrice as long as deep; one præ- and two postoculars; nine upper labials, fifth and sixth entering the eye. Scales smooth, in 15 rows. Above uniform deep green; an inconspicuous temporal streak; a very delicate black line traverses the centre of each of the two rows of scales that bound the vertebral row; upper lip and lower parts light green.

Total length 650 millim.

Cocuyas de Veraguas, Colombia.

3. Leptophis modestus.

Ahætulla modesta, *Günth. Ann. & Mag. N. H.* (4) ix. 1872, p. 26, pl. vi. fig. C.
Philothamnus modestus, *Cope, Proc. Amer. Philos. Soc.* xxiii. 1886, p. 279.
Leptophis modestus, *Günth. Biol. C.-Am., Rept.* p. 129, pl. xlviii. (1894).

Eye a little shorter than its distance from the nostril. Rostral broader than deep, just visible from above; internasals shorter than the præfrontals; frontal once and one third as long as broad, as long as its distance from the rostral, shorter than the parietals; nasal elongate, divided; loreal present, once and a half as long as deep; one præocular, not reaching the frontal; two postoculars; temporals 1+2; eight upper labials, fourth and fifth entering the eye; five lower labials in contact with the anterior chin-shields, which are shorter than the posterior. Scales in 15 rows, all except the outer strongly keeled; caudal scales smooth, except the basal ones. Ventrals without trace of lateral keel, 171; anal divided; subcaudals 171. Uniform olive-green above, pale green beneath; lips and throat yellowish; a blackish streak on each side of the head, behind the eye.

Total length 1300 millim.; tail 540.

Guatemala.

a. ♂ (V. 171; C. 171).   Rio Chisoy, below the   O. Salvin, Esq. [P.].
                         town of Cubulco.   (Type.)

4. Leptophis mexicanus.

Leptophis mexicanus, *Dum. & Bibr.* vii. p. 536 (1854); *Jan, Icon. Gén.* 49, pl. vi. fig. 3 (1879).
Ahætulla mexicana, *Günth. Cat.* p. 154 (1858).
Thrasops mexicanus, *Cope, Proc. Ac. Philad.* 1860, p. 557.
Hapsidophrys mexicanus, *Cope, Proc. Am. Philos. Soc.* xxii. 1885, p. 382.

Rostral broader than deep, just visible from above; internasals as long as or a little shorter than the præfrontals; frontal once and one third to once and two thirds as long as broad, as long as its distance from the end of the snout, a little shorter than the parietals; nasal elongate, divided; loreal present, twice to thrice as long as deep; one præocular, not reaching the frontal; two postoculars; temporals 1+2; eight upper labials, fourth and fifth entering the eye; four or five (rarely six) lower labials in contact with the anterior chin-shields, which are shorter than the posterior. Scales in 15 rows, striated, all except the outer row strongly keeled; caudal scales keeled. Ventrals feebly angulate laterally, 154–174; anal divided; subcaudals 141–172.

Total length 1340 millim.; tail 520.

Northern Mexico to Costa Rica.

A. Pale bronzy brown or golden above, head blue or green; a black lateral stripe on the head and body, passing through the eye, enclosing some bluish-green spots in the anterior half of the body; upper lip and lower parts mother-of-pearl or pale golden.

a. Hgr. (V.166; C.162).   Amula, Guerrero.   Mr. H. H. Smith [C.];
                                             F. D. Godman, Esq.
                                             [P.].
b. Hgr. (V.164; C.156).   Teapa, Tabasco.   F. D. Godman, Esq.
                                             [P.].

c. ♂ (V. 161; C. ?).	Mexico.	Mr. H. Finck [C.].
d. ♂ (V. 159; C. 165).	Belize.	J. Smith, Esq. [P.].
e, f. ♂ (V. 154; C. 166) & ♀ (V. 165; C. 152).	Stann Creek, Brit. Honduras.	Rev. J. Robertson [C.].
g, h, i. ♂ (V. 161; C. 157), ♀ (V. 166; C. 154), & yg. (V. 173; C. 167).	Honduras.	

B. Two broad black stripes along the body (continued from the sides of the head) separated by three series of pale reddish-brown scales; upper surface of head and nape olive; lips and throat yellowish; belly and lower surface of tail pale brown.

a. ♂ (V. 163; C. 166). Yucatan.

C. Pale bronzy olive above, head and neck green; the black lateral stripe of the head continued on the body and tail as a broad dark green stripe, some of the scales on which are tipped or edged with black; upper lip and lower parts white.

a. ♀ (V. 162; C. 151).	Jalisco, N. of Rio de Santiago.	F. D. Godman, Esq. [P.].
b. ♂ (V. 154; C. 148).	Atoyac, Guerrero.	Mr. H. H. Smith [C.]; F. D. Godman, Esq. [P.].
c. ♂ (V. 165; C. 152).	Santo Domingo de Guzman, Mexico.	Dr. A. C. Buller [C.].

D. Bright green above, the lateral scales on the anterior half of the body tipped or edged with black; a black streak on each side of the head, passing through the eyes; lips and throat bright yellow; belly and lower surface of tail pale green.

a. ♂ (V. 164; C. 162).	Tampico.	F. D. Godman, Esq. [P.].
b–g. ♂ (V. 167, 167, 173; C. 152, 162, 141) & ♀ (V. 169, 174, 173; C. 146, 172, 157).	Mexico.	

5. Leptophis cupreus.

Thrasops cupreus, *Cope, Proc. Ac. Philad.* 1868, p. 106.

Head unusually short and broad. Rostral much broader than deep, little visible from above; nasal long and narrow; loreal present, absent on one side in the type specimen; one præ- and two postoculars; temporals 1+2; eight upper labials, fourth and fifth entering the eye. Scales in 15 rows, the five median feebly keeled. Ventrals 152; anal divided; subcaudals 136. Above metallic copper colour; a dark streak on the side of the head, passing through the eye; labials and chin yellowish white; ventrals brown, copper-coloured, with darker dashes.

Total length 520 millim.; tail 205.

Upper Amazon.

6. Leptophis saturatus.

Leptophis saturatus, *Cope, Journ. Ac. Philad.* (2) viii. 1876, p. 133.
Hapsidophrys saturatus, *Cope, Proc. Amer. Philos. Soc.* xxiii. 1886, p. 279.

Head short and wide; snout not longer than the diameter of the eye. Frontal rather broad; nasals not elongate; loreal present, nearly twice as long as deep; one præocular, nearly reaching the frontal; two postoculars; temporals $1+2$; nine upper labials, fifth and sixth entering the eye. Scales in 15 rows, one on each side of the median vertebral weakly keeled. Ventrals not angulate, 160; anal divided; subcaudals 133. Indigo-blue (in spirits), very dark on head and spine; lips dark green; a blackish shade above the labials behind the eye.

Total length 880 millim.; tail 340.

Sipurio, Costa Rica.

7. Leptophis diplotropis.

Ahætulla diplotropis, *Günth. Ann. & Mag. N. H.* (4) ix. 1872, p. 25, pl. vi. fig. A.
Hapsidophrys diplotropis, *Cope, Proc. Amer. Philos. Soc.* xxiii. 1886, p. 279.
Leptophis diplotropis, *Günth. Biol. C.-Am., Rept.* p. 130 (1894).

Rostral broader than deep, just visible from above; internasals shorter than the præfrontals; frontal once and a half to once and two thirds as long as broad, as long as its distance from the end of the snout, shorter than the parietals; nasal elongate, divided; loreal usually present, once and a half to thrice as long as deep; one præocular, not reaching the frontal; two postoculars; temporals $1+2$; eight upper labials, fourth and fifth entering the eye; four or five lower labials in contact with the anterior chin-shields, which are shorter than the posterior. Scales in 15 rows, those of the series on each side of the vertebral with a strong keel. Ventrals feebly angulate laterally, 165–186; anal divided; subcaudals 126–166. Olive or pale green above, with or without oblique black cross bars; scales of the vertebral series yellowish or pale olive; a black stripe on each side of the head and neck, passing through the eye; two black lines following the dorsal keels; upper lip and lower parts pale greenish yellow.

Total length 1250 millim.; tail 440.

Western Mexico.

a–c. Hgr. ♀ (V. 181, 175, 179; C. 145, 133, 146). Tehuantepec. F. Sumichrast [C.]. (Types.)
d. Hgr. ♀ (V. 179; C. 153). Tapana, Tehuantepec. F. Sumichrast [C.]; F. D. Godman & O. Salvin, Esqs. [P.].

e. ♀ (V. 180 ; C. ?).	Santo Domingo de Guzman.	Dr. A. C. Buller [C.].
f. Head and neck.	La Laguna, Sierra de Alica.	Dr. A. C. Buller [C.].
g-i. ♂ (V. 171, 167; C. 141, 144) & ♀ (V. 165; C. 126).	Amula, Guerrero.	Mr. H. H. Smith [C.]; F. D. Godman, Esq. [P.].
k-l. ♂ (V. 178; C. ?) & ♀ (V. 178; C. 149).	Presidio, near Mazatlan.	Mr. A. Forrer [C.].
m-n. ♂ (V. 185; C. 160) & hgr. (V. 186; C. 166).	Tres Marias Ids.	Mr. A. Forrer [C.].

8. Leptophis bilineatus.

Diplotropis bilineata, *Günth. Ann. & Mag. N. H.* (4) ix. 1872, p. 24, pl. vi. fig. B.
Leptophis bilineatus, *Cope, Proc. Amer. Philos. Soc.* xxiii. p. 279; *Günth. Biol. C.-Am., Rept.* p. 130 (1894).

Rostral broader than deep, just visible from above; internasals as long as the præfrontals; frontal once and a half as long as broad, as long as its distance from the end of the snout, as long as the parietals; nasal moderately elongate, divided; loreal present, twice as long as deep; one præocular, not quite reaching the frontal; two postoculars; temporals 1+2; eight or nine upper labials, fourth and fifth (or fifth and sixth) entering the eye; five lower labials in contact with the anterior chin-shields, which are shorter than the posterior. Scales in 15 rows, those of the series on each side of the vertebral with a strong keel. Ventrals feebly angulate laterally, 144; anal divided. Bright green above, light greenish beneath; two black lines following the dorsal keels; a very indistinct dark streak on each side of the head, passing through the eye.

Total length 910 millim.

Costa Rica, Nicaragua, Tehuantepec.

a. ♂ (V. 144; C. ?).	Costa Rica.	O. Salvin, Esq. [P.]. (Type.)

9. Leptophis occidentalis. (PLATE III. fig. 2.)

Dendrophis liocercus, part., *Schleg. Phys. Serp.* ii. p. 224 (1837); *Guichen. in Gay, Hist. de Chile, Zool.* ii. p. 88 (1854).
—— liocercus, *Berthold, Abh. Ges. Götting.* iii. 1847, p. 11.
Leptophis liocercus, part., *Dum. & Bibr.* vii. p. 533 (1854).
Ahætulla liocercus, part., *Günth. Cat.* p. 153 (1858).
—— occidentalis, *Günth. Proc. Zool. Soc.* 1859, p. 412; *Peters, Mon. Berl. Ac.* 1873, p. 607.
Thrasops occidentalis, *Cope, Proc. Ac. Philad.* 1860, p. 557.
—— præstans, *Cope, Proc. Ac. Philad.* 1868, p. 309.
Leptophis præstans, *Cope, Journ. Ac. Philad.* (2) viii. 1876, p. 133; *Günth. Biol. C.-Am., Rept.* p. 130 (1894).
Thrasops (Ahætulla) sargii, *Fischer, Arch. f. Nat.* 1881, p. 229, pl. xi. figs. 7-9.
Leptophis sargii, *Cope, Proc. Amer. Philos. Soc.* xxiii. 1886, p. 279.
—— occidentalis, *Cope, Proc. Amer. Philos. Soc.* xxiii. 1886, p. 279.

Rostral broader than deep, just visible from above; internasals shorter than the præfrontals; frontal once and one fourth to once and a half as long as broad, as long as its distance from the rostral or the end of the snout, as long as or a little shorter than the parietals; nasal elongate, divided; no loreal; præfrontal in contact with the second and third, or second, third, and fourth labials; one præocular, in contact with or not quite reaching the frontal; two or three postoculars; temporals 1+2; eight or nine upper labials, fourth and fifth or fifth and sixth entering the eye; four or five lower labials in contact with the anterior chin-shields, which are much shorter than the posterior. Scales in 15 rows, all except the outer row more or less distinctly keeled; scales of middle row more feebly keeled than the one or two on each side of it; scales on the neck and tail smooth. Ventrals 152–185, feebly angulate laterally; anal divided; subcaudals 133–176.

Total length 2000 millim.; tail 730.

Central America, Colombia, Ecuador, Peru.

A. Green above, the keels brown or black; yellowish or pale green beneath; a black streak on each side of the head behind the eye, disappearing in old specimens.

a. ♂ (V. 152; C. ?).	W. Ecuador.	Mr. Fraser [C.].	(Type.)
b–d. Hgr. ♂ (V. 178; C. 147) & yg. (V. 174, 167; C. ?, 159).	Cartagena.	Capt. Garth [P.].	
e. ♀ (V. 173; C. 155).	Brazil?		
f. ♀ (V. 185; C. 170).	British Honduras.	F. D. Godman, Esq. [P.].	
g. ♂ (V. 172; C. 169).	Stann Creek, British Honduras.	Rev. J. Robertson [C.].	
h. ♀ (V. 182; C. 172).	Tabasco.	F. D. Godman, Esq. [P.].	
i. ♂ (V. 180; C. ?).	Amula, Guerrero.	Mr. H. H. Smith [C.]; F.D.Godman,Esq.[P.].	

B. Black above and beneath, chin and throat white.

a. Hgr. ♀ (V. 165; C. 133).	Guayaquil.	(Type.)
b. Hgr. ♀ (V. 163; C. 140).	——?	Royal College of Surgeons.

10. Leptophis nigromarginatus. (PLATE III. fig. 3.)

Ahætulla nigromarginata, *Günth. Ann. & Mag. N. H.* (3) xviii. 1866, p. 28.

Leptophis marginatus, *Cope, Journ. Ac. Philad.* (2) viii. 1876, p. 177.

Rostral broader than deep, just visible from above; internasals as long as or a little shorter than the præfrontals; frontal once and a half to once and three fourths as long as broad, as long as its distance from the end of the snout, as long as or slightly shorter than the parietals; nasal elongate, entire or divided; no loreal;

præfrontal in contact with labials ; one præocular, in contact with or narrowly separated from the frontal; temporals 1+2; eight or nine upper labials, fourth and fifth or fifth and sixth entering the eye; five or six lower labials in contact with the anterior chin-shields, which are shorter than the posterior. Scales in 15 rows, all except the outer row more or less distinctly keeled; the keel on the middle vertebral series more feeble, or obsolete ; scales on neck and tail smooth. Ventrals distinctly angulate laterally, 149–161; anal divided; subcaudals 106–163. Green above, head-shields and scales broadly edged with black ; keels black ; ventrals and subcaudals yellowish or pale green, uniform or edged with dark green.

Total length 1010 millim.; tail 390.

Upper Amazons.

a. ♀ (V. 158; C. 115).	Upper Amazons.	Mr. E. Bartlett [C.]. (Type.)
b–c. ♂ (V. 155; C. 142) & ♀ (V. 161; C. 163).	Pebas.	
d–e. ♀ (V. 158, 149; C. 153, 150).	Peruvian Amazons.	
f–g. ♂ (V. 155; C. 117) & ♀ (V. 153; C. 144).	Moyobamba, N.E. Peru.	Mr. A. H. Roff [C.].
h. ♀ (V. 151; C. 106).	Sarayacu, N.E. Peru.	Mr. W. Davis [C.]; Messrs. Veitch [P.].

11. Leptophis liocercus.

Coluber ahætulla, part., *Gmel. S. N.* i. p. 1116 (1788); *Bonnat. Encycl. Méth., Ophiol.* p. 28 (1789); *Lacép. Serp.* p. 223, pl. xi. fig. 1 (1789).
—— boiga, part., *Daud. Rept.* vii. p. 63, pl. lxxxiv. (1803).
—— liocercus, *Wied, Abbild.* (1824), *and Beitr. Nat. Bras.* i. p. 265 (1825).
—— richardii, *Bory de St. Vinc. Ann. Sc. Nat.* i. 1824, p. 408, pl. xxiv.
Ahætula liocercus, *Gray, Griff. A. K.* ix. *Syn.* p. 93 (1831).
—— richardi, *Gray, l. c.*
Leptophis ahætulla, part., *Bell, Zool. Journ.* ii. 1825, p. 328.
Dendrophis liocercus, part., *Schleg. Phys. Serp.* ii. p. 224, pl. ix. figs. 1 & 2 (1837).
Leptophis liocercus, part., *Dum. & Bibr.* vii. p. 533 (1845).
Ahætulla liocercus, part., *Günth. Cat.* p. 153 (1858).
—— liocercus, *Günth. Proc. Zool. Soc.* 1859, p. 413.
Thrasops ahætulla, *Cope, Proc. Ac. Philad.* 1860, p. 557.
—— marginatus, *Cope, Proc. Ac. Philad.* 1862, p. 349.
Ahætulla ahætulla, *Peters, Mon. Berl. Ac.* 1873, p. 607.
Leptophis liocercus, *Jan, Icon. Gén.* 49, pl. vi. fig. 1 (1879).

Rostral broader than deep, just visible from above; internasals as long as or shorter than the præfrontals; frontal once and a half to twice as long as broad, as long as its distance from the rostral or the end of the snout, as long as or a little shorter than the parietals; nasal elongate, entire or divided; no loreal; præfrontal in contact with labials; one præocular (rarely divided) in contact with or narrowly separated from the frontal; two (rarely three) postoculars; temporals 1+2; eight or nine upper labials, fourth and fifth or fifth and sixth entering the eye; five or six (rarely seven) lower labials in contact with the anterior chin-shields, which are shorter than the posterior. Scales in 15 rows, all except the outer strongly keeled on the body, smooth on the neck and tail. Ventrals strongly angulate laterally, 151-167; anal divided; subcaudals 140-173. Bronzy or golden above, head, neck, and usually vertebral region bright green; or bright green above and on the sides; the keels on the scales dark brown or black; a black streak on each side of the head, passing through the eye; scales and head-shields sometimes black-edged; upper lip and lower parts white or yellow.

Total length 1720 millim.; tail 590.

South America east of the Andes.

a.	♂ (V. 167; C. 173).	Tobago.	A. Ludlam, Esq. [P.].
b.	♀ (V. 159; C. ?).	Trinidad.	Rev. C. Kingsley [C.].
c-d.	♂ (V. 160; C. 144) & hgr. (V. 156; C. 123).	Trinidad.	F. W. Urich, Esq. [P.].
e.	♂ (V. 164; C. 161).	Demerara.	Capt. Friend [P.].
f-h.	♂ (V. 161, 153; C. 154, 147) & ♀ (V. 166, 161; C. 153, 148).	Berbice.	
i.	♂ (V. 156; C. 163).	Bahia.	Mr. Ker [P.].
k.	♀ (V. 151; C. 150).	Brazil.	Lord Stuart [P].
l.	♀ (V. 160; C. ?).	Asuncion, Paraguay.	Dr. J. Bohls [C.].
m.	Skull.	Caracas.	
n.	Skull.	Cayenne.	

12. Leptophis ortonii.

Leptophis ortonii, *Cope, Journ. Ac. Philad.* (2) viii. 1876, p. 177.

Nearly allied to *L. nigromarginatus*, but scales smooth and ventrals non-angulate. Blue above; a coppery golden spot within the apex of many of the scales, which extends on those of the external two rows so as to cover the scale except at its base; ventrals coppery golden, the front margin sea-green; the blue scales have a black tip, and often a narrow border; head uniform green; labials yellow, with a narrow black line along the upper margin of the posterior ones.

Total length 965 millim.; tail 390.

Middle Amazon, Peru.

13. Leptophis urostictus.

Ahætulla urosticta, *Peters, Mon. Berl. Ac.* 1873, p. 606.

No loreal; nine upper labials. Scales in 13 rows, all except the outer keeled; smooth on the tail. Ventrals 163. Olive-green above, the keels black; vertebral scales dotted with black on the sides; a black streak on each side of the head, behind the eye; pale green beneath.

Bogota.

72. UROMACER.

Dendrophis, part., *Schleg. Phys. Serp.* ii. p. 220 (1837).
Uromacer, *Dum. & Bibr. Mém. Ac. Sc.* xxiii. 1853, p. 478, *and Erp. Gén.* vii. p. 719 (1854); *Jan, Elenco sist. Ofid.* p. 87 (1863).
Ahætulla, part., *Günth. Cat. Col. Sn.* p. 151 (1858).

Maxillary teeth 16 to 20, increasing in size posteriorly, the two last very large and separated from the others by a wide interspace; anterior mandibular teeth longest. Head elongate, distinct from neck; eye moderate, with round pupil. Body slender, very elongate; scales narrow, lanceolate, smooth, with apical pits, in 17 or 19 rows; ventrals without or with a very obtuse lateral keel. Tail very long; subcaudals in two rows.

Santo Domingo, West Indies.

Synopsis of the Species.

Snout twice as long as the eye, truncate at the end; rostral twice as broad as deep; scales in 17 rows; ventrals 163–176 .... 1. *catesbyi*, p. 115.
Snout twice and a half as long as the eye, subacuminate; rostral as broad as deep; scales in 17 rows; ventrals 182–193 .... 2. *frenatus*, p. 116.
Snout thrice as long as the eye, acutely pointed; rostral deeper than broad; scales in 19 rows; ventrals 192–210 .... 3. *oxyrhynchus*, [p. 116.

1. Uromacer catesbyi.

Dendrophis catesbyi, *Schleg. Phys. Serp.* ii. p. 226 (1837).
Uromacer catesbyi, *Dum. & Bibr.* vii. p. 721, pl. lxxxiii. fig. 2 (1854); *Jan, Icon. Gén.* 33, pl. iii. fig. 2 (1869).
Ahætulla catesbyi, *Günth. Cat.* p. 154 (1858).
Leptophis catesbyi, *Cope, Proc. Am. Philos. Soc.* xviii. 1879, p. 273.

Snout about twice as long as the eye, much depressed and truncate at the end. Rostral twice as broad as deep, scarcely visible from above; internasals longer than broad, shorter than the præfrontals; frontal twice as long as broad, as long as its distance from the end of the snout, slightly shorter than the parietals; nasal long and narrow, divided or semidivided; præfrontal in contact with the second labial; a small loreal between the præocular and the præfrontal; one or two præ- and two or three postoculars; temporals 1+2; eight upper labials, third, fourth, and fifth entering the eye; five lower labials in contact with the anterior chin-shields, which are shorter than the posterior. Scales in 17 rows. Ventrals 163-176; anal divided; subcaudals 162-206. Olive-green above, each scale black-edged; upper lip, two or three outer rows of scales, and lower parts greenish white; a more or less distinct white lateral streak; a blackish streak on each side of the head and neck, passing through the eye.

Total length 1310 millim.; tail 580.

Santo Domingo.

a, b, c. ♂ (V. 170, 163; C. 181, 170) & ♀ (V. 165; C. 172).	S. Domingo.	
d. ♀ (V. 163; C. 162).	Hayti.	Zoological Society.
e. ♀ (V. 170; C. 184).	Cape Hayti.	Hr. Rolle [C.].

2. Uromacer frenatus.

Ahætulla frenata, *Günth. Ann. & Mag. N. H.* (3) xv. 1865, p. 94, pl. ii. fig. B.
Uromacer inornatus, *Garm. Proc. Am. Philos. Soc.* xxiv. 1887, p. 284.

Head very long and narrow, body extremely slender; snout rather pointed, projecting, twice and a half as long as the eye. Rostral nearly as deep as broad, scarcely visible from above; internasals much longer than broad, shorter than the præfrontals; frontal rather more than twice as long as broad, a little shorter than its distance from the end of the snout or than the parietals; nasal long and narrow, divided; præfrontal in contact with the second labial; a small loreal between the præocular and the præfrontal; one præ- and two postoculars; temporals 1+2; eight upper labials, third, fourth, and fifth entering the eye; four lower labials in contact with the anterior chin-shields, which are shorter than the posterior. Scales in 17 rows. Ventrals 182-193; anal divided; subcaudals 201-205. Greyish olive above and below, punctulated or powdered with darker and with a few widely scattered small black spots; upper surface of head and nape green with a blackish lateral streak, passing through the eye, edged below by a whitish streak.

Total length 1250 millim.; tail 550.

Santo Domingo.

a. ♂ (V. 191; C. ?).	——?	(Type.)
b. ♂ (V. 190; C. ?).	Jeremie, Hayti.	Dr. J. G. Fischer.
c. ♂ (V. 193; C. 205).	S. Domingo.	

3. Uromacer oxyrhynchus.

Uromacer oxyrhynchus, *Dum. & Bibr.* vii. p. 722, pl. lxxxiii. fig. 1 (1854); *Garman, Proc. Am. Philos. Soc.* xxiv. 1887, p. 284; *Fischer, Jahrb. Hamb. Wiss. Anst.* v. 1888, p. 41, pl. iii. fig. 6.

Ahætulla oxyrhyncha, *Günth. Cat.* p. 154 (1858).
Leptophis oxyrhynchus, *Cope, Proc. Am. Philos. Soc.* xviii. 1879, p. 273.

Head very long and narrow, body extremely slender; snout acutely pointed, strongly projecting, about thrice as long as the eye. Rostral deeper than broad, scarcely visible from above; internasals much longer than broad, shorter than the præfrontals; frontal twice and a half as long as broad, shorter than its distance from the end of the snout or than the parietals; nasal long and narrow, divided or semidivided; præfrontal in contact with the second labial; a small loreal between the præocular and the præfrontal; one præ- and two postoculars; temporals 1+2; eight upper labials, third, fourth, and fifth entering the eye; five lower labials in contact with the anterior chin-shields, which are much shorter than the posterior. Scales in 19 rows. Ventrals 192-210; anal divided; subcaudals 185-202. Emerald-green above, pale green below; a white and yellow lateral streak, running along the lower series of scales; a black line, edged with white below, on each side of the head, passing through the eye.

Total length 1270 millim.; tail 550.

Santo Domingo.

a. ♀ (V. 210; C. 185).	S. Domingo.	
b. ♂ (V. 210; C. 185).	Cape Hayti.	Hr. Rolle [C.].
c. ♂ (V. 197; C. 200).	———?	

73. HYPSIRHYNCHUS.

Hypsirhynchus, *Günth. Cat. Col. Sn.* p. 48 (1858).

Maxillary teeth 14 or 15, slightly increasing in size posteriorly; anterior mandibular teeth much larger than the posterior. Head slightly distinct from neck; eye moderate or rather small, with round or subelliptic pupil. Body cylindrical; scales smooth, with apical pits, in 19 rows; ventrals rounded. Tail moderate; subcaudals in two rows.

Santo Domingo, West Indies.

1. Hypsirhynchus ferox. (PLATE VI. fig. 1.)

Hypsirhynchus ferox, *Günth. l. c.* p. 49; *Cope, Proc. Ac. Philad.* 1871, p. 218, *and Proc. Amer. Philos. Soc.* xviii. 1879, p. 273; *Fischer, Jahrb. Hamb. Wiss. Anst.* v. 1888, p. 41.
—— scalaris, *Cope, Proc. Ac. Philad.* 1862, p. 72; *Garm. Proc. Am. Philos. Soc.* xxiv. 1887, p. 284.

Snout rather elongate, acuminate, obliquely truncate, strongly projecting, and slightly turned up at the end. Rostral broader than deep, not or but scarcely visible from above; internasals as long as broad or a little longer, shorter than the præfrontals, which may be fused to a single shield; frontal not broader than the supraocular, twice as long as broad, as long as its distance from the end of the snout, as long as or a little shorter than the parietals; loreal longer than deep; one præ- and two postoculars; temporals 1+2; eight upper labials, third, fourth, and fifth entering the eye; five lower labials in contact with the anterior chin-shields, which are shorter than the posterior. Scales in 19 rows. Ventrals 166-177; anal divided; subcaudals 72-88. Pale grey-brown above, with a dorsal series of rhombic or arrow-headed dark brown markings; head with dark undulous markings above; lips and chin grey or brown; a whitish streak from the eye to the side of the neck; a more or less distinct dark lateral stripe; lower parts spotted or powdered with brown.

Total length 700 millim.; tail 140.

Santo Domingo.

a. ♀ (V. 174; C. 72).	S. Domingo *.	(Type.)	
b. ♂ (V. 167; C. 86).	Cape Hayti.	Hr. Rolle [C.].	Fischer Coll.
c. Yg. (V. 167; C. 88).	Gonaives.	Hr. Rolle [C.].	
d. Yg. (V. 172; C. 78).	Hayti.	Hr. Rolle [C.].	
e. ♀ (V. 177; C. 75).	Hayti.	Christiania Museum.	

74. DROMICUS.

Chironius, part., *Fitzing. N. Class. Rept.* p. 29 (1826).
Psammophis, part., *Schleg. Phys. Serp.* ii. p. 201 (1837).
Dromicus, part., *Bibr. in R. de la Sagra, Hist. Cuba, Erp.* p. 221 (1843); *Dum. & Bibr. Erp. Gén.* vii. p. 646 (1854); *Günth. Cat. Col. Sn.* p. 126 (1858); *Jan, Elenco sist. Ofid.* p. 66 (1863); *Bocourt, Miss. Sc. Mex., Rept.* p. 707 (1890).
Tæniophis, *Girard, Proc. Ac. Philad.* 1854, p. 226.
Liophis, part., *Cope, Proc. Ac. Philad.* 1862, p. 76.
Alsophis, *Cope, l. c.*
Ocyophis, *Cope, Proc. Am. Philos. Soc.* xxiii. 1886, p. 491.
Orophis, *Garman, Bull. Essex Inst.* xxiv. 1892, p. 86.

Fig. 11.

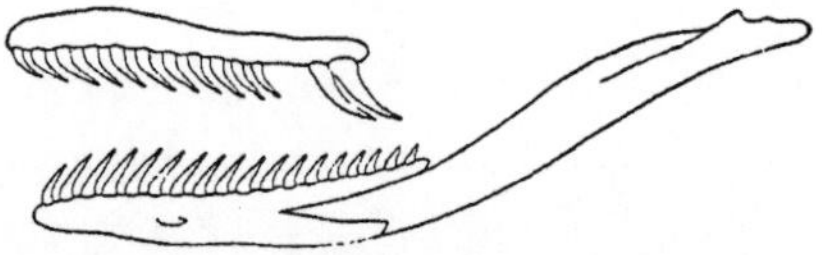

Maxillary and mandible of *Dromicus angulifer*.

Maxillary teeth 11 to 17, increasing in size posteriorly, followed, after a considerable interspace, by two strongly enlarged ones; anterior mandibular teeth much larger than the posterior. Head

---

* Purchased of H. Cuming as from Barbados. I find that all the specimens marked "Barbados, Cuming's Coll.," belong to species found in S. Domingo.

slightly distinct from neck; eye moderate or rather large, with round pupil. Body cylindrical; scales smooth, with apical pits, in 17 to 23 rows; ventrals not or but obtusely angulate laterally. Tail long; subcaudals in two rows.

West Indies, Chili and Peru.

Synopsis of the Species.

I. Subcaudals 100 or more.

    A. Fourth and fifth upper labials entering the eye; scales in 19 rows .... 1. *chamissonis*, p. 119.

    B. Third, fourth, and fifth upper labials entering the eye; anterior chin-shields shorter than the posterior.

        1. Scales in 17 rows.

Loreal present, tetragonal .... 2. *angulifer*, p. 120.
Loreal absent.................... 3. *ater*, p. 121.
Loreal present, pentagonal .... 4. *sanctæ-crucis*, p. 122.

        2. Scales in 19 rows.

Ventrals 172–189 .... 5. *antillensis*, p. 123.
Ventrals 191–210 .... 6. *leucomelas*, p. 123.

        3. Scales in 21 or 23 rows ...... 7. *rufiventris*, p. 124.

    C. Fourth labial entering the eye; posterior chin-shields no longer than the anterior; scales in 21 rows.

        8. *anomalus*, p. 125.

II. Subcaudals 80–83; scales in 19 rows. 9. *exiguus*, p. 126.

1. Dromicus chamissonis.

Coronella chamissonis, *Wiegm. N. Acta Ac. Leop.-Carol.* xvii. i. 1835, p. 246, pl. xix.
Psammophis temminckii, *Schleg. Phys. Serp.* ii. p. 218, pl. viii. figs. 14 & 15 (1837); *Guichen. in Gay, Hist. Chile, Zool.* ii. p. 84 (1848).
Dromicus temminckii, *Dum. & Bibr.* vii. p. 663 (1854); *Günth. Cat.* p. 131 (1858); *Girard, U.S. Explor. Exped., Herp.* p. 161 (1858).
Tæniophis tantillus, *Girard, Proc. Ac. Philad.* 1854, p. 227, *and in Gillis, U.S. Nav. Astr. Exped.* ii. p. 215, pl. xxxvii. figs. 7–12 (1855).
Herpetodryas biserialis, *Günth. Proc. Zool. Soc.* 1860, p. 97.
Liophis temminckii, *Cope, Proc. Ac. Philad.* 1862, p. 76, *and Journ. Ac. Philad.* (2) viii. 1876, p. 181.
Dromicus chamissonis, *Steind. Novara, Rept.* p. 65 (1867); *Peters, Mon. Berl. Ac.* 1869, p. 719; *Günth. Zool. Rec.* 1869, p. 115.
—— chamissonis, vars. dorsalis *et* habeli, *Steind. Festschr. zool.-bot. Ges. Wien*, 1876, p. 306, pl. i.

Opheomorphus chamissonis, *Cope, Proc. U.S. Nat. Mus.* xii. 1889, p. 144.
Orophis biserialis, *Garman, Bull. Essex Inst.* xxiv. 1892, p. 85.

Eye moderate, superciliary border projecting; snout moderately elongate, obtuse. Rostral broader than deep, just visible from above; internasals as long as broad, shorter than the præfrontals; frontal twice as long as broad, longer than its distance from the end of the snout, as long as the parietals; loreal as long as deep or longer; one præ- and two (rarely three) postoculars; temporals 1+2 or 2+2; eight upper labials, fourth and fifth entering the eye; four or five lower labials in contact with the anterior chin-shields, which are as long as or a little shorter than the posterior. Scales in 19 rows. Ventrals 179-225; anal divided; subcaudals 100-122. Length of tail $3\frac{1}{2}$ to 4 times in the total length. Brown or olive above, with or without black spots; a dark vertebral stripe, four or five scales wide, bordered on each side by a more or less distinct yellowish streak which begins on the canthus rostralis; this dorsal stripe may be broken up into two series of spots; labials yellowish, usually with dark borders; belly yollowish or pale olive, dark-dotted or with a light dark-edged longitudinal streak on each side; chin and throat, in the young, brown, variegated with yellow.

Total length 1160 millim.; tail 310.

Chili, Peru, Galapagos Islands.

a. ♀ (V. 193; C. 103). Chili.
b. ♂ (V. 188; C. 108). Chili. Mr. Bridge [C.].
c, d. ♀ (V. 200, 197; Colchagua.
  C. 104, 100).
e. Yg. (V. 182; C. 118). Luco Bay. Dr. Cunningham [P.].
f. Yg. (V. 209; C. 110). Charles Id., Gala- C. Darwin [P.].
  pagos. (Type of *H. biserialis*.)
g. ♂ (V. 179; C. 110). ——?
h. Skull. Chili.

2. Dromicus angulifer.

Dromicus angulifer, *Bibr. in R. de la Sagra, Hist. Cuba, Rept.* p. 222, pl. xxvii. [Coluber cantherigerus] (1843); *Dum. & Bibr.* vii. p. 670 (1854); *Günth. Cat.* p. 129 (1858); *Jan, Icon. Gén.* 23, pl. vi. (1867); *Peters, Mon. Berl. Ac.* 1869, p. 440.
—— unicolor, *Dum. & Bibr.* p. 658; *Jan, op. cit.* 24, pl. vi. fig. 2.
Alsophis vudii, *Cope, Proc. Ac. Philad.* 1862, p. 74.
—— angulifer, *Cope, l. c.* p. 76; *Garm. Proc. Am. Philos. Soc.* xxiv. 1887, p. 282.
Dromicus (Alsophis) angulifer, var. adspersus, *Gundl. & Peters, Mon. Berl. Ac.* 1864, p. 388.
—— adspersus, *Gundl. Repert. fis. Cuba*, ii. 1868, p. 116, and *Erp. Cubana*, p. 78 (1880).
Diadophis rubescens, *Cope, Proc. Am. Philos. Soc.* xxii. 1885, p. 403.
Alsophis caymanus, *Garm. l. c.* p. 276.
—— angulifer caymanus, *A. E. Brown, Proc. Ac. Philad.* 1893, p. 433.

Eye rather large, superciliary border projecting; snout rather elongate. Rostral broader than deep, just visible from above; internasals as long as broad, shorter than the præfrontals; frontal twice as long as broad, longer than its distance from the end of the snout, shorter than the parietals; loreal longer than deep; one præ- and two postoculars; temporals 1+2, anterior very large; eight upper labials, third, fourth, and fifth entering the eye; five lower labials in contact with the anterior chin-shields, which are shorter than the posterior. Scales in 17 rows. Ventrals obtusely angulate laterally, 164–180; anal usually divided; subcaudals 102–129. Length of tail 3 to $3\frac{1}{2}$ times in the total length. Olive or pale brown above, uniform or with small dark spots, or dark brown with small yellow spots; a more or less distinct dark T-shaped marking on the head, the longitudinal branch on the parietal suture; young with a dark streak on each side of the head, passing through the eye; yellowish or pale olive inferiorly.

Total length 1150 millim.

Bahamas, Cuba, Caymans.

a. ♀ (V. 173; C. ?).	Cuba.	M. Sallé [C.].
b. Yg. (V. 176; C. 108).	Cuba.	
c. ♀ (V. 180; C. 113).	Little Cayman.	Lieut. A. Carpenter [P.].
d, e. Skulls.	Cuba.	

3. Dromicus ater.

Natrix atra, *Gosse, Nat. Soj. Jamaica*, p. 228 (1851).
—— capistrata, *Gosse, l. c.* p. 371.
Dromicus ater, *Günth. Cat.* p. 127 (1858); *Jan, Icon. Gén.* 23, pl. iv. (1867).
Alsophis ater, *Cope, Proc. Ac. Philad.* 1862, p. 76; *Garm. Proc. Am. Philos. Soc.* xxiv. 1887, p. 282.
Dromicus (Ocyophis) ater, *Bocourt, Miss. Sc. Mex., Rept.* p. 713, pl. l. fig. 1 (1890).
—— ater, var. inconstans (*Jan*), *Bocourt, l. c.* p. 714, fig. 2.

Eye moderate, with projecting superciliary border; snout moderate. Rostral twice as broad as deep, scarcely visible from above; internasals as long as broad, shorter than the præfrontals; frontal once and a half to once and two thirds as long as broad, longer than its distance from the end of the snout, shorter than the parietals; no loreal, præfrontal in contact with the second and third labials; one præ- and two postoculars; temporals 1+2; eight upper labials, third, fourth, and fifth entering the eye; five lower labials in contact with the anterior chin-shields, which are shorter than the posterior. Scales in 17 rows. Ventrals 171–185; anal divided; subcaudals 144–162. Length of tail $2\frac{1}{2}$ to $2\frac{2}{3}$ times in the total length. Olive above and below, uniform or dotted with black; a more or less distinct dark line on each side of the head, passing through the eye; some specimens uniform black.

Total length 1100 millim.; tail 430.

Jamaica.

A. Uniform black or blackish olive; chin sometimes yellowish. (*N. atra*, Gosse.)

a, b. ♀ (V. 175, 181; C. 160, ?).	Jamaica.	P. H. Gosse, Esq. [P.]. (Types.)
c. ♀ (V. 183; C. 162).	Jamaica.	R. Heward, Esq. [P.].
d. ♂ (V. 178; C. ?).	Bluefields, Jamaica.	
e. ♀ (V. 173; C. 158).	W. Indies.	

B. Pale olive, with small brown or black spots, a dark temporal streak, and usually a dark streak on the suture between the parietals. (*N. capistrata*, Gosse.)

a, b, c. ♀ (V. 183; C. ?), hgr. (V. 176; C. 156), & yg. (V. 174; C. 158).	Jamaica.	P. H. Gosse, Esq. [P.]. (Types of *N. capistrata*.)
d. ♀ (V. 179; C. 148).	Jamaica.	R. Heward, Esq. [P.].
e. Hgr. (V. 178; C. 147).	Jamaica.	Capt. Parry [P.].
f. ♂ (V. 179; C. 150).	Cinchona, Jamaica.	T. D. A. Cockerell, Esq. [P.].

4. Dromicus sanctæ-crucis.

Psammophis antillensis, part., *Schleg. Phys. Serp.* ii. p. 214 (1837).
Dromicus antillensis, part., *Dum. & Bibr.* vii. p. 659 (1854).
—— antillensis, *Günth. Ann. & Mag. N. H.* (3) iv. 1859, p. 210.
Alsophis sancticrucis, *Cope, Proc. Ac. Philad.* 1862, p. 76; *Reinh. & Lütk. Vidensk. Meddel.* 1862, p. 218; *Cope, Proc. Ac. Philad.* 1868, p. 312.
—— melanichnus, *Cope, Proc. Ac. Philad.* 1862, p. 76; *Garm. Proc. Am. Philos. Soc.* xxiv. 1887, p. 283.
—— portoricensis, *Reinh. & Lütk. l. c.* p. 221.

Eye moderate; snout moderate, somewhat projecting. Rostral twice as broad as deep, not or scarcely visible from above; internasals as long as broad, shorter than the præfrontals; frontal once and three fourths or twice as long as broad, as long as or a little longer than its distance from the end of the snout, shorter than the parietals; loreal pentagonal, as long as deep; one præ- and two postoculars; temporals 1+2; eight upper labials, third, fourth, and fifth entering the eye; five lower labials in contact with the anterior chin-shields, which are shorter than the posterior. Scales in 17 rows. Ventrals 170–198; anal divided; subcaudals 120–147. Length of tail about one third of the total. Dark brown above, powdered with black, and with small yellow spots or lines which may form more or less regular cross bands; flanks and ends of ventrals blackish; a black streak, bordered with yellow, on each side of the head, passing through the eye; upper lip yellow, speckled with brown; lower parts yellow, uniform or with black dots or fine black edges to the shields.

Total length 1240 millim.; tail 410.

West Indies (St. Croix, Porto Rico, Santo Domingo).

a, b–c. ♂ (V. 195; C. 145) & ♀ (V. 195, 195; C. ?, ?).	St. Croix.	A. Newton, Esq. [P.].
d. ♂ (V. 191; C. 147).	St. Croix.	Hr. Riise [C.].

5. Dromicus antillensis.

Psammophis antillensis, part., *Schleg. Phys. Serp.* ii. p. 214 (1837).
Dromicus antillensis, part., *Dum. & Bibr.* vii. p. 659 (1854).
—— antillensis, *Günth. Cat.* p. 129 (1858); *Jan, Icon. Gén.* 25, pl. i. fig. 1 (1867).
Alsophis antillensis, *Cope, Proc. Ac. Philad.* 1862, p. 76; *Reinh. & Lütk. Vidensk. Meddel.* 1862, p. 218; *Garm. Proc. Am. Philos. Soc.* xxiv. 1887, p. 282.

Eye moderate; snout moderate, somewhat projecting. Rostral twice as broad as deep, scarcely visible from above; internasals as long as broad, shorter than the præfrontals; frontal once and two thirds or once and three fourths as long as broad, as long as or a little longer than its distance from the end of the snout, shorter than the parietals; loreal pentagonal, as long as deep or a little longer; one præ- and two postoculars; temporals $1+2$; eight upper labials, third, fourth, and fifth entering the eye; five lower labials in contact with the anterior chin-shields, which are shorter than the posterior. Scales in 19 rows. Ventrals 172–189; anal divided; subcaudals 116–142. Length of tail $2\frac{3}{4}$ to 3 times in the total length. Pale brown above, with small darker spots; a dark brown streak on each side of the head, passing through the eye; frequently a dark streak along each side of the body; upper labials and lower parts yellowish, usually brown-dotted; sometimes a brown line along each side of the belly.

Total length 1000 millim.; tail 350.

West Indies (Martinique, Dominica, Guadeloupe, St. John, St. Thomas, Vieque, Santo Domingo).

a–c. ♂ (V. 178, 175; C. ?, 121) & ♀ (V. 181; C. 119).	Dominica.	
d–f. ♂ (V. 179; C. 116) & ♀ (V. 189, 179; C. ?, 129).	Vieque.	Hr. Riise [C.].
g–n. ♂ (V. 177, 177; C. 126, 129), ♀ (V. 180; C. ?), hgr. (V. 181, 182, 184; C. 134, 133, 135), & yg. (V. 174; C. 126).	St. Thomas.	Hr. Riise [C.].
o. ♂ (V. 178; C. 122).	St. Thomas.	M. Sallé [C.].
p–q. ♂ (V. 178, 174; C. 131, 131).	St. Thomas.	
r. Skull.	W. Indies.	

6. Dromicus leucomelas.

Dromicus leucomelas, *Dum. & Bibr.* vii. p. 666 (1854); *Jan, Icon. Gén.* 25, pl. ii. fig. 1 (1867); *Günth. Ann. & Mag. N. H.* (6) ii. 1888, p. 366.
Alsophis leucomelas, *Cope, Proc. Ac. Philad.* 1862, p. 76.
—— sibonius, *Cope, Proc. Am. Philos. Soc.* xviii. 1879, p. 275; *Garm. Proc. Am. Philos. Soc.* xxiv. 1887, p. 283.

Eye moderate; snout moderate, somewhat projecting. Rostral

twice as broad as deep, scarcely visible from above; internasals as long as broad or a little longer, as long as or shorter than the præfrontals; frontal once and a half to once and three fourths as long as broad, longer than its distance from the end of the snout, as long as or a little shorter than the parietals; loreal tetra- or pentagonal, longer than deep; one (rarely two) præ- and two postoculars; temporals 1+2 or 1+1+2; eight upper labials, third, fourth, and fifth entering the eye; five lower labials in contact with the anterior chin-shields, which are shorter than the posterior. Scales in 19 rows. Ventrals obtusely angulate laterally, 191-210; anal divided; subcaudals 103-144. Length of tail $2\frac{2}{3}$ to $3\frac{2}{3}$ times in the total length. Coloration very variable: some specimens pale brown with dark brown spots or with a dark zigzag vertebral band, others yellow with black spots or black with yellow spots; a dark band on each side of the head, passing through the eye, sometimes continued along the body; belly yellow, uniform or with small black spots; hinder half of belly, and tail, frequently uniform black.

Total length 1300 millim.; tail 370.

West Indies (Antigua, Montserrat, Guadeloupe, Marie-Galante, Dominica).

a-b. Yg. (V. 196, 195; C. 105, 103).	Antigua.	
c-h. ♂ (V. 200, 205; C. 125, 132) & ♀ (V. 206, 203, 199, 203; C. 121, ?, 113, 118).	Montserrat.	J. S. Hollings, Esq. [P.].
i. ♂ (V. 203; C. 139).	Guadeloupe.	
k-n, o-q. ♂ (V. 195, 200; C. 114, 120), ♀ (V. 194, 200, 202; C. 115, ?, ?), hgr. (V. 200; C. 115), & yg. (V. 191; C. 120).	Dominica.	G. A. Ramage, Esq. [C.].

7. Dromicus rufiventris.

Dromicus rufiventris, *Dum. & Bibr.* vii. p. 668 (1854); *Günth. Cat.* p. 130 (1858); *Jan, Icon. Gén.* 25, pl. iii. fig. 1 (1867).
Alsophis rijersmæi, *Cope, Proc. Am. Philos. Soc.* xi. 1869, p. 154.
—— rufiventris, *Garm. Proc. Am. Philos. Soc.* xxiv. 1887, p. 282.
—— cinereus, *Garm. l. c.*

Eye moderate; snout projecting. Rostral much broader than deep, not or but scarcely visible from above; internasals as long as broad or a little longer, as long as or shorter than the præfrontals; frontal about twice as long as broad; loreal longer than deep; one (sometimes two) præ- and two postoculars; temporals 1+2; eight upper labials, third, fourth, and fifth entering the eye; five lower labials in contact with the anterior chin-shields, which are shorter than the posterior. Scales in 21 or 23 rows. Ventrals obtusely angulate laterally, 200-220; anal divided; subcaudals 91-122. Length of tail $3\frac{1}{4}$ to $4\frac{1}{2}$ times in the total length. Yellowish brown above, with dark brown spots or interrupted longitudinal streaks;

the spots vary exceedingly, and may be very small or so large as to nearly cover the ground-colour; a dark band on each side of the head, passing through the eye, widening on the temple, and sometimes continued along the body; lower parts yellow, more or less dotted with brown; in some specimens the hinder part of the body and the tail blackish brown.

Total length 1230 millim.; tail 270.

West Indies (Anguilla, St. Martin, St. Bartholomew, Saba, St. Eustatius, St. Kitts, Nevis).

A. Scales in 23 rows.

a. ♂ (V. 210; C. 114).	St. Eustatius.	Commodore Markham [P.].
b–h. ♀ (V. 210, 215; C. 97,97),hgr.(V.211.215, 213, 204; C. 91, 95, ?, 100), & yg. (V. 213; C. 94).	St. Kitts.	— Hogan, Esq. [P.].
i. ♀ (V. 214; C. 96).	St. Kitts.	Mus. Comp. Zoology [E.].
k. ♀ (V. 201; C. 99).	St. Bartholomew.	Mus. Comp. Zoology [E.].
l. ♂ (V. 214; C. 119).	Nevis.	Mus. Comp. Zoology [E.].
m. ♀ (V. 217; C. 98).	—— ?	

B. Scales in 21 rows.

n. ♂ (V. 202; C. 112).	Anguilla.	Mus. Comp. Zoology [E.]. (One of the types of *A. cinereus*.)
o–p. ♂ (V. 208; C. ?) & hgr. (V. 205; C. 114).	Anguilla.	W. R. Elliott [C.].

8. Dromicus anomalus.

Zamenis anomalus, *Peters, Mon. Berl. Ac.* 1863, p. 282.
Dromicus (Alsophis) anomalus, *Fischer, Jahrb. Hamb. Wiss. Anst.* v. 1888, p. 37.

Eye moderate; snout projecting, rather pointed. Rostral nearly as deep as broad, well visible from above; internasals as long as broad, a little shorter than the præfrontals; frontal once and one third to once and two thirds as long as broad, as long as or shorter than its distance from the end of the snout, shorter than the parietals; loreal longer than deep; one præ- and two postoculars; a large subocular usually separating the eye from the fifth and sixth labials; temporals $1+2$; eight upper labials, fourth entering the eye, third sometimes just touching the eye; five lower labials in contact with the anterior chin-shields, which are as long as the posterior. Scales in 21 rows. Ventrals obtusely angulate laterally, 205–219; anal divided; subcaudals 118–130. Length of tail $3\frac{2}{3}$ to $4\frac{1}{5}$ times in the total length. Pale brown above, the head-shields and scales edged with darker; young with numerous, more or less distinct, irregular dark cross bands; ventrals yellowish, with the outer ends brown.

Total length 1950 millim.; tail 550.

Santo Domingo.

a, *b*. ♀ (V. 211; C. 120) & yg.   Hayti.   Hr. Rolle [C.].
 (V. 213; C. 118).
c. ♀, head imperfect (V. 214; C.?).   ———?   Zoological Soc.

9. Dromicus exiguus.

Dromicus exiguus, *Cope, Proc. Ac. Philad.* 1862, p. 79; *Garm. Proc. Am. Philos. Soc.* xxiv. 1887, p. 282.

Eye moderate; snout prominent. Rostral nearly twice as broad as deep, scarcely visible from above; frontal once and two thirds as long as broad, longer than its distance from the end of the snout, shorter than the parietals; loreal very small; one præ- and two postoculars; eight upper labials, third, fourth, and fifth entering the eye; anterior chin-shields shorter than the posterior. Scales in 19 rows. Ventrals 137–140; anal divided; subcaudals 80–83. Above light brown or yellowish, densely punctulated with darker; the median dorsal region of a deeper shade; distant dark brown dots sometimes forming two parallel series along the back; a dark brown stripe along the fourth row of scales nearly to the end of the tail, continuous with a head-band which passes through the eye; beneath yellowish, dotted with brown, especially towards the extremities of the ventrals.

Total length 433 millim.; tail 132 (*Cope*).
West Indies (St. Thomas, St. John).

a. Yg. (V. 128; C. 83).   St. Thomas.   H.M.S. 'Challenger.'

75. LIOPHIS.

Liophis, part., *Wagl. Syst. Amph.* p. 187 (1830); *Dum. & Bibr. Erp. Gén.* vii. p. 697 (1854); *Günth. Cat. Col. Sn.* p. 42 (1858); *Jan, Arch. Zool. Anat. Phys.* ii. 1863, p. 287; *Cope, Proc. Ac. Philad.* 1862, p. 76.
Coronella, part., *Schleg. Phys. Serp.* ii. p. 50 (1837).
Xenodon, part., *Schleg. l. c.* p. 80; *Dum. & Bibr. t. c.* p. 753; *Günth. l. c.* p. 53.
Herpetodryas, part., *Schleg. l. c.* p. 173.
Dromicus, part., *Bibr. in R. de la Sagra, Hist. Cuba, Erp.* p. 221 (1843); *Dum. & Bibr.* vii. p. 646; *Günth. l. c.* p. 126; *Cope, Proc. Ac. Philad.* 1862, p. 76; *Jan, Elenco sist. Ofid.* p. 66 (1863).

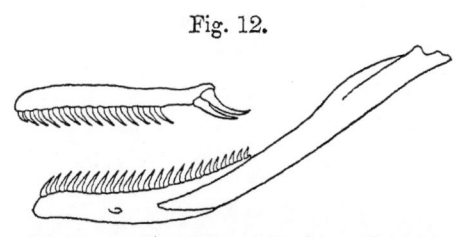

Fig. 12.

Maxillary and mandible of *Liophis pœcilogyrus*.

Maxillary teeth 14 to 24, followed, after a considerable interspace, by two strongly enlarged ones; mandibular teeth subequal. Head slightly distinct from neck; eye moderate or rather large, in one species rather small, with round pupil. Body cylindrical; scale smooth, with apical pits, in 17 or 19 rows; ventrals not or but obtusely angulate laterally. Tail moderate or long; subcaudals in two rows.

South and Central America, West Indies, South-eastern North America.

Synopsis of the Species.

I. Tail 4 to 7 times in the total length; eye moderate or rather small, its length not twice its distance from the oral border.

   A. Scales in 17 rows (exceptionally 19 in *L. tæniurus*).

      1. Seven upper labials, third and fourth entering the eye; ventrals 128, subcaudals 31 .... 1. *pygmæus*, p. 129.

      2. Eight upper labials, fourth and fifth entering the eye.

        a. Internasals as long as broad.

          α. Eye rather small; ventrals 187–195.
                                                   2. *triscalis*, p. 129.

          β. Eye moderate; ventrals 148–185.

Anterior chin-shields as long as the posterior; subcaudals 53–57 .... 3. *tæniurus*, p. 130.
Anterior chin-shields shorter than the posterior; subcaudals 70 .... 5. *fraseri*, p. 131.

        b. Internasals broader than long.

Frontal once and a half as long as broad, shorter than the parietals .... 4. *albiventris*, p. 130.
Frontal once and two thirds to twice as long as broad, as long as or a little shorter than the parietals .... 8. *melanotus*, p. 134.

   B. Scales in 19 rows.

      1. Snout not pointed.

Subcaudals 40–63 .... 6. *pœcilogyrus*, p. 131.
Subcaudals 75–80 .... 7. *perfuscus*, p. 133.

      2. Snout rather pointed; rostral largely exposed above; subcaudals 58–73.

Ventrals 149–168 .... 9. *almadensis*, p. 134.
Ventrals 169–197; uniform green above.. 10. *viridis*, p. 135.

II. Tail 4¼ to 6 times in the total length; eye rather large.

Scales in 19 rows .... 11. *typhlus*, p. 136.
Scales in 17 rows .... 12. *epinephelus*, p. 137.

III. Tail not more than 4 times in the total length; eye large or rather large; eight or nine upper labials.

A. Fourth and fifth upper labials entering the eye.

Ventrals 136–150; subcaudals 58–81 .... 13. *reginæ*, p. 137.
Ventrals 157–164; subcaudals 64–85 .... 14. *juliæ*, p. 139.
Ventrals 185–200; subcaudals 82–105 .. 15. *cursor*, p. 139.

B. Third, fourth, and fifth (or fourth, fifth, and sixth) upper labials entering the eye.

1. Eye not twice as long as its distance from the mouth.

Ventrals 140–150; subcaudals 85–106 .. 16. *andreæ*, p. 140.
Ventrals 145–165; subcaudals 110–130 .. 17. *parvifrons*, p. 141.

2. Eye twice as long as its distance from the mouth. [p. 142.
18. *melanostigma*,

IV. Tail not 4 times in the total length; eye moderate; seven upper labials, third and fourth entering the eye.

Scales in 19 rows; ventrals 131–134 .... 19. *callilæmus*, p. 142.
Scales in 17 rows; ventrals 167 .... 20. *temporalis*, p. 143.
Scales in 17 rows; ventrals 126 .... 21. *flavilatus*, p. 143.

TABLE SHOWING NUMBERS OF SCALES AND SHIELDS.

	Sc.	V.	C.	Lab.
pygmæus	17	128	31	7
triscalis	17	187–195	62–85	8
tæniurus	17–19	148–185	53–57	8
albiventris	17	138–158	50–66	8
fraseri	17	160	70	8
pœcilogyrus	19	143–179	40–63	8
perfuscus	19	182–194	75–80	8–9
melanotus	17	144–157	53–65	8
almadensis	19	149–168	58–68	8
viridis	19	169–197	60–78	8
typhlus	19	136–171	46–57	8
epinephelus	17	137–151	51–59	8
reginæ	17	136–150	58–81	8
juliæ	17	157–164	64–85	8
cursor	17	185–200	80–105	8
andreæ	17	140–150	85–106	8
parvifrons	17–19	145–165	110–130	8
melanostigma	17	150–160	90–107	8
callilæmus	19	130–134	70–110	7
temporalis	17	167	?	7
flavilatus	17	126	77	7

1. Liophis pygmæus.

Liophis pygmæus, *Cope, Proc. Ac. Philad.* 1868, p. 103.

Head slightly distinct from neck, ovate, narrowed in front; snout slightly prominent. Rostral much broader than deep, just visible from above; internasals broader than long; frontal large, elongate, longer than its distance from the tip of the snout; supraoculars rather narrow; loreal subquadrate; one præ- and two postoculars; temporals 1+2; seven upper labials, third and fourth entering the eye; chin-shields equal. Scales in 17 rows. Ventrals 128; anal divided; subcaudals 31. Above deep olive; leaden on the sides and the ends of the ventral shields; a black line from the eye to the angle of the mouth; a black occipito-nuchal collar, directed backwards; a series of black spots on each side, confluent into a band from the middle of the body to the end of the tail; below uniform yellow.

Total length 185 millim.; tail 30.

Napo or Upper Maranon.

2. Liophis triscalis.

Seba, ii. pl. xxxviii. fig. 3 (1735).
Coluber triscalis, *Linn. S. N.* i. p. 385 (1766); *Daud. Rept.* vi. p. 377 (1803).
Dromicus triscalis, *Dum. & Bibr.* vii. p. 672 (1854); *Günth. Cat.* p. 131 (1858).
Liophis triscalis, *Jan, Arch. Zool. Anat. Phys.* ii. 1863, p. 299, *and Icon. Gén.* 18, pl. iv. fig. 1 (1866).
? Liophis rufus, *Jan, l. c.* p. 301.

Eye rather small. Rostral a little broader than deep, well visible from above; internasals as long as broad, as long as or shorter than the præfrontals; frontal about once and a half as long as broad, as long as or a little longer than its distance from the end of the snout and a little shorter than the parietals; loreal as long as deep or a little deeper than long; one or two præ- and two postoculars; temporals 1+2; eight upper labials, fourth and fifth entering the eye; five lower labials in contact with the anterior chin-shields, which are as long as or a little longer than the posterior. Scales in 17 rows. Ventrals 187–195; anal divided; subcaudals 62–85. Length of tail 4½ to 6 times in the total length. Pale brown or greyish above, with three more or less distinct dark brown longitudinal streaks on the back and two on the tail; lower parts whitish, uniform or dotted with brown.

Total length 710 millim.; tail 120.

Curaçao, Venezuela, Surinam.

a. ♀ (V. 189; C. 84).	Curaçao.	E. Hartert, Esq. [C.].
b, c. ♂ (V. 191, 192; C. ?, 85).	Surinam (?).	Van Lidth de Jeude Collection.
d. ♀ (V. 192; C. 62).	Paraguay (?).	

3. Liophis tæniurus.

Liophis tæniurus, *Tschudi, Faun. Per., Herp.* p. 51, pl. v. (1845);
Jan, Arch. Zool. An. Phys. ii. 1863, p. 296, *and Icon. Gén.* 18, pl. iii. fig. 1 (1866).
Aporophis tæniurus, *Cope, Proc. Amer. Philos. Soc.* xviii. 1879, p. 277.

Eye moderate. Rostral broader than deep, scarcely visible from above; internasals as long as broad, as long as the præfrontals; frontal once and a half or once and two thirds as long as broad, a little longer than its distance from the end of the snout, a little shorter than the parietals; loreal deeper than long; one præ- and two postoculars; temporals 1+2; eight upper labials, fourth and fifth entering the eye; five lower labials in contact with the anterior chin-shields, which are as long as the posterior. Scales in 17 or 19 rows. Ventrals 148–185; anal divided; subcaudals 53–57. Length of tail 4½ to 5 times in the total length. Brown above, anteriorly with black transverse spots or cross-bands, posteriorly with two or four black longitudinal bands; labials whitish, a black lateral streak passing through the eye; a black ∧-shaped nuchal collar; belly white with black spots, some of which extend over a whole ventral shield; lower surface of tail black, with or without a light streak along each side.

Total length 630 millim.; tail 125.

Peru and Ecuador.

a. ♀ (Sc. 17; V. 164; C. 53).	Muña, Peru.	Mr. W. Davis [C.]; Messrs. Veitch [P.].
b. ♂ (Sc. 17; V. 148; C. 57).	Moyobamba, Peru.	Mr. A. H. Roff [C.].
c. ♀ (Sc. 19; V. 157; C. 55).	Guayaquil.	H. B. James, Esq. [P.].

4. Liophis albiventris.

Liophis reginæ, vars. albiventris *et* quadrilineata, *Jan, Arch. Zool. Anat. Phys.* ii. 1863, p. 294, *and Icon. Gén.* 16, pl. vi. figs. 2 & 3 (1866); *Bocourt, Miss. Sc. Mex., Rept.* p. 633, pl. xli. fig. 4 (1886).
Opheomorphus alticolus, *Cope, Proc. Ac. Philad.* 1868, p. 102.
Liophis alticolus, *Bouleng. Ann. & Mag. N. H.* (5) ix. 1882, p. 460.

Eye moderate; snout short. Rostral nearly twice as broad as deep, just visible from above; internasals broader than long, shorter than the præfrontals; frontal once and a half as long as broad, as long as or slightly longer than its distance from the end of the snout, shorter than the parietals; loreal deeper than long, sometimes very small or absent; one præ- and one or two postoculars; temporals 1+2; eight upper labials, fourth and fifth entering the eye; four or five lower labials in contact with the anterior chin-shields, which are as long as the posterior. Scales in 17 rows. Ventrals 138–158; anal divided; subcaudals 50–66. Length of tail 4 to 5½ times in the total length. Olive or greenish above, uniform or spotted with black; a dark streak on each side of the head, passing through the eye; labials white; young with a pale black-edged nuchal collar; a more or less distinct black streak, sometimes edged above and below by a series of whitish dots, along each side of the posterior half of the body to the end of the tail; sometimes two other dark lines on the posterior part of the back, confluent into one on the tail; lower parts white, uniform or black-spotted.

Total length 750 millim.; tail 140.

Ecuador.

A. Belly unspotted.

a.	♀ (V. 158; C. 60).	Olalla, 8490 ft.	E. Whymper, Esq. [C.].
b.	♂ (V. 152; C. 66).	Quito.	
c–d.	♂ (V. 151; C. 61) & ♀ (V. 155; C. 62).	W. Ecuador.	Mr. Fraser [C.].

B. Belly black-spotted.

e.	♀ (V. 156; C. 57).	Intac.	Mr. Buckley [C.].
f.	Yg. (V. 138; C. 55).	Guayaquil.	Mr. Fraser [C.].

5. Liophis fraseri. (PLATE VI. fig. 2.)

Eye moderate; snout rather elongate. Rostral twice as broad as deep, just visible from above; internasals as broad as long, a little shorter than the præfrontals; frontal nearly twice as long as broad, as long as its distance from the end of the snout, shorter than the parietals; loreal a little deeper than long; one præ- and two post-oculars; temporals 1+2; eight upper labials, fourth and fifth entering the eye; five lower labials in contact with the anterior chin-shields, which are shorter than the posterior. Scales in 17 rows. Ventrals 160; anal divided; subcaudals 70. Pale olive above, with small black spots; a dark streak on each side of the head, passing through the eye; upper lip white; four black streaks along the posterior part of the body, the two upper confluent into one on the tail; lower parts white, with scattered small black spots.

Total length 690 millim.; tail 150.

Ecuador.

a. ♀ (V. 160; C. 70).   W. Ecuador.   Mr. Fraser [C.].

6. Liophis pœcilogyrus.

Coluber M-nigrum, *Raddi, Mem. Soc. Ital. Modena*, xviii. (Fis.) 1820, p. 338.
? Coluber alternans, *Licht. Verz. Doubl. Mus. Berl.* p. 104 (1823).
Coluber pœcilogyrus, *Wied, Beitr. Nat. Bras.* i. p. 371, *and Abbild.* (1825).
—— doliatus (*non* L.), *Wied, Beitr.* p. 368, *and Abbild.*
—— bicolor, *Reuss, Mus. Senckenb.* i. 1834, p. 145, pl. viii. fig. 1.
Coronella merremii, part., *Schleg. Phys. Serp.* ii. p. 58 (1837).
Liophis merremii, part., *Dum. & Bibr.* vii. p. 708 (1854); *Günth. Cat.* p. 44 (1858); *Hensel, Arch. f. Nat.* 1868, p. 324.

Liophis merremi, *Girard*, *U.S. Explor. Exped.*, *Herp.* p. 159, pl. xi. figs. 1–6 (1858).
—— merremii, var. sublineatus, *Cope*, *Proc. Ac. Philad.* 1860, p. 252.
—— reginæ, *Burm. Reise La Plata*, i. p. 528 (1861).
Ophiomorphus doliatus, *Cope*, *Proc. Ac. Philad.* 1862, pp. 75 & 348.
Liophis subfasciatus, *Cope*, *l. c.* p. 77.
—— pœcilogyrus, *Jan*, *Arch. Zool. Anat. Phys.* ii. 1863, p. 291, *and Icon. Gén.* 17, pl. vi. fig. 1 (1866); *Bouleng. Ann. & Mag. N. H.* (5) xviii. 1886, p. 432.
—— reginæ, var. viridicyanea, *Jan*, *ll. cc.*, *Icon.* 18, pl. ii. fig. 1.
Opheomorphus meleagris, *Cope*, *Proc. Am. Philos. Soc.* xxii. 1885, p. 191.
—— meleagris doliatus, *Cope*, *Proc. Am. Philos. Soc.* xxiv. 1887, p. 57.

Eye moderate. Rostral broader than deep, visible from above; internasals as long as broad or a little broader, shorter than the præfrontals; frontal once and a half to once and two thirds as long as broad, as long as or a little longer than its distance from the end of the snout, as long as or shorter than the parietals; loreal as long as deep or deeper than long; one præ- and two postoculars; temporals 1+2; eight upper labials, fourth and fifth entering the eye; four or five lower labials in contact with the anterior chin-shields, which are as long as or longer than the posterior. Scales in 19 rows. Ventrals 143–179; anal divided; subcaudals 40–63. Length of tail 5 to 6 times in the total length. Pale brownish, olive, greenish, or red above, spotted and reticulated with black, or with more or less distinct dark cross bands, or with alternating red and pale olive transverse bands; some specimens blackish, with small pale olive spots. Young with well-marked black spots arranged symmetrically, or with broad black cross bands, continuous across the back or interrupted and alternate, and separated by narrow pale interspaces; a broad black occipito-nuchal collar. Lower parts yellowish or red, usually spotted with black.
Total length 660 millim.; tail 120.
Brazil, Uruguay, Paraguay, Argentine Republic.

a. Yg. (V. 145; C. 50).	Santarem, Amazon.	H. W. Bates, Esq. [C.].
b. ♂ (V. 179; C. 62).	Rio Janeiro.	J. MacGillivray, Esq. [C.].
c. ♂ (V. 176; C. 57).	Rio Janeiro.	Haslar Collection.
d. Hgr. (V. 174; C. 63).	Tijuca R.	R. Bennett, Esq. [P.].
e. ♀ (V. 175; C. 62).	Tijuca R., near Rio Janeiro.	W. Schaus, Esq. [P.].
f–p, q. ♂ (V. 162, 157, 162; C. 55, ?, 62), ♀ (V. 157, 161; C. 59, 56), hgr. (V. 160; C. 56), & yg. (V. 153, 152, 156, 156; C. 59, ?, 58, 62).	Porto Real, Prov. Rio Janeiro.	M. F. Hardy du Dréneuf [C.].
r–t. Hgr. ♂ (V. 146; C. 50), ♀ (V. 153; C. 52), & yg. (V. 143; C. 49).	S. José dos Campos, Prov. S. Paulo.	Mr. A. Thomson [P.].

u. ♀ (V. 154; C. 47).	Santa Cruz, Prov. Matto Grosso.	S. Moore, Esq. [P.].
v, w-y. ♀ (V. 144; C. 53) & yg. (V. 144, 150, 148; C. 50, 55, 53).	Rio Grande do Sul.	Dr. H. v. Ihering [C.].
z-a. Hgr. (V. 172, 173; C. ?, 62).	Brazil.	G. L. Conyngham, Esq. [P.].
β-γ. Hgr. (V. 155, 154; C. 50, 49).	Paraguay.	Prof. Grant [P.].
δ-θ. ♀ (V. 149, 155, 147; C. 45, 43, 43), hgr. (V. 157; C. 44), & yg. (V. 152; C. 43).	Asuncion, Paraguay.	Dr. J. Bohls [C.].
ι. ♀ (V. 151; C. ?).	Dept. of Soriano, Uruguay.	R. Havers, Esq. [C.].
κ-μ. ♂ (V. 155; C. 52) & ♀. (V. 156, 157; C. 48, 51).	Uruguay.	
ν-ξ. ♀ (V. 150, 155; C. 46, 46).	Buenos Ayres.	E. W. White, Esq. [C.].
o-σ. ♂ (V. 158; C. 50), ♀ (V. 149; C. 44), & yg. (V. 152, 149; C. 42, 46).	S. of Rio de la Plata.	Lieut. Gairdner [P.].
τ. ♀ (V. 154; C. 44).	Argentina.	
υ. Skull.	Uruguay.	

7. Liophis perfuscus.

Liophis perfuscus, *Cope, Proc. Ac. Philad.* 1862, p. 77.
Dromicus perfuscus, *Günth. Ann. & Mag. N. H.* (3) xii. 1863, p. 349.

Eye moderate; snout short. Rostral broader than deep, visible from above; internasals as long as broad, as long as or a little shorter than the præfrontals; frontal about once and a half as long as broad, as long as or a little longer than its distance from the end of the snout, shorter than the parietals; loreal deeper than long; one præ- and two postoculars; temporals $1+1$ or $1+2$; eight or nine upper labials, fourth and fifth or fifth and sixth entering the eye; five lower labials in contact with the anterior chin-shields, which are as long as or longer than the posterior. Scales in 19 rows. Ventrals obtusely angulate laterally, 182–194; anal divided; subcaudals 75–80. Length of tail $4\frac{3}{4}$ to 5 times in the total length. Brown above, vertebral region darker; lower parts greyish or yellowish brown [*].

Total length 1000 millim.; tail 200.

Barbados.

a-d. ♂ (V. 192; C. 78), ♀ (V. 189; C. 75), & yg. (V. 182, 194; C. 77, 77).	Barbados.	Col. Feilden [P.].
e. ♂ (V. 193; C. 80).	——?	College of Surgeons [P.].

---

[*] A young specimen, preserved in the Christiania Museum, is entirely black above and beneath.

8. Liophis melanotus.

Coluber melanotus, *Shaw, Zool.* iii. p. 534 (1802).
—— raninus, *Merr. Tent.* p. 106 (1820).
—— vittatus, *Hallow. Proc. Ac. Philad.* 1845, p. 242.
Dromicus melanotus, *Günth. Cat.* p. 133 (1858); *Garm. Proc. Am. Philos. Soc.* xxiv. 1887, p. 282.
Liophis vittatus, *Cope, Proc. Ac. Philad.* 1859, p. 297.
—— melanonotus, *Cope, Proc. Ac. Philad.* 1860, p. 253.
—— melanotus, *Jan, Arch. Zool. Anat. Phys.* ii. 1863, p. 298, *and Icon. Gén.* 18, pl. iii. fig. 4 (1866).

Eye moderate; snout short. Rostral nearly twice as broad as deep, just visible from above; internasals broader than long, shorter than the præfrontals; frontal once and two thirds to twice as long as broad, longer than its distance from the end of the snout, as long as or a little shorter than the parietals; loreal deeper than long; one præ- and two postoculars; temporals 1+2; eight upper labials, fourth and fifth entering the eye; four or five lower labials in contact with the anterior chin-shields, which are as long as or a little shorter than the posterior. Scales in 17 rows. Ventrals obtusely angulate laterally, 144–157; anal divided; subcaudals 53–65. Length of tail 4⅓ to 5 times in the total length. Yellowish; vertebral region (4 or 5 scales wide) black, with a white and black streak on each side; the vertebral stripe narrower on the nape, and sometimes broken up into spots; the dorso-lateral streaks usually broken up into spots on the nape; head olive above, with a black lateral streak passing through the eye; upper lip and lower parts uniform yellowish white.

Total length 610 millim.; tail 140.

Colombia, Venezuela, Trinidad, Tobago, Grenada.

a–d. ♂ (V. 148, 151; C. 64, 61) & ♀ (V. 151, 150; C. 65, 65).	Rosario de Cucuta, Colombia.	Mr. C. Webber [C.].
e. ♀ (V. 144; C. 58).	Cartagena.	Capt. Barth [P.].
f. ♂ (V. 148; C. 57).	Sta. Marta.	
g. Yg. (V. 144; C. 61).	Colombia.	
h. ♂ (V. 153; C. 53).	Venezuela.	Mr. Dyson [C.].
i. ♀ (V. 145; C. 56).	Tobago.	W. J. A. Ludlam, Esq. [P.].

9. Liophis almadensis.

Natrix almadensis, *Wagl. in Spix, Serp. Bras.* p. 30, pl. x. fig. 3 (1824).
Liophis conirostris, *Günth. Cat.* p. 46 (1858), *and Zool. Rec.* 1866, p. 126.
—— almadensis, *Cope, Proc. Ac. Philad.* 1862, p. 78; *Boettg. Zeitschr. f. ges. Naturw.* lviii. 1885, p. 228; *Bouleng. Ann. & Mag. N. H.* (5) xv. 1885, p. 194, & xviii. 1886, p. 432.
Lygophis conirostris, *Cope, Proc. Ac. Philad.* 1862, p. 76.
Liophis wagleri, part., *Jan, Arch. Zool. Anat. Phys.* ii. 1863, p. 297, *and Icon. Gén.* 18, pl. iii. figs. 2 & 3 (1866).
? Liophis verecundus, *Jan, Arch. Zool. Anat. Phys.* ii. p. 300.

Liophis (Lygophis) y-græcum, *Peters, Sitzb. Ges. nat. Fr.* 1882, p. 129.
Aporophis conirostris, *Cope, Proc. Am. Philos. Soc.* xxii. 1885, p. 191.

Eye moderate; snout short and rather pointed. Rostral broader than deep, well visible from above; internasals as long as broad or broader than long, as long as or shorter than the præfrontals; frontal nearly twice as long as broad, as long as or longer than its distance from the end of the snout, as long as the parietals; loreal deeper than long; one præ- and two postoculars; temporals 1+2; eight upper labials, fourth and fifth entering the eye; five lower labials in contact with the anterior chin-shields, which are as long as the posterior. Scales in 19 rows. Ventrals 149–168; anal divided; subcaudals 58–68. Length of tail 4½ to 5 times in the total length. Olive above, with four longitudinal series of dark spots; a light streak along the posterior half (sometimes the whole) of the body and tail; head brown, with a light V- or Y-shaped marking on the parietals; upper lip yellowish, the shields usually with black sutures; belly yellowish, spotted or checkered with black.

Total length 500 millim.; tail 105.

Brazil, Paraguay.

a.	♀ (V. 155; C. 68).	Brazil.	J. S. Bowerbank, Esq. [P.].	(Types of *L. conirostris*.)
b.	♀ (V. 160; C. 68).	Bahia.	Haslar Collection.	
c, d, e.	♀ (V. 162; C. 63) & yg. (V. 155, 157; C. 64,?).	Bahia.	Dr. O. Wucherer [C.].	
f-g.	♂ (V. 149; C. ?) & ♀ (V. 163; C. 62).	S. José dos Campos, Prov. S. Paulo.	Mr. A. Thomson [P.].	
h-i.	♀ (V. 156, 154; C. 58, ?).	Rio Grande do Sul.	Dr. H. v. Ihering [C.].	
k.	Yg. (V. 156; C. 63).	Paraguay.	Lord Dormer [P.].	

10. Liophis viridis.

Liophis viridis, *Günth. Ann. & Mag. N. H.* (3) ix. 1862, p. 58, pl. ix. fig. 2.
—— typhlus, var. prasina, *Jan, Arch. Zool. Anat. Phys.* ii. 1863, p. 288, *and Icon. Gén.* 18, pl. iv. fig. 3 (1866).
Opheomorphus viridis, *Cope, Proc. Ac. Philad.* 1868, p. 102.

Eye moderate; snout short and rather pointed. Rostral broader than deep, well visible from above; internasals as long as broad, as long as or shorter than the præfrontals; frontal once and two thirds to twice as long as broad, longer than its distance from the end of the snout, as long as or a little longer than the parietals; loreal as long as deep or deeper than long; one præ- and two postoculars; temporals 1+2; eight upper labials, fourth and fifth entering the eye; five lower labials in contact with the anterior chin-shields, which are as long as or a little longer than the posterior. Scales in 19 rows. Ventrals obtusely angulate laterally, 169-197; anal divided; subcaudals 60-78. Length of tail 4 to 5 times in the total length. Uniform green above, upper lip and lower parts uniform yellowish white; a black nuchal band may be present in the young.

Total length 550 millim.; tail 110.

Brazil, Argentine Republic.

a. ♀ (V. 175; C. 60).	S. America.	(Types.)
b. ♀ (V. 178; C. 66).	Pernambuco.	
c. Yg. (V. 169; C. 73).	Pernambuco.	W. A. Forbes, Esq. [P.].
d. Yg. (V. 182; C. 78).	Rio Janeiro.	D. Wilson Barker, Esq. [P.].
e. ♀ (V. 197; C. ?).	Salta, Buenos Ayres.	Herbert Druce, Esq. [P.].

11. Liophis typhlus.

Coluber typhlus, *Linn. S. N.* i. p. 378 (1766); *Daud. Rept.* vii. p. 135 (1803).
Natrix forsteri, *Wagl. in Spix, Serp. Bras.* p. 16, pl. iv. fig. 1 (1824).
Xenodon typhlus, *Schleg. Phys. Serp.* ii. p. 94 (1837); *Dum. & Bibr.* vii. p. 760 (1854); *Günth. Cat.* p. 57 (1858).
Liophis typhlus, *Jan, Arch. Zool. Anat. Phys.* ii. 1863, p. 300, *and Icon. Gén.* 18, pl. iv. fig. 2 (1866); *Boettg. Zeitschr. f. ges. Naturw.* lviii. 1885, p. 229.
Opheomorphus typhlus, *Cope, Proc. Ac. Philad.* 1868, p. 102.
Xenodon isolepis, *Cope, Proc. Am. Philos. Soc.* xi. 1869, p. 155.
Opheomorphus brachyurus, *Cope, Proc. Am. Philos. Soc.* xxiv. 1887, p. 57.

Eye rather large; snout short. Rostral broader than deep, visible from above; internasals as long as broad or slightly broader than long, as long as or shorter than the præfrontals; frontal once and a half as long as broad, as long as or a little longer than its distance from the end of the snout, as long as the parietals; loreal deeper than long; one præ- and two postoculars; temporals 1+2; eight upper labials, fourth and fifth entering the eye; five lower labials in contact with the anterior chin-shields, which are as long as the posterior. Scales in 19 rows. Ventrals obtusely angulate laterally, 136-171; anal divided; subcaudals 46-57. Length of tail 4½ to 6 times in the total length. Olive or green above, uniform or with small black spots, or some or all of the scales with darker or lighter edges; young with a Λ-shaped black nuchal band; upper lip and lower parts white; belly usually uniform, rarely with very small dark spots.

Total length 650 millim.; tail 120.

South America east of the Andes.

a. ♂ (V. 144; C. 55).	British Guiana.	
b. ♀ (V. 147; C. 54).	Berbice.	Lady Essex [P.].
c. Hgr. (V. 148; C. 55).	Vryheids Lust, Demerara.	Rev. W. Turner [C.].

d. ♂ (V. 155; C. 55).	Bahia.	Dr. O. Wucherer [C.].
e. ♀ (V. 171; C. 52).	Corumba, Prov. Matto Grosso.	S. Moore, Esq. [P.].
f. ♂ (V. 156; C. 54).	Chapada Plateau, Prov. Matto Grosso.	S. Moore, Esq. [P.].
g. ♂ (V. 150; C. 46).	S. Lorenzo, Rio Grande do Sul.	Dr. H. v. Ihering [C.].
h–i. Hgr. (V. 149; C. 56) & yg. (V. 153; C. 48).	Moyobamba, N.E. Peru.	Mr. A. H. Roff [C.].
k. Skull.	Cayenne.	

12. Liophis epinephelus.

Liophis epinephelus, *Cope, Proc. Ac. Philad.* 1862, p. 78, *and Journ. Ac. Philad.* viii. 1876, p. 136; *Günth. Biol. C.-Am., Rept.* p. 107, pl. xxxvii. fig. B (1893).

Eye rather large; snout moderate. Rostral broader than deep, visible from above; internasals as long as broad or a little broader than long, as long as or shorter than the præfrontals; frontal once and one third to once and two thirds as long as broad, as long as or longer than its distance from the end of the snout, as long as or a little shorter than the parietals; loreal deeper than long; one præ- and two postoculars; temporals 1+2; eight upper labials, fourth and fifth entering the eye; five lower labials in contact with the anterior chin-shields, which are as long as the posterior. Scales in 17 rows. Ventrals 137–151; anal divided; subcaudals 51–59. Length of tail 4¼ to 5 times in the total length. Olive, greyish, or purplish above, with black cross bands; the scales between the black cross bands edged with white or red; upper surface of head dark olive or dark grey, usually with a black spot on each parietal shield; upper lip white; belly white, usually with black spots; outer ends of ventrals grey.

Total length 710 millim.; tail 150.

Costa Rica, Colombia.

a. ♂ (V. 143; C. 58).	Cartago, Costa Rica.	
b–g. ♀ (V. 145, 147, 146, 149, 149; C. 57, 57, 52, 51, 59) & hgr. (V. 141; C. 55).	Irazu, Costa Rica.	F. D. Godman Esq. [P.].
h. ♂ (V. 151; C. 53).	Rio Fusio, Costa Rica.	Mr. H. Rogers [C.]; F. D. Godman, Esq. [P.].
i–n. ♂ (V. 137, 142; C. 53, 55), ♀ (V. 139, 140; C. 59, 59), & yg. (V. 142, 147; C. 59, 58).	Chiriqui.	G. C. Champion, Esq. [C.].

13. Liophis reginæ.

Coluber reginæ, *Linn. Mus. Ad. Frid.* p. 24, pl. xiii. fig. 3 (1754) *and S. N.* i. p. 378 (1766); *Daud. Rept.* vii. p. 172 (1803).
? Coluber violaceus, *Lacép. Serp.* pp. 116, 172, pl. viii. fig. 1 (1789).
Coluber graphicus, *Shaw, Zool.* iii. p. 474 (1802).

Natrix semilineata, *Wagl. in Spix, Serp. Bras.* p. 33, pl. xi. fig. 2 (1824).
Coronella reginæ, part., *Schleg. Phys. Serp.* ii. p. 61 (1837).
Liophis reginæ, part., *Dum. & Bibr.* vii. p. 704 (1854); *Jan, Arch. Zool. Anat. Phys.* ii. 1863, p. 293, *and Icon. Gén.* 16, pl. vi. fig. 1 (1866).
—— reginæ, *Günth. Cat.* p. 46 (1858).
—— tæniurus (*non Tsch.*), *Günth. l. c.* pp. 46, 246.
—— merremii, part., *Günth. l. c.* p. 44.
—— reginæ, var. ornata, *Jan, ll. cc.* 18, pl. ii. fig. 2.

Eye rather large; snout short. Rostral broader than deep, visible from above; internasals broader than long, shorter than the præfrontals; frontal once and two thirds to twice as long as broad, much longer than its distance from the end of the snout, as long as the parietals or slightly shorter; loreal deeper than long; one præ- and two postoculars; temporals 1+2; eight upper labials, fourth and fifth entering the eye; five lower labials in contact with the anterior chin-shields, which are as long as or a little longer than t h posterior. Scales in 17 rows. Ventrals 136–150; anal divided; subcaudals 58–81. Length of tail 3 to 4 times in the total length. Olive or greenish above, which colour extends to the ends of the ventrals; usually, each scale edged with black, forming a reticulate pattern; a more or less distinct black streak along each side of the tail; upper lip yellow; young with a yellowish collar and an oblique yellowish, black-edged band on each side of the head, from the parietal shield to the last upper labial; traces of these markings are usually to be found in the adult; belly yellowish, usually with black spots.

Total length 650 millim.; tail 165.

Tropical South America.

a. ♂ (V.139; C.78).	Trinidad.	J. H. Hart, Esq. [P.]
b, c–g, h, i. ♂ (V.139,144, 141,140; C. 72, 71, ?,?), ♀ (V. 144, 140, 139; C. ?, 74, 76), & yg. (V. 136; C. 73).	Demerara.	
k–m, n. ♂ (V.143; C.77) & ♀ (V.144,146,141; C. 72, 68, ?).	Berbice.	
o–p. ♂ (V.148; C.71) & ♀ (V. 140; C. 74).	Para.	
q. ♂ (V. 145; C. ?).	Para.	R. Graham, Esq. [P.].
r–s. ♂ (V. 149, 138; C. 80, 62).	Bahia.	Dr. O. Wucherer [C.].
t, u. ♀ (141, 141; C. 70, 58).	Moyobamba, N.E. Peru.	Mr. A. H. Roff [C.].
v. ♀ (V.144; C.72).	Sarayacu, N.E. Peru.	Mr. W. Davis [C.]; Messrs. Veitch [P.].
w–x. ♀ (V. 148; C. ?) & yg. (V. 142; C. 71).	Guayaquil.	Mr. Fraser [C.].
y. ♀ (V. 140; C. 76).	S. America.	
z. ♂ (V. 139; C. 75).	——?	(Type of *C. graphicus.*)

Zamenis ater, Günth. Ann. & Mag. N. H. (4) ix. 1872, p. 22, stated by the donor to have been obtained at, and described as from, Biskra, Algeria, is a snake closely allied to *Liophis reginæ*, of which it is probably a melanotic (insular?) variety, distinguished by the somewhat more elongate body, the blackish colour of the upper parts, and the uniform yellowish belly.

a-c. ♂ (V. 156, 148; C. ?, 60) J. Brenchley, Esq. [C.].
& ♀ (V. 150; C. 62). (Types of *Z. ater*.)

14. Liophis juliæ.

Aporophis juliæ, *Cope, Proc. Am. Philos. Soc.* xviii. 1879, p. 274.
Dromicus juliæ, *Garm. Proc. Am. Philos. Soc.* xxiv. 1887, p. 281.
Liophis juliæ, *Günth. Ann. & Mag. N. H.* (6) ii. 1888, p. 365.

Eye rather large; snout short. Rostral broader than deep, visible from above; internasals broader than long, shorter than the præfrontals; frontal once and a half to once and two thirds as long as broad, longer than its distance from the end of the snout, shorter than the parietals; loreal deeper than long; one præ- and two postoculars; temporals 1+2; eight upper labials, fourth and fifth entering the eye; five lower labials in contact with the anterior chin-shields, which are as long as or a little longer than the posterior. Scales in 17 rows. Ventrals 157-164; anal divided; subcaudals 64-85. Length of tail $3\frac{1}{3}$ to $3\frac{2}{3}$ times in the total length. Black above, each scale with a pale olive or whitish spot; head black, with pale brownish variegations; no light nuchal collar in the young; upper labials white, lower parts white, uniform or spotted or checkered with black.

Total length 610 millim.; tail 170.

West Indies (Dominica and Marie Galante).

a-h. ♂ (V. 163; C. 64), ♀ (V. Dominica. G. A. Ramage, Esq. [C.].
157, 163, 161, 160, 163; C.
79, 80, 75, 79, 82), hgr. (V.
162; C. 79), & yg. (V. 158;
C. 85).

15. Liophis cursor.

Coluber cursor, *Lacép. Serp.* pp. 96, 281, pl. xiv. (1789); *Bonnat. Encycl. Méth., Ophiol.* p. 27 (1789); *Daud. Rept.* vi. p. 411 (1803); *Moreau de Jonnès, Bull. des Sc. Soc. Philom.* 1818, p. 111.
—— fugitivus, *Donnd. Zool. Beitr.* iii. p. 206 (1798).
—— ornatus, *Shaw, Zool.* iii. p. 477 (1802).
Herpetodryas cursor, part., *Schleg. Phys. Serp.* ii. p. 199 (1837).
Dromicus cursor, part., *Dum. & Bibr.* vii. p. 650 (1854); *Jan, Icon. Gén.* 23, pl. v. fig. 3 (1867).
—— fugitivus, part., *Günth. Cat.* p. 132 (1858); *Bocourt, Miss. Sc. Mex., Rept.* p. 708, pl. l. fig. 4 (1890).
? Liophis putnamii, *Cope, Proc. Ac. Philad.* 1862, p. 78.
Dromicus cursor, *Garm. Proc. Am. Philos. Soc.* xxiv. 1887, p. 280.
—— ornatus, *Garm. l. c.* p. 281.
Liophis ornatus, *Cope, Proc. U.S. Nat. Mus.* xii. 1889, p. 141.
—— fugitivus, *Bouleng. Proc. Zool. Soc.* 1891, p. 353.

Eye rather large; snout moderate. Rostral broader than deep, just visible from above; internasals a little broader than long, shorter than the præfrontals; frontal twice as long as broad, as long as or longer than its distance from the end of the snout, a little shorter than the parietals; loreal deeper than long; one præ- and two postoculars; temporals 1+2; eight upper labials, fourth and fifth entering the eye; five lower labials in contact with the anterior chin-shields, which are as long as the posterior. Scales in 17 rows. Ventrals obtusely angulate laterally, 185–200; anal divided; subcaudals 80–105. Length of tail 3¾ to 4 times in the total length. Black and olive or black and yellow above, usually with a light streak along each side of the hinder part of the body, sometimes extending to the nape; upper lip and lower parts yellow; usually a black spot or border at the outer end of each ventral.

Total length 830 millim.; tail 220.

West Indies (Guadeloupe, Martinique, St. Lucia).

a–b. ♀ (V. 187, 188; C. 97, ?).	Guadeloupe.	
c–f. ♀ (V. 192, 188, 196; C. 88, 89, 88) & hgr. (V. 193; C. 82).	St. Lucia.	Miss Alexander [P.].
g. ♀ (V. 191; C. 84).	St. Lucia.	Mus. Comp. Zool. [E.].
	(One of the types of *Dromicus ornatus*.)	
h. ♂ (V. 191; C. 86).	St. Lucia.	G. A. Ramage, Esq. [C.].
i. ♀ (V. 189; C. 86).	St. Lucia.	Haslar Collection.
k. ♂ (V. 194; C. 85).	St. Lucia.	
l. ♂ (V. 194; C. ?).	Caracas (??).	
m. ♀ (V. 191; C. 85).	——?	(Type of *Coluber ornatus*.)
n. ♀ (V. 189; C. 82).	——?	Dr. Günther [P.].
o. ♂ (V. 200; C. 86).	——?	
p. Skull.	Martinique.	

16. Liophis andreæ.

Herpetodryas cursor, part., *Schleg. Phys. Serp.* ii. p. 199 (1837).
Dromicus cursor, *Bibron, in R. de la Sagra, Hist. Cuba, Erp.* p. 225, pl. xxviii. (1843).
—— cursor, part., *Dum. & Bibr.* vii. p. 650 (1854); *Jan, Icon. Gén.* 23, pl. v. fig. 2 (1867).
—— fugitivus, part., *Günth. Cat.* p. 132 (1858); *Cope, Proc. Am. Philos. Soc.* xi. 1869, p. 164.
Liophis andreæ, *Reinh. & Lütk. Vidensk. Meddel.* 1862, p. 214; *Bouleng. Proc. Zool. Soc.* 1891, p. 354.
Dromicus fugitivus, *Gundlach, Erp. Cubana,* p. 79 (1880).
—— cubensis, *Garm. Proc. Am. Philos. Soc.* xxiv. 1887, p. 281.

Eye rather large; snout moderate. Rostral broader than deep, just visible from above; internasals as long as broad, shorter than the præfrontals; frontal twice as long as broad, longer than its distance from the end of the snout, as long as the parietals; loreal deeper than long; one præ- and two postoculars; temporals 1+2; eight upper labials, third, fourth, and fifth entering the eye; four or five lower labials in contact with the anterior chin-shields, which are much shorter than the posterior. Scales in 17 rows. Ventrals 140-150; anal divided; subcaudals 85-106. Blackish above, with an interrupted yellow line or series of yellow spots on each side, commencing from the tip of the snout; upper lip white; ventrals, subcaudals, and outer row of scales white, edged with black.

Total length 655 millim.; tail 215.

Cuba.

a. ♀ (V. 143; C. 89).	Cuba.	M. Sallé [C.].
b-c. ♀ (V. 150, 148; C. 90, 92).	Cuba.	

17. Liophis parvifrons.

Dromicus fugitivus, part., *Günth. Cat.* p. 132 (1858).
—— lineatus, part., *Günth. l. c.* p. 134.
—— parvifrons, *Cope, Proc. Ac. Philad.* 1862, p. 79, *and Proc. Amer. Philos. Soc.* xviii. 1879, p. 273; *Fischer, Jahrb. Hamb. Wiss. Anst.* v. 1888, p. 40.
—— protenus, *Jan, Icon. Gén.* 25, pl. iii. fig. 2 (1867).
Leptophis frenatus, *Fischer, Oster-Progr. Akad. Gymn. Hamb.* 1883, p. 8, pl. —. figs. 9-11 *.

Eye rather large; snout rather short. Rostral broader than deep, just visible from above; internasals as long as broad or a little broader than long, shorter than the præfrontals; frontal twice as long as broad, longer than its distance from the end of the snout, as long as or a little shorter than the parietals; loreal nearly as long as deep; one præ- and two postoculars; temporals 1+1 or 1+2; eight upper labials, third, fourth, and fifth entering the eye; five lower labials in contact with the anterior chin-shields, which are shorter than the posterior. Scales in 19 (exceptionally 17) rows. Ventrals 145-165; anal divided; subcaudals 110-130. Length of tail $2\frac{1}{3}$ to $2\frac{3}{4}$ times in the total length. Brown or dark olive above, with two light longitudinal streaks and a more or less distinct darker vertebral line; the olive of the sides extending to the ends of the ventrals; upper lip and lower parts whitish, the labials, the gular region, and the anterior ventrals usually dotted with black. Specimen *h* uniform black.

Total length 670 millim.; tail 250.

Santo Domingo.

a-b. ♂ (V. 158, 156; C. 123, 120).	Hayti.	Zoological Society.
c-f, g. ♀ (V. 152, 155, 155, 153; C. 110, 114, ?, 112), & yg. (V. 150; C. 128).	Hayti.	Hr. Rolle [C.].
h. ♂ (V. 145; C. 123).	S. Domingo.	M. Sallé [C.].
i. ♀ (V. 155; C. ?).	—? [Barbados].	

---

* Through the kindness of Dr. F. Müller, I have examined one of the type specimens, stated to be from Sierra Leone, preserved in the Basle Museum.

18. Liophis melanostigma.

Natrix melanostigma, *Wagl. in Spix, Serp. Bras.* p. 17, pl. iv. fig. 2 (1824).
Dromicus pleii (*non D. & B.*), *Günth. Cat.* p. 128 (1858).
—— melanostigma, *Jan, Icon. Gén.* 24, pl. v. fig. 3 (1867); *Bouleng. Ann. & Mag. N. H.* (5) xviii. 1886, p. 433.
Liophis reginæ, part., *Steind. Sitzb. Ak. Wien,* lv. 1867, p. 268.
Aporophis cyanopleurus, *Cope, Proc. Am. Philos. Soc.* xxii. 1885, p. 191.

Eye large; snout moderate. Rostral broader than deep, just visible from above; internasals a little broader than long, shorter than the præfrontals; frontal about once and two thirds as long as broad, as long as its distance from the end of the snout or a little longer, as long as the parietals; loreal a little deeper than long; one præ- and two postoculars; temporals 1+2; eight (rarely nine) upper labials, third, fourth, and fifth (or fourth, fifth, and sixth) entering the eye; four or five lower labials in contact with the anterior chin-shields, which are much shorter than the posterior. Scales in 17 rows. Ventrals 150–160; anal divided; subcaudals 90–107. Tail about one third of the total length. Olive above; nape with two more or less distinct yellowish longitudinal bands; a lateral series of light and black dots on the fourth row of scales; sides and ends of ventrals plumbeous grey; upper lip yellowish, edged with black above; ventrals yellowish, edged with grey or black, and with a black dot or short streak on each side.

Total length 770 millim.; tail 250.

Brazil.

a. ♂ (V. 160; C. 107).	Rio Janeiro.	A. Fry, Esq. [P.].
b. ♂ (V. 153; C. 100).	Rio Janeiro.	Dr. Cunningham [P.].
c. ♀ (V. 152; C. 98).	Rio Grande do Sul.	Dr. H. v. Ihering [C.].

19. Liophis callilæmus. (PLATE VI. fig. 3.)

Natrix callilæma, *Gosse, Nat. Soj. Jamaica,* p. 384 (1851).
Dromicus callilæma, *Günth. Cat.* p. 131 (1858), *and Ann. & Mag. N. H.* (3) xii. 1863, p. 357.
? Alsophis funereus, *Cope, Proc. Ac. Philad.* 1862, p. 77.

Eye moderate; snout slightly prominent. Rostral broader than deep, scarcely visible from above; internasals as long as broad or a little broader, shorter than the præfrontals; frontal a little longer than broad, nearly twice as broad as the supraocular, as long as its distance from the end of the snout, shorter than the parietals; loreal small, square; one præ- and two postoculars; temporals 1+2; seven upper labials, third and fourth entering the eye; four lower labials in contact with the anterior chin-shields, which are as long as or a little shorter than the posterior. Scales in 19 rows. Ventrals 130–134; anal divided; subcaudals 70–110. Length of tail 2½ to 3¼ times in the total length. Brown above, with darker dots;

a blackish streak on the middle of the head and along the anterior part of the back and a blackish lateral streak on the head and body, passing through the eye; lips and gular region yellowish, variegated with brown; belly uniform yellowish. Full-grown specimens nearly uniform brown.

Total length 410 millim.; tail 125.

Jamaica.

a, b–c. Hgr. ♂ (V. 132; C. 110) & yg. (V. 131, 134; C. 97, 109).	Bluefields, Jamaica.	P. H. Gosse, Esq. [P.]. (Types.)
d. ♀ (V. 132; C. 70).	Jamaica.	R. Heward, Esq. [P.].
e, f. Hgr. ♂ (V. 133; C. 108) & yg. (V. 130; C. 107).	W. Indies.	

20. Liophis temporalis.

Dromicus temporalis, *Cope, Proc. Ac. Philad.* 1860, p. 370; *Garm. Proc. Am. Philos. Soc.* xxiv. 1887, p. 281.

Mouth very inferior; rostral plate prominent, but barely visible from above; nasals distinct; internasals small; frontal broad; loreal confluent with præocular; one or two postoculars; seven upper labials, third and fourth entering the eye; chin-shields equal. Scales in 17 rows. Ventrals 167; anal divided; (tail mutilated). Chocolate-brown above, with a yellow lateral streak, edged with blackish above, running along the sutures of the two outer rows of scales; a blackish plumbeous band along the outer row of scales and the outer ends of the ventrals; indications of a yellow collar; belly yellow.

Length of head and body 430 millim.

Cuba.

21. Liophis flavilatus.

Dromicus flavilatus, *Cope, Proc. Ac. Philad.* 1871, p. 222, *and Proc. Am. Philos. Soc.* xiv. 1877, p. 65; *Garm. N. Am. Rept.* p. 58 (1883).

Closely allied to *L. callilæmus*, but scales in 17 rows. Ventrals 126; subcaudals 77. Colour above a rich golden brown, the scales of the two lower rows on each side broadly gold-edged; head dark brown, darkest behind, with numerous but obscure paler vermiculations; sides of head paler, with a reddish-brown band from the rostral plate through the eye to the middle of the last labial; labials whitish, with black dots on the posterior; below white, lower labials sparsely black-dotted; a pair of pale dots on the common occipital suture.

S.E. United States (N. Carolina to Florida).

76. CYCLAGRAS.

Xenodon, part., *Dum. & Bibr. Erp. Gén.* vii. p. 753 (1854).
Leiosophis, part., *Jan, Arch. Zool. Anat. Phys.* ii. 1863, p. 320.
Cyclagras, part., *Cope, Proc. Am. Philos. Soc.* xxii. 1885, p. 185.

Maxillary teeth 18, the last two very strongly enlarged and separated from the others by a short interspace; mandibular teeth subequal. Head distinct from neck; eye moderate, with round pupil; a series of suboculars. Body moderately elongate, cylindrical; scales smooth, with apical pits, in 19 rows; ventrals rounded. Tail rather long; subcaudals in two rows.

South America.

1. Cyclagras gigas.

Xenodon gigas, *Dum. & Bibr.* vii. p. 761 (1854).
Leiosophis gigas, *Jan, Arch. Zool. Anat. Phys.* ii. 1863, p. 320, *and Icon. Gén.* 48, pl. iii. fig. 6 (1876), & 50, pl. ii. figs. 25-27 (1881).
Cyclagras gigas, *Cope, Proc. Am. Philos. Soc.* xxii. 1885, p. 185.

Rostral nearly as deep as broad, just visible from above; internasals much shorter than the præfrontals; frontal as long as broad, much shorter than the parietals; loreal nearly as long as deep; a præocular, two postoculars, and three suboculars, separating the eye from the labials; temporals 2+2; eight upper labials; four or five lower labials in contact with the anterior chin-shields, which are a little longer than the posterior. Scales in 19 rows. Ventrals 153-170; anal entire; subcaudals 60-87. Yellowish or reddish brown above, with broad black cross bands or rings; a black stripe from the eye to the side of the neck; anterior part of belly with three longitudinal series of brown dots or small round spots.

Total length 2050 millim.; tail 550.

Argentina, Brazil, Bolivia.

a. ♂ (V. 170; C. 87).	Para.	
b. ♂ (V. 158; C. 81).	Bolivia.	M. Suarez [P.].

77. XENODON.

Ophis, *Wagl. in Spix, Serp. Bras.* p. 47 (1824), *and Syst. Amph.* p. 172 (1830).
Xenodon, part., *Boie, Isis,* 1827, p. 540; *Schleg. Phys. Serp.* ii. p. 80 (1837); *Dum. & Bibr. Erp. Gén.* vii. p. 753 (1854); *Günth. Cat. Col. Sn.* p. 53 (1858); *Jan, Arch. Zool. Anat. Phys.* ii. 1863, p. 316.
Xenodon, *Günth. Ann. & Mag. N. H.* (3) xii. 1863, p. 353.
Acanthophallus, *Cope, Am. Nat.* 1893, p. 482.

Maxillary short, with 6 to 15 teeth, followed, after an interspace, by a pair of very strongly enlarged fangs; mandibular teeth subequal. Head distinct from neck; eye large, with round pupil. Body cylindrical or depressed; scales disposed obliquely, smooth, with apical pits, in 19 or 21 rows; ventrals rounded or obtusely angulate. Tail short or moderate; subcaudals in two rows.

Tropical America.

Fig. 13.

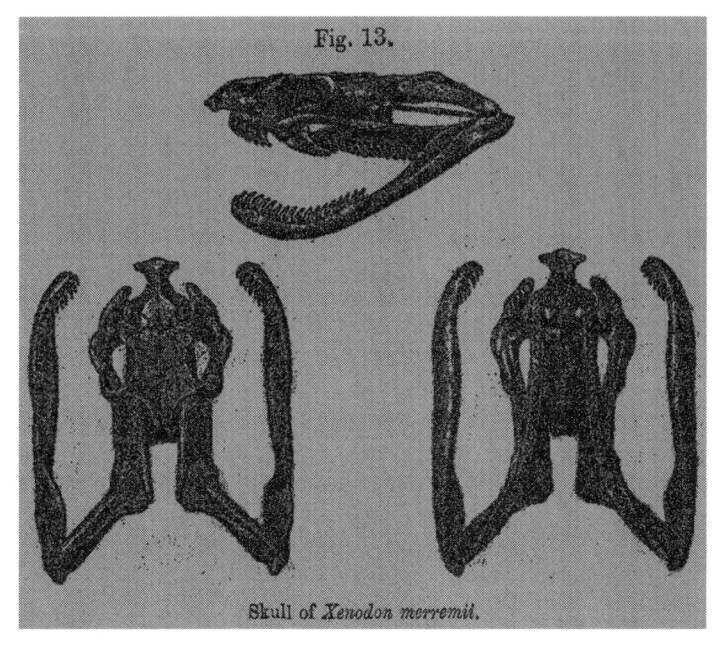

Skull of *Xenodon merremii*.

Synopsis of the Species.

I. Maxillary teeth 13 to 15+2; rostral shield not twice as broad as deep.

    A. Ventrals not angulate laterally, 130–157; anal entire; subcaudals 35–50; scales in 19 rows (rarely 21).

Rostral once and a half to once and two
  thirds as broad as deep .... 1. *colubrinus*, p. 146.
Rostral once and one third as broad as deep. 2. *suspectus*, p. 147.

    B. Ventrals obtusely angulate laterally, 152–176; anal usually divided; subcaudals 50–69.

Scales in 19 rows; rostral as deep as broad. 3. *guentheri*, p. 147.
Scales in 21 rows; rostral broader than deep. 4. *neuwiedii*, p. 148.

II. Maxillary teeth 6 to 12+2; rostral shield twice as broad as deep; anal usually divided.

Scales in 21 rows; 8 upper labials; maxillary teeth 10 to 12+2 .... 5. *severus*, p. 149.
Scales in 19 rows; usually 7 upper labials;
  maxillary teeth 6 or 7+2.. .... 6. *merremii*, p. 150.

1. Xenodon colubrinus.

Coluber rabdocephalus, part., *Wied, Beitr. Nat. Bras.* i. p. 351 (1825), *and Abbild.* (1831).
Xenodon rhabdocephalus, part., *Schleg. Phys. Serp.* ii. p. 87 (1837); *Dum. & Bibr.* vii. p. 758 (1854); *Jan, Arch. Zool. Anat. Phys.* ii. 1863, p. 317; *Günth. Biol. C.-Am., Rept.* p. 114 (1894).
—— colubrinus, *Günth. Cat.* p. 55 (1858), *and Ann. & Mag. N. H.* (3) xii. 1863, p. 353, pl. v. fig. E, *and Zool. Rec.* 1865, p. 126.
—— severus, part., *Jan, l. c.* p. 317, *and Icon. Gén.* 19, pl. iii. fig. A (1866).
—— bertholdi, *Jan, ll. cc.* p. 318, pl. iv. fig. 2.
—— angustirostris, *Peters, Mon. Berl. Ac.* 1864, p. 390, & 1873, p. 607.
——, sp., *F. Müller, Verh. nat. Ges. Basel,* vi. 1878, p. 663.
—— severus, var. angustirostris, *Bocourt, Miss. Sc. Mex., Rept.* p. 638, pl. xxxviii. fig. 3 (1886).
—— bipræocularis, *Cope, Proc. Am. Philos. Soc.* xxii. 1886, p. 103.

Maxillary teeth 13–15 + 2. Head moderately depressed. Rostral shield once and a half to once and two thirds as broad as deep, the portion visible from above measuring one fifth to one third its distance from the frontal; internasals as long as broad or a little broader than long, as long as or a little shorter than the præfrontals; frontal as long as broad or a little longer, as long as its distance from the rostral, as long as or a little shorter than the parietals; loreal as long as deep or deeper than long; one or two præ- and two or three postoculars; temporals 1 + 2 (rarely 2 + 2); eight upper labials (exceptionally seven), fourth and fifth (or third and fourth) entering the eye; four or five lower labials in contact with the anterior chin-shields, which are larger than the posterior. Scales in 19 rows (rarely 21). Ventrals not angulate laterally, 131–153; anal entire; subcaudals 36–50. Pale brown above, with broad brown dark-edged cross bands constricted or interrupted in the middle; a dark marking on the crown and nape, usually bifurcate behind, separated from a dark band, extending from the eye to the angle of the mouth, by a pale oblique temporal streak; lower parts yellowish, more or less obscured by brown dots, or with dark cross bars in the young.

Total length 1100 millim.; tail 150.

Central and Tropical South America.

a. ♀ (V. 151; C. 44).	Para.	R. Graham, Esq. [P.]. (Type.)
b, c. ♀ (V. 147, 146; C. 47, 45).	Bahia.	Dr. Wucherer [C.].
d. Yg. (V. 142; C. 45).	Bahia.	Haslar Collection.
e. ♀ (V. 148; C. 43).	Moyobamba, N.E. Peru.	Mr. A. H. Roff [C.].
f. Yg. (V. 141; C. 40).	Sarayacu, N.E. Peru.	Mr. W. Davies [C.]; Messrs. Veitch [P.].
g–i. ♀ (V. 153, 148; C. 49, 45) & yg. (V. 148; C. 42).	Medellin.	

k, l–m. ♀ (V. 149, 147; C. 46, 44) & yg. (V. 143; C. 49).	Panama.	Christiania Museum.
n. ♀ (V. 144; C. 39).	San Gerónimo, Guatemala.	O. Salvin, Esq. [C.].
o–p. Yg. (V. 131, 132; C. 43, 41).	British Honduras.	Dr. Günther [P.].
q. Yg. (V. 132; C. 42).	Amula, Guerrero.	F. D. Godman, Esq. [P.].

2. Xenodon suspectus.

Xenodon suspectus, *Cope, Proc. Ac. Philad.* 1868, p. 133.

Maxillary teeth 14+2. Head moderately depressed. Rostral shield once and one third as broad as deep, the portion visible from above measuring one fourth its distance from the frontal; internasals a little broader than long, shorter than the præfrontals; frontal a little longer than broad, as long as its distance from the end of the snout, as long as the parietals; loreal deeper than long; one præ- and two postoculars; temporals 1+2; eight upper labials, fourth and fifth entering the eye; four lower labials in contact with the anterior chin-shields, which are a little larger than the posterior. Scales in 19 rows. Ventrals not angulate laterally, 134–141; anal entire; subcaudals 35–40. Olive above, with broad dark black-edged cross bars contracted or interrupted in the middle; upper surface of head with black dots; sides of head black, with a few yellow dots; chin and throat black, spotted with yellow; belly dark brown, with yellow spots on the sides.

Total length 650 millim.; tail 100.

Upper Amazon.

a. ♂ (V. 141; C. 40). Moyobamba, N.E. Peru. Mr. A. H. Roff [C.].

3. Xenodon guentheri. (PLATE VII. fig. 1.)

Maxillary teeth 13+2. Head moderately depressed. Rostral shield as deep as broad, the portion visible from above as long as half its distance from the frontal; internasals broader than long, shorter than the præfrontals; frontal longer than broad, as long as its distance from the end of the snout, as long as the parietals; loreal deeper than long; one præ- and two postoculars; temporals 1+2; eight upper labials, fourth and fifth entering the eye; four lower labials in contact with the anterior chin-shields, which are larger than the posterior. Scales in 19 rows. Ventrals obtusely angulate laterally, 170; anal divided; subcaudals 57. Above with dark brown, blacked-edged, angular, transverse spots separated by narrow pale interspaces; nape with an elongate rhomboidal dark brown spot with a central longitudinal light streak; head with three dark brown angular transverse bands, the middle one across the interorbital space and passing through the eye; belly yellowish with large black transverse spots.

Total length 430 millim.; tail 70.
Southern Brazil.

a. Hgr. ♂ (V. 170; C. 57). Lagos, Sta. Catharina. Hr. Michaëlis [C.].

4. Xenodon neuwiedii.

Xenodon rhabdocephalus, part., *Dum. & Bibr.* vii. p. 758 (1854);
  Günth. Cat. p. 56 (1858); *Jan, Arch. Zool. Anat. Phys.* ii. 1863,
  p. 317, and *Icon. Gén.* 19, pl. v. fig. 1 (1866).
—— neuwiedii, *Günth. Ann. & Mag. N. H.* (3) xii. 1863, p. 354,
  pl. v. fig. C; *Hensel, Arch. f. Nat.* 1868, p. 228.
—— severus, part., *Jan, Arch. Zool. Anat. Phys.* ii. 1863, p. 317.
—— neovidii, *Cope, Proc. Ac. Philad.* 1868, p. 133.

Maxillary teeth 13–15+2. Head moderately depressed. Rostral shield once and a half to once and two thirds as broad as deep, the portion visible from above measuring about one fourth to one third its distance from the frontal; internasals broader than long (if but very slightly), shorter than the præfrontals; frontal longer than broad, as long as its distance from the end of the snout, as long as or a little shorter than the parietals; loreal as long as deep or deeper than long; one præ- and two postoculars; temporals 1+2; eight upper labials, fourth and fifth entering the eye; five lower labials in contact with the anterior chin-shields, which are larger than the posterior. Scales in 21 rows. Ventrals obtusely angulate laterally, 152–176; anal divided (exceptionally entire); subcaudals 50–69. Above with more or less distinct, dark brown, black-edged, angular cross bars separated by narrow yellowish or pale olive interspaces; usually a large ∧ or ⋋-shaped dark marking on the nape; a ∩-shaped dark band round the head, passing through the eyes, and a dark cross bar between the eyes; belly yellowish, usually with brown dots which crowd together to form cross bands; usually a few deep black spots on each side of the anterior part of the body.

Total length 800 millim.; tail 130.
Southern Brazil.

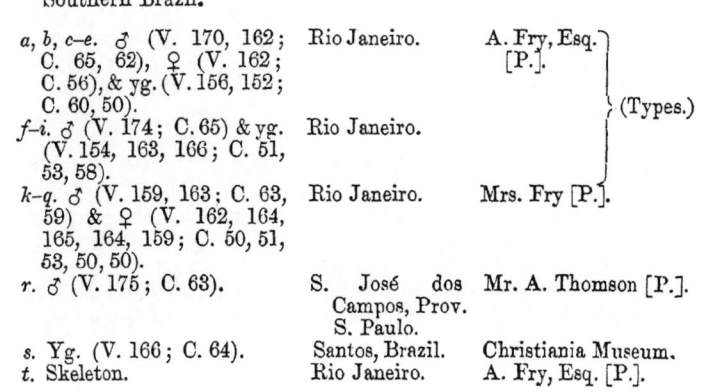

a, b, c–e. ♂ (V. 170, 162; C. 65, 62), ♀ (V. 162; C. 56), & yg. (V. 156, 152; C. 60, 50). Rio Janeiro. A. Fry, Esq. [P.].
f–i. ♂ (V. 174; C. 65) & yg. (V. 154, 163, 166; C. 51, 53, 58). Rio Janeiro. } (Types.)
k–q. ♂ (V. 159, 163; C. 63, 59) & ♀ (V. 162, 164, 165, 164, 159; C. 50, 51, 53, 50, 50). Rio Janeiro. Mrs. Fry [P.].
r. ♂ (V. 175; C. 63). S. José dos Campos, Prov. S. Paulo. Mr. A. Thomson [P.].
s. Yg. (V. 166; C. 64). Santos, Brazil. Christiania Museum.
t. Skeleton. Rio Janeiro. A. Fry, Esq. [P.].

5. Xenodon severus.

Coluber severus, *Linn. Mus. Ad. Frid.* p. 25, pl. viii. fig. 1 (1754), and *S. N.* i. p. 379 (1766).
Cerastes severus, *Laur. Syn. Rept.* p. 81 (1768).
Coluber breviceps, *Shaw, Zool.* iii. p. 430 (1802).
—— versicolor, *Merr. Tent.* p. 95 (1820).
—— saurocephalus, *Wied, Beitr. Nat. Bras.* i. p. 359 (1825).
Xenodon severus, *Schleg. Phys. Serp.* ii. p. 83, pl. iii. figs. 1–5 (1837); *Dum. & Bibr.* vii. p. 756 (1854); *Günth. Cat.* p. 54 (1858).
—— severus, part., *Jan, Arch. Zool. Anat. Phys.* ii. 1863, p. 317, and *Icon. Gén.* 19, pl. iii. fig. B (1866).

Maxillary teeth 10–12+2. Head short, much depressed; body more or less depressed. Rostral twice as broad as deep, the portion visible from above measuring one fourth to one third its distance from the frontal; internasals nearly as long as broad, as long as the præfrontals or a little longer; frontal as long as broad or slightly longer, as long as or a little shorter than its distance from the rostral, shorter than the parietals; loreal deeper than long; one præ- and two or three (rarely four) postoculars; temporals 1+2 or 1+3; eight upper labials, fourth and fifth entering the eye; five or six lower labials in contact with the anterior chin-shields, which are larger than the posterior. Scales in 21 rows. Ventrals not angulate laterally, 128–149; anal divided; subcaudals 33–42. Young with angular, brown, dark-edged bands separated by narrow pale brown interspaces; usually a round white spot on the nape, and light and dark curved bands on the head; belly brown or black, with large yellowish spots on the sides. These markings usually become indistinct or disappear in the adult, some of which are pale brownish or greyish above and yellowish inferiorly; others have broad black bands across the body, or black edges to some or all of the scales, or the upper parts may be entirely black with a few small yellowish spots.

Total length 1500 millim.; tail 190.

Tropical South America.

a. Yg. (V. 149; C. 36).	Rosario de Cucuta, Colombia.	Mr. C. Webber [C.].
b. ♂ (V. 135; C. 36).	Venezuela.	Mr. Dyson [C.].
c, d. ♀ (V. 140, 139; C. 38, 38).	Demerara.	
e, f. ♂ (V. 131; C. 40) & ♀ (V. 140; C. 37).	Berbice.	Lady Essex [P.].
g. Yg. (V. 134; C. 41).	Berbice.	
h. ♀ (V. 140; C. 36).	British Guiana.	
i. Yg. (V. 136; C. 38).	Surinam.	
k, l. Hgr. ♀ (V. 139; C. 37) & yg. (V. 142; C. 40).	Para.	R. Graham, Esq. [P.].
m, n. ♂ (V. 134; C. 36) & yg. (V. 137; C. 34).	Para.	
o, p. ♀ (V. 136; C. 34) & hgr. (V. 131; C. 36).	Moyobamba, N.E. Peru.	Mr. A. H. Roff [C.].

q. ♀ (V. 136 ; C. 39).	Cayaria, N.E. Peru.	Mr. W. Davies [C.] ; Messrs. Veitch [P.].
r. ♀ (V. 138 ; C. 35).	Yurimaguas, N.E. Peru.	Dr. Hahnel [C.].
s, t. ♀ (V. 144 ; C. 38) & yg. (V. 134 ; C. 37).	Guayaquil.	Mr. Fraser [C.].
u. ♀, skel.	Demerara.	Demerara Museum.

6. Xenodon merremii.

Ophis merremii, *Wagl. in Spix, Serp. Bras.* p. 47, pl. xvii. (1824).
Coluber rabdocephalus, part., *Wied, Beitr. Nat. Bras.* i. p. 351, *and Abbild.* (1825).
? Xenodon ocellatus, *Boie, Isis,* 1827, p. 541.
? Xenodon æneus, *Boie, l. c.*
Xenodon rhabdocephalus, part., *Schleg. Phys. Serp.* ii. p. 87 (1837) ; *Dum. & Bibr.* vii. p. 758, pl. lxxvi. fig. 5 (1854) ; *Günth. Cat.* p. 56 (1858); *Jan, Arch. Zool. Anat. Phys.* ii. 1863, p. 317, *and Icon. Gén.* 13, pl. iv. fig. 1 (1866).
—— rhabdocephalus, *Günth. Ann. & Mag. N. H.* (3) xii. 1863, p. 353, pl. v. fig. B ; *Hensel, Arch. f. Nat.* 1868, p. 325.
—— irregularis, *Günth. l. c.* p. 354, fig. D.
—— severus, *Boettg. Zeitschr. f. ges. Naturw.* lviii. 1885, p. 232.

Maxillary teeth 6 or 7+2. Head short, much depressed ; body more or less depressed. Rostral twice as broad as deep, the portion visible from above measuring one third to one half its distance from the frontal; internasals as long as broad or a little broader than long, a little shorter than the præfrontals ; frontal longer than broad, as long as or a little longer than its distance from the rostral, as long as or a little longer than the parietals ; loreal deeper than long ; one or two præ- and two or three postoculars ; one or two small suboculars sometimes present ; temporals 1+2 ; seven (rarely eight) upper labials, third and fourth (or third or fourth only) entering the eye; five or six lower labials in contact with the anterior chin-shields, which are larger than the posterior. Scales in 19 rows. Ventrals not angulate laterally, 132–157 ; anal divided (rarely entire) ; subcaudals 33–48. Pale brown above, with broad brown, black-edged cross bands constricted or sometimes even interrupted in the middle; head with dark curved bands ; these markings may disappear or be reduced to blackish cross bars in old specimens ; belly yellowish, uniform or dotted with brown.

Total length 1040 millim. ; tail 130.

Guianas, Brazil, Paraguay.

a. Yg. (V. 146 ; C. 48).	Demerara.	Capt. E. Sabine [P.].
b. ♀ (V. 153 ; C. 36).	Para.	
		(Types of *X. irregularis*.)
c. Yg. (V. 142 ; C. 42).	Pernambuco.	J. P. G. Smith, Esq. [P.].
d. ♀ (V. 144 ; C. 37).	Pernambuco.	W. A. Forbes, Esq. [P.].
e. Yg. (V. 132 ; C. 44).	Iguarasse, Pernambuco.	G. A. Ramage, Esq. [C.].
f. ♀ (V. 150 ; C. 35).	Bahia.	

g. Yg. (V. 147; C. 41).	Corumba, Mato Grosso.	S. Moore, Esq. [P.]
h–i. ♂ (V. 137; C. 41) & ♀ (V. 147; C. 38).	San Lorenzo, Rio Grande do Sul.	Dr. H. v. Ihering [C.].
k. Yg. (V. 157; C. 47).	Brazil.	
l–m. Hgr. ♂ (V. 131; C. 36) & ♀ (V. 144; C. 34).	Asuncion, Paraguay.	Dr. J. Bohls [C.].
n. ♀ (V. 146; C. 39).	S. America.	Zoological Society.
o. Skull.	Bahia.	
p. Skull.	S. America.	

78. LYSTROPHIS.

Heterodon, part., *Dum. & Bibr. Erp. Gén.* vii. p. 764 (1854); *Günth. Cat. Col. Sn.* p. 82 (1858); *Jan, Arch. Zool. Anat. Phys.* ii. 1864, p. 218.
Lystrophis, *Cope, Proc. Am. Philos. Soc.* xxii. 1885, p. 193.

Maxillary very short, shorter than ectopterygoid, with 4 or 5 small teeth, followed, after an interspace, by a pair of strongly enlarged fangs; mandibular teeth subequal. Head scarcely distinct from neck; snout very short, projecting, cuneiform; rostral very large, trihedral, with sharp Λ-shaped edge, produced posteriorly in a narrow branch separating the internasals; eye moderate, with round pupil. Body subcylindrical; scales slightly oblique, smooth, with apical pits, in 19 or 21 rows; ventrals obtusely angulate laterally. Tail short; subcaudals in two rows.

Southern parts of South America, east of the Andes.

Synopsis of the Species.

Eye usually separated from the labials by a series of suboculars; scales in 21 rows; ventrals 131–147; tail pointed	1. *dorbignyi*, p. 151.
Eye in contact with two labials; scales in 19 rows; ventrals 133–144; tail pointed	2. *histricus*, p. 152.
Eye in contact with two labials; scales in 21 rows; ventrals 153–173; tail rounded off at the end	3. *semicinctus*, p. 153.

1. Lystrophis dorbignyi.

Heterodon dorbignyi, *Dum. & Bibr.* vii. p. 772 (1854); *Günth. Cat.* p. 83 (1858); *Jan, Arch. Zool. Anat. Phys.* ii. 1863, p. 221; *Hensel, Arch. f. Nat.* 1868, p. 329; *Jan, Icon. Gén.* 48, pl. iii. figs. 3 & 4 (1876); *Boettg. Zeitschr. f. ges. Naturw.* lviii. 1885, p. 233.
Lystrophis dorbignyi, *Cope, Proc. Am. Philos. Soc.* xxii. 1885, p. 193.

Frontal once and one third to once and a half as long as broad, nearly as long as or shorter than its distance from the end of the snout, as long as or longer than the parietals; an azygous shield separates the præfrontals; eye usually encircled by three to five shields, in addition to the supraocular; temporals 1+2; seven upper labials, third or fourth rarely entering the eye; four or five lower labials in contact with the anterior chin-shields; posterior chin-shields small. Scales in 21 rows. Ventrals 131-147; anal divided; subcaudals 29-44. Yellowish or pale brown above, with three series of large black or brown black-edged spots; three black Λ-shaped bands on the head and a larger one on the nape; lower parts black and yellow (red?), the two colours in nearly equal proportion.

Total length 560 millim.; tail 80.

Southern Brazil, Paraguay, Uruguay, Argentina, Southern Chili.

a-c. ♂ (V. 136; C. 40), ♀ (V. 131; C. 29), & yg. (V. 145; C. 40).	Rio Grande do Sul.	Dr. H. v. Ihering [C.].
d. ♂ (V. 134; C. 41).	San Lorenzo, Rio Grande do Sul.	Dr. H. v. Ihering [C.].
e-g. ♂ (V. 143; C. 44) & yg. (V. 136, 145; C. 35, 40).	Dept. of Soriano, Uruguay.	R. Havers, Esq. [P.].
h-i. ♂ (V. 141; C. 43) & ♀ (V. 132; C. 33).	Montevideo	
k. ♀ (V. 135; C. 33).	Buenos Ayres.	G. Wilkes, Esq. [P.].
l. ♀ (V. 138; C. 36).	Buenos Ayres.	
m-n. ♂ (V. 147; C. 39) & ♀ (V. 135; C. 36).	Argentina.	Mr. Wilson [C.].
o. ♀ (V. 133; C. 34).	S. Chili.	A. Lane, Esq. [C.]; H. B. James, Esq. [P.].
p. Skull.	Argentina.	

2. Lystrophis histricus.

Heterodon histricus, *Jan, Arch. Zool. Anat. Phys.* ii. 1863, p. 224, and *Icon. Gén.* 11, pl. iv. fig. 2 (1865); *Steindachn. Verh. zool.-bot. Ges. Wien,* 1864, p. 232, pl. vi.
—— nattereri, *Steindachn. Novara, Rept.* p. 90 (1867).

Frontal as long as broad, shorter than its distance from the end of the snout, as long as the parietals; no azygous shield separating the præfrontals; one (or two) præ- and two (or one) postoculars; temporals 1+2; seven upper labials, third and fourth entering the eye; five lower labials in contact with the anterior chin-shields; posterior chin-shields small. Scales in 19 rows. Ventrals 133-144; anal divided; subcaudals 27-37. Reddish above, with more or less regular black cross bars edged with yellowish; sides yellowish, spotted with black; three black Λ-shaped bands on the head, and a larger one on the nape; yellowish-white inferiorly, spotted with black.

Total length 285 millim.; tail 45.
Southern Brazil.

a-b. ♂ (V. 138; C. 33) & hgr. (V. 133; C. 27).		Cio Cahy, near S. João de Monte Negro.	Dr. H. v. Ihering [C.].

3. Lystrophis semicinctus.

Heterodon semicinctus, *Dum. & Bibr.* vii. p. 774 (1854); *Jan, Arch. Zool. Anat. Phys.* ii. 1863, p. 224, *and Icon. Gén.* 48, pl. iii. fig. 5 (1876).
—— pulcher, *Jan, ll. cc.* p. 222, *Icon.* 11, pl. iv. fig. 1 (1865).

Frontal as long as broad, shorter than its distance from the end of the snout, as long as the parietals; usually an azygous shield separating the præfrontals; one or two præ- and two postoculars; temporals 1+2; eight upper labials (exceptionally seven), fourth and fifth entering the eye; four or five lower labials in contact with the anterior chin-shields; posterior chin-shields small. Scales in 21 rows. Ventrals 153–173; anal divided; subcaudals 25–41. End of tail rounded off. Above with black cross bars disposed in pairs, the interspace between the two yellow, that between the pairs red; the red scales, and sometimes also the yellow ones, with a black spot or a black edge; head variegated with black; a black band between the eyes; belly black, usually yellow on the sides.

Total length 660 millim.; tail 80.
Uruguay, Bolivia, Argentina, Northern Patagonia.

a. ♂ (V. 156; C. 30).	Uruguay.	
b. ♀ (V. 153; C. 26).	Rio de Cordova, Argentina.	E. Fielding, Esq. [P.].
c. ♂ (V. 162; C. 36).	Catamarca.	Lord Dormer [P.].
d–e. Yg. (V. 159, 154; C. 33, 30).	Argentina.	E. W. White, Esq. [C.].
f–g. ♀ (V. 160; C. 27) & hgr. (V. 162; C. 25).	Patagonia.	

79. HETERODON.

Heterodon, *Latreille, Hist. Rept.* iv. p. 32 (1800); *Baird & Gir. Cat. N. Am. Rept.* p. 51 (1853); *Cope, Proc. U.S. Nat. Mus.* xiv. 1892, p. 642.
Heterodon, part., *Schleg. Phys. Serp.* ii. p. 96 (1837); *Dum. & Bibr. Erp. Gén.* vii. p. 764 (1854); *Günth. Cat. Col. Sn.* p. 82 (1858); *Jan, Arch. Zool. Anat. Phys.* ii. 1863, p. 218; *Bocourt, Miss. Sc. Mex., Rept.* p. 603 (1886).

Maxillary very short, shorter than the ectopterygoid, with 6 to 11 teeth, followed, after an interspace, by a pair of strongly enlarged fangs; mandibular teeth subequal. Head scarcely distinct from neck; snout very short, projecting, cuneiform; rostral very large, trihedral, with sharp Λ-shaped edge; eye moderate, with round pupil. Body stout, subcylindrical; scales keeled, with apical pits, in 23 to 27 rows; ventrals rounded. Tail short; subcaudals in two rows.

North America.

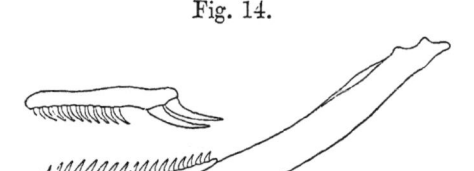

Fig. 14.

Maxillary and mandible of *Heterodon platyrhinus*.

Synopsis of the Species.

Maxillary teeth 9 to 11+2; præfrontals usually in contact with each other; frontal longer than broad; rostral narrower than the distance between the eyes .... 1. *platyrhinus*, p. 154.

Maxillary teeth 8 or 9+2; internasals and præfrontals separated by small irregular shields; frontal as broad as long or slightly longer; rostral narrower than the distance between the eyes .... 2. *simus*, p. 156.

Maxillary teeth 6 to 8+2; internasals and præfrontals separated by small irregular shields; frontal as broad as long or broader; rostral as broad as the distance between the eyes .... 3. *nasicus*, p. 156.

1. Heterodon platyrhinus.

Catesby, Nat. Hist. Carol. ii. pls. xliv., lvi. (1743).
? Coluber leberis, *Linn. S. N.* i. p. 375 (1766).
Heterodon platyrhinus, *Latr. Rept.* iv. p. 32 (1800); *Storer, Rep. Fish. & Rept. Mass.* p. 231 (1839); *Holbr. N. Am. Rept.* iv. p. 67, pl. xvii. (1842); *Dekay, Faun. N. York, Rept.* p. 51, pl. xiii. fig. 28 (1842); *Baird & Gir. Cat. N. Am. Rept.* p. 51 (1853); *Günth. Cat.* p. 82 (1858); *Jan, Arch. Zool. Anat. Phys.* ii. 1863, p. 220, and *Icon. Gén.* 48, pl. iii. fig. 2 (1876); *Garm. N. Am. Rept.* p. 75, pl. vi. fig. 5 (1883); *Cope, Proc. U.S. Nat. Mus.* xiv. 1892, p. 643; *H. Garm. Bull. Illin. Lab.* iii. 1892, p. 302; *Hay, Batr. & Rept. Indiana*, p. 102 (1893).
Coluber cacodæmon, *Shaw, Zool.* iii. p. 377, pl. cii. (1802).
Scytale niger, *Daud. Rept.* v. p. 342 (1803).
Coluber heterodon, part., *Daud. Rept.* vii. p. 153 (1803); *Say, Amer. Journ.* i. 1818, p. 261.

Coluber heterodon, *Harlan, Phys. Med. Res.* p. 120 (1835).
—— thraso, *Harl. l. c.*
Heterodon niger, *Troost, Ann. Lyc. N. York*, iii. 1836, p. 186; *Holbr. l. c.* p. 63, pl. xvi.; *Baird & Gir. Cat.* p. 55; *Dum. & Bibr.* vii. p. 769; *Günth. Cat.* p. 83.
—— annulatus, *Troost, l. c.* p. 188.
—— tigrinus, *Troost, l. c.* p. 189.
—— cognatus, *Baird & Gir. l. c.* p. 54; *Garm. Bull. Essex Inst.* xxiv. 1892, p. 106.
—— atmodes, *Baird & Gir. l. c.* p. 57.

Maxillary teeth 9 to 11 + 2. Rostral considerably narrower than the distance between the eyes; an azygous shield separating the internasals; præfrontals usually in contact with each other; frontal longer than broad, shorter than its distance from the end of the snout, as long as the parietals; loreal deeper than long; eye surrounded by 10 or 11 shields in addition to the supraocular; temporals small, scale-like; seven or eight upper labials; one pair of chin-shields, in contact with three lower labials. Scales strongly keeled, of outer row smooth, in 23 to 27 rows. Ventrals 120–150; anal divided; subcaudals 37–60.

Total length 800 millim.; tail 130.

United States, east of the Rocky Mountains.

A. Brown or red above, with a dorsal series of large square or transverse dark brown or black spots, separated from each other by light interspaces, and a lateral alternating series of smaller spots; an elongate black blotch or band on each side of the head, sometimes extending forwards to the frontal, a transverse dark band between the eyes, and an oblique one from the eye to the angle of the mouth; belly yellowish or reddish, clouded with brown.

a, b. ♂ (Sc. 25; V. 127; C. 50) & yg. (Sc. 25; V. 128; C. 53).	Bloomington, Indiana.	C. H. Bollman, Esq. [C.].
c. ♂ (Sc. 25; V. 127; C. 52).	Pennsylvania.	Smithsonian Instit. [P.].
d. Yg. (Sc. 23; V. 120; C. 52).	Charleston.	
e. Yg. (Sc. 25; V. 128; C. 49).	S. Carolina.	Smithsonian Instit. [P.]. (As typical of *H. atmodes.*)
f–h, i. ♂ (Sc. 25; V. 129, 130; C. 48, 48), ♀ (Sc. 25; V. 134; C. 37), & yg. (Sc. 25; V. 128; C. 48).	Duval Co., Texas.	W. Taylor, Esq. [C.].
k. ♀ (Sc. 25; V. 141; C. 44).	Texas.	
l. Skeleton.	N. America.	
m. Skull.	N. America.	

B. Uniform black or blackish above.

n. ♂ (Sc. 25; V. 120; Georgia. Smithsonian Instit.
C. 56). [P.].
o, p. ♂ (Sc. 27; V. 131; N. America.
C. 51) & ♀ (Sc. 25;
V. 139; C. 42).

2. Heterodon simus.

Coluber simus, *Linn. S. N.* i. p. 375 (1766).
—— heterodon, part., *Daud. Rept.* vii. p. 153 (1803).
Heterodon platyrhinos, *Schleg. Phys. Serp.* ii. p. 97, pl. iii. figs. 20-22 (1837).
—— simus, *Holbr. N. Am. Herp.* iv. p. 57, pl. xv. (1842); *Baird & Gir. Cat. N. Am. Rept.* p. 59 (1853); *Garm. N. Am. Rept.* p. 76, pl. vi. fig. 4 (1883); *Cope, Proc. U.S. Nat. Mus.* xiv. 1892, p. 643; *Hay, Batr. & Rept. Indiana,* p. 105 (1893).
—— platyrhinus, part., *Dum. & Bibr. Erp. Gén.* vii. p. 766 (1854).
—— catesbyi, part., *Günth. Cat.* p. 83 (1858).
—— nasicus, part., *Jan, Arch. Zool. Anat. Phys.* ii. 1863, p. 220.

Maxillary teeth 8 or 9+2. Rostral narrower than the distance between the eyes; several small shields separating the internasals and the præfrontals; frontal as broad as long or slightly longer, shorter than its distance from the end of the snout, as long as the parietals; eye surrounded by 10 or 11 shields in addition to the supraocular; eight upper labials; one pair of chin-shields, in contact with three or four lower labials. Scales strongly keeled, of two or three outer rows smooth, in 25 or 27 rows. Ventrals 114-132; anal divided; subcaudals 30-55. Coloration as in *H. platyrhinus*.

Total length 470 millim.; tail 90.

Mississippi and South Carolina.

a. ♀ (Sc. 25; V. 121; Charleston.
C. 33).
b-c. ♂ (Sc. 25; V. 114; N. America. College of Surgeons [P.].
C. 40) & yg. (Sc. 25; (Types of *H. catesbyi*.)
V. 124; C. 34).

3. Heterodon nasicus.

Heterodon nasicus, *Baird & Gir. in Stansbury, Explor. Great Salt Lake,* p. 352 (1852), *and Cat. N. Am. Rept.* p. 61 (1853), *and in Marcy, Explor. Red Riv.* p. 222, pl. iv. (1853); *Hallow. Proc. Ac. Philad.* 1856, p. 249; *Baird, U.S. Mex. Bound. Surv.* ii., *Rept.* p. 18, pl. xi. fig. 1 (1859); *Jan. Icon. Gén.* 10, pl. v. (1865); *Bocourt, Miss. Sc. Mex., Rept.* p. 604, pl. xxxviii. figs. 1 & 2 (1886); *Cope, Proc. U.S. Nat. Mus.* xiv. 1892, p. 644.
—— catesbyi, part., *Günth. Cat.* p. 83 (1858).
—— kennerlyi, *Kennicott, Proc. Ac. Philad.* 1860, p. 336.
—— nasicus, part., *Jan, Arch. Zool. Anat. Phys.* ii. 1863, p. 220.
—— simus nasicus, *Cope, Check-list N. Am. Rept.* p. 43 (1875); *Coues & Yarrow, Bull. U.S. Geol. Surv.* iv. 1878, p. 270; *Garm.*

N. Am. Rept. p. 77, pl. vi. fig. 6 (1883); *H. Garm. Bull. Illin. Lab.* iii. 1892, p. 305.
Heterodon simus kennerlyi, *Coues & Yarrow, l. c.* p. 271.

Maxillary teeth 6 to 8+2. Rostral as broad as the distance between the eyes; several small shields separating the internasals and the præfrontals; frontal as broad as long or broader, shorter than its distance from the end of the snout, as long as or longer than the parietals; eye surrounded by 10 or 11 shields in addition to the supraocular; eight upper labials; one pair of chin-shields, in contact with three or four lower labials. Scales all, except the outer row, strongly keeled, in 23 rows. Ventrals 126-150; anal divided; subcaudals 32-45. Pale brown or yellowish above, with three alternating series of brown spots, the median largest and usually broadest; a dark cross band between the eye, a broad oblique band from the eye to the mouth, and three elongate blotches behind the head; belly yellowish white, with deep black blotches, or entirely black in the middle.

Total length 600 millim.; tail 70.

United States, west of the Mississippi, North Mexico.

a. ♂ (V. 148; C. 36).	Kansas.	Smithsonian Instit. [P.].
b-c, d. ♀ (V. 138, 138, 141; C. 35, 38, 38).	Duval Co., Texas.	W. Taylor, Esq. [C.].
e. Yg. (V. 126; C. 37).	Texas.	Smithsonian Instit. [P.]. (As typical of *H. kennerlyi*.)
f. ♀ (V. 138; C. 32).	Texas.	(One of the types of *H. catesbyi*.)

80. APOROPHIS.

Herpetodryas, part., *Schleg. Phys. Serp.* ii. p. 173 (1837).
Dromicus, part., *Dum. & Bibr. Erp. Gén.* vii. p. 646 (1854); *Günth. Cat. Col. Sn.* p. 126 (1858); *Jan, Elenco sist. Ofid.* p. 66 (1863).
Lygophis (non *Tsch.*), part., *Cope, Proc. Ac. Philad.* 1862, p. 75.
Aporophis, part., *Cope, Proc. Am. Philos. Soc.* xvii. 1877, p. 34.

Maxillary teeth 18 to 22, followed after an interspace by two enlarged ones; mandibular teeth subequal. Head narrow, slightly distinct from neck; eye large, with round pupil. Body cylindrical; slender; scales smooth, without apical pits, in 17 or 19 rows. ventrals rounded. Tail rather long; subcaudals in two rows.

South America.

Synopsis of the Species.

Scales in 19 rows; ventrals 165-178 ..	1. *lineatus*, p. 158.
Scales in 17 rows; ventrals 157-178; frontal as long as the parietals......	2. *flavifrenatus*, p. 158.
Scales in 17 rows; ventrals 155; frontal a little shorter than the parietals.........................	3. *coralliventris*, p. 159.
Scales in 17 rows; ventrals 137-143; frontal as long as the parietals......	4. *amœnus*, p. 160.

1. Aporophis lineatus.

Coluber lineatus, *Linn. Mus. Ad. Frid.* p. 30, pls. xii. fig. 1, and xx.
fig. 1 (1754), *and S. N.* i. p. 382 (1766); *Daud. Rept.* vii. p. 25
(1803).
Herpetodryas lineatus, *Schleg. Phys. Serp.* ii. p. 191 (1837).
Dromicus lineatus, part., *Dum. & Bibr.* vii. p. 655 (1854); *Günth.
Cat.* p. 134 (1858).
—— lineatus, *Jan, Icon. Gén.* 24, pl. vi. fig. 4 (1867) ; *Fischer, Arch.
f. Nat.* 1882, p. 285.
Lygophis lineatus, *Cope, Proc. Ac. Philad.* 1862, p. 76.
—— dilepis, *Cope, l. c.* p. 81.
Aporophis lineatus, *Cope, Proc. Am. Philos. Soc.* xxii. 1885, p. 191.
—— dilepis, *Cope, l. c.*

Rostral broader than deep, visible from above; posterior border of internasals more than twice as broad as the anterior; suture between the internasals as long as or shorter than that between the præfrontals ; frontal narrow, twice to twice and one third as long as broad, longer than its distance from the end of the snout, as long as the parietals ; loreal as long as deep or deeper than long ; one or two præ- and two postoculars ; temporals 1+2 ; eight upper labials, fourth and fifth entering the eye ; five or six lower labials in contact with the anterior chin-shields, which are as long as or a little longer than the posterior. Scales in 19 rows. Ventrals 165–178 ; anal divided ; subcaudals 79–93. Yellowish or pale olive above, with a brown black-and-yellow-edged vertebral band commencing on the snout ; a black or brown lateral line, widening on the side of the head and passing through the eye ; lower parts yellowish white.

Total length 650 millim.; tail 190.

Guianas, Brazil, Paraguay.

a–h. ♂ (V. 165; C. 93), ♀ (V. 167, 169, 164, 166, 173; C. 90, 90, 82, 88, 90), & hgr. (V. 170, 168; C. 87, 87).	Demerara.	
i. ♂ (V. 175; C. ?).	Berbice.	
k. ♀ (V. 170; C. 79).	Para.	R. Graham, Esq. [P.].
l–m. ♀ (V. 178, 173; C. 79, 81).	Asuncion, Paraguay.	Dr. J. Bohls [C.].

2. Aporophis flavifrenatus.

Lygophis flavifrenatus, *Cope, Proc. Ac. Philad.* 1862, p. 80.
Dromicus amabilis, *Jan, Icon. Gén.* 24, pl. v. fig. 2 (1867).
Aporophis flavifrenatus, *Cope, Proc. Am. Philos. Soc.* xxii. 1885, p. 191.
Dromicus flavifrenatus, *Bouleng. Ann. & Mag. N. H.* (5) xviii. 1886, p. 433.

Rostral a little broader than deep, just visible from above ; posterior border of internasals more than twice as broad as the anterior ; suture between the internasals as long as or shorter than that between the præfrontals; frontal narrow, twice as long as broad, considerably longer than its distance from the end of the snout, as long as the parietals; loreal as long as deep or deeper than long; one præ- and two postoculars; temporals 1+2; eight upper labials, fourth and fifth entering the eye; five lower labials in contact with the anterior chin-shields, which are as long as or a little longer than the posterior. Scales in 17 rows. Ventrals 157-178; anal divided; subcaudals 76-94. Olive-brown above, with two yellow longitudinal lines, commencing on the snout, and separated by a dark brown band three scales wide; a black lateral streak or series of spots; the two outer rows of scales pale olive; upper lip and præ- and postoculars yellowish; lower parts greenish white, the anterior border of each ventral black on each side.

Total length 700 millim.; tail 190.

Southern Brazil to North-eastern Argentina.

a-b. ♀ (V. 166; C. 78) & yg. (V. 160; C. 94).	Rio Grande.	Dr. H. v. Ihering [C.].
c-d. ♀ (V. 160; C. 86) & yg (V. 157; C. 86).	S. Lorenzo, Rio Grande do Sul.	Dr. H. v. Ihering [C.].
e. ♀ (V. 166; C. 76).	Colonia Resistencia, Central Chaco.	Prof. Spegazzini [C.]; Marquis G. Doria [P.].
f. Hgr. (V. 178; C. ?).	Candelaria, Prov. Missiones.	Marquis G. Doria [P.].

3. Aporophis coralliventris. (PLATE VII. fig. 2.)

Aporophis coralliventris, *Bouleng. Ann. & Mag. N. H.* (6) xiii. 1894, p. 346.

Rostral broader than deep, just visible from above; internasals broader than long, shorter than the præfrontals; frontal twice as long as broad, longer than its distance from the end of the snout, a little shorter than the parietals; loreal deeper than long; one præ- and two postoculars; temporals 1+2; eight upper labials, fourth and fifth entering the eye; five lower labials in contact with the anterior chin-shields, which are as long as the posterior. Scales in 17 rows. Ventrals 155; anal divided; subcaudals 71. Olive-brown above, darker along the five median rows of scales, bluish grey on the sides (three rows of scales); head without streaks or markings; upper lip white; throat and anterior ventral region white, rest of belly and tail coral-red, the shields edged with black.

Total length 300 millim.; tail 80.

Paraguay.

a. ♂ (V. 155; C. 71).	Island north of Concepcion, near S. Salvador.	Dr. J. Bohls [C.]. (Type.)

4. Aporophis amœnus.

Enicognathus amœnus, *Jan, Arch. Zool. Anat. Phys.* ii. 1863, p. 270, and *Icon. Gén.* 16, pl. ii. fig. 1 (1866).

Rostral broader than deep, just visible from above; internasals nearly as long as broad, a little shorter than the præfrontals; frontal twice as long as broad, considerably longer than its distance from the end of the snout, as long as the parietals; loreal a little deeper than long: one præ- and two postoculars; temporals 1+2; eight upper labials, fourth and fifth (or third, fourth, and fifth) entering the eye; four lower labials in contact with the anterior chin-shields, which are shorter than the posterior. Scales in 17 rows. Ventrals 137–143; anal divided; subcaudals 92–94. Olive-brown above, with a blackish vertebral line, which is indistinct or absent on the anterior half of the body: two outer rows of scales grey-brown, third row black, with a round white spot in the centre of each scale; a black streak on each side of the head, passing through the eye; upper lip white; lower parts white, with a lateral series of black dots.

Total length 590 millim.; tail 210. The specimen in the collection measures 235, tail 75.

Brazil.

a. Yg. (V. 137; C. 92). Theresopolis, Prov. Rio Janeiro. Dr. E. A. Göldi [P.].

81. RHADINÆA.

Liophis, part., *Wagl. Syst. Amph.* p. 187 (1830); *Dum. & Bibr. Erp. Gén.* vii. p. 697 (1854); *Günth. Cat. Col. Sn.* p. 42 (1858); *Jan, Arch. Zool. Anat. Phys.* ii. 1863, p. 287.
Coronella, part., *Schleg. Phys. Serp.* ii. p. 50 (1837); *Günth. l. c.* p. 34; *Jan, l. c.* p. 236.
Dromicus, part., *Dum. & Bibr. t. c.* p. 646; *Günth. l. c.* p. 126; *Cope, Proc. Ac. Philad.* 1862, p. 76; *Jan, Elenco sist. Ofid.* p. 66 (1863); *Bocourt, Miss. Sc. Mex., Rept.* p. 707 (1890).
Ophiomorphus (*non Dej.*), *Cope, l. c.* p. 75, and *Proc. U.S. Nat. Mus.* xii. 1889, p. 144.
Lygophis (*non Tsch.*), part., *Cope, l. c.*
Rhadinæa, *Cope, Proc. Ac. Philad.* 1863, p. 100, and 1868, p. 132.
Calonotus (*non Hübn.*), *Jan, l. c.* p. 239.
Diadophis, part., *Jan, l. c.* p. 261; *Bocourt, Miss. Sc. Mex., Rept.* p. 618 (1886).
Enicognathus, part., *Jan, l. c.* p. 266; *Bocourt, l. c.* p. 625.
Ablabes, *Jan, l. c.* p. 279.
Aporophis, part., *Cope, Proc. Am. Philos. Soc.* xvii. 1877, p. 34.

Maxillary teeth 14 to 24, posterior largest, forming an uninterrupted series, or the two or three hindermost separated from the others by a very short interspace; mandibular teeth subequal. Head but slightly distinct from neck; eye moderate or rather small, with round pupil. Body cylindrical; scales smooth, without apical pits, in 15 to 21 rows; ventrals not or but obtusely angulate laterally. Tail moderate or long; subcaudals in two rows.

Central and South America.

Fig. 15.

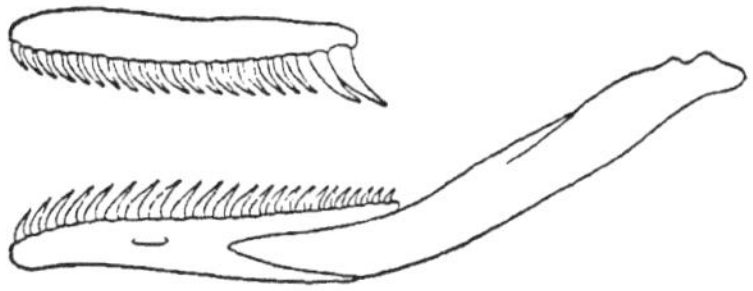

Maxillary and mandible of *Rhadinæa merremii*.

Synopsis of the Species.

I. Scales in 15 to 19 rows.

  A. Anterior chin-shields at least as long as the posterior; tail not more than one fourth of the total length.

    1. Six or seven upper labials, third and fourth entering the eye.

      a. Scales in 17 rows.

One postocular	1. *leucogaster*, p. 163.
Two postoculars; nostril between two nasals........................	2. *breviceps*, p. 164.
Two postoculars; nasal single	3. *calligaster*, p. 164.

      b. Scales in 15 rows .... 4. *mimus*, p. 164.

    2. Eight upper labials, fourth and fifth entering the eye.

      a. Scales in 19 rows.

        α. Ventrals obtusely angulate laterally.

Ventrals 147–159; subcaudals 50–73.	5. *anomala*, p. 165.
Ventrals 186–198; subcaudals 75–92.	6. *sagittifera*, p. 165.

        β. Ventrals not angulate laterally.

Suture between the frontal and the præfrontals not longer than that between the frontal and the supraocular	13. *obtusa*, p. 171.
Frontal wide, supraocular suture shorter than anterior...........	14. *serperastra*, p. 172.

    b. Scales in 17 rows.

        a. Ventrals not more than 185.

            * Parietals a little longer than their distance from the internasals; ventrals 143–163.
            7. *cobella*, p. 166.

            ** Parietals as long as their distance from the internasals.

Ventrals 150–170, with black spots .. 8. *purpurans*, p. 167.
Ventrals 141–164, unspotted, uniform, or edged with black .... 9. *merremii*, p. 168.
Ventrals 165–183, unspotted, edged with black .... 10. *fusca*, p. 169.
Ventrals 147–165, unspotted, outer ends green .... 11. *jægeri*, p. 170.

        β. Ventrals more than 185 . 12. *genimaculata*, p. 170.

    B. Anterior chin-shields shorter than the posterior, or tail more than one fourth of the total length.

        1. Scales in 17 rows.

            a. Seven upper labials, third and fourth entering the eye.

Frontal once and a half as long as broad; subcaudals 58–71 .... 15. *affinis*, p. 172.
Frontal once and two thirds as long as broad; subcaudals 66–75 ...... 16. *pœcilopogon*, p. 173.
Frontal once and a half to once and two thirds as long as broad; subcaudals 85–95 .... 25. *laureata*, p. 179.

            b. Eight upper labials, fourth and fifth entering the eye.

Ventrals 173; subcaudals 78 ...... 17. *lachrymans*, p. 174.
Ventrals 117–137; subcaudals 90–120. 21. *decorata*, p. 176.
Ventrals 117; subcaudals 79 ...... 22. *vermiculaticeps*, p. 177.
Ventrals 145–183; subcaudals 88–132 .... 24. *vittata*, p. 178.

            c. Eight upper labials, third, fourth, and fifth entering the eye.
            18. *undulata*, p. 174.

        2. Scales in 19 rows .... 23. *clavata*, p. 177.

        3. Scales in 15 rows.

Three postoculars .... 19. *melanauchen*, p. 175.
Two postoculars .... 20. *occipitalis*, p. 175.

II. Scales in 21 rows .... 26. *godmani*, p. 179.

TABLE SHOWING NUMBERS OF SCALES AND SHIELDS.

	Scales.	Ventrals.	Subcaudals.	Labials.
leucogaster	17	140	30	6
breviceps	17	154	48–54	6–7
calligaster	17	152	46	7
mimus	15	?	?	7
anomala	19	147–159	50–73	7–8
sagittifera	19	186–198	75–92	8
cobella	17	143–163	45–57	8
purpurans	17	150–170	51–65	8
merremii	17	141–164	47–58	8
fusca	17	165–183	51–60	8
jægeri	17–19	147–165	55–62	8
genimaculata	17	191–208	51–66	8
obtusa	19	174–192	65–79	8
serperastra	19	164	78	8
affinis	17	147–185	58–71	7
pœcilopogon	17	141–164	66–75	7
lachrymans	17	173	78	8
undulata	17	149–167	52–85	8
melanauchen	15	148	60	8
occipitalis	15	152–183	64–86	8
decorata	17	117–137	90–120	8
vermiculaticeps	17	117	79	8
clavata	19	127	89	8
vittata	17	145–183	88–132	8
laureata	17	159–164	85–95	7
godmani	21	168–176	72–92	8

1. Rhadinæa leucogaster.

Liophis leucogaster, *Jan, Arch. Zool. Anat. Phys.* ii. 1863, p. 289, and *Icon. Gén.* 13, pl. vi. fig. 1 (1865).

Eye rather small; rostral broader than deep, just visible from above; internasals broader than long; suture between the internasals a little shorter than that between the præfrontals; frontal about once and a half as long as broad, longer than its distance from the end of the snout, as long as the parietals; latter as long as their distance from the internasals; loreal square; one præ- and one postocular; temporals 1+2; six upper labials, third and fourth entering the eye; four lower labials in contact with the anterior chin-shields, which are as long as the posterior. Scales in 17 rows. Ventrals 140; anal divided; subcaudals 30. Brown above, with black cross bands, some of the scales on the middle line between the bands edged with white; a black lateral stripe on the hinder part of the body, extending to the ends of the ventrals; lower surface white.

Total length 180 millim.; tail 28.

Habitat unknown.

2. Rhadinæa breviceps.

Liophis breviceps, *Cope, Proc. Ac. Philad.* 1860, p. 252.
Ophiomorphus breviceps, *Cope, Proc. Ac. Philad.* 1862, p. 75.

Eye rather small; rostral broader than deep, just visible from above; internasals broader than long, a little shorter than the præfrontals; frontal not quite once and a half as long as broad, longer than its distance from the end of the snout, slightly shorter than the parietals; latter as long as their distance from the rostral; loreal deeper than long; one præ- and two postoculars; temporals 1+2; six or seven upper labials, third and fourth entering the eye; four lower labials in contact with the anterior chin-chields, which are as long as or a little longer than the posterior. Scales in 17 rows. Ventrals 154; anal divided; subcaudals 48-54. Length of tail $5\frac{1}{2}$ to $5\frac{2}{3}$ times in the total length. Brown above, the scales edged with black, with more or less regular narrow black cross bands; below with black and white (red?) cross bands of subequal width.

Total length 450 millim.; tail 80.

Ecuador; Surinam.

a. ♂ (V. 154; C. 48).	W. Ecuador.	Mr. Fraser [C.].
b. ♀ (V. 154; C. 48).	Palltanga, Ecuador.	Mr. Buckley [C.].

3. Rhadinæa calligaster.

Contia calligaster, *Cope, Journ. Ac. Philad.* viii. 1876, p. 146, pl. xxviii. fig. 12.

Snout narrowed; one nasal shield; loreal subquadrate; one præ- and two postoculars; temporals 1+1 or 1+2; seven upper labials, third and fourth entering the eye; anterior and posterior chin-shields equal. Scales in 17 rows. Ventrals 152; anal divided; subcaudals 46. Dark brown above, with a narrow black vertebral stripe; two lateral paler stripes; a black stripe along the ends of the ventrals; labials broadly black bordered; belly yellow, with a series of black crescents on the median front of each suture; middle line of tail below black.

Pico Blanco, Costa Rica.

4. Rhadinæa mimus.

Opheomorphus mimus, *Cope, Proc. Ac. Philad.* 1868, p. 307.

Snout short; internasals broader than long; frontal shorter than parietals; loreal very small or absent; one præ- and two postoculars; temporals 1+2; seven upper labials, third and fourth entering the eye. Scales in 15 rows. Crimson, each scale with a brown spot near its tip; ten black rings on the body, complete across the belly; head above and spot below eye black.

Total length 340 millim.; tail 60.

Andes of Ecuador or Colombia.

5. Rhadinæa anomala.

Coronella anomala, *Günth. Cat.* p. 37 (1858), *and Zool. Rec.* 1866, p. 126; *Bouleng. Ann. & Mag. N. H.* (5) xviii. 1886, p. 431.
Lygophis rutilus, *Cope, Proc. Ac. Philad.* 1862, p. 80, *and* 1863, p. 101.
Coronella pulchella, *Jan, Arch. Zool. Anat. Phys.* ii. 1863, p. 251, *and Icon. Gén.* 17, pl. iii. fig. 4 (1866).
Aporophis anomalus, *Cope, Proc. Am. Philos. Soc.* xvii. 1877, p. 93.

Eye moderate. Rostral broader than deep, just visible from above; internasals as long as broad, as long as or a little shorter than the præfrontals *; frontal narrow, twice as long as broad, as long as or a little longer than its distance from the end of the snout, as long as the parietals; loreal as long as deep or a little longer †; one præ- and two postoculars; temporals 1+2; eight (exceptionally seven) upper labials, fourth and fifth entering the eye; four or five lower labials in contact with the anterior chin-shields, which are as long as or a little longer than the posterior. Scales in 19 rows. Ventrals indistinctly angulate laterally, 147–159; anal divided; subcaudals 50–73. Length of tail $4\frac{1}{3}$ to $5\frac{1}{2}$ times in the total length. Brown or olive above, with large black spots dotted with yellow, and two yellow longitudinal lines or series of spots; the black spots forming symmetrical vertical bars on the sides; belly yellow.

Total length 570 millim.; tail 105.

From Southern Brazil and Paraguay to Buenos Ayres.

a. ♂ (V. 155; C.?).	Parana.	Haslar Collection. (Type.)
b. ♂ (V. 158; C. 69).	Paraguay.	Prof. Grant [P.].
c–d. ♂ (V. 159; C. 69) & ♀ (V. 155; C. 66).	Rio Grande do Sul.	Dr. H. v. Ihering [C.].
e–f. ♀ (V. 147, 154; C. 66, 65).	Dept. of Soriano, Uruguay.	R. Havers, Esq. [P.].
g–i. ♂ (V. 152; C. 58) & ♀ (V. 155, 155; C. 53, 63).	Buenos Ayres.	G. Wilks, Esq. [P.].
k–l. ♂ (V. 151; C. 59) & ♀ (V. 147; C. 50).	South of Rio de la Plata.	Lieut. Gairdner [P.].
m. Skull.	Rio Grande do Sul.	

6. Rhadinæa sagittifera.

Chlorosoma sagittifer, *Jan, in Burm. Reise La Plata,* ii. p. 530 (1861) [no description].
Liopeltis sagittifer, *Jan, Elenco,* p. 82 (1863), *and Icon. Gén.* 31, pl. v. fig. 2 (1869).
Liophis pulcher, *Steind. Sitzb. Ak. Wien,* lv. 1867, p. 267, pl. ii.

Eye moderate; rostral a little broader than deep, well visible

---

* The type specimen is anomalous in having an azygous shield separating the pair of præfrontals.
† Fused with the præfrontal in the type specimen.

from above; nasals sometimes entire or semidivided; internasals nearly as long as broad, as long as the præfrontals; frontal narrow, nearly twice as long as broad, with concave sides, slightly longer than its distance from the end of the snout, as long as the parietals; loreal a little longer than deep; one or two præ- and two postoculars; temporals 1+2; eight upper labials, fourth and fifth entering the eye; five lower labials in contact with the anterior chin-shields, which are as long as or a little longer than the posterior. Scales in 19 rows. Ventrals 186–198; anal divided; subcaudals 75–92. Length of tail 4 to 4½ times in the total length. Pale buff above, with two alternating series of large black spots; a broad black stripe from the eye to the side of the neck; upper lip, two or three outer rows of scales, and lower surface white.

Total length 740 millim.; tail 170.

Western Argentina.

a–b. ♂ (V. 198; C. 86) & Mendoza.
♀ (V. 193; C. 75).
c–f. ♂ (V. 196, 197, 186; Tucuman.
C. 86, 92, 82) & yg.
(V. 187; C. 76).

7. Rhadinæa cobella.

Coluber cobella, *Linn. Mus. Ad. Frid.* p. 24 (1754), *and S. N.* i. p. 378 (1766); *Merr. Beitr.* i. p. 16, pl. iv., and ii. p. 32, pl. viii. (1790).
Elaps cobella, *Schneid. Hist. Amph.* ii. p. 296 (1801).
Coluber cenchrus, *Daud. Rept.* vii. p. 139 (1803).
—— serpentinus, part., *Daud. l. c.* p. 87.
Coronella cobella, *Schleg. Phys. Serp.* ii. p. 63, pl. i. figs. 4 & 5 (1837).
Liophis cobella, *Dum. & Bibr.* vii. p. 698 (1854); *Günth. Cat.* p. 43 (1858); *Jan, Icon. Gén.* 16, pl. v. fig. 1 (1866).
Ophiomorphus cobella, *Cope, Proc. Ac. Philad.* 1862, p. 75.
Liophis tæniogaster, *Jan, Arch. Zool. Anat. Phys.* ii. 1863, p. 292, and *Icon. Gén.* 18, pl. i. (1866).

Eye rather small. Rostral broader than deep, visible from above; internasals as long as broad or a little broader, as long as or a little shorter than the præfrontals; frontal once and a half to once and three fourths as long as broad, as long as or slightly longer than its distance from the end of the snout, a little shorter than the parietals; loreal as long as deep or deeper than long; one præ- and two postoculars; temporals 1+2; eight upper labials, fourth and fifth entering the eye; five lower labials in contact with the anterior chin-shields, which are as long as or a little longer than the posterior. Scales in 17 rows. Ventrals 143–163; anal divided; subcaudals 45–57. Length of tail 4⅔ to 6 times in the total length. Brown or blackish above, with whitish lines or cross bands; belly yellowish (coral-red in life) with transverse black spots or cross bands, always more or less marked in the young, which have a blackish nuchal collar edged behind with whitish, and a pair of whitish dots close together on the parietal shields; upper labials yellowish.

Total length 730 millim.; tail 125.

Guianas, Brazil.

A. Brown above, with darker spots and white edges to some of the scales, the assemblage of short white lines sometimes forming irregular cross bands. (*C. cobella.*)

a. ♀ (V. 150; C. 61).	Trinidad.	F. W. Urich, Esq. [P.].
b–n. ♂ (V. 154, 151, 153; C. 57, 54, 51), ♀ (V. 151, 155, 151, 153; C. 56, 46, 52, 52), hgr. (V. 151, 156, 154; C. 54, 53, 51), & yg. (V. 150, 148; C. 51, 54).	Demerara.	
o. ♀ (V. 153; C. 55).	Demerara.	J. Henslow, Esq. [P.].
p–u. ♂ (V. 155, 154, 153; C. 54, 53, 53) & ♀ (V. 148, 150, 151; C. 52, 50, ?).	Berbice.	
v. ♀ (V. 163; C. 45).	Cayenne.	H. C. Rothery, Esq. [P.].
w. ♀ (V. 154; C. 56).	Brazil.	

B. Dark brown or black above, with regular transverse white bands, widening towards the body. (*L. tæniogaster.*)

a. ♂ (V. 154; C. 52).	Bahia.	Dr. Wucherer [C.].
b. ♀ (V. 157; C. 52).	Bahia.	D. Wilson Barker, Esq. [P.].
c–d. ♂ (V. 150; C. 49) & ♀ (V. 154; C. 53).	Bahia.	Haslar Collection.
e–f. Yg. (V. 157, 155; C. 51, 50).	Pernambuco.	J. P. G. Smith, Esq. [P.].
g. ♀ (V. 154; C. 56).	Para.	R. Graham, Esq. [P.].

A variety without spots on the belly is named by Jan (*ll. cc.,* Icon. 16, pl. v. fig. 2) var. *flaviventris.*

---

a. Skeleton.	S. America.
b, c. Skulls.	Brazil.

8. Rhadinæa purpurans.

Ablabes purpurans, *Dum. & Bibr.* vii. p. 312 (1854).
Diadophis purpurans, *Jan, Arch. Zool. Anat. Phys.* ii. 1863, p. 265, *and Icon. Gén.* 15, pl. v. fig. 5 (1866).
Coronella orientalis, *Günth. Rept. Brit. Ind.* p. 236 (1864).
? Liophis cobella, var. collaris, *Jan, ll. cc.* p. 293, *Icon.* 16, pl. v. fig. 3 (1866).
Rhadinæa chrysostoma, *Cope, Proc. Ac. Philad.* 1868, p. 104.

Coronella pœcilolæmus, *Günth. Ann. & Mag. N. H.* (4) ix. 1872, p. 19.
Liophis purpurans, *Günth. l. c.*

Eye rather small. Rostral broader than deep, scarcely visible from above; internasals a little broader than long, as long as or a little shorter than the præfrontals; frontal nearly twice as long as broad, longer than its distance from the end of the snout, slightly shorter than the parietals; loreal as long as deep or deeper than long; one præ- and two postoculars; temporals 1+2; eight upper labials, fourth and fifth entering the eye; five lower labials in contact with the anterior chin-shields, which are as long as or a little longer than the posterior. Scales in 17 rows. Ventrals 150–170; anal divided; subcaudals 51–65. Length of tail 5 to 5½ times in the total length. Dark brown above, with a lighter, black-edged streak along each side of the posterior half of the body, most distinct on the tail; sometimes a white line across the nape; upper lip white; below white, with square or transverse black spots, which may be confluent on the anterior portion of the body, more sparse posteriorly, and absent on the tail.

Total length 460 millim.; tail 90.

From the Guianas to the Upper Amazon.

a. Hgr. (V. 150; C. 53).	Demerara.	Zool. Society.
b, c. ♂ (V. 156; C. 62) & ♀ (V. 170; C. 52).	Upper Amazon.	Mr. E. Bartlett [C.]. (Types of *C. pœcilolæmus*.)
d. Hgr. (V. 163; C. 65).	—— ?	E. India Company. (Type of *C. orientalis*.)

9. Rhadinæa merremii.

? Coluber miliaris, *Linn. Mus. Ad. Frid.* p. 27 (1754), *and S. N.* i. p. 380 (1766).
Coluber merremii, *Wied, Reise Bras.* ii. p. 121 (1821), *Beitr. Nat. Bras.* i. p. 332, *and Abbild.* (1825).
—— miliaris, *Raddi, Mem. Soc. Ital. Modena*, xix. (*Fis.*) 1823, p. 65.
Natrix chiametla, *Wagl. in Spix, Serp. Bras.* p. 14, pl. ii. b. fig. 2 (1824).
Coluber bicolor, *Reuss, Mus. Senckenb.* i. p. 145, pl. viii. fig. 1 (1833).
Coronella merremii, part., *Schleg. Phys. Serp.* ii. p. 58 (1837).
Liophis merremii, part., *Dum. & Bibr.* vii. p. 708 (1854); *Günth. Cat.* p. 44 (1858); *Jan, Arch. Zool. Anat. Phys.* ii. 1863, p. 291.
Coronella australis, *Günth. l. c.* p. 40.
Ophiomorphus merremii, *Cope, Proc. Ac. Philad.* 1862, p. 75.
Liophis bicolor, *Jan, l. c.* p. 296.

Eye rather small. Rostral broader than deep, visible from above; internasals as long as broad or a little broader, shorter than the præfrontals; frontal once and two thirds to twice as long as broad; frontal as long as or a little longer than its distance from the end of the snout, slightly shorter than the parietals; loreal as long as deep or deeper than long; one præ- and two postoculars; temporals 1+2; eight upper labials, fourth and fifth entering the eye; five lower labials in contact with the anterior chin-shields, which are as long as or a little longer than the posterior. Scales in 17 rows. Ventrals 141–164; anal divided; subcaudals 47–58. Length of tail 5 to 5½ times in the total length. Yellowish or pale olive-brown above, each scale edged with black; lower parts uniform yellow, the shields edged with black at the sides. Young with a light nuchal collar.

Total length 500 millim.; tail 95.

Brazil.

a–d, e–f. ♂ (V. 144, 141, 150, 143; C. 53, ?, 49, 51), ♀ (V. 146; C. 49), & hgr. (V. 148; C. 50).	Brazil.	
g. Yg. (V. 143; C. 50).	Brazil.	Lord Stuart [P.].
h. ♀ (V. 148; C. ?).	Tijuca, near Rio Janeiro.	W. Schaus, Esq. [P.].
i. Yg. (V. 160; C. 56).	Colonia Alpina, Theresopolis, Prov. Rio Janeiro.	Dr. E. A. Göldi [P.].
k–l. ♀ (V. 152, 164; C. 47, 56).	S. José dos Campos, Prov. S. Paulo.	Mr. A. Thomson [P.].
m. ♀ (V. 145; C. 52).	——?	Sir J. Richardson [P.]. (Type of *C. australis*.)
n. Skull.	Brazil.	

10. Rhadinæa fusca.

Liophis merremii, *Burm. Reise La Plata*, i. p. 528 (1861).
? Opheomorphus merremii, var. semiaureus, *Cope, Proc. Ac. Philad.* 1862, p. 348.
Liophis merremii, part., *Jan, Arch. Zool. Anat. Phys.* ii. 1863, p. 291, and *Icon. Gén.* 17, pl. v. (1866); *Hensel, Arch. f. Nat.* 1868, p. 324.
Opheomorphus fuscus, *Cope, Proc. Am. Philos. Soc.* xxii. 1885, p. 190; *Bouleng. Ann. & Mag. N. H.* (5) xvi. 1885, p. 297.
Liophis fuscus, *Bouleng. Ann. & Mag. N. H.* (5) xviii. 1886, p. 431.
Rhadinæa fusca, *Bouleng. Ann. & Mag. N. H.* (6) xiii. 1894, p. 346.

Eye rather small. Rostral broader than deep, visible from above; internasals broader than long, shorter than the præfrontals; frontal once and two thirds to twice as long as broad, longer than its distance from the end of the snout, as long as or a little shorter than the parietals; loreal deeper than long; one præocular, sometimes with a small subocular below it; two postoculars; temporals 1 + 2; eight upper labials, fourth and fifth entering the eye; five lower labials in contact with the anterior chin-shields, which are as long as or a little longer than the posterior. Scales in 17 rows. Ventrals 165–183; anal divided; subcaudals 51–60. Length of tail 5 to 6 times in the total length. Yellowish or olive-brown above, each scale edged with darker; ventrals and subcaudals yellow, black-edged. Young with a Λ-shaped black band on the occiput, another on the nape, and a series of large black spots along each side, which unite into a stripe on the posterior part of the body and on the tail; these markings may be more or less distinctly preserved in the adult.

Total length 1070 millim.; tail 175.

Southern Brazil to Buenos Ayres.

a. Hgr. (V. 174; C. 59).	Brazil.	Capt. J. Parish [P.].
b, c–d. ♀ (V. 180, 175; C. 58, 58) & yg. (V. 169; C. 53).	Rio Grande do Sul.	Dr. H. v. Ihering [C.].
e. ♀ (V. 183; C. 51).	Asuncion, Paraguay.	Dr. J. Bohls [C.].
f–m. ♂ (V. 172, 168, 167, 170; C. 58, 57, 56, 51) & ♀ (V. 170, 170, 166; C. 60, 59, 57).	Uruguay.	

This form may have to be regarded as merely a variety of the preceding.

11. Rhadinæa jægeri. (PLATE VII. fig. 3.)

Coronella jægeri, *Günth. Cat.* p. 37 (1858); *Bouleng. Ann. & Mag. N. H.* (5) xviii. 1886, p. 431.
Liophis (Ophiomorphus) dorsalis, *Peters, Mon. Berl. Ac.* 1863 p. 283; *Hensel, Arch. f. Nat.* 1868, p. 325.
Opheomorphus dorsalis, *Cope, Proc. Ac. Philad.* 1868, p. 102.

Eye moderate. Rostral broader than deep, just visible from above; internasals as long as broad or a little broader, as long as or shorter than the præfrontals; frontal not twice as long as broad, a little longer than its distance from the end of the snout, a little shorter than the parietals; loreal deeper than long; one præ- and two postoculars; temporals 1+2; eight upper labials, fourth and fifth entering the eye; five lower labials in contact with the anterior chin-shields, which are as long as or a little longer than the posterior. Scales in 17 rows (exceptionally 19). Ventrals 147–165; anal divided; subcaudals 55–62. Length of tail 4½ to 5 times in the total length. Olive above, this colour extending to the sides of the ventrals; vertebral region brown; lower surface yellowish or salmon-pink, uniform or obscured with olive.

Total length 560 millim.; tail 115.

Southern Brazil and Uruguay.

a. ♀ (V. 153; C. 56).	Brazil.	Dr. Gardiner [C.].	Types.
b. ♀ (V. 147; C. 62).	Brazil.	M. Clausen [C.].	
c–e. ♂ (V. 157; C. 62), ♀ (V. 165; C. 60), & yg. (V. 164; C. 59).	Rio Grande do Sul.	Dr. H. v. Ihering [C.].	
f. ♀ (V. 164; C. 55).	Uruguay.		

12. Rhadinæa genimaculata.

Dromicus lineatus, part., *Dum. & Bibr.* vii. p. 155 (1854).
Liophis (Lygophis) genimaculata, *Boettg. Zeitschr. f. ges. Naturw.* lviii. 1885, p. 229.

Rhadinæa genimaculata, *Bouleng. Ann. & Mag. N. H.* (6) xiii. 1894, p. 347.

Eye moderate. Rostral slightly broader than deep, scarcely visible from above; internasals broader than long, as long as the præfrontals; frontal twice as long as broad, not much wider than the supraocular, much longer than its distance from the end of the snout, a little shorter than the parietals; loreal as long as deep; one præ- and two postoculars; temporals 1+2; eight upper labials, fourth and fifth entering the eye; five lower labials in contact with the anterior chin-shields, which are a little longer than the posterior. Scales in 17 rows. Ventrals 191–208; anal divided; subcaudals 51–66. Pale grey-brown above, with three dark brown light-edged longitudinal streaks; the vertebral streak widest, hardly two scales wide, and commencing on the occiput; the lateral streaks widening abruptly at the angle of the mouth and extending, through the eye, to the end of the snout; upper lip and lower surface uniform white; a pink stripe along the middle of the belly and under the tail.

Total length 390 millim.; tail 75.

Paraguay.

a.	♂ (V. 191; C. 55).	Paraguay.	Hr. H. Rohde [C.]. (One of the types.)
b.	♀ (V. 199; C. 55).	Asuncion.	Dr. J. Bohls [C.].

13. Rhadinæa obtusa.

Rhadinæa obtusa, *Cope, Proc. Ac. Philad.* 1863, p. 101.
Enicognathus, sp., *F. Müll. Verh. nat. Ges. Basel*, vii. 1882, p. 144.
Coronella obtusa, *Bouleng. Ann. & Mag. N. H.* (5) xv. 1885, p. 194.

Eye rather small. Rostral a little broader than deep, scarcely visible from above; internasals broader than long, as long as the præfrontals; frontal once and a half as long as broad, twice as broad as the supraocular, longer than its distance from the end of the snout and a little shorter than the parietals; loreal square; one præ- and two postoculars; temporals 1+2; eight upper labials, fourth and fifth entering the eye; five lower labials in contact with the anterior chin-shields, which are as long as or a little longer than the posterior. Scales in 19 rows. Ventrals 174–192; anal divided; subcaudals 65–79. Pale brown above, with a dark brown vertebral streak, occupying one scale and two halves, commencing on the snout; a dark brown lateral streak, extending round the snout and passing through the eye; sides and outer ends of ventrals dark olive-grey; upper lip and lower surface yellowish, unspotted.

Total length 380 millim.; tail 95.

Southern Brazil, Uruguay, Argentina.

a.	♂ (V. 187; C. 79).	Rio Grande do Sul.	Dr. H. v. Ihering [C.].
b.	♀ (V. 185; C. 74).	Dept. of Soriano, Uruguay.	R. Havers, Esq. [P.].
c.	♂ (V. 174; C. 65).	Colonia Resistencia, C. Chaco.	Prof. Spegazzini [C.]; Marquis G. Doria [P.].

14. Rhadinæa serperastra.

Rhadinæa serperastra, *Cope, Proc. Ac. Philad.* 1871, p. 212, *and Journ. Ac. Philad.* viii. 1876, p. 140.
Ablabes serperastra, *Günth. Biol. C.-Am., Rept.* p. 105 (1893).

Internasals transverse, narrow; frontal wide, supraciliary suture shorter than anterior, total length exceeding that of common parietal suture; loreal square; one præ- and two postoculars; temporals 1+2; eight upper labials, fourth and fifth entering the eye; chin-shields subequal. Scales in 19 rows. Ventrals 164; anal divided; subcaudals 78. Dark brown, with six longitudinal yellow or white lines; head dark brown above, with a pale shade across frontal and two just behind parietals; labials black, yellow-spotted; belly yellowish, with a series of dark spots on the ends of the ventrals.

Costa Rica.

15. Rhadinæa affinis.

Dromicus affinis, part., *Günth. Cat.* p. 128 (1858).
? Enicognathus melanocephalus, part., *Jan, Arch. Zool. Anat. Phys.* ii. 1863, p. 269, *and Icon. Gén.* 16, pl. 1. fig. 4* (1866).
Coronella iheringii, *Bouleng. Ann. & Mag. N. H.* (5) xv. 1885, p. 194.

Eye moderate. Snout short; rostral broader than deep, scarcely visible from above; internasals broader than long, shorter than the præfrontals; frontal about once and a half as long as broad, a little longer than its distance from the end of the snout and shorter than the parietals; loreal as long as deep or a little deeper than long; one præ- and two postoculars; temporals 1+2, the second upper very long; seven upper labials, third and fourth entering the eye; four lower labials in contact with the anterior chin-shields, which are shorter than the posterior. Scales in 17 rows. Ventrals 147–185; anal divided; subcaudals 58–71. Length of tail $3\frac{1}{2}$ to 5 times in the total length. Grey-brown above; a black band on each side of the head, passing through the eye, uniting with a broad transverse band on the occiput, covering the posterior half of the parietals; the rest of the upper surface of the head with black variegations, or almost entirely black; a triangular light spot behind the eye and two roundish ones close together behind the point of the frontal; the black occipital band edged with lighter behind; a blackish longitudinal line on the nape, sometimes continued along the back as a vertebral series of small spots; a roundish dark spot on each side of the nape; pale yellow inferiorly, with one or two black dots on the side of each ventral and one on each caudal; a few other minute dots may be scattered on the ventrals; chin and throat brown, with yellowish, black-edged spots or vermiculations.

Total length 710 millim.; tail 145.
Brazil.

a. Hgr. ♀ (V. 170 ; C. 64).	Rio Janeiro.	A. Fry, Esq. [P.].
b. Hgr. ♂ (V. 147 ; C. 71).	Rio Janeiro.	G. Busk, Esq. [P.].
		(Types.)
c-e. ♂ (V. 178; C. ?) & yg. (V. 172, 176; C. 58, ?).	Rio Grande do Sul.	Dr. H. v. Ihering [C.]. (Types of *C. iheringii*.)
f. ♀ (V. 185 ; C. 65).	Theresopolis.	Fischer Collection.

16. Rhadinæa pœcilopogon.

Dromicus affinis, part., *Günth. Cat.* p. 128 (1858).
Rhadinæa pœcilopogon, *Cope, Proc. Ac. Philad.* 1863, p. 100; *Günth. Zool. Rec.* 1866, p. 125.
Enicognathus elegans, *Jan, Arch. Zool. Anat. Phys.* ii. 1863, p. 268, and *Icon. Gén.* 16, pl. i. fig. 3 (1866).
Dromicus melanocephalus, *Peters, Mon. Berl. Ac.* 1863, p. 277.
? Liophis persimilis, *Cope, Proc. Ac. Philad.* 1868, p. 308.
Coronella pœcilopogon, *Bouleng. Ann. & Mag. N. H.* (5) xv. 1885, p. 194, and xviii. 1886, p. 431.
Enicognathus bilineatus, *Fischer, Jahrb. Hamb. Wiss. Anst.* ii. 1885, p. 98, pl. iii. fig. 5.

Eye rather small. Rostral broader than deep, scarcely visible from above; internasals broader than long, shorter than the præfrontals; frontal once and two thirds as long as broad, much longer than its distance from the end of the snout, and a little shorter than the parietals; loreal as long as deep or deeper than long; one præ- and two postoculars; temporals 1+2; seven upper labials, third and fourth entering the eye; four lower labials in contact with the anterior chin-shields, which are shorter than the posterior. Scales in 17 rows. Ventrals 141-164; anal divided; subcaudals 66-75. Length of tail $3\frac{1}{3}$ to 4 times in the total length. Back brown, with three black longitudinal lines or series of dots; sides usually dark grey, which colour extends upon the sides of the ventrals; head dark brown or blackish, with a white line round the snout, along the canthus rostralis, and from the eye to the second upper temporal; upper lip white, brown-dotted; ventral region yellowish in the middle, with a lateral series of small black spots or short lines, which may be confluent into a streak on the hinder half of the body.

Total length 440 millim.; tail 120.

Southern Brazil, Paraguay, and Uruguay.

a-b. ♂ (V. 151 ; C. 75) & ♀ (V. 160 ; C. 69).	Rio Grande do Sul.	Dr. H. v. Ihering [C.].
c. ♀ (V. 163 ; C. 70).	Sta. Catharina.	Dr. H. v. Ihering [C.].
d. ♂ (V. 141 ; C. 72).	Santos.	Dr. J. G. Fischer's Collection. (Type of *E. bilineatus*.)
e. ♂ (V. 155 ; C. 66).	Brazil.	Haslar Collection. (One of the types of *D. affinis*.)
f. ♂ (V. 154 ; C. 72).	Paraguay.	Prof. Grant [P.].

17. Rhadinæa lachrymans.

Lygophis lachrymans, *Cope, Proc. Amer. Philos. Soc.* xi. 1869, p. 164.
Rhadinæa lachrymans, *Cope, Journ. Ac. Philad.* viii. 1876, p. 140.
Dromicus lachrymans, *Günth. Biol. C.-Am., Rept.* p. 114 (1894).

Snout short; rostral broader than deep; internasals broader than long; frontal broad, shorter than the parietals; loreal longer than deep; one præ- and two postoculars; temporals 1+2; eight upper labials, fourth and fifth entering the eye; anterior chin-shields shorter than the posterior. Scales in 17 rows. Ventrals 173; anal divided; subcaudals 78. Chestnut-brown above; end of ventrals and first three and a half rows of scales blackish, yellowish-margined above from side of neck to end of tail; on the anterior half the body is divided by a yellowish band on the first and second rows of scales; head above brown; a deep brown band from eye across sixth labial, another across seventh, and a black spot on each side of neck; below and labials bright yellow, the anterior upper labials brown margined.

Habitat unknown.

18. Rhadinæa undulata.

Coluber undulatus, *Wied, Beitr. Nat. Bras.* i. p. 329, and *Abbild.* (1825).
Enicognathus melanocephalus, part., *Dum. & Bibr.* vii. p. 330 (1854); *Jan, Arch. Zool. Anat. Phys.* ii. 1863, p. 269, and *Icon. Gén.* 16, pl. i. fig. 4 (1866).
Coronella decorata (*non Günth.* 1858), *Günth. Proc. Zool. Soc.* 1859, p. 412.
Dromicus brevirostris, *Peters, Mon. Berl. Ac.* 1863, p. 280, and 1871, p. 400.
—— undulatus, *Peters, l. c.* 1863, p. 281.
Enicognathus tæniolatus, *Jan, Il. cc.* p. 272, 16, pl. ii. f. 4.
Dromicus boursieri, *Jan, Icon. Gén.* 25, pl. ii. fig. 2 (1867).
—— viperinus, *Günth. Ann. & Mag. N. H.* (4) i. 1868, p. 418.
Lygophis nicagus, *Cope, Proc. Ac. Philad.* 1868, p. 132.
Rhadinæa tæniolata, *Cope, Proc. Am. Philos. Soc.* xi. 1869, p. 154.
Coronella whymperi, *Bouleng. Ann. & Mag. N. H.* (5) ix. 1882, p. 460, fig.
? Enicognathus joberti, *Sauvage, Bull. Soc. Philom.* (7) viii. 1884, p. 146.
Rhadinæa nicaga, *Cope, Proc. Am. Philos. Soc.* xxiii. 1886, p. 102.
Coronella tæniolata, *Boettg. Ber. Senck. Ges.* 1888, p. 195.

Snout very short; rostral broader than deep, scarcely visible from above; internasals as long as broad, or a little broader, shorter than the præfrontals; frontal once and a half to once and two thirds as long as broad, longer than its distance from the end of the snout, shorter than the parietals; loreal as long as deep or deeper than long; one præ- and two postoculars; temporals 1+2 (rarely 2+2); eight upper labials, third, fourth, and fifth entering the eye; four lower labials in contact with the anterior chin-shields, which are shorter than the posterior. Scales in 17 rows. Ventrals 149–167;

anal divided; subcaudals 52–85. Length of tail 3 to 4⅓ times in the total length. Brown above, with a blackish vertebral streak or zigzag band, becoming indistinct on the anterior part of the body in the adult; a black streak from the eye to the side of the neck, and another along each side of the tail; upper labials yellowish, uniform or spotted with black, or blackish spotted with yellow; a light, black-edged spot on each side of the nape, and usually a pair of light dots close together on the parietals; yellowish inferiorly, uniform or the edges of the ventrals blackish; a black spot at the outer end of each ventral; chin and throat usually dotted or spotted with black.

Total length 410 millim.; tail 135.

Brazil, Ecuador, Guiana *.

a–b. ♂ (V. 156; C. 66) & ♀ (V. 154; C. ?)	Milligalli, Ecuador, 6200 feet.	E. Whymper, Esq. [C.]. (Types of *C. whymperi.*)
c. Hgr. (V. 165; C. 57).	W. Ecuador.	Mr. Fraser [C.].
d–e. Hgr. (V. 160, 161; C. 60, ?).	Pebas.	Mr. J. Hauxwell [C.]. (Types of *D. viperinus.*)

19. Rhadinæa melanauchen.

Enicognathus melanauchen, *Jan, Arch. Zool. Anat. Phys.* ii. 1863, p. 267, *and Icon. Gén.* 16, pl. i. fig. 2 (1866).

Eye rather small. Rostral much broader than deep, scarcely visible from above; internasals as long as broad, nearly as long as the præfrontals; frontal once and a half as long as broad, a little longer than its distance from the end of the snout, a little shorter than the parietals; loreal square; one præ- and three postoculars; temporals 1+2; eight upper labials, third, fourth, and fifth entering the eye; four lower labials in contact with the anterior chin-shields, which are much shorter than the posterior. Scales in 15 rows. Ventrals 148; anal divided; subcaudals 60. Reddish brown above, with blackish cross bands; a large blackish, white-edged transverse spot on the nape; lips and chin white, speckled with black; lower surface uniform white.

Total length 350 millim.; tail 95.

Bahia.

20. Rhadinæa occipitalis.

Coronella elegans (*non Tsch.*), *Günth. Cat.* p. 38 (1858).
Enicognathus occipitalis, *Jan, Arch. Zool. Anat. Phys.* ii. 1863, p. 267, *and Icon. Gén.* 16, pl. i. fig. 1 (1866).
Dromicus (Lygophis) wuchereri, *Günth. Ann. & Mag. N. H.* (3) xii. 1864, p. 225, fig.
Liophis reginæ, part., *Steind. Sitz. Ak. Wien,* lv. 1867, p. 268.
Dromicus miolepis, *Boettg. Zool. Anz.* 1891, p. 395.
Rhadinæa occipitalis, *Bouleng. Ann. & Mag. N. H.* (6) xiii. 1894, p. 347.

---

* I have examined a specimen from Cayenne, preserved in the Basle Museum.

Rostral broader than deep, scarcely visible from above; internasals broader than long, shorter than the præfrontals; frontal nearly twice as long as broad, much longer than its distance from the end of the snout, and slightly shorter than the parietals; loreal as long as deep, or a little deeper than long; one præ- and two postoculars; temporals 1+2 or 2+2; eight upper labials, third, fourth, and fifth entering the eye; four or five lower labials in contact with the anterior chin-shields, which are shorter than the posterior. Scales in 15 rows. Ventrals 152–183; anal divided; subcaudals 64–86. Tail $3\frac{2}{3}$ to $4\frac{1}{2}$ times in the total length. Back olive-brown, with two alternating series of blackish, light-edged spots, which are largest and sometimes confluent into cross bands on the anterior part of the body, and may be lost altogether posteriorly; sides usually darker, sometimes blackish grey; head brown above, with a blackish lateral streak, passing through the eyes, and a white line round the snout, along the canthus rostralis and supraciliary edge; usually a pair of white dots close together on the parietals; sometimes a pair of larger light spots on the occiput; upper lip white; belly white, with a lateral series of black dots, which may be confluent into a longitudinal line.

Total length 475 millim.; tail 105.

Brazil, Paraguay, North Argentina, Bolivia, Eastern Peru.

a. ♀ (V. 160; C. 66).	Bahia.	Dr. O. Wucherer [P.]. (Type of *D. wuchereri*).
b. ♀ (V. 183; C. 64).	Rio Grande do Sul.	Dr. H. v. Ihering [C.].
c. Yg. (V. 175; C. 68).	Asuncion, Paraguay.	Dr. J. Bohls (C.).
d. ♂ (V. 178; C. ?).	Candelaria, Argentina.	M. J. Bove [C.]; Marquis G. Doria [P.].
e. ♂ (V. 172; C. 73).	Cashiboya, N.E. Peru.	Mr. W. Davis [C.]; Messrs. Veitch [P.].
f. ♂ (V. 165; C. 70).	Peru ?	

21. Rhadinæa decorata.

Coronella decorata, *Günth. Cat.* p. 35 (1858); *Zool. Rec.* 1866, p. 125, *and Biol. C.-Am., Rept.* p. 111 (1893).

Diadophis decoratus, *Cope, Proc. Ac. Philad.* 1860, p. 250; *Bocourt, Miss. Sc. Mex., Rept.* p. 624, pl. xl. fig. 3 (1886).

Enicognathus vittatus, part., *Jan, Arch. Zool. Anat. Phys.* ii. 1863, p. 271, *and Icon. Gén.* 16, pl. ii. fig. 2 (1866).

Rhadinæa decorata, *Cope, Proc. Ac. Philad.* 1863, p. 101, *and Journ. Ac. Philad.* viii. 1876, p. 138.

Dromicus ignitus, *Cope, Proc. Ac. Philad.* 1871, p. 201.

Rhadinæa ignita, *Cope, Journ. Ac. Philad.* viii. 1876, p. 140, *and Proc. Amer. Philos. Soc.* xxxi. 1893, p. 344.

Coronella ignita, *Günth. Biol. C.-Am., Rept.* p. 111 (1893).

Eye moderate. Rostral broader than deep, scarcely visible from above; internasals broader than long, shorter than the præfrontals; frontal once and a half as long as broad, longer than its distance from the end of the snout, shorter than the parietals; loreal square; one or two præ- and two postoculars; a small subocular may be present below the præocular; temporals 1+2; eight upper labials, fourth and fifth entering the eye; four or five lower labials in contact with the anterior chin-shields, which are shorter than the posterior. Scales in 17 rows. Ventrals 117-137; anal divided; subcaudals 90-120. Length of the tail more than one third of the total. Pale brown above, with a yellowish black-edged streak along each side of the back, an oblique yellow black-edged streak on the outer border of the parietal, and a yellow black-edged spot on each side of the nape; upper lip yellow, black-edged; sides darker than back; yellow below, with or without a black spot or dot at the outer end of each ventral.

Total length 410 millim.; tail 180.

Central America.

a, b. ♂ (V. 128, 128; C. 120, ?). Mexico. M. Sallé [C.]. (Types.)
c-e. ♂ (V. 122; C. ?), ♀ (V. 129; Atoyac. Mr. H. H. Smith [C.];
  C. 102), & hgr. (V. 122; C. ?).   F. D. Godman, Esq. [P.].

22. Rhadinæa vermiculaticeps.

Tæniophis vermiculaticeps, *Cope, Proc. Ac. Philad.* 1860, p. 249.
Rhadinæa vermiculaticeps, *Cope, Proc. Ac. Philad.* 1863, p. 101.
Coronella vermiculaticeps, *Günth. Biol. C.-Am., Rept.* p. 111 (1893).

Snout short, eye large; frontal elongate; loreal as deep as long; one præ- and two postoculars; eight upper labials, fourth and fifth entering the eye; posterior chin-shields longer than the anterior. Scales in 17 rows. Ventrals 117; anal divided; subcaudals 79. Yellowish brown above, with two deep brown dorsal streaks, separated by the width of one scale, diverging on the neck and extending to the outer posterior angle of the supraocular shield; another dark streak on each side of head and body, passing through the eye; head brown above, vermiculated with yellowish; labials whitish, narrowly edged with brown; chin and belly yellowish white, each ventral with a deep brown dot at each end near the posterior border.

Total length 330 millim.; tail 115.

Veragua, Costa Rica.

23. Rhadinæa clavata.

Dromicus clavatus, *Peters, Mon. Berl. Ac.* 1864, p. 388; *Bocourt, Miss. Sc. Mex., Rept.* p. 711, pl. xlv. fig. 2 (1890).

Rostral visible from above; frontal once and a half as long as broad, a little shorter than the parietals; loreal a little longer than deep; one præ- and two postoculars; temporals 1+2; eight upper labials, fourth and fifth entering the eye; four lower labials in contact with the anterior chin-shields. Scales in 19 rows. Ventrals 127; anal divided; subcaudals 89. Pale brown above, with three dark longitudinal lines; head brown above, with a dark longitudinal streak on the frontal, and a yellow line on each side beginning above the nostril and terminating club-shaped behind the eye; a short yellowish-brown longitudinal band on each side of the nape; labials yellowish white, edged above by a black line; beneath yellow.

Total length 240 millim.; tail 87.

Mexico.

24. Rhadinæa vittata.

Enicognathus vittatus, part., *Jan, Arch. Zool. Anat. Phys.* ii. 1863, p. 271, *and Icon. Gén.* 16, pl. ii. fig. 3 (1866); *Bocourt, Miss. Sc. Mex., Rept.* p. 630, pl. xli. fig. 1 (1886).
Dromicus tæniatus, *Peters, Mon. Berl. Ac.* 1863, p. 275; *Günth. Biol. C.-Am., Rept.* p. 113 (1894).
Rhadinæa fulvivittis, *Cope, Journ. Ac. Philad.* viii. 1876, p. 139.
—— tæniata, *Cope, l. c.* p. 140.
Diadophis fulvivittis, *Garm. N. Am. Rept.* p. 158 (1883).
Rhadinæa quinquelineata, *Cope, Proc. Amer. Philos. Soc.* xxiii. 1866, p. 277.
—— vittata, *Cope, Bull. U.S. Nat. Mus.* no. 32, 1887, p. 80.
Diadophis decoratus, *Garm. Bull. Essex Inst.* xix. 1888, p. 127.
Coronella quinquelineata, *Günth. Biol. C.-Am., Rept.* p. 111 (1893).
Dromicus omiltemanus, *Günth. l. c.* p. 113, pl. xl. fig. B (1894).
—— fulvivittis, *Günth. l. c.* p. 113.

Eye rather small. Rostral broader than deep, just visible from above; internasals broader than long, shorter than the præfrontals; frontal once and a half to once and two thirds as long as broad, longer than its distance from the end of the snout, shorter than the parietals; loreal as long as or a little longer than deep; one præ- and two postoculars; sometimes a small subocular below the præ-ocular; temporals 1+2 or 1+1; eight upper labials, fourth and fifth entering the eye; five lower labials in contact with the anterior chin-shields, which are as long as or a little shorter than the posterior. Scales in 17 rows. Ventrals 145–183; anal divided; subcaudals 88–132. Length of tail 3 to $3\frac{1}{3}$ times in the total length. Dark brown above, with two yellow longitudinal streaks commencing on the snout, or yellowish brown with three dark brown or black stripes; upper lip and lower parts yellowish white.

Total length 700 millim.; tail 210.

Mexico.

A. A black dot at the outer end of each ventral.

a–b. Hgr. ♀ (V. 182, 183; C. 106, 100).	City of Mexico.	Mr. Doorman [C.].
c, d–e. ♂ (V. 172; C. ?), ♀ (V. 182; C. 88), & yg. (V. 165; C. 106).	La Cumbre de los Arrastrados, Jalisco.	Dr. A. C. Buller [C.].
f. Hgr. ♀ (V. 153; C. 88).	Omilteme, Guerrero.	F. D. Godman, Esq. [P.]. (Type of *D. omiltemanus.*)
g. Hgr. ♂ (V. 175; C. 90).	Mexico.	

B. Ventrals without black dots; two outer rows of scales yellowish.

h. ♂ (V. 163; C. 99).	Xantipa, Guerrero.	Mr. H. H. Smith [C.]; F. D. Godman, Esq. [P.].
i. Yg. (V. 162; C. 132).	Amula, Guerrero.	Mr. H. H. Smith [C.]; F. D. Godman, Esq. [P.].

C. Ends of ventrals and four outer rows of scales closely speckled with blackish; yellow lateral streak interrupted on the temple.

k. Yg. (V. 166; C. 130).	Amula, Guerrero.	Mr. H. H. Smith [C.]; F. D. Godman, Esq. [P.].
l, m. Hgr. ♂ (V. 156; C. 125) & ♀ (V. 169; C. ?).	S. Mexico.	F. D. Godman, Esq. [P.].

Rhadinœa fulviceps, Cope, Proc. Amer. Philos. Soc. xxiii. 1886, p. 279, from Panama, appears to be very closely allied to *R. vittata*.

25. Rhadinæa laureata.

Dromicus laureatus, *Günth. Ann. & Mag. N. H.* (4) i. 1868, p. 419, pl. xix. fig. E; *Bocourt, Miss. Sc. Mex., Rept.* p. 710, pl. xlv. fig. 1 (1890); *Günth. Biol. C.-Am., Rept.* p. 112, pl. xl. fig. A (1893).
Rhadinæa loreata, *Cope, Journ. Ac. Philad.* viii. 1876, p. 140.

Eye rather small. Rostral twice as broad as deep, scarcely visible from above; internasals much broader than long, much shorter than the præfrontals; frontal once and a half to once and two thirds as long as broad, longer than its distance from the end of the snout, shorter than the parietals; loreal as long as or a little longer than deep; one præ- and two postoculars; temporals 1+2; seven upper labials, third and fourth entering the eye; four lower labials in contact with the anterior chin-shields, which are as long as or a little shorter than the posterior. Scales in 17 rows. Ventrals 159-164; anal divided; subcaudals 85-95. Head and vertebral region dark brown, sides yellowish brown or pale reddish, with an ill-defined darker lateral streak, on and below which the scales are speckled with black; a fine yellow line round the snout on the canthus rostralis, extending posteriorly to the second temporal or to the last labial shield; a crescentic white black-edged line may border the head posteriorly; a white streak on the upper lip, edged with black above; labials speckled with brown; lower parts uniform white or with a few minute blackish dots.

Total length 510 millim.; tail 170.

Mexico.

a. ♂ (V. 163; C. 95).	City of Mexico.	Mr. Doorman [C.]. (Type.)
b-e. ♂ (V. 164, 163, 159; C. 87, 90, 90) & hgr. (V. 160; C. 85).	La Cumbre de los Arrastrados, Jalisco.	Dr. A. C. Buller [C.].

26. Rhadinæa godmani.

Dromicus godmanni, *Günth. Ann. & Mag. N. H.* (3) xv. 1865, p. 94.
Rhadinæa godmanii, *Cope, Journ. Ac. Philad.* viii. 1876, p. 139.
Henicognathus godmanii, *Bocourt, Miss. Sc. Mex., Rept.* p. 631, pl. xl. fig. 5 (1886).

Coronella godmani, *Günth. Biol. C.-Am., Rept.* p. 110, pl. xxxix. fig. B (1893).

Eye rather small; snout very short. Rostral broader than deep, scarcely visible from above; internasals much broader than long, about half as long as the præfrontals; frontal once and one third to once and a half as long as broad, a little longer than its distance from the end of the snout, shorter than the parietals; loreal square or longer than deep; one præ- and two postoculars; temporals 1+2; eight upper labials, fourth and fifth entering the eye; four lower labials in contact with the anterior chin-shields, which are as long as or a little longer than the posterior. Scales in 21 rows. Ventrals 168–176; anal divided; subcaudals 72–92. Length of tail three and a half to four times in the total length. Pale brown above, with three or five dark brown longitudinal streaks; head dark brown, bordered with yellow spots posteriorly; upper labials dark brown with yellow spots; yellowish beneath.

Total length 445 millim.; tail 130.

Guatemala.

a–d. ♂ (V. 176, 173; C. 88, 92), ♀ (V. 176; C. 72), & yg. (V. 170; C. 81).	Dueñas.	Messrs. Salvin & Godman [C.]. (Types.)

82. UROTHECA.

Xenodon, part., *Schleg. Phys. Serp.* ii. p. 80 (1837).
Urotheca, *Bibr. in R. de la Sagra, Hist. Cuba, Rept.* p. 217 (1843).
Liophis, part., *Dum. & Bibr. Erp. Gén.* vii. p. 697 (1854); *Günth. Cat. Col. Sn.* p. 42 (1858); *Bocourt, Miss. Sc. Mex., Rept.* p. 633 (1886).
Pliocercus, *Cope, Proc. Ac. Philad.* 1860, p. 253.
Elapochrous, *Peters, Mon. Berl. Ac.* 1860, p. 294.
Cosmiosophis, *Jan, Arch. Zool. Anat. Phys.* ii. 1863, p. 289.
Leiosophis, part., *Jan, l. c.* p. 320.
Cyclagras, part., *Cope, Proc. Am. Philos. Soc.* xxiii. 1886, p. 488.

Maxillary teeth 11 to 14, increasing in size posteriorly, followed after a short interspace by two enlarged ones; mandibular teeth subequal. Head scarcely distinct from neck; eye moderate or rather small, with round pupil; usually one or more subocular shields. Body cylindrical; scales smooth, without apical pits, in 17 to 21 rows; ventrals rounded. Tail long, thick throughout, ending obtusely; subcaudals in two rows.

Cuba, Central and South America.

Synopsis of the Species.

I. Eye in contact with two labials; scales in 17 rows.

A. Four lower labials in contact with the anterior chin-shields.

Anterior chin-shields much shorter than
  the posterior; temporals 1+2 .... 1. *dumerilii*, p. 181.
Anterior chin-shields as long as the pos-
  terior; temporals 1+1 .... 2. *lateristriga*, p. 181.

B. Five or six lower labials in contact with the anterior chin-shields.

Temporals 1+2; black above, with narrow bright transverse lines .... 3. *euryzona*, p. 182.
Temporals usually 1+1; red with black or black and yellow rings .... 4. *elapoides*, p. 182.

II. Eye separated from the labials by a series of suboculars; scales in 19 or 21 rows .... 5. *bicincta*, p. 184.

1. Urotheca dumerilii.

Urotheca dumerilii, *Bibr. in R. de la Sagra, Hist. Cuba, Rept.* p. 218, pl. xxvi. (1843).
Dromicus dumerilii, *Garm. Proc. Am. Philos. Soc.* xxiv. 1887, p. 280.

Eye moderate; snout short, truncate. Rostral just visible from above; internasals half as long as præfrontals; frontal once and two thirds as long as broad, longer than its distance from the end of the snout, shorter than the parietals; loreal as long as deep; two præoculars, with a small subocular below them; two postoculars; temporals 1+2; eight upper labials, fourth and fifth entering the eye; four lower labials in contact with the anterior chin-shields, which are much shorter than the posterior. Scales in 17 rows. Ventrals 128-153; anal divided. The tail of the type specimen is probably mutilated, hence the small number (42) of subcaudals. Brown above; a blackish lateral streak, running along the outer ends of the ventrals; lower parts orange.

Total length 344 millim.

Cuba.

2. Urotheca lateristriga.

Liophis lateristriga, *Berth. Götting. Anz.* iii. 1859, p. 180; *Jan, Arch. Zool. Anat. Phys.* ii. 1863, p. 303, *and Icon. Gén.* 18, pl. v. fig. 2 (1866).
Dromicus frenatus, *Peters, Mon. Berl. Ac.* 1863, p. 278.
—— multilineatus, *Peters, l. c.* p. 279.
—— nuntius, *Jan, Icon.* 24, pl. vi. fig. 1 (1867).
—— lateristriga, *Cope, Proc. Ac. Philad.* 1868, p. 103.
Ablabes decipiens, *Günth. Biol. C.-Am., Rept.* p. 105, pl. xxxvii. fig. A (1893).

Eye rather small; snout very short, rounded. Rostral much broader than deep, just visible from above; internasals much shorter than the præfrontals; frontal about once and a half as long as broad, longer than its distance from the end of the snout, shorter than the parietals; loreal nearly as long as deep (fused with the lower præ-ocular in one of the specimens in the collection); one or two præoculars, usually with a small subocular below; two postoculars; temporals 1+1; eight upper labials, fourth and fifth entering the eye; four lower labials in contact with the anterior chin-shields, which are nearly as long as the posterior. Scales in 17 rows.

Ventrals 133–163; anal divided; subcaudals 83–110. Brown above, with one or two yellowish streaks along each side, the lowermost bordered inferiorly by a dark purplish-brown streak running along the ends of the ventrals and the lower half of the outer row of scales; this dark streak may be replaced by a series of black spots along the outer ends of the ventrals; a dark brown streak on each side of the head, passing through the eye; upper labials yellowish above, brown inferiorly; lower parts uniform yellowish.

Total length 580 millim.; tail 200.

Costa Rica, Colombia, Ecuador, Venezuela.

a. ♀ (V. 162; C. 83).	Intac, Ecuador.	Mr. Buckley [C.].
b-d. ♂ (V. 144, 133; C. ?, 110) & ♀ (V. 151; C. ?).	Irazu, Costa Rica.	F. D. Godman, Esq. [P.]. (Types of *Ablabes decipiens*.)

3. Urotheca euryzona.

Pliocercus euryzonus, *Cope, Proc. Ac. Philad.* 1862, p. 72, and 1865, p. 190.
Liophis splendens, *Jan, Arch. Zool. Anat. Phys.* ii. 1863, p. 302, *and Icon. Gén.* 18, pl. v. fig. 1 (1866).
Elapochrus euryzona, *Günth. Biol. C.-Am., Rept.* p. 197 (1893).

Eye moderate; snout short and broad. Rostral twice as broad as deep, just visible from above; internasals half as long as the præfrontals; frontal about once and a half as long as broad, slightly longer than its distance from the end of the snout, shorter than the parietals; loreal as long as deep; one præocular, with a small subocular below it; two postoculars; temporals 1+2; eight upper labials, fourth and fifth entering the eye; five lower labials in contact with the anterior chin-shields, which are a little shorter than the posterior. Scales in 17 rows. Ventrals 120–138; anal divided; subcaudals 116. Black above, with narrow pale (red?) transverse lines, about half the length of a scale, and equidistant; these lines widen into irregular broad cross bars on the abdomen, where the red and black are in nearly equal proportions; a few small pale spots on the head.

Total length 560 millim.; tail 260.

Colombia and Ecuador.

a. ♂ (V. 120; C. 116).	Nanegal, Ecuador, 3000 feet.	E. Whymper, Esq. [C.].

4. Urotheca elapoides.

Pliocercus elapoides, *Cope, Proc. Ac. Philad.* 1860, p. 253; *Salvin, Proc. Zool. Soc.* 1861, p. 227; *Cope, Proc. Ac. Philad.* 1865, p. 190; *Peters, Mon. Berl. Ac.* 1869, p. 876; *F. Müll. Verh. nat. Ges. Basel*, vi. 1878, p. 660.
Elapochrous deppii, *Peters, Mon. Berl. Ac.* 1860, p. 294, pl. — fig. 2.
Pleiocercus æqualis, *Salvin, l. c.*; *Cope, Proc. Ac. Philad.* 1865, p. 190; *F. Müll. l. c.* p. 662, pl. ii. fig. A.

Liophis tricinctus, *Jan, Arch. Zool. Anat. Phys.* ii. 1863, p. 301, *and Icon. Gén.* 18, pl. iv. figs. 4-6 (1866).
Pliocercus dimidiatus, *Cope, Proc. Ac. Philad.* 1865, p. 190, *and Journ. Ac. Philad.* viii. 1876, p. 138, *and Proc. Amer. Philos. Soc.* xxii. 1885, p. 183.
—— sargii, *Fisch. Arch. f. Nat.* 1881, p. 225, pl. xi. figs. 1-3.
Liophis elapoides, *Garm. N. Am. Rept.* p. 69 (1883); *Bocourt, Miss. Sc. Mex., Rept.* p. 635, pl. xli. fig. 6 (1886).
—— elapoides, vars. diastema *et* æqualis, *Bocourt, l. c.* figs. 7 & 8.
Elapochrus æqualis, *Günth. Biol. C.-Am., Rept.* p. 106, pl. xxxvi. fig. A (1893).
—— dimidiatus, *Günth. l. c.* p. 107.

Eye rather small; snout short, rounded. Rostral nearly twice as broad as deep, just visible from above; internasals shorter than the præfrontals; frontal as long as or a little longer than its distance from the end of the snout, shorter than the parietals; loreal as long as deep; one or two præoculars and a small subocular; two postoculars; temporals 1+1 (rarely 1+2); eight (rarely nine) upper labials, fourth and fifth (or fifth and sixth) entering the eye; five lower labials in contact with the anterior chin-shields, which are a little longer than the posterior. Scales in 17 rows. Ventrals 124-143; anal divided; subcaudals 85-127. Red above, the scales usually tipped with black, with single or triple black annuli, the annuli when triad separated by yellow interspaces; some large black spots may be present on the red areas; head black in front and behind, with a yellow band across the parietal shields and the temples; belly yellowish between the rings.

Total length 550 millim.; tail 240.

Mexico, Guatemala, Costa Rica.

The individuals of this species vary immensely in the number and arrangement of the annuli, as may be seen by referring to the descriptions and figures quoted in the synonymy. I am therefore disposed to regard the forms *diastema* (5 triad annuli on the body) and *æqualis* as extreme colour-varieties of one species. I have dealt in the same manner with *Atractus elaps* and *A. latifrons*, which present analogous variations.

A. Black annuli triad, with yellow interspaces.

  a. 7 to 10 annuli on the body; irregular black blotches on the red areas.

a.	♂ (V. 126; C. 97).	Mexico.	
b.	♂ (V. 128; C. ?).	Teapa, Tabasco.	F. D. Godman, Esq. [P.].
c.	♀ (V. 133; C. 110).	Jalisco.	F. D. Godman, Esq. [P.].

  b. 8 annuli on the body; no black spots.

d.	Yg. (V. 130; C. 104).	Dueñas, Guatemala.	O. Salvin, Esq. [C.].

B. 25 to 27 equidistant black annuli on the body.

e.	♀ (V. 133; C. 92).	S. Gerónimo, Guatemala.	Robert Owen, Esq. [C.]. (Type of *P. æqualis*.)
f.	Yg. (V. 131; C. 112).	Vera Paz, low forest.	O. Salvin, Esq. [C.].

5. Urotheca bicincta.

Coluber bicinctus, *Hermann, Obs. Zool.* p. 276 (1804).
Elaps schranckii, *Wagl. in Spix, Serp. Bras.* p. 1, pl. i. (1824).
Xenodon bicinctus, *Schleg. Phys. Serp.* ii. p. 95 (1837).
Liophis bicinctus, *Dum. & Bibr.* vii. p. 716 (1854); *Günth. Cat.* p. 43 (1858).
Leiosophis bicinctus, *Jan, Arch. Zool. Anat. Phys.* ii. 1863, p. 321.

Eye rather small; snout short, rounded. Rostral much broader than deep, just visible from above; internasals as long as or shorter than the præfrontals; frontal as long as broad or a little longer, as long as or shorter than its distance from the end of the snout, shorter than the parietals; loreal nearly as long as deep or longer than deep; six scales surround the eye in addition to the supraocular, the labials being excluded from the eye; temporals 2+2 or 3+3; eight or nine upper labials; chin-shields short. Scales in 19 or 21 rows. Ventrals 168–192; anal entire; subcaudals 69–94. Reddish brown above, with black annuli arranged in pairs, each pair bordered with yellow in front and behind; a black band behind the eye, confluent with another across the occiput and descending to the gular region; belly yellow, spotted with black; tail with complete annuli.

Total length 1950 millim.; tail 650.

Guianas, Brazil.

a. ♂ (Sc. 19; V. 171; C. 90).	British Guiana.	Demerara Museum.
b. ♀ (Sc. 21; V. 176; C. 72).	S. America.	
c. Yg. (Sc. 19; V. 169; C. 87).	S. America.	Zool. Soc.

83. TRIMETOPON.

Trimetopon, *Cope, Proc. Amer. Philos. Soc.* xxii. 1885, p. 177.

Maxillary short, with 12 teeth, which gradually increase in length to the last; mandibular teeth subequal. Head scarcely distinct from neck; eye small, with round pupil; a single præfrontal. Body elongate, cylindrical; scales smooth, with very indistinct apical pits, in 15 rows; ventrals rounded. Tail moderate; subcaudals in two rows.

Central America.

1. Trimetopon gracile.

Ablabes gracilis, *Günth. Ann. & Mag. N. H.* (4) ix. 1872, p. 18, pl. iii. fig. D.
Trimetopon gracile, *Cope, Proc. Amer. Philos. Soc.* xxii. 1885, p. 177.

Snout short, rounded. Rostral more than twice as broad as deep, scarcely visible from above; internasals small, much broader than deep; præfrontal nearly twice as broad as deep; frontal once and one third to once and a half as long as deep, a little longer than its distance from the end of the snout, much shorter than the parietals; loreal small, longer than deep; one præ- and one postocular; temporals 1+1; seven or eight upper labials, third and fourth or fourth and fifth entering the eye; four or five lower labials in contact with the anterior chin-shields, which are longer than the posterior. Scales in 15 rows. Ventrals 141–149; anal divided; subcaudals 60–65. Dark brown above, with five blackish longitudinal lines; a more or less distinct yellowish collar; lateral scales lighter in the centre; labials yellowish, spotted with black; lower parts uniform whitish.

Total length 290 millim.; tail 77.

Costa Rica.

a–b. ♂ (V. 149, 141; C. 65, 60).   Cartago, Costa Rica.   (Types.)

84. HYDROMORPHUS.

Hydromorphus, *Peters, Mon. Berl. Ac.* 1859, p. 276.

Maxillary short, with 14 teeth, which gradually increase in length to the last; mandibular teeth subequal. Head scarcely distinct from neck; eye very small, with round pupil; a single præfrontal; nostril directed upwards, in a single shield; loreal entering the eye. Body rather elongate, cylindrical; scales smooth, in 17 rows; ventrals rounded. Tail short; subcaudals in two rows.

Central America.

1. Hydromorphus concolor.

Hydromorphus concolor, *Peters, l. c.* p. 277, pl. —. fig. 3.

Head much depressed, with broadly rounded snout. Nostrils directed upwards, the nasals separated by a pair of very small internasals; præfrontal twice as broad as long; frontal slightly longer than broad, as long as its distance from the rostral, much shorter than the parietals; loreal twice as long as deep; a small præocular between the supraocular and the loreal; two postoculars; temporals 1+2; six upper labials, third bordering the eye; four lower labials in contact with the anterior chin-shields; posterior chin-shields small. Scales in 17 rows. Ventrals 175; anal divided; subcaudals 31. Dark greyish brown above; sides and lower parts paler, the scales and shields being partly yellowish.

Total length 850 millim.; tail 86.

Costa Rica.

85. DIMADES.

Pseudoeryx, part., *Fitzing. N. Class. Rept.* p. 29 (1826).
Helicops, part., *Wagler, Syst. Amph.* p. 170 (1830).
Homalopsis, part., *Schleg. Phys. Serp.* ii. p. 297 (1837).
Dimades, part., *Gray, Zool. Misc.* p. 65 (1842).
Dimades, *Gray, Cat. Sn.* p. 76 (1849).
Colopisma, part., *Dum. & Bibr. Erp. Gén.* vii. p. 336 (1854); *Jan, Arch. Zool. Anat. Phys.* iii. 1865, p. 241.

Maxillary teeth 15 to 17, gradually increasing in size; mandibular teeth subequal. Head small, not distinct from neck; eye rather small, with round pupil; nostril directed upwards, in a semi-divided nasal; a single internasal; no loreal. Body cylindrical, stout; scales as broad as long, smooth, without apical pits, in 15 rows; ventrals rounded. Tail rather short; subcaudals in two rows. South America.

1. Dimades plicatilis.

Coluber plicatilis, *Linn. Mus. Ad. Frid.* p. 23, pl. vi. fig. 1 (1754), and *S. N.* i. p. 376 (1766); *Daud. Rept.* vii. p. 193 (1803).
Cerastes plicatilis, *Laur. Syn. Rept.* p. 81 (1768).
Elaps plicatilis, *Schneid. Hist. Amph.* ii. p. 294 (1801).
Pseudoeryx daudini, *Fitzing. N. Class. Rept.* p. 55 (1826).
Homalopsis plicatilis, *Boie, Isis,* 1827, p. 551.
—— plicatilis, part., *Schleg. Phys. Serp.* ii. p. 353, pl. xiii. figs. 21 & 22 (1837).
Dimades plicatilis, *Gray, Zool. Misc.* p. 65 (1842), and *Cat.* p. 76 (1849).
Calopisma plicatile, *Dum. & Bibr.* vii. p. 344 (1854); *Jan, Arch. Zool. Anat. Phys.* iii. 1865, p. 242, and *Icon. Gén.* 29, pl. v. figs. 2 & 3 (1868).

Snout very short, rounded; rostral broader than deep, just visible from above, in contact with the nasal fissure; nasals in contact behind the rostral; internasals small, rhomboidal, at least twice as broad as long; frontal twice to twice and a half as long as broad, longer than its distance from the end of the snout, shorter than the parietals; one præ- and two postoculars; temporals 1+2; eight upper labials, third and fourth entering the eye; four lower labials in contact with the anterior chin-shields, which are as long as or a little longer than the posterior. Scales in 15 rows. Ventrals 129–140; anal divided; subcaudals 32–51. Brown above, uniform or with two rows of small black spots; a black lateral band, with a series of small whitish spots; a black streak on each side of the head, passing through the eye; lips yellowish, spotted or vermiculated with brown; belly yellowish with black dots, which may form regular longitudinal series; a series of large black dots runs along each side of the belly, on the first row of scales, and is continued on the subcaudals.

Total length 920 millim.; tail 120.

Guianas, Brazil.

a. ♀ (V. 137; C. 36).	British Guiana.	
b, c. ♂ (V. 135; C. 46) & ♀ (V. 140; C. 36).	British Guiana.	Demerara Museum.
d. ♂ (V. 132; C. 47).	Berbice.	
e. Hgr. (V. 129; C. 51).	Para.	
f. Hgr. (V. 131; C. 46).	——?	College of Surgeons [P.].

86. HYDROPS.

Hydrops, *Wagler, Syst. Amph.* p. 170 (1830); *Gray, Cat. Sn.* p. 75 (1849); *Dum. & Bibr. Erp. Gén.* vii. p. 482 (1854).
Homalopsis, part., *Schleg. Phys. Serp.* ii. p. 297 (1837).
Higina, *Gray, Zool. Misc.* p. 67 (1842), and *Cat. Sn.* p. 75.
Calopisma, part., *Jan, Arch. Zool. Anat. Phys.* iii. 1865, p. 241.

Maxillary teeth 14 or 15, gradually increasing in size; mandibular teeth subequal. Head small, not distinct from neck; eye very small, with round pupil; nostril directed upward, in a semidivided nasal; a single internasal; no loreal. Body cylindrical, slender; scales short, smooth, without apical pits, in 15 or 17 rows; ventrals rounded. Tail rather short; subcaudals in two rows. South America.

1. Hydrops triangularis.

Elaps triangularis, *Wagl. in Spix, Serp. Bras.* p. 5, pl. 2 a. fig. 1 (1824).
Homalopsis martii, *Schleg. Phys. Serp.* ii. p. 356, pl. xiii. figs. 19 & 20 (1837).
Hydrops martii, *Gray, Zool. Misc.* p. 68 (1842), *and Cat.* p. 75 (1849); *Dum. & Bibr.* vii. p. 484 (1854); *Günth. Ann. & Mag. N. H.* (4) i. 1868, p. 421.
Higina fasciata, *Gray, ll. cc.* pp. 67, 75.
Calopisma martii, *Jan, Arch. Zool. Anat. Phys.* iii. 1865, p. 242, *and Icon. Gén.* 29, pl. iv. fig. 1 (1868).

Snout broadly rounded. Rostral much broader than deep, scarcely visible from above; nasals in contact with each other behind the rostral; internasal rhomboidal, nearly twice as broad as long; frontal a little longer than broad, as long as or longer than its distance from the end of the snout, shorter than the parietals; one præ- and two postoculars; temporals 1+1; eight upper labials, fourth entering the eye; four lower labials in contact with the anterior chin-shields, which are as long as or shorter than the posterior. Scales in 15 rows. Ventrals 145-176; anal divided; subcaudals 40-66. Purplish brown above, red on the sides, white below; with black annuli, which may be interrupted and alternate on the middle dorsal and ventral lines.

Total length 780 millim.; tail 110.

Guianas, Brazil.

a-b, c-e. Hgr. (V. 165, 170, 168, 176, 168; C. 56, 58, ?, 56, 54).	Demerara.	(Types of *H. fasciata*.)
f. Yg. (V. 167; C. 46).	Demerara.	Zoological Society.
g-i. ♀ (V. 169, 161, 166; C. 45, 43, 45).	Berbice.	
k. Hgr. (V. 166; C. 64).	Surinam.	W. Carruthers, Esq. [P.].
l. Yg. (V. 175; C. 42).	Surinam.	
m, n. Hgr. (V. 159; C. 60) & yg. (V. 157; C. 58).	——?	

2. Hydrops martii.

Elaps martii, *Wagler, in Spix, Serp. Bras.* p. 3, pl. ii. fig 2 (1824).
Hydrops callostictus, *Günth. Ann. & Mag. N. H.* (4) i. 1868, p. 421, pl. xvii. fig. B.

Very closely allied to the preceding, but scales in 17 rows. Ventrals 168-179; subcaudals 71-76. Red, back purplish, with black annuli, which are bordered on the upper surface of the body with round whitish spots.

Total length 265 millim.; tail 60.

Brazil and N.E. Peru.

a. Hgr. (V. 168; C. 71).	Chyavetas, Upper Amazon.	Mr. E. Bartlett [C.]. (Type of *H. callostictus*.)

87. SYMPHOLIS.

Sympholis, *Cope, Proc. Ac. Philad.* 1861, p. 524 (1862); *Bocourt, Miss. Sc. Mex., Rept.* p. 555 (1883).
Cheilorhina (*non Lütk.*), *Jan, Arch. Zool. Anat. Phys.* ii. 1862, p. 57.

Maxillary teeth small, 12, increasing in length posteriorly. Head not distinct from neck; eye very small, with round pupil; internasals fused with the præfrontals; nasal fused with the first labial; loreal and præocular present. Body cylindrical; scales very short, smooth, without pits, in 19 rows; ventrals rounded. Tail very short, obtuse; subcaudals in two rows.

Mexico.

1. Sympholis lippiens.

Sympholis lippiens, *Cope, l. c., and Am. Journ. Sc. & Arts*, xxxv. 1863, p. 458; *Bocourt, l. c.* pl. xxxiv. fig. 5.
Cheilorhina villarsii, *Jan. l. c., and Icon. Gén.* 48, pl. i. fig. 5 (1876).
Geophis lippiens, *Garm. N. Am. Rept.* p. 103 (1883).

Snout rounded, feebly prominent. Rostral as deep as broad, visible from above; frontal longer than broad, pointed in front, longer than its distance from the end of the snout, as long as the parietals; supraocular nearly as broad as the frontal; a small loreal and a small præocular; one postocular, which may be fused with the supraocular; one temporal; five upper labials, first fused with the nasal, third, or second and third, entering the eye; a single pair of small chin-shields. Scales in 19 rows. Ventrals 213; anal entire; subcaudals 15. Yellow, with black annuli, which are more or less incomplete inferiorly; snout black.

Total length 500 millim.

Guadalajara.

88. CORONELLA.

Coronella, part., *Laurenti, Syn. Rept.* p. 84 (1768); *Boie, Isis*, 1827, p. 519; *Schleg. Phys. Serp.* ii. p. 50 (1837); *Dum. & Bibr. Erp. Gén.* vii. p. 607 (1854); *Günth. Cat. Col. Sn.* p. 34 (1858); *Jan, Arch. Zool. Anat. Phys.* ii. 1863, p. 236; *Bocourt, Miss. Sc. Mex., Rept.* p. 607 (1886).
Zacholus, *Wagl. Syst. Amph.* p. 190 (1860).
Calamaria, part., *Schleg. l. c.* p. 25.
Herpetodryas, part., *Schleg. l. c.* p. 173.
Ophibolus, *Baird & Gir. Cat. N. Am. Rept.* p. 82 (1853); *Cope, Proc. U.S. Nat. Mus.* xiv. 1892, p. 607.
Diadophis, *Baird & Gir. l. c.* p. 112; *Cope, l. c.* p. 614.
Osceola, *Baird & Gir. l. c.* p. 133; *Cope, l. c.* p. 606.
Ablabes, part., *Dum. & Bibr. t. c.* p. 304; *Günth. l. c.* p. 27.
Meizodon, *Fischer, Abh. Nat. Hamb.* iii. 1856, p. 112.
Coryphodon, part., *Günth. l. c.* p. 107.
Lampropeltis, *Cope, Proc. Ac. Philad.* 1860, p. 254, and 1861, p. 302.
Diadophis, part., *Jan, l. c.* p. 261; *Bocourt, l. c.* p. 618.
Bellophis, *Lockington, Proc. Cal. Acad.* vii. 1877, p. 52.
Coronella, *Bouleng. Faun. Ind., Rept.* p. 308 (1890).

Maxillary teeth 12 to 20, increasing in size posteriorly, sometimes very slightly; mandibular teeth subequal. Head not or but slightly distinct from neck; eye moderate or rather small, with round pupil. Body cylindrical or slightly compressed; scales smooth, with apical pits, in 15 to 25 rows; ventrals not or but obtusely angulate laterally. Tail moderate; subcaudals in two rows.

Europe, South-western Asia, India, Africa, America north of the Equator.

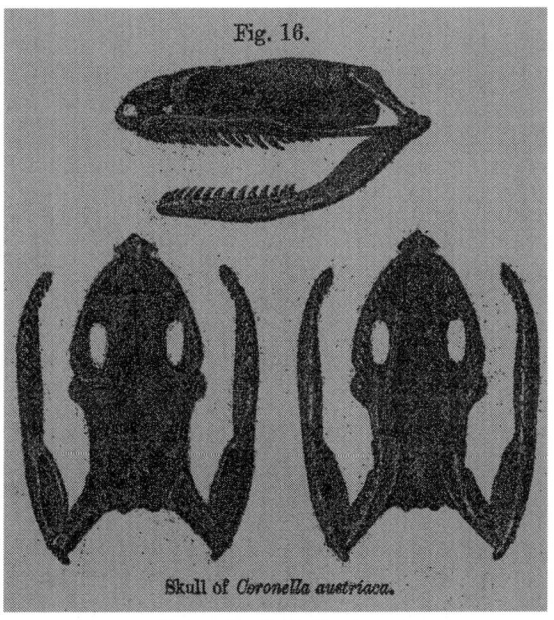

Fig. 16.

Skull of *Coronella austriaca*.

Synopsis of the Species.

I. One præocular; scales in 17 rows or more.

  A. Anal divided; scales in 19–21 rows.

    1. Rostral as deep as broad; usually two superposed anterior temporals.

Third and fourth (rarely fourth and fifth) labials entering the eye; scales in 19 rows	1. *austriaca*, p. 191.
Fourth and fifth labials entering the eye; scales in 21 rows	2. *amaliæ*, p. 193.

    2. Rostral much broader than deep; usually two or three superposed anterior temporals; scales usually in 21 rows.

Supraocular extensively in contact with the præfrontal	3. *girondica*, p. 194.
Præocular reaching or nearly reaching the frontal	4. *semiornata*, p. 195.

3. Rostral a little broader than deep; a single anterior temporal; scales in 19 rows.

Four lower labials in contact with the anterior chin-shields; belly yellow ....   5. *coronata*, p. 196.
Five lower labials in contact with the anterior chin-shields; belly blackish ....   6. *regularis*, p. 196.

B. Anal undivided.

1. Upper portion of rostral broad; normally seven upper labials.

    a. Scales in 19–25 rows; ventrals 198–240, not angulate laterally; rostral once and a half as broad as deep.

Eye moderate; scales in 19–23 rows; dark brown or black above, with yellow or white markings ....   7. *getula*, p. 197.
Eye small; scales in 21–25 rows ....   8. *calligaster*, p. 198.

    b. Scales in 21–23 rows; ventrals 188–214; rostral twice as broad as deep.

Portion of rostral visible from above measuring less than one fourth its distance from the frontal...................   9. *leonis*, p. 199.
Portion of rostral visible from above measuring one third to one half its distance from the frontal...................   10. *triangulum*, p. 200.

    c. Scales in 19–25 rows, usually 21 or 23; ventrals angulate laterally; rostral once and a half as broad as deep.

Ventrals 179–210; first black band on nape only ....   11. *gentilis*, p. 201.
Ventrals 198–224; body compressed; first black band on nape only ....   12. *zonata*, p. 202.
Ventrals 204–240; first black band forming a complete ring, extending across the throat..........................   13. *micropholis*, p. 203.

    d. Scales in 17–19 rows; ventrals 161–180.
                     14. *doliata*, p. 205.

2. Upper portion of rostral not broader than the anterior nasal; eight upper labials ....   15. *brachyura*, p. 206.

II. Usually two præoculars; anal divided; scales in 15–17 rows.

Scales in 15 rows; ventrals 136–160 ....   16. *punctata*, p. 206.
Scales in 15 rows; ventrals 182–222 ....   17. *amabilis*, p. 207.
Scales in 17 rows; ventrals 183–237 ....   18. *regalis*, p. 208.

88. CORONELLA.

TABLE SHOWING NUMBERS OF SCALES AND SHIELDS.

	Sc.	V.	A.	C.	Lab.	Pr.oc.	Ant. temp.
austriaca	19	153-199	2(1)	42-70	7(8)	1(2)	2(1)
amaliæ	21	190-198	2	52-64	8	1	2
girondica	21(19)	170-200	2	55-72	8	1	2-3
semiornata	21	176-204	2	63-88	8	1	2(1)
coronata	19	180-205	2	63-75	8	1	1
regularis	19	186-200	2	58-73	8	1	1
getula	19-23	198-240	1	41-65	7	1	2
calligaster	21-25	192-213	1	40-52	7	1	1-2
leonis	23	200	1	51	7	1	2
triangulum	21(19)	184-214	1	43-55	7	1	2(1)
gentilis	21(23)	179-210	1	44-53	7(8)	1	2
zonata	21-23	198-224	1	45-66	7	1	2
micropholis	19-25	204-240	1	46-60	7(8)	1	2(1)
doliata	17-19	161-180	1	31-54	7	1	1
brachyura	23	213-223	1	46-53	8	1	2
punctata	15	136-160	2	36-62	7-8	2(1)	1
amabilis	15	182-210	2	53-63	7-8	2	1
regalis	17	183-237	2	56-77	7-8	2	1

1. Coronella austriaca.

Coronella austriaca, *Laur. Syn. Rept.* p. 84, pl. v. fig. 1 (1768); *Lindaker, Abh. böhm. Ges. Wiss.* i. 1791, p. 123; *De Betta, Erp. Veron., Mem. Accad. Verona*, xxxv. 1857, p. 183; *Günth. Cat.* p. 34 (1858); *Gray, The Zool.* 1859, p. 6730; *F. Bond, t. c.* p. 6787; *Jan, Arch. Zool. Anat. Phys.* ii. 1863, p. 250; *De Betta, Atti Ist. Ven.* (3) xiii. 1868, p. 920; *Strauch, Schlang. Russ. R.* p. 43 (1873); *Schreib. Herp. Eur.* p. 303 (1875); *De Betta, Faun. Ital., Rett. Anf.* p. 36 (1876); *Boettg. Zeitschr. ges. Nat.* xlix. 1877, p. 286; *Collett, Vidensk. Selsk. Forh. Christ.* 1878, no. 3, p. 2; *Leydig, Abh. Senck. Ges.* xiii. 1883, p. 183; *Boettg. in Radde, Faun. Flor. Casp.-Geb.* p. 67 (1886); *Camerano, Mon. Ofid. Ital., Colubr.* p. 59 (1891); *Werner, Verh. zool.-bot. Ges. Wien*, xli. 1891, p. 764; *Méhely, Beitr. Mon. Kronstadt, Herp.* p. 26 (1892); *Bouleng. The Zool.* 1894, p. 10; *Dürigen, Deutschl. Amph. u. Rept.* p. 321, pl. viii. fig. 2 (1894).

Coluber lævis, *Lacép. Serp.* pp. 98, 158 (1789); *Hermann, Observ. Zool.* p. 278 (1804).

—— austriacus, *Gmel. S. N.* i. p. 1114 (1789); *Sturm, Deutschl. Faun.* iii. Heft 2 (1799); *Daud. Rept.* vii. p. 19 (1803); *Metaxa, Mon. Serp. Rom.* p. 39 (1823); *Lenz, Schlangenk.* p. 500, pl. vii. fig. 10 (1832); *Bonap. Icon. Faun. Ital.* (1836).

—— versicolor, *Razoum. Hist. Nat. Jorat*, p. 122, pl. ii. fig. 6 (1789); *Daud. t. c.* p. 96.

—— coronella, *Bonnaterre, Encycl. Méth., Ophid.* p. 31, pl. xxxvi. fig. 2 (1790).

—— ferruginosus, *Sparrman, Vetensk. Ak. Handl. Stockh.* xvi. 1795, p. 180, pl. vii.

Natrix coronilla, *Schranck, Faun. Boica*, i. p. 291 (1798); *Reider & Hahn, Faun. Boica*, iii. pl. — (1832).

Coluber cupreus, *Georgi, Beschr. Russ. R.* iii. p. 1884 (1800); *Pall. Zoogr. Ross.-As.* iii. p. 45 (1811).

Coluber alpinus, *Georgi, l. c.*
? Coluber ponticus, *Georgi, l. c.*
Coluber gallicus, *Herm. l. c.* p. 281.
—— caucasius, *Pall. l. c.* p. 46.
Coronella lævis, *Boie, Isis,* 1827, p. 539; *Eichw. Zool. Spec.* iii. p. 175 (1831); *Nordm. in Demid. Voy. Russ. Mér.* iii. p. 350, *Rept.* pls. xii. & xiii. (1840); *Dum. & Bibr.* vii. p. 610 (1854); *Jan, Icon. Gén.* 14, pl. vi. fig. 4 (1865); *Cooke, Our Rept.* p. 53, pl. iv. (1865); *Viaud-Grandm. Et. Serp. Vend.* p. 12 (1868); *Fatio, Vert. Suisse,* iii. p. 177 (1872); *Lataste, Herp. Gir.* p. 145 (1876); *Cambridge, Proc. Dorset N. H. Club,* vii. 1886, p. 84, pl. vi.; *Tomasini, Wiss. Mitth. aus Bosn. u. Herzeg.* ii. p. 622 (1894).
Zacholus austriacus, *Fitzing. Beitr. Landesk. Oesterr.* i. p. 326 (1832); *Bonap. Mem. Acc. Tor.* (2) ii. 1839, p. 431; *Glückselig, Lotos,* i. 1851, p. 198.
Coluber nebulosus, *Ménétr. Cat. Rais.* p. 73 (1832).
Coronella lævis, part., *Schleg. Phys. Serp.* ii. p. 65, pl. ii. figs. 12 & 13 (1837).
Zacholus fitzingeri, *Bonap. l. c.*
—— lævis, *Eichw. Faun. Casp.-Cauc.* p. 118 (1841).
Coronella lævis, vars. caucasica *et* ægyptiaca, *Jan, Arch. Zool. Anat. Phys.* ii. 1863, p. 238.
—— austriaca, var. italica, *Schreib. l. c.* fig.
—— lævis, var. leopardina, *F. Müll. Verh. nat. Ges. Basel,* vii. 1884, p. 283.

Snout obtuse or more or less prominent: rostral at least as deep as broad, more or less produced posteriorly between the internasals, the portion visible from above at least half as long (in some specimens quite as long) as its distance from the frontal, rarely separating the internasals, which are shorter than the præfrontals; frontal as long as or longer than its distance from the end of the snout, shorter than the parietals; loreal longer than deep; one (rarely two) præ- and two postoculars; temporals 2+2 or 2+3 (rarely 1+2); seven (rarely eight) upper labials, third and fourth (or fourth and fifth) entering the eye; four (rarely three) lower labials in contact with the anterior chin-shields, which are as long as or longer than the posterior. Scales in 19 rows. Ventrals 153-199; anal divided (rarely entire); subcaudals 42-70. Brown or reddish above, often with one or three lighter stripes, with small dark brown or brick-red spots usually disposed in pairs; frequently two dark brown or brick-red stripes on the nape, usually confluent with a large dark blotch on the occiput; a dark streak on each side of the head, from the nostril to the angle of the mouth, passing through the eye, sometimes extending along each side of the neck; lower parts red, orange, brown, grey, or blackish, uniform or speckled with black and white. Spec. *u* is pale brown above, with four black lines along the anterior part of the body, and two small, yellowish, dark-edged spots close together on the back of the head, separated by the parietal suture.

Total length 720 millim.; tail 140.

Europe, as far north as $62\frac{1}{2}°$ (in Great Britain only in Hampshire and Dorsetshire, perhaps also in Surrey), Transcaucasia, Talysch, Syria.

a. ♂ (V. 158; C. 61).	Bournemouth.	Lord A. Russell [P.].
b. Hgr. ♀ (V. 165; C. 48).	Bournemouth.	Master J. L. Monk [P.].
c. ♂ (V. 154; C. 58).	Ringwood.	F. Bond, Esq. [P.].
d–f. ♂ (V. 153, 163; C. 57, 54) & hgr. ♀ (V. 177; C. 55).	Hampshire.	Zoological Society.
g. ♂ (V. 154; C. 56).	Hampshire.	F. Beckford, Esq. [P.].
h. ♀ (V. 173; C. 45).	Argenton, Indre.	M. R. Parâtre [E.].
i–l. ♀ (V. 181, 174, 175; C. 51, 47, 54).	Houffalize, Belgium.	Mdlle. L. Héger [P.].
m–n. ♂ (V. 161; C. 53) & ♀ (V. 187; C. 45).	Hanover.	Dr. J. E. Gray [P.].
o–q. ♀ (V. 178, 185, 184; C. 45, 47, 51).	Kahlenberg, near Brandis, Saxony.	Dr. W. Wolterstorff [P.].
r. ♂ (V. 170; C. 53).	Walkmühle, near Eisenberg, Altenburg.	Dr. W. Wolterstorff [P.].
s. ♂ (V. 165; C. 57).	Jocketa, near Plauen, Saxony.	Dr. W. Wolterstorff [P.].
t. Yg. (V. 190; C. 51).	Saxony.	Dr. A. B. Meyer [P.].
u. ♂ (V. 172; C. 47).	Between Kierling and Weidling, near Vienna.	Dr. F. Werner [E.].
v, w–a, β–γ, δ, ε–ζ, η–ι. ♂ (V. 175, 171, 175, 168, 168, 188; C. 57, 57, 54, ?, 58, 44), ♀ (V. 187, 187, 183, 181; C. 51, 52, 49, 53), & yg. (V. 176, 175, 191, 177; C. 60, 58, 51, ?).	Near Vienna.	Dr. F. Werner [E.].
κ. ♂ (V. 164; C. 62).	Admont, Upper Styria.	Hr. F. Henkel [E.].
λ–ν. ♂ (V. 174; C. 56), ♀ (V. 186; C. 50), & yg. (V. 190; C. 47).	Carinthia.	Hr. F. Henkel [E.].
ξ–ο. ♂ (V. 166; C. 54) & ♀ (V. 189; C. 48).	Transylvania.	Dr. F. Werner [E.].
π. ♀ (V. 179; C. ?).	Turin.	Count M. G. Peracca [P.].
ρ. Yg. (V. 161; C. 53).	Bologna.	Prof. J. J. Bianconi [P.].
σ. Yg. (V. 167; C. 47).	Ferrieare, Apennines.	Prof. G. B. Howes [P.].
τ. ♂ (V. 166; C. 61).	Corunna.	M. V. L. Seoane [P.].
υ. ♀ (V. 177; C. 50).	Pontevedra.	M. V. L. Seoane [P.].
φ. Yg. (V. 168; C. 57).	Travnik, Bosnia.	Dr. F. Werner [P.].
χ. ♀ (V. 177; C. 53).	Saagdan, Kuban, Ciscaucasia.	St. Petersburg Mus. [E.].
ψ. Yg. (V. 198; C. 44).	Tiflis.	St. Petersburg Mus. [E.].
ω, aa. Skulls.	France.	

2. Coronella amaliæ.

Rhinechis amaliæ, *Boettg. Zool. Anz.* 1881, p. 570, *and Abh. Senck. Ges.* xiii. 1883, p. 98, pl. i. fig. 1.
Coronella amaliæ, *Bouleng. Ann. & Mag. N. H.* (6) iii. 1889, p. 305, *and Tr. Zool. Soc.* xiii. 1891, p. 144, pl. xviii. fig. 1.

Intermediate between *C. austriaca* and *C. girondica*. Snout prominent; rostral as deep as broad, produced between the internasals, the portion seen from above about half as long as its distance from the frontal; suture between the internasals one third the length of that between the præfrontals; frontal a little longer than its distance from the end of the snout, a little shorter than the parietals; loreal longer than deep; one præ- and two postoculars; temporals 2+3; eight upper labials, fourth and fifth entering the eye; four lower labials in contact with the anterior chin-shields; posterior chin-shields three fourths the length of the anterior. Scales in 21 rows. Ventrals 190-198; anal divided; subcaudals 52-64. Grey-brown above, with reddish-brown spots and four rather indistinct dark longitudinal bands; vertebral region light; a pair of elongate dark brown spots on the nape; a black streak on each side of the head, from the nostril, through the eye, to the angle of the mouth; a dark band between the eyes, crossing the præfrontals; a black line below the eye; lower parts coral-red, with quadrangular black spots.

Total length 390 millim.; tail 72.

Morocco and Algeria.

a-b. ♀ (V. 193; C. 63)   Benider Hills, near   M. H. Vaucher [C.].
& yg. (V. 190; C. 64).   Tangier.

3. Coronella girondica.

Coluber girondicus, *Daud. Rept.* vi. p. 432 (1803).
—— meridionalis, *Daud.* vii. p. 158 (1803).
—— riccioli, *Metaxa, Mon. Serp. Rom.* p. 40, pl. —. figs. 3 & 4 (1823); *Bonap. Icon. Faun. Ital.* (1822).
Coronella meridionalis, *Boie, Isis,* 1827, p. 539.
Coluber rubens, *Gachet, Actes Soc. Linn. Bord.* iii. 1829, p. 255.
Coronella lævis, part., *Schleg. Phys. Serp.* ii. p. 65 (1837).
Zamenis riccioli, *Bonap. Mem. Acc. Tor.* (2) ii. 1839, p. 432.
Coronella girondica, *Dum. & Bibr.* vii. p. 612 (1854); *Günth. Cat.* p. 35 (1858); *Jan, Arch. Zool. Anat. Phys.* ii. 1863, p. 251, *and Icon. Gén.* 17, pl. iii. figs. 1-3 (1866); *Boettg. Abh. Senck. Ges.* ix. 1873, p. 150; *Schreib. Herp. Eur.* p. 299 (1875); *De Betta, Faun. d'Ital., Rett. Anf.* p. 37 (1875); *Lataste, Herp. Gir.* p. 151 (1876); *Bedriaga, Amph. et Rept. de Portug.* p. 65 (1889); *Camerano, Mon. Ofid. Ital., Col.* p. 64 (1891); *Bouleng. Tr. Zool. Soc.* xiii. 1891, p. 145; *Caruccio, Boll. Soc. Rom.* i. 1892, p. 55; *Kolombatović, N. Nadod. Kralj. Dalm.* p. 12 (1893).
—— riccioli, *De Betta, Mem. Acc. Verona,* xxxv. 1857, p. 191, *and Atti Ist. Ven.* (3) xiii. 1868, p. 921.
—— lævis, var. hispanica, *Boetty. Ber. Offenb. Ver. Nat.* x. 1869, p. 55.

Snout scarcely prominent; rostral much broader than deep, just visible from above; suture between the internasals not more than half as long as that between the præfrontals; frontal as long as or a little longer than its distance from the end of the snout, shorter than the parietals; loreal longer than deep; one præ- and two postoculars; temporals 2+3 or 3+3; eight upper labials, fourth and fifth entering the eye; four lower labials in contact with the anterior chin-shields, which are as long as or longer than the posterior. Scales in 21 (rarely 19) rows. Ventrals 170–200; anal divided; subcaudals 55–72. Brown, greyish, or reddish above, with dark brown or blackish transverse spots or bars; a pair of elongate blackish spots or a U-shaped mark on the nape; a black streak from the eye to the angle of the mouth, and a dark band between the eyes, crossing the præfrontals; a black line below the eye; lower parts yellowish or red, with quadrangular black spots or with two black longitudinal bands or series of spots.

Total length 620 millim.; tail 125.

South of France, Italy, Pyrenean Peninsula, Morocco, Algeria.

a. ♂ (V. 186; C. 65).	Montpellier.	
b. ♀ (V. 196; C. 56).	Maritime Alps.	Count M. G. Peracca [P.].
c. Hgr. ♀ (V. 183; C. 61).	Genoa.	Count M. G. Peracca [P.].
d. Yg. (V. 194; C. 59).	Ferrieare, Apennines.	Prof. G. B. Howes [P.].
e. ♂ (V. 185; C. 64).	Pisa.	Prof. Della Torre [E.].
f. ♀ (V. 194; C. 58).	Italy.	Prof. J. J. Bianconi [P.].
g–h. ♂ (V. 187; C. 65) & yg. (V. 190; C. 58).	Serra de Gerez, Portugal.	Dr. Gadow [C.].
i. ♂ (V. 180; C. 67).	Coimbra.	Dr. J. L. Vieira [P.].
k. ♀ (V. 191; C. 64).	Morocco.	M. Curty [C.], Héron-Royer Collection.
l. Skull.	Montpellier.	

4. Coronella semiornata.

Coronella semiornata, *Peters, Mon. Berl. Ac.* 1854, p. 622; *Günth. Cat.* p. 39 (1858); *Peters, Reise n. Mossamb.* iii. p. 116, pl. xvii. fig. 2 (1882).
Zamenis fischeri, *Peters, Mon. Berl. Ac.* 1879, p. 777.
Coronella inornata, *Fischer, Jahrb. Hamb. Wiss. Anst.* i. 1884, p. 6, pl. i. fig. 2.
—— plumbiceps, *Boettg. Zool. Anz.* 1893, p. 117.

Snout not prominent; rostral much broader than deep, just visible from above; internasals as long as the præfrontals; frontal as long as its distance from the end of the snout, shorter than the parietals; loreal longer than deep; one præocular, reaching or nearly reaching the frontal; two postoculars; temporals 2+2 or 2+3 (rarely 1+2); eight upper labials, fourth and fifth entering the eye: four or five lower labials in contact with the anterior chin-shields, which are as long as the posterior. Scales in 21 rows. Ventrals 176–204; anal divided; subcaudals 63–88. Olive-brown above, with black transverse lines on the anterior portion of the body; these lines indistinct or broken up in the adult; upper lip and præ- and postoculars yellowish; ventrals yellowish, uniform or edged with black.

Total length 610 millim.; tail 150.

East Africa.

A. Head olive above.

a. ♂ (V. 185; C. 83). E. Africa. Sir J. Kirk [C.].
b. Yg. (V. 186; C. 86). Coast of Zanzibar.

B. Head black above.

c. Yg. (V. 189; C. 87). Mombasa. D. J. Wilson, Esq. [P.].

5. Coronella coronata.

Calamaria coronata, *Schleg. Phys. Serp.* ii. p. 46 (1837).
Coronella (Meizodon) bitorquata, *Günth. Proc. Zool. Soc.* 1860, p. 428, fig.
—— coronata, *Jan, Arch. Zool. Anat. Phys.* ii. 1863, p. 254, *and Icon. Gén.* 15, pl. iii. fig. 1 (1866).
Mizodon coronatus, *Steind. Sitzb. Akad. Wien*, lxii. i. 1870, p. 332.
Meizodon bitorquatum, *Matschie, Zool. Jahrb.* v. 1890, p. 332.

Snout scarcely prominent. Rostral a little broader than deep, just visible from above; internasals as long as or shorter than the præfrontals; frontal once and a half as long as broad, a little longer than its distance from the end of the snout, a little shorter than the parietals; loreal a little longer than deep; one præ- and two postoculars; temporals 1+2; eight upper labials, fourth and fifth entering the eye; four lower labials in contact with the anterior chin-shields, which are as long as the posterior. Scales in 19 rows. Ventrals 180–205; anal divided; subcaudals 63–75. Olive above, which colour extends to the outer ends of the ventrals; a black band across the occiput, another across the nape; a black spot on the loreal region and another on each side of the neck; the space between the black bands and spots yellowish; belly yellowish.

Total length 520 millim.; tail 115.
West Africa.

a. Yg. (V. 205; C. 75). Senegal. (Type of *C. bitorquata.*)
b. ♀ (V. 188; C. 68). McCarthy Island. Officers of the Chatham Museum [P.].
c. ♂ (V. 180; C. 70). W. Africa.

6. Coronella regularis.

Meizodon regularis, *Fischer, Abh. Nat. Hamb.* iii. 1856, p. 112, pl. iii. fig. 3; *Günth. Cat.* p. 250 (1858); *Matschie, Zool. Jahrb.* v. 1890, p. 614.
Coryphodon margaritiferus, *Günth. Cat.* p. 109.
Coronella (Meizodon) regularis, *Günth. Proc. Zool. Soc.* 1862, p. 428, fig.; *Jan, Icon. Gén.* 15, pl. iii. figs. 2 & 3 (1866); *Fischer, Oster-Progr. Ak. Gymn. Hamb.* 1883, p. 15.
—— elegans, *Jan, Arch. Zool. Anat. Phys.* ii. 1863, p. 255.

Very closely allied to the preceding. Differs in the rather shorter frontal shield, the presence of five lower labials in contact with the anterior chin-shields, and in the coloration. Dark olive, each scale black in the centre and with a white dot; three black bands across the head, separated by narrow pale lines and followed by one or more black bands across the nape; lips and gular region yellowish; lower parts whitish, the borders of the shields blackish.

West Africa.

a. Yg. (V. 187; C. 58).   W. Africa.   Sir J. Richardson [P.].
(Type of *Coryphodon margaritiferus*.)

7. Coronella getula.

Coluber getulus, *Linn. S. N.* i. p. 382 (1766); *Catesby, Nat. Hist. Carol.* pl. lii. (1771); *Daud. Rept.* vi. p. 314, pl. lxxvii. fig. 1 (1803); *Harl. Journ. Ac. Philad.* v. 1827, p. 358, *and Med. Phys. Res.* p. 122 (1835); *DeKay, N. Y. Faun.* iii. p. 37, pl. x. fig. 21 (1842); *Günth. Cat.* p. 249 (1858).

? Coluber eximius, *Harl. ll. cc.* p. 360.

Coluber (Ophis) californiæ, *Blainv. N. Ann. Mus.* iv. 1835, p. 292, pl. xxvii. fig. 1.

Herpetodryas getulus, *Schleg. Phys. Serp.* ii. p. 198 (1837).

Coronella getulus, *Holbr. N. Am. Herp.* iii. p. 95, pl. xxi. (1842); *Dum. & Bibr.* vii. p. 616 (1854); *Jan, Arch. Zool. Anat. Phys.* ii. 1863, p. 244, *and Icon. Gén.* 12, pl. vi. & 14, pl. v. (1865).

—— sayi, *Holbr. l. c.* p. 99, pl. xxii.; *Dum. & Bibr.* p. 619; *Günth. Cat.* p. 41; *Jan, ll. cc.*

Ophibolus boylii, *Baird & Gir. Cat. N. Am. Rept.* p. 82 (1853).

—— splendidus, *Baird & Gir. l. c.* p. 83.

—— sayi, *Baird & Gir. l. c.* p. 84, *and in Marcy's Explor. Red Riv.* p. 228, pl. vii. (1853).

—— getulus, *Baird & Gir. l. c.* p. 85; *Cope, Check-list N. Am. Rept.* p. 37 (1875); *Garm. N. Am. Rept.* p. 68, pl. v. figs. 3 & 4 (1883); *Cope, Proc. U.S. Nat. Mus.* xiv. 1892, p. 611; *H. Garm. Bull. Illin. Lab.* iii. 1892, p. 297; *Hay, Batr. & Rept. Indiana,* p. 110 (1893).

Coronella balteata, *Hallow. Proc. Ac. Philad.* 1853, p. 226, *and Rep. Explor. Surv. R. R.* x. pt. 4, pl. v. (1857).

—— californiæ, *Dum. & Bibr.* p. 623.

Lampropeltis sayi, *Cope, Proc. Ac. Philad.* 1860, p. 254

—— splendida, *Cope, l. c.* p. 255.

—— getula, *Cope, l. c.*

—— boylii, *Cope, l. c.*

—— boylii, var. conjuncta, *Cope, Proc. Ac. Philad.* 1861, p. 301.

Coronella pseudogetulus, *Jan, ll. cc.* p. 247, *Icon.* 12, pl. vi. fig. 2.

Ophibolus californiæ, *Cope, Check-list,* p. 37.

—— getulus niger, eiseni, multicinctus, *Yarrow, Proc. U.S. Nat. Mus.* v. 1882, p. 438.

Snout slightly prominent; rostral about once and a half as broad as deep, the portion visible from above measuring about one third its distance from the frontal; suture between the internasals about half as long as that between the præfrontals; frontal considerably longer than broad, as long as its distance from the end of the snout, shorter than the parietals; loreal as long as deep, or deeper than long, or longer than deep, sometimes absent; one præ- and two (rarely three) postoculars; temporals 2+2 or 2+3; seven upper labials, third and fourth entering the eye; four lower labials in contact with the anterior chin-shields, which are as long as or longer than the posterior. Scales in 21 or 23 (rarely 19) rows. Ventrals 198-240; anal entire; subcaudals 41-65. Dark brown or black above, with yellow or white markings; labials and lower parts black and white or black and yellow.

Total length 1780 millim.; tail 200.

North America.

A. Above with small yellow spots or narrow yellow transverse bands restricted to the back. (*C. sayi.*)

a. ♂ (Sc. 19; V. 209; C. 39). Indiana.
b. ♂ (Sc. 21; V. 211; C. 46). Arkansas.
c. ♀ (Sc. 21; V. 216; C. 45). Louisiana.
d, e-f. ♀ (Sc. 21; V. 207; C. 49) New Orleans. M. Sallé [C.].
  & yg. (Sc. 21, 21; V. 201, 198; C. 50, 53).
g. ♂ (Sc. 23; V. 218; C. 53). Duval County, Texas. W. Taylor, Esq. [C.].
h, i. ♂ (Sc. 21; V. 209; C. 54) Texas.
  & ♀ (Sc. 21; V. 206; C. 50).
k. Skull. New Orleans.

B. Above with yellow transverse bands connected with their fellows on the sides. (*C. getula.*)

a. Hgr. (Sc. 21; V. 210; C. 39). District of Columbia. Smithsonian Inst.
b-d. ♂ (Sc. 23; V. 216; C. 52), Marion Co., Florida. A. Erwin Brown, Esq. [P.].
  ♀ (Sc. 23; V. 218; C. ?), & hgr. (Sc. 23; V. 214; C. 47).
e, f. ♂ (Sc. 21; V. 217; C. 50) N. America.
  & yg. (Sc. 21; V. 211; C. 50).

C. More or less complete whitish annuli round the body, widening on the belly. (*C. californiæ, C. boylii.*)

a. ♂ (Sc. 23; V. 225; C. 56). Camp Taylor, Marion Co., Cal. Prof. C. Eigenmann [C.].
b-d. ♂ (Sc. 23; V. 228; C. 46), San Diego, California. Christiania Mus.
  ♀ (Sc. 23; V. 235; C. 54), & yg. (Sc. 23; V. 226; C. 54).
e. Yg. (Sc. 23; V. 228; C. 54). California. Smithsonian Inst.
f. Yg. (Sc. 21; V. 218; C. 55). California. Lord Walsingham [P.].

8. Coronella calligaster.

Coluber calligaster, *Harlan, Journ. Ac. Philad.* v. 1827, p. 359, *and Med. Phys. Res.* p. 122 (1835).
—— guttatus (*non L.*), *Schleg. Phys. Serp.* ii. p. 168 (1837).
Coronella rhombomaculata, *Holbr. N. Am. Herp.* iii. p. 103, pl. xxiii. (1842); *Jan, Arch. Zool. Anat. Phys.* ii. 1863, p. 243, *and Icon. Gén.* 17, pl. ii. figs. 1 & 2 (1866).
Ophibolus rhombomaculatus, *Baird & Gir. Cat. N. Am. Rept.* p. 86 (1853).
Ablabes triangulum, var. calligaster, *Hallow. Proc. Ac. Philad.* 1856, p. 244.

Ophibolus evansii, *Kennicott, Proc. Ac. Philad.* 1859, p. 99.
Lampropeltis calligaster, *Cope, Proc. Ac. Philad.* 1860, p. 255.
—— rhombomaculata, *Cope, l. c.*; *Stejneger, Proc. U.S. Nat. Mus.* xiv. 1891, p. 503.
Coronella evansii, *Jan, ll. cc.* p. 243, fig. 3.
—— tigrina, *Jan, Arch. Zool. Anat. Phys.* ii. p. 244.
Ophibolus calligaster, *Cope, Check-list N. Am. Rept.* p. 37 (1875), and *Proc. U.S. Nat. Mus.* xiv. 1892, p. 610; *H. Garm. Bull. Illin. Lab.* iii. 1892, p. 293.
—— triangulus, vars. calligaster *et* rhombomaculatus, *Garm. N. Am. Rept.* pp. 66 & 156 (1883).

Eye small. Snout scarcely prominent. Rostral about once and a half as broad as deep, the portion visible from above measuring about one third its distance from the frontal; suture between the internasals shorter than that between the præfrontals; frontal considerably longer than broad, as long as or longer than its distance from the end of the snout, shorter than the parietals; loreal longer than deep; one præ- and two postoculars; temporals 1+2 or 2+3; seven upper labials, third and fourth entering the eye; four lower labials in contact with the anterior chin-shields, which are longer than the posterior. Scales in 21 to 25 rows. Ventrals 192-213; anal entire; subcaudals 40-52. Pale brown above, with large, transversely elliptic, dark, black-edged spots on the back and an alternating series of smaller ones along the side; a pair of elongate dark spots or bands on the nape, and a dark oblique streak behind the eye; red inferiorly, with brown spots.

Total length 790 millim.; tail 110.

North America east of the Rocky Mountains, as far north as Virginia.

a. ♀ (Sc. 25; V. 210; C. 49).	Missouri.	Smithsonian Institution [P.].
b. Yg. (Sc. 21; V. 199; C. 49).	Florida.	Dr. Guillemard [P.].
c. ♀ (Sc. 25; V. 204; C. 40).	N. America.	Dr. Rüppell.

9. Coronella leonis.

Coronella leonis, *Günth. Biol. C.-Am., Rept.* p. 110, pl. xxxix. fig. A (1893).

Snout longer and eye larger than in *C. triangulum*. Rostral twice as broad as deep, just visible from above; suture between the internasals about two thirds the length of that between the præfrontals; frontal a little longer than broad, as long as its distance from the rostral, shorter than the parietals; loreal twice as long as deep; one præ- and two postoculars; temporals 2+3; seven upper labials, third and fourth entering the eye; four lower labials in contact with the anterior chin-shields, which are longer than the posterior. Scales in 23 rows. Ventrals 200; anal entire; subcaudals 51. Grey above, with a dorsal series of ∞-shaped red, black-edged spots; a ⌒-shaped marking on the nape; a black blotch with red centre on the frontal and another on each parietal and a small oblique black spot on the temple; belly yellowish, with a few black blotches; an interrupted black stripe along the lower surface of the tail.

Total length 610 millim.; tail 95.

Nuevo Leon, North Mexico.

a. ♀ (V. 200; C. 51).   N. Leon.   W. Taylor, Esq. [C.].   (Type.)

10. Coronella triangulum.

Coluber triangulum, *Daud. Rept.* vi. p. 322 (1803).
—— eximius (*non Harl.*), *Holbr. N. Am. Herp.* iii. p. 69, pl. xv. (1842); *DeKay, N. Y. Faun.* iii. p. 38, pl. xii. fig. 25 (1842); *Günth. Cat.* p. 91 (1858).
Pseudoelaps Y, *Berth. Abh. Ges. Wiss. Götting.* i. 1843, p. 67, pl. i. figs. 11 & 12.
Ophibolus eximius, *Baird & Gir. Cat. N. Am. Rept.* p. 87 (1853).
—— clericus, *Baird & Gir. l. c.* p. 88.
Ablabes triangulum, *Dum. & Bibr.* vii. p. 315 (1854); *Hallow. Proc. Ac. Philad.* 1856, p. 244.
Lampropeltis triangula, *Cope, Proc. Ac. Philad.* 1860, p. 256.
Coronella eximia, *Jan, Arch. Zool. Anat. Phys.* ii. 1863, p. 242, and *Icon. Gén.* 17, pl. i. fig. 3 (1866).
—— doliata (typica), *Jan, ll. cc.* p. 241, *Icon.* 14, pl. iv. fig. A (1865); *Bocourt, Miss. Sc. Mex., Rept.* p. 609, pl. xxxix. fig. 2 (1886).
Ophibolus doliatus triangulum, *Cope, Check-list N. Am. Rept.* p. 4 (1875); *and Proc. U.S. Nat. Mus.* 1888, p. 381 (1889), *and Am. Nat.* 1893, p. 1067, pl. xxiv. fig. 1.
—— triangulus, *Garm. N. Am. Rept.* p. 65, pl. v. fig. 1 (1883); *H. Garm. Bull. Illin. Lab.* iii. 1892, p. 295.
—— doliatus parallelus, clericus, *et* collaris, *Cope, Proc. U.S. Nat. Mus.* xi. 1888, p. 383, & xiv. 1892, p. 608, *and Am. Nat.* 1893, p. 1067, pls. xxiv., xxv., & xxvii.
—— doliatus temporalis, *Cope, Am. Nat.* 1893, p. 1068, pl. xxv. fig. 4.
—— doliatus, *Hay, Batr. & Rept. Indiana,* p. 107 (1893).

Snout not prominent; rostral twice as broad as deep, the portion visible from above measuring one third to one half its distance from the frontal; suture between the internasals one third to two thirds the length of that between the præfrontals; frontal nearly as broad as long or little longer than broad, as long as its distance from the end of the snout, shorter than the parietals; loreal longer than deep; one præ- and two postoculars; temporals 2+2, 2+3, or, rarely, 1+2; seven upper labials, third and fourth entering the eye; four lower labials in contact with the anterior chin-shields, which are longer than the posterior. Scales in 21 (rarely 19) rows. Ventrals 184–214; anal entire; subcaudals 43–55. Above with large brown or red, black-edged, elliptic, transverse spots separated by narrow yellowish or greyish cross bars, which widen towards the abdomen; a lateral series of smaller spots, alternating with the dorsals; a black oblique streak from the eye to the angle of the mouth; often one or two small light spots on the parietal suture behind the frontal; lower parts white, usually largely spotted with black.

Total length 760 millim.; tail 120.
North America east of the Mississippi.

A. A light V- or Y-shaped marking on the occiput.   (*C. triangulum*.)

a. ♂ (V. 207 ; C. 50).	Canada.	Chatham Mus.
b-c. ♀ (V. 198; C. 52) & yg. (V. 196; C. 43).	Bloomington, Indiana.	C. H. Bollman, Esq. [C.].
d-e. Hgr. (V. 197; C. 46) & yg. (V. 190; C. 47).	New York.	Mr. J. Murray [P.].
f. Yg. (V. 206; C. 53).	N. America.	E. Doubleday, Esq. [P.].
g, h, i, k. ♂ (V. 195; C. 49), ♀ (V. 193; C. 49), hgr. (V. 202 ; C. 51), & yg. (V. 202; C. 51).	N. America.	

B. A light transverse collar behind the parietals.
(Var. *collaris*, Cope.)

a. Hgr. (V. 195; C. 45).	N. America.	G. Henshaw, Esq. [P.].
b. Hgr. (V. 198 ; C. 46).	N. America.	Dr. Jacob Green [P.].
c. Hgr. (V. 189; C. 51).	N. America.	

This variety connects *C. triangulum* with *C. gentilis* and *C. doliata*.

11. Coronella gentilis.

Ophibolus doliatus, *Baird & Gir. Cat. N. Am. Rept.* p. 89 (1853); *Dugès, La Naturaleza*, iii. 1875, p. 222, pl. — (?); *Cope, Proc. U.S. Nat. Mus.* xiv. 1892, p. 608.
—— gentilis, *Baird & Gir. l. c.* p. 90, *and Marcy's Explor. Red Riv.* p. 229, pl. viii. (1853).
Coronella doliata, part., *Dum. & Bibr.* vii. p. 621 (1854); *Günth. Cat.* p. 41 (1858).
—— doliata, *Hallow. Proc. Ac. Philad.* 1856, p. 247; *Wied, N. Acta Ac. Leop.-Carol.* xxxii. i. 1865, p. 99.
Lampropeltis doliata, *Cope, Proc. Ac. Philad.* 1860, p. 256.
—— annulata, *Cope, l. c.* p. 257.
—— multistriata, *Kennicott, Proc. Ac. Philad.* 1860, p. 328; *Stejneger, Proc. U.S. Nat. Mus.* xiv. 1891, p. 502.
—— annulata, part., *Kennicott, l. c.* p. 329.
Coronella doliata, var. gentilis, *Jan, Arch. Zool. Anat. Phys.* ii. 1863, p. 241, *and Icon. Gén.* 17, pl. i. fig. 2 (1866); *Bocourt, Miss. Sc. Mex., Rept.* p. 610, pl. xxxix. fig. 5 (1886).
Ophibolus doliatus coccineus, gentilis, annulatus, *et* doliatus, *Cope, Check-list N. Am. Rept.* pp. 4, 36 (1875), *and Proc. U.S. Nat. Mus.* xi. 1888, p. 381, *and Am. Nat.* 1893, p. 1067, pls. xxvi. & xxviii.
—— multistratus, *Cope, Check-list*, p. 37.
—— triangulus, vars. mexicanus (?), doliatus, *et* gentilis, *Garman, N. Am. Rept.* p. 66 (1883).
—— doliatus syspylus, *Cope, Proc. U.S. Nat. Mus.* xi. 1888, p. 383, *and Am. Nat.* 1893, p. 1067, pl. xxvii. fig. 1.
? Coronella mexicana, *Günth. Biol. C.-Am., Rept.* p. 110 (1893).

Snout slightly prominent; rostral about once and a half as broad as deep, the portion visible from above measuring about one third its distance from the frontal; suture between the internasals much shorter than that between the præfrontals; frontal a little longer than broad, as long as its distance from the end of the snout, a little shorter than the parietals; loreal longer than deep; one præ- and two postoculars; temporals 2+2 or 2+3; seven (rarely eight) upper labials, third and fourth entering the eye; four lower labials in contact with the anterior chin-shields, which are usually longer than the posterior. Scales in 21 (rarely 23) rows. Ventrals obtusely angulate laterally, 179–210; anal entire; subcaudals 44–53. Scarlet, with pairs of black rings enclosing a yellow one; these rings may be interrupted on the belly, and the black nuchal band never extends across the throat; the greater part of the upper surface of the head black.

Total length 490 millim.; tail 80.

Southern United States, North Mexico.

a–f. ♂ (V. 191, 188; C. 52, 49), ♀ (V. 191, 202; C. 46, 50), & hgr. (V. 188, 191; C. 53, 52).	New Orleans.	M. Sallé [C.].
g. Hgr. (V. 188; C. 48).	Texas.	
h. Hgr. (V. 179; C. 49).	N. America.	Sir R. Murchison [P.].

12. Coronella zonata.

Coluber (Zacholus) zonatus, *Blainv. N. Ann. Mus.* iv. 1835, p. 293.
Ophibolus pyromelanus, *Cope, Proc. Ac. Philad.* 1866, p. 305.
—— pyrrhomelas, *Cope, Check-list N. Am. Rept.* p. 37 (1875); *Yarrow, Wheeler's Rep. Explor. Surv. W.* 100*th Mer.* v. pl. xix. (1875); *Cope, Proc. U.S. Nat. Mus.* xiv. 1892, p. 610.
Bellophis zonatus, *Lockington, Proc. Cal. Acad.* vii. 1877, p. 52.
Coronella multifasciata, *Bocourt, Miss. Sc. Mex., Rept.* p. 616, pl. xl. fig. 2 (1886).

Body compressed in its posterior half. Snout slightly prominent; rostral about once and a half as broad as deep, the portion visible from above hardly one third its distance from the frontal; suture between the internasals shorter than that between the præfrontals; frontal a little longer than broad, as long as its distance from the end of the snout, shorter than the parietals; loreal longer than deep (rarely absent); one præ- and two postoculars; temporals 2+3; seven upper labials, third and fourth entering the eye; four lower labials in contact with the anterior chin-shields, which are longer than the posterior. Scales in 21 (or 23) rows. Ventrals obtusely angulate laterally, 198–224; anal entire; subcaudals 45–66. Scarlet, with pairs of blacks rings enclosing a yellow one; these rings very numerous (47–55) and close together on the posterior half of the body, where the red interspaces are scarcely wider than the yellow rings or may disappear altogether; head black anteriorly, yellow posteriorly; a black band across the nape, not extending to the throat.

Total length 780 millim.; tail 125.
Arizona and California.

a. ♀ (V. 209; C. 60).   Sta. Cruz, California.   Hr. A. Forrer [C.].

13. Coronella micropholis.

Coronella doliata, part., *Dum. & Bibr.* vii. p. 621 (1854); *Günth. Cat.* p. 41 (1858).
Lampropeltis micropholis, *Cope, Proc. Ac. Philad.* 1860, p. 257.
—— polyzona, *Cope, l. c.* p. 258.
—— amaura, *Cope, l. c.*
Coronella formosa, *Jan, Arch. Zool. Anat. Phys.* ii. 1863, p. 241, *and Icon. Gén.* 14, pl. iv. fig. B (1865); *Bocourt, Miss. Sc. Mex., Rept.* p. 612, pl. xxxix. fig. 3 (1886).
—— doliata, var. conjuncta, *Jan, ll. cc.* p. 242, fig. C; *Bocourt, l. c.* p. 611, fig. 6.
—— doliata, var. formosa, *Bouleng. Bull. Soc. Zool. France*, 1880, p. 44.
Ophibolus polyzonus, *Sumichrast, Bull. Soc. Zool. France*, 1880, p. 181.
—— triangulus, var. zonatus, *Garm. N. Am. Rept.* p. 67 (1883).
Coronella formosa, vars. anomala, oligozona, polyzona, *et* abnorma, *Bocourt, l. c.* figs. 4, 7, 8.
Ophibolus doliatus polyzonus, occipitalis, *et* annulatus, *Cope, Proc. U.S. Nat. Mus.* 1888, p. 382, *and Am. Nat.* 1893, p. 1067, pl. xxviii. fig. 1.
Lampropeltis annulata, *Stejneger, Proc. U.S. Nat. Mus.* xiv. 1891, p. 503.
Coronella annulata, *Günth. Biol. C.-Am., Rept.* p. 109, pl. xxxviii. (1893).

Snout slightly prominent; rostral about once and a half as deep as broad, the portion visible from above measuring one third to two thirds its distance from the frontal; suture between the internasals shorter than that between the præfrontals; frontal as long as or a little longer than broad, as long as its distance from the end of the snout, as long as or a little shorter than the parietals; loreal usually longer than deep; one præ- and two postoculars; temporals 2+2 or 2+3 (exceptionally 1+2); seven (rarely eight) upper labials, third and fourth (or fourth and fifth) entering the eye; four lower labials in contact with the anterior chin-shields, which are as long as or longer than the posterior. Scales in 21 or 23 (exceptionally 19 or 25) rows. Ventrals obtusely angulate laterally, 204–240; anal entire; subcaudals 46–60. Scarlet, usually with pairs of black rings enclosing yellow ones; the red scales may be tipped with black; head black above, usually with a continuous or interrupted yellow band across the snout and another across the occiput; the first black ring usually extends across the throat.

Total length 1370 millim.; tail 170.
America, from Texas and Mexico to the Equator.

A. Annuli separated by broad red interspaces; the red scales not tipped with black.

a. Yg. (Sc. 21; V. 221; C. 58). Mazatlan. Hr. A. Forrer [C.].
b. ♀ (Sc. 21; V. 215; C. 53). Presidio, nr. Mazatlan. Hr. A. Forrer [C.].
c-e. ♂ (Sc. 23; V. 230; C. 56) & yg. (Sc. 23, 23; V. 232, 231; C. 57, 56). Tres Marias Ids. Hr. A. Forrer [C.].
f. ♀ (Sc. 21; V. 215; C. 52). Mezquital del Oro, Zacatecas. Dr. A. C. Buller [C.].
g. ♀ (Sc. 23; V. 228; C. 59). Para. J. P. G. Smith, Esq. [P.].

B. Annuli separated by broad red interspaces, the scales of which are tipped with black.

a. ♂ (Sc. 21; V. 231; C. ?). City of Mexico. Mr. Doorman [C.].
b. ♂ (Sc. 21; V. 218; C. 55). Tierra Colorado, Guerrero. Mr. H. H. Smith [C.]; F. D. Godman, Esq. [P.].
c. Yg. (Sc. 21; V. 209; C. 50). Amula, Guerrero. Mr. H. H. Smith [C.]; F. D. Godman, Esq. [P.].
d. Hgr. ♂ (Sc. 21; V. 229; C. 59). Huatuzco. F. D. Godman, Esq. [P.].
e-f. Hgr. (Sc. 25; V. 231; C. 56) & yg. (Sc. 23; V. 227; C. 51). Teapa, Tabasco. F. D. Godman, Esq. [P.].
g. Yg. (Sc. 23; V. 213; C. 49). Yucatan.
h. ♀ (Sc. 21; V. 230; C. ?). S. Mexico. F. D. Godman, Esq. [P.].
i-k. ♂ (Sc. 21; V. 223; C. 58) & hgr. (Sc. 23; V. 226; C. 49). Mexico. Mr. Warwick [C.].
l, m-n. ♂ (Sc. 23, 21; V. 224, 221; C. 55, 54) & hgr. (Sc. 21; V. 223; C. 51). Mexico. Mr. H. Finck [C.].
o. Yg. (Sc. 23; V. 230; C. 56). Belize. Dr. Günther [P.].
p-q. Yg. (Sc. 23, 23; V. 228, 226; C. 56, 50). Dueñas, Guatemala. O. Salvin, Esq. [C.].
r. Hgr. ♀ (Sc. 19; V. 230; C. 51). Irazu, Costa Rica. F. D. Godman, Esq. [P.].
s. Yg. (Sc. 19; V. 232; C. 53). Chiriqui. G. Champion, Esq. [C.]; F. D. Godman, Esq. [P.].
t. ♂ (Sc. 21; V. 231; C. ?). —— ? Dr. J. E. Gray [P.].

C. Black rings irregular; belly black.

a. Hgr. ♀ (Sc. 21; V. 204; C. 51). Mexico. Mr. Warwick [C.].

D. Black, with yellow rings, the red colour appearing on
each side as rounded spots.

a. Yg. (Sc. 21; V. 218; Tehuantepec.
C. 47).

E. Red above, each scale tipped with black, the rings reduced
to mere traces here and there.

a. Hgr. (Sc. 23; V. 218; Yucatan.
C. 55).

F. Almost uniform black, with very indistinct traces
of the light annuli.

a. ♂ (Sc. 21; V. 225; C. 51). Irazu, Costa   F. D. Godman, Esq.
Rica.   [P.].

14. Coronella doliata.

Coluber doliatus, *Linn. S. N.* i. p. 379 (1766); *Harl. Journ. Ac. Philad.* v. 1827, p. 362, *and Phys. Med. Res.* p. 125 (1835).
—— doliatus, part., *Daud. Rept.* vii. p. 74 (1803).
—— dumfriesiensis, *Sowerby, Brit. Misc.* p. 5, pl. iii. (1804).
Coronella coccinea, *Schleg. Phys. Serp.* ii. p. 57 (1837); *Jan, Arch. Zool. Anat. Phys.* ii. 1863, p. 239, *and Icon. Gén.* 17, pl. i. fig. 1 (1866); *Bocourt, Miss. Sc. Mex., Rept.* p. 608, pl. xxxix. fig. 1 (1886).
—— doliata, *Holbr. N. Am. Herp.* iii. p. 105, pl. xxiv. (1842).
Calamaria elapsoidea, *Holbr. l. c.* p. 119, pl. xxviii.
Osceola elapsoidea, *Baird & Gir. Cat. N. Am. Rept.* p. 133 (1853), *and Rep. U.S. Surv. R. R.* x. pt. iii. pl. xxxiii. fig. 97 (1859); *Cope, Proc. U.S. Nat. Mus.* xiv. 1892, p. 606.
Lampropeltis coccinea, *Cope, Proc. Ac. Philad.* 1860, p. 257.
Ophibolus doliatus, *Garm. N. Am. Rept.* p. 64, pl. v. fig. 2 (1883); *H. Garm. Bull. Illin. Lab.* iii. 1892, p. 298.
—— elapsoideus, *H. Garm. l. c.*

Snout much depressed, slightly prominent; rostal once and a half as broad as deep, the portion visible from above measuring about half its distance from the frontal; suture between the internasals shorter than that between the præfrontals; frontal a little longer than broad, as long as its distance from the end of the snout, a little shorter than the parietals; loreal usually absent; one præ- and two postoculars; temporals 1+2; seven upper labials, third and fourth entering the eye; four lower labials in contact with the anterior chin-shields, which are longer than the posterior. Scales in 17 or 19 rows. Ventrals 161–180; anal entire; subcaudals 31–54. Head and body scarlet, with pairs of black rings enclosing each a white one; the first white ring on the nape; these rings may be interrupted on the ventral surface.

Total length 230 millim.; tail 35.

South-eastern United States.

a. ♀ (Sc. 17; V. 161; C. 42). N. America.   E. Doubleday, Esq. [P.].

15. Coronella brachyura.

Zamenis brachyurus, *Günth. Ann. & Mag. N. H.* (3) xviii. 1866,
p. 27, pl. vi. fig. A.; *Blanf. Journ. As. Soc. Beng.* xxxix. 1870,
p. 372; *Anders. Proc. Zool. Soc.* 1871, p. 176; *Theob. Cat. Rept.
Brit. Ind.* p. 171 (1876).
Coronella brachyura, *Bouleng. Faun. Ind., Rept.* p. 309 (1890).

Rostral broader than deep, visible from above; suture between the internasals shorter than that between the præfrontals; frontal once and a half as long as broad, as long as its distance from the end of the snout, a little shorter than the parietals; loreal as long as deep; one præocular; two postoculars; temporals 2+2; eight upper labials, fourth and fifth entering the eye; five lower labials in contact with the anterior chin-shields, which are slightly longer than the posterior; the latter widely separated from each other by three series of scales. Scales in 23 rows. Ventrals 213–223; anal entire; subcaudals 46–53. Olive-brown above, with rather indistinct light variegations on the anterior half of the body; lower surface whitish.

Total length 445 millim.; tail 55.

Deccan.

a. ♀ (V. 223; C. 46).   Poonah.   Dr. Leith [P.]. (Type.)

16. Coronella punctata.

Coluber punctatus, *Linn. S. N.* i. p. 376 (1766); *Daud. Rept.* vii.
p. 178 (1803); *Harlan, Journ. Ac. Philad.* v. 1827, p. 354, *and
Med. Phys. Res.* p. 117 (1835); *Storer, Rep. Fish. & Rept. Mass.*
p. 225 (1839); *Holbr. N. Am. Herp.* iii. p. 81, pl. xviii. (1842);
DeKay, N. Y. Faun. iii. p. 39, pl. xiv. fig. 29 (1842).
—— torquatus, *Shaw, Zool.* iii. p. 553 (1802).
Calamaria punctata, *Schleg. Phys. Serp.* ii. p. 39 (1837).
Diadophis punctatus, *Baird & Gir. N. Am. Rept.* p. 112 (1853);
Cope, Proc. Ac. Philad. 1860, p. 250; *Jan, Arch. Zool. Anat. Phys.*
ii. 1863, p. 263, *and Icon. Gén.* 15, pl. vi. fig. 1 (1866); *Garm. N.
Am. Rept.* p. 72, pl. ii. fig. 2 (1883); *Bocourt, Miss. Sc. Mex., Rept.*
p. 618, pl. xl. fig. 1 (1886); *Cope, Proc. U.S. Nat. Mus.* xiv. 1892,
p. 617; *H. Garm. Bull. Illin. Lab.* iii. 1892, p. 300.
Ablabes punctatus, *Dum. & Bibr.* vii. p. 310 (1854); *Günth. Cat.*
p. 28 (1858).
—— occipitalis, *Günth. l. c.* p. 29.
Diadophis occipitalis, *Cope, Proc. Ac. Philad.* 1860, p. 250.
—— punctatus, var. pallidus, *Cope, l. c.*; *Bocourt, l. c.* p. 621.
—— dysopes, *Cope, l. c.* p. 251.

Snout not prominent; rostral nearly twice as broad as deep, just visible from above; suture between the internasals shorter than that between the præfrontals; frontal a little longer than its distance from the end of the snout, about two thirds the length of the parietals; loreal as long as deep or a little deeper than long; two (rarely one) præ- and two postoculars; temporals 1+1; seven upper labials, third and fourth entering the eye, or eight, fourth and fifth entering the eye; five lower labials in contact with the anterior chin-shields, which are longer than the posterior. Scales in 15 rows. Ventrals 136–160; anal divided; subcaudals 36–62. Dark grey above, with the upper lip and a nuchal collar yellow and black-edged; yellow or orange inferiorly, each ventral with a small black spot at its outer extremity.

Total length 360 millim.; tail 65.

North America, east of the Rocky Mountains; Mexico.

A. A median row of black spots along the belly.

a–b. ♀ (V. 151; C. 42) & hgr. (V. 149; C. 41).	Raleigh, N. Carolina.	Messrs. Brimley [C.].
c. ♂ (V. 155; C. 56).	Wisconsin.	hristiania Museum.
d. ♀ (V. 151; C. 40).	United States.	
e. ♀ (V. 149; C. 41).	Mexico.	(Type of *A. occipitalis*.)

B. No black spots along the middle of the belly.

f. ♂ (V. 158; C. 50).	Nova Scotia.	J. M. Jones, Esq. [P.].
g. Yg. (V. 146; C. 62).	Chippeway, Canada.	
h. Yg. (V. 159; C. 51).	Delaware.	E. Doubleday, Esq. [P.].
i. ♀ (V. 160; C. 47).	Bloomington, Indiana.	C. Bollman, Esq. [C.].
k. ♂ (V. 155; C. 58).	Kansas.	

17. Coronella amabilis.

Diadophis amabilis, *Baird & Gir. Cat. N. Am. Rept.* p. 113 (1853), and *Rep. U.S. Surv. R. R.* x. pt. iii. pl. xxxiii. fig. 83 (1859); *Cope, Proc. U.S. Nat. Mus.* xiv. 1892, p. 615.
—— docilis, *Baird & Gir. ll. cc.* p. 114, fig. 84.
—— pulchellus, *Baird & Gir. ll. cc.* p. 115, fig. 85; *Cooper, Rep. U.S. Surv. R. R.* xii. pt. ii. p. 302 (1860).
—— punctatus, var. stictogenys, *Cope, Proc. Ac. Philad.* 1860, p. 250.
—— texensis, *Kennicott, Proc. Ac. Philad.* 1860, p. 328.
—— punctatus, var. amabilis, *Jan, Arch. Zool. Anat. Phys.* ii. 1863, p. 265, and *Icon. Gén.* 15, pl. vi. fig. 4 (1866); *Garm. N. Am. Rept.* p. 159 (1883); *Bocourt, Miss. Sc. Mex., Rept.* p. 620 (1886); *H. Garm. Bull. Illin. Lab.* iii. 1892, p. 300.
—— punctatus, var. docilis, *Jan, ll. cc.* fig. 2; *Garm. l. c.* p. 72; *Bocourt, l. c.* p. 619.
—— punctatus, var. pulchellus, *Jan, ll. cc.* fig. 3; *Cope, Proc. Ac. Philad.* 1883, p. 27; *Bocourt, l. c.* p. 620.

Differs from the preceding in the more elongate body, the ventrals numbering 182 to 210, and the subcaudals 53 to 63. Dark grey or blackish above, with a yellow nuchal collar; belly and the scales of the outer row orange, usually with numerous round black spots.

Total length 470 millim.; tail 80.

United States west of the Mississippi.

a. ♂ (V. 194; C. 59). San Diego, California. Prof. C. Eigenmann [C.].
b. ♀ (V. 210; C. 58). San Bernardino, California. Hr. A. Forrer [C.].
c. Hgr. (V. 195; C. 54). Santa Cruz, California. Hr. A. Forrer [C.].
d. ♀ (V. 205; C. 62). Upper Sacramento River Valley. Mr. Gruber [C.].
e. ♂ (V. 196; C. 63). California. Lord Walsingham [P.].
f. ♀ (V. 198; C. 53). California. Christiania Museum.

18. Coronella regalis.

Diadophis regalis, *Baird & Gir. Cat. N. Am. Rept.* p. 115 (1853), and *Rep. U.S. Surv. R. R.* x. pt. iii. pl. xxxiii. fig. 86 (1859), and *Rep. U.S. Mex. Bound. Surv.* ii. *Rept.* pl. xix. fig. 2 (1859); *Garm. N. Am. Rept.* p. 73 (1883); *Cope, Proc. Ac. Philad.* 1883, p. 12, and *Proc. U.S. Nat. Mus.* xiv. 1892, p. 615.
—— arnyi, *Kennicott, Proc. Ac. Philad,* 1859, p. 99.
? Diadophis docilis, *Kennicott, in Baird, Rep. U.S. Mex. Bound. Surv.* ii., *Rept.* p. 22, pl. xxi. fig. 3 (1859).
Diadophis punctatus, var. arnyi, *Jan, Arch. Zool. Anat. Phys.* ii. 1863, p. 265, and *Icon. Gén.* 15, pl. vi. fig. 5 (1866); *Bocourt, Miss. Sc. Mex., Rept.* p. 622 (1886); *H. Garm. Bull. Illin. Lab.* iii. 1892, p. 302.
—— punctatus, var. lætus, *Jan, ll. cc.* fig. 6; *Bocourt, l. c.* p. 622, pl. xl. fig. 4.
Liophis (Diadophis) arnyi, *Günth. Ann. & Mag. N. H.* (4) i. 1868, p. 413.
Diadophis punctatus, var. dougesii, *Villada, La Naturaleza,* iii. 1875, p. 226, pl. —.
—— punctatus, vars. modestus *et* regalis, *Bocourt, l. c.* p. 623.
—— texensis, *Garm. Bull. Essex Inst.* xix. 1888, p. 127.
Liophis regalis, *Günth. Biol. C.-Am., Rept.* p. 108 (1893).
—— lætus, *Günth. l. c.*

Distinguished from *C. punctata* and *C. amabilis* by having 17 rows of scales. Ventrals 183 to 237; subcaudals 56 to 77. Temporals 1+1 or 1+2. Olive-brown, bluish grey, or blackish above, with or without a yellow nuchal collar; belly yellow or orange, with numerous black spots.

Total length 490 millim.; tail 100.

From Illinois and Kansas to Mexico.

a. ♂ (V. 183; C. 56).   City of Mexico.   Mr. Doorman [C.].

89. HYPSIGLENA.

Hypsiglena, *Cope, Proc. Ac. Philad.* 1860, p. 246, and *Proc. U.S. Nat. Mus.* xiv. 1892, p. 617.
Pseudodipsas, *Peters, Mon. Berl. Ac.* 1860, p. 521.
Comastes, *Jan, Elenco sist. Ofid.* p. 102 (1863).

Maxillary teeth 12 to 16, the two posterior strongly enlarged and separated from the rest by an interspace; mandibular teeth subequal. Head distinct from neck; eye rather small, with vertically elliptic pupil. Body cylindrical; scales smooth, with apical pits, in 19 or 21 rows; ventrals rounded. Tail moderate; subcaudals in two rows.

Southern parts of North America, Central America, Venezuela.

Synopsis of the Species.

I. Subcaudals 39-55.

A. Scales in 21 rows; usually a subocular below the præocular.

Rostral once and a half as broad as deep, the portion visible from above nearly half as long as its distance from the frontal	1. *ochrorhynchus*, p. 209.
Rostral nearly twice as broad as deep, the portion visible from above one fourth to one third its distance from the frontal	2. *torquata*, p. 210.
B. Scales in 19 rows; no subocular..	3. *affinis*, p. 210.

II. Subcaudals 80-93.

Scales in 19 rows..................	4. *discolor*, p. 211.
Scales in 21 rows; body cylindrical ..	5. *latifasciata*, p. 211.
Scales in 21 rows; body feebly compressed.......................	6. *ornata*, p. 211.

1. Hypsiglena ochrorhynchus.

Hypsiglena ochrorhyncha, *Cope, Proc. Ac. Philad.* 1860, p. 246; *Garm. N. Am. Rept.* p. 80 (1883); *Cope, Proc. U.S. Nat. Mus.* xiv. 1892, p. 617; *Stejneger, N. Am. Faun.* no. 7, pt. ii. p. 204 (1893).

—— chlorophæa, *Cope, Proc. Ac. Philad.* 1860, p. 246; *Stejneger, l. c.*

—— texana, *Stejneger, l. c.* p. 205.

—— torquata, var., *Günth. Biol. C.-Am., Rept.* p. 137 (1894).

Snout conical, prominent. Rostral once and a half as broad as deep, the portion visible from above nearly half as long as its distance from the frontal; internasals shorter than the præfrontals; frontal once and one third to one and a half as long as broad, as long as or a little longer than its distance from the end of the snout, much shorter than the parietals; loreal longer than deep; one præocular, usually with a subocular below it; two postoculars; temporals 1+2; eight upper labials, fourth and fifth entering the eye; four or five lower labials in contact with the anterior chinshields, which are as long as or a little shorter than the posterior. Scales in 21 rows. Ventrals 167-175; anal divided; subcaudals 41-55. Grey or pale buff above, with a dorsal series of large transverse dark brown or blackish spots and two alternating series of smaller spots on each side; three elongate dark blotches on the nape, the lateral ones produced forwards as a lateral streak, which passes through the eye and is white-edged beneath; upper surface of head and lips finely dotted with dark brown; lower parts uniform white.

Total length 225 millim.; tail 57.

Lower California, California, Arizona, Texas, North Mexico.

a–b. ♀ (V. 172, 171; C. 41, 43).		Duval Co., Texas.	W. Taylor, Esq. [P.].
c. ♂ (V. 168; C. 45).		Nuevo Leon, Mexico.	W. Taylor, Esq. [P.].

2. Hypsiglena torquata.

Leptodira torquata, *Günth. Ann. & Mag. N. H.* (3) v. 1860, p. 170, pl. x. fig. A; *Peters, Mon. Berl. Ac.* 1860, p. 521.
Comastes quincunciatus, *Jan, Elenco*, p. 102 (1863), *and Icon. Gén.* 38, pl. i. fig. 1 (1871).
Hypsiglena torquata, *Cope, Bull. U.S. Nat. Mus.* no. 32, 1887, p. 78.
—— torquata, part., *Günth. Biol. C.-Am., Rept.* p. 137 (1894).

Snout rounded, obtuse. Rostral nearly twice as broad as deep, the portion visible from above one fourth to one third as long as its distance from the frontal; internasals shorter than the præfrontals; frontal once and one third to once and a half as long as broad, as long as or a little longer than its distance from the end of the snout, much shorter than the parietals; loreal longer than deep; one præ-ocular, with a subocular below it; two postoculars; temporals 1+2; eight (exceptionally seven) upper labials, fourth and fifth (or third and fourth) entering the eye; five lower labials in contact with the anterior chin-shields, which are a little shorter than the posterior. Scales in 21 rows. Ventrals 169–174; anal divided; subcaudals 41–55. Greyish or pale brown above, with a dorsal or two alternating dorsal series of large dark brown spots separated by yellowish interspaces; sides with two or three alternating series of small dark brown spots; a large dark brown blotch on the nape, preceded by a yellow collar; a dark brown streak on each side of the head, passing through the eye; head speckled with dark brown; lower parts white.

Total length 400 millim.; tail 60.

Mexico to Venezuela.

a. ♂ (V. 169; C. 53).	Ventanas, Durango.	Hr. A. Forrer [C.].
b–c. ♂ (V. 170, 171; C. 49, 50).	Presidio, near Mazatlan.	Hr. A. Forrer [C.].
d. ♀ (V. 174; C. 41).	Nicaragua.	Derby Museum. (One of the types.)

3. Hypsiglena affinis. (PLATE VIII. fig. 1.)

Hypsiglena torquata, var., *Günth. Biol. C.-Am., Rept.* p. 137 (1894).

Agrees in structure and coloration with *H. torquata*, except that the subocular is absent, the upper labials are seven in number, third and fourth entering the eye, and the scales form only 19 rows. Ventrals 162–168; subcaudals 39–51.

Total length 310 millim.; tail 52.

Mexico.

a. ♂ (V. 164; C. 43).	Zacatecas.	J. M. Cameron, Esq. [P.].

b. Hgr. ♀ (V. 168; C. 39).   Mezquital del Oro,   Dr. A. C. Buller [C.].
                              Zacatecas.
c. ♀ (V. 162; C. 51).   Jalisco.   F. D. Godman, Esq. [P.].

4. Hypsiglena discolor.

Leptodira discolor, *Günth. Proc. Zool. Soc.* 1860, p. 317.
Hypsiglena discolor, *Cope, Bull. U.S. Nat. Mus.* no. 32, 1887, p. 78; *Günth. Biol. C.-Am., Rept.* p. 137, pl. xlix. fig. 1 (1894).

Rostral nearly twice as broad as deep, just visible from above; internasals shorter than the præfrontals; frontal slightly longer than broad; loreal as long as deep or slightly longer than deep; one præ- and two postoculars; temporals 1+2; seven or eight upper labials, third and fourth or fourth and fifth entering the eye; anterior chin-shields as long as or a little shorter than the posterior. Scales in 19 rows. Ventrals 173–180; anal divided; subcaudals 85–89. Above with blackish cross bars separated by pale yellowish-brown interspaces; upper surface of head dark, nape yellowish; labials yellowish with blackish sutures; uniform yellowish white beneath.

Total length 560 millim.; tail 155.

Mexico.

a–b. ♂ (V. 180, 173; C. 85, 89).   Oaxaca.   M. Sallé [C.].   (Types.)

5. Hypsiglena latifasciata.

Hypsiglena latifasciata, *Günth. Biol. C.-Am., Rept.* p. 138, pl. xlix. fig. 2 (1894).

Rostral nearly thrice as broad as deep, scarcely visible from above; internasals a little shorter than the præfrontals; frontal nearly twice as long as broad, with concave sides; loreal deeper than long; one præocular, with a small subocular below it; two postoculars; temporals 1+2; eight upper labials, fourth and fifth entering the eye; anterior chin-shields shorter than the posterior. Scales in 21 rows. Ventrals 186; anal divided; subcaudals *circa* 80 (end of tail absent in the specimen). Dark brown above, with ten widely separated whitish cross bands; head whitish, spotted with brown, except on the occiput; beneath dirty white.

Total length 350 millim.

South Mexico.

a. ♀ (V. 186; C. ?).   S. Mexico.   F. D. Godman, Esq. [P.].   (Type.)

6. Hypsiglena ornata.

Comastes ornatus, *Bocourt, Bull. Soc. Philom.* (7) viii. 1884, p. 141.
Hypsiglena ornata, *Cope, Bull. U.S. Nat. Mus.* no. 32, 1887, p. 78.

Scaling as in *H. torquata*, but body feebly compressed, ventrals and subcaudals more numerous, rostral smaller, and eye larger. Eight upper labials, fourth and fifth entering the eye. Ventrals 185–189; subcaudals 84–93. Yellowish above, with a dorsal series of large brown spots, and smaller spots alternating with them on the sides.

Total length 260 millim.; tail 87.

Isthmus of Darien.

90. RHINOCHILUS.

Rhinocheilus, *Baird & Gir. Cat. N. Am. Rept.* p. 120 (1853); *Cope, Proc. Ac. Philad.* 1860, p. 244; *Jan, Arch. Zool. Anat. Phys.* ii. 1863, p. 217; *Bocourt, Miss. Sc. Mex., Rept.* p. 602 (1886); *Cope, Proc. U.S. Nat. Mus.* xiv. 1892, p. 605.

Maxillary teeth 16 to 19, the two posterior largest; anterior mandibular teeth a little larger than the posterior. Head not distinct from neck, with pointed, strongly projecting snout; eye moderate, pupil round; rostral large. Body much elongate, cylindrical; scales smooth, with apical pits*, in 17 to 23 rows. Tail moderate; subcaudals entire, or only the posterior divided.

North America, Mexico, Venezuela.

Synopsis of the Species.

Scales in 23 rows .... 1. *lecontii*, p. 212.
Scales in 19 rows .... 2. *thominoti*, p. 213.
Scales in 17 rows .... 3. *antonii*, p. 213.

1. Rhinochilus lecontii.

Rhinocheilus lecontei, *Baird & Gir. Cat.* p. 120 (1853), *and Rep. U.S. Explor. R. R.* x. pt. iii. pl. xxxiii. fig. 90 (1859), *and U.S. Mex. Bound. Surv.* ii. *Rept.* pl. xx. (1859); *Jan, Arch. Zool. Anat. Phys.* ii. 1863, p. 217; *Cope, Proc. Ac. Philad.* 1866, p. 304; *Jan, Icon. Gén.* 48, pl. iii. fig. 1 (1876); *Garm. N. Am. Rept.* p. 73 (1883); *Bocourt, Miss. Sc. Mex., Rept.* p. 602, pl. xl. fig. 7 (1886); *Cope, Proc. U.S. Nat. Mus.* xiv. 1892, p. 606; *Günth. Biol. C.-Am., Rept.* p. 100 (1893).

Rostral as deep as broad, the portion visible from above at least half as long as its distance from the frontal; suture between the internasals as long as or a little longer than that between the præfrontals †; frontal a little longer than its distance from the end of the snout or than the parietals, twice as broad as the supraocular; loreal longer than deep; one or two præ- and two or three postoculars; temporals 2+2 or 2+3; eight upper labials, fourth and fifth entering the eye; three or four lower labials in contact with the anterior chin-shields; posterior chin-shields small, and separated from each other by two or three series of scales. Scales in 23 rows. Ventrals 191–206; anal entire; subcaudals 40–55. Black above, with brick-red square spots or cross bands; head and sides black and yellow; lower parts yellowish white, with black spots, at least on the sides.

Total length 940 millim.; tail 140.

California to Texas.

a, b, c. ♂ (V. 191, 205;   Duval Co., Texas.   W. Taylor, Esq. [C.].
C. 49, 50, 6 or 11 last
divided) & hgr. (V. 201;
C. 42, 8 last divided).

---

* Stated to be absent in *R. thominoti* and *antonii*.
† In spec. *c* the præfrontals are separated by the internasals, which are in contact with the point of the frontal.

d. Hgr. ♀ (V. 204; C. 46, San Diego, California. Prof. C. Eigenmann 11 last divided). [C.].
e. ♀ (V. 194; C. 42). California.

Var. *tessellatus*, Garm. *l. c.* p. 74.
Ventrals 178, subcaudals 37 entire and 14 pairs.
Coahuila, Mexico.

2. Rhinochilus thominoti.

Rhinocheilus thominotii, *Bocourt, Le Naturaliste*, (2) i. 1887, p. 45, figs.

Portion of rostral visible from above about half as long as its distance from the frontal; suture between the internasals shorter than that between the præfrontals: frontal a little shorter than its distance from the end of the snout; loreal longer than deep; one præ- and two postoculars; temporals 2+3; eight upper labials, fourth and fifth entering the eye; four lower labials in contact with the anterior chin-shields; posterior chin-shields nearly as large as the anterior and in contact with each other in front. Scales in 19 rows. Ventrals 181; anal entire; subcaudals 82, all single. Reddish yellow above; upper surface of head and a broad nuchal band brown; lower parts yellow.
Total length 280 millim.; tail 67.
Venezuela.

3. Rhinochilus antonii.

Rhinocheilus antonii, *Dugès, Proc. Am. Phil. Soc.* xxiii. 1886, p. 290, fig., and *Le Naturaliste*, (2) i. 1887, p. 46, and *La Naturaleza*, (2) i. 1888, p. 66, pl. vii.; *Bocourt, Miss. Sc. Mex., Rept.* pl. xlv. fig. 4 (1888).

Head-shields as in *R. lecontii*. Scales in 17 rows. Ventrals 200; anal entire; subcaudals 41, last three divided. Black above, with narrow yellow (red?) cross bands; lips yellow; belly yellow, checkered with black.
Total length 310 millim.; tail 35.
San Blas, near Mazatlan, Mexico.

91. CEMOPHORA.

Heterodon, part., *Schleg. Phys. Serp.* ii. p. 96 (1837).
Simotes, part., *Dum. & Bibr. Erp. Gén.* vii. p. 624 (1854); *Günth. Cat. Col. Sn.* p. 23 (1858).
Cemophora, *Cope, Proc. Ac. Philad.* 1860, p. 244; *Jan, Arch. Zool. Anat. Phys.* ii. 1863, p. 230; *Bocourt, Miss. Sc. Mex., Rept.* p. 567 (1883); *Cope, Proc. U.S. Nat. Mus.* xiv. 1892, p. 602.
Stasiotes, *Jan, Arch. Zool. Anat. Phys.* ii. 1862, p. 75.

Maxillary teeth 8 to 11, posterior much longer than anterior; mandibular teeth subequal. Head not distinct from neck, with pointed, strongly projecting snout; eye small, pupil round; rostral and frontal large; nostril in a single or divided nasal. Body elongate, cylindrical; scales smooth, with apical pits, in 19 rows. Tail rather short; subcaudals in two rows.
North America.

1. Cemophora coccinea.

Coluber coccineus, *Blumenb., Voigt's Mag. Phys. u. Naturg.* v. 1788, p. 11, pl. i.; *Daud. Rept.* vii. p. 43, pl. lxxxiii. fig. 1 (1803); *Harlan, Journ. Ac. Philad.* v. 1827, p. 356, *and Phys. Med. Res.* p. 119 (1835).
Elaps coccineus, *Merr. Tent.* p. 145 (1820).
Heterodon coccineus, *Schleg. Phys. Serp.* ii. p. 102, pl. iii. figs. 15 & 16 (1837).
Rhinostoma coccineus, *Holbr. N. Am. Herp.* iii. p. 125, pl. xxx. (1842); *Baird & Gir. Cat. N. Am. Rept.* p. 118 (1853), *and Rep. U.S. Explor. Surv. R. R.* x. pt. iii. pl. xxxiii. fig. 89 (1859).
Simotes coccineus, *Dum. & Bibr.* vii. p. 637, pl. lxxxii. fig. 2 (1854); *Günth. Cat.* p. 26 (1858).
Cemophora coccinea, *Cope, Proc. Ac. Philad.* 1860, p. 244; *Jan, Arch. Zool. Anat. Phys.* ii. 1863, p. 230, *and Icon. Gén.* 11, pl. v. figs. 1 & 2 (1865); *Garm. N. Am. Rept.* p. 78, pl. vi. fig. 1 (1883); *Bocourt, Miss. Sc. Mex., Rept.* p. 567, pl. xxxv. fig. 6 (1883); *Cope, Proc. U.S. Nat. Mus.* xiv. 1892, p. 602.
—— copii, *Jan, ll. cc.* fig. 3.

Portion of rostral visible from above as long as or a little longer than its distance from the frontal; suture between the internasals usually shorter than that between the præfrontals; frontal as long and as broad or a little longer, longer than its distance from the end of the snout, as long as the parietals, at least twice as broad as the supraocular; loreal longer than deep; one or two præ- and one or two postoculars; temporals 1+2 (rarely 2+2); six or seven upper labials, second and third or third and fourth, or third only entering the eye; three or four lower labials in contact with the anterior chin-shields; posterior chin-shields small. Scales in 19 rows. Ventrals 157-188; anal entire; subcaudals 35-45. Red above (in life), with yellow or pale olive transverse bands between pairs of crescentic black bands; lower parts white.

Total length 660 millim.; tail 80.

South-eastern United States.

a. ♂ (V. 166; C. 44).	Charleston.	
b. ♂ (V. 163; C. 41).	Raleigh, N. Carolina.	Messrs. Brindley [C.].
c. ♂ (V. 174; C. 45).	Lake Kerr, Florida.	A. Erwin Brown, Esq. [P.].
d-e. ♀ (V. 169, 177; C. 39, 43).	Marion Co., Florida.	A. Erwin Brown, Esq. [P.].
f. ♀ (V. 168; C. 37).	Florida.	Dr. Guillemard [P.].
g-h, i. ♂ (V. 174, 157; C. 43, 39) & ♀ (V. 174; C. 40).	N. America.	

92. SIMOTES.

Coronella, part., *Schleg. Phys. Serp.* ii. p. 50 (1837).
Xenodon, part., *Schleg. t. c.* p. 80.
Simotes, part., *Dum. & Bibr. Mém. Ac. Sc.* xxiii. 1853, p. 472, *and Erp. Gén.* vii. p. 624 (1854); *Günth. Cat. Col. Sn.* p. 23 (1858).
Simotes, *Jan, Arch. Zool. Anat. Phys.* ii. 1863, p. 232; *Günth. Rept. Brit. Ind.* p. 212 (1864); *Bouleng. Faun. Ind., Rept.* p. 309 (1890).

92. SIMOTES.

Holarchus, *Cope, Proc. Amer. Philos. Soc.* xxiii. 1886, p. 488.
Dicraulax, *Cope, Am. Nat.* 1893, p. 480.

Maxillary teeth 8 to 12, posterior very strongly enlarged and compressed; mandibular teeth subequal. Head short, not distinct from neck; eye rather small, with round pupil; rostral large. Body cylindrical; scales smooth or feebly keeled, in 13 to 21 rows, with or without apical pits; ventrals rounded or obtusely keeled laterally. Tail short or moderate; subcaudals in two rows.

Southern China, East Indian continent and archipelago.

Fig. 17.

Skull of *Simotes albocinctus*.

Synopsis of the Species.

I. Anal entire.

  A. Portion of rostral seen from above as long as or a little shorter than its distance from the frontal.

    1. Scales in 19 or 21 rows.

      a. Two superposed anterior temporals; fifth or fourth and fifth labials usually entering the eye.

        α. Four internasals; ventrals not angulate laterally.
                              1. *splendidus*, p. 217.

        β. Two internasals; ventrals angulate laterally.

Suture between the internasals as long as or longer than that between the præfrontals; frontal large, considerably longer than its distance from the end of the snout .... 2. *purpurascens*, p. 218.

Suture between the internasals as long
as or shorter than that between the
præfrontals; frontal moderate ....   3. *cyclurus*, p. 219.

    b. A single anterior temporal.

Portion of rostral seen from above
shorter than its distance from the
frontal ....   4. *albocinctus*, p. 220.

Portion of rostral seen from above as
long as its distance from the frontal.   5. *formosanus*, p. 222.

   2. Scales in 17 rows.

      a. Fourth and fifth labials entering the eye; a single anterior temporal........   6. *violaceus*, p. 222.

      b. Fourth or third and fourth labials entering the eye.

         α. A single labial entering the eye; a single anterior temporal ....   7. *woodmasoni*, p. 223.

         β. Two labials entering the eye; two superposed anterior temporals ....   8. *octolineatus*, p. 224.

         γ. Two labials entering the eye; a single anterior temporal.

         * Two postoculars.

Loreal longer than deep ....   9. *phænochalinus*, p. 225.
Loreal rather deeper than long ......   10. *forbesii*, p. 225.

        ** One postocular.

Scales smooth ....   11. *signatus*, p. 226.
Scales feebly keeled ....   12. *subcarinatus*, p. 226.

   3. Scales in 15 rows ....   13. *annulifer*, p. 226.

  B. Portion of rostral visible from above much shorter than its distance from the frontal; scales in 17 or 19 rows.

Temporals 1+1+2; ventrals 146-169;
subcaudals 30-47 ....   14. *tæniatus*, p. 227.
Temporals 1+2 or 2+2; ventrals 170-
191; subcaudals 55-60 ....   15. *chinensis*, p. 228

II. Anal divided.

  A. Scales in 19 rows ....   16. *vaillanti*, p. 228.

  B. Scales in 17 rows.

    1. Nasal undivided; fourth and fifth labials entering the eye; ventrals 167-168; subcaudals 43-46.

                  17. *beddomii*, p. 229.

    2. Nasal divided, rarely semidivided.

Third and fourth labials entering the
eye; ventrals 170-202; subcaudals
41-59 ....   18. *arnensis*, p. 229.

Fourth and fifth labials entering the eye;
 ventrals 171–180; subcaudals 34–42. 19. *theobaldi*, p. 230.
Fourth and fifth (rarely third and fourth)
 labials entering the eye; ventrals
 148–173; subcaudals 27–37 ...... 20. *cruentatus*, p. 231.
C. Scales in 15 rows .... 21. *torquatus*, p. 232.
D. Scales in 13 rows .... 22. *planiceps*, p. 232.

TABLE SHOWING NUMBERS OF SCALES AND SHIELDS.

	Sc.	V.	A.	C.	Lab.	Ant. temp.
splendidus	21	193	1	41	8	2
purpurascens	19–21	160–210	1	40–60	8 (7)	2
cyclurus	19–21	156–210	1	37–58	8 (7)	2
albocinctus	19	177–205	1	51–69	7 (8)	1
formosanus	19	162–173	1	46–55	7–8	1
violaceus	17	160–196	1	33–41	8	1
woodmasoni	17	180–186	1	57	6	1
octolineatus	17	156–197	1	43–61	6	2
phænochalinus	17	156–173	1	36–45	7	1 (2)
forbesii	17	155–165	1	45	7	1
signatus	17	141–157	1	47–59	7	1
subcarinatus	17	155	1	50	7	1
annulifer	15	153	1	49	7	1
tæniatus	17–19	146–169	1	30–47	8	1
chinensis	17	170–191	1	55–60	8	1–2
vaillanti	19	?	2	?	8	1
beddomii	17	167–168	2	43–46	8	1
arnensis	17	170–202	2	41–59	7	1
theobaldi	17	171–180	2	34–42	8	1
cruentatus	17	148–173	2	27–37	8 (7)	1
torquatus	15	144–159	2	27–34	7	1
planiceps	13	132	2	27	5	1

1. Simotes splendidus.

Simotes splendidus, *Günth. Proc. Zool. Soc.* 1875, p. 231, pl. xxxiii.;
 Bouleng. Faun. Ind., Rept. p. 310 (1890).

Nasal divided; portion of rostral seen from above as long as its distance from the frontal; each of the internasals broken up into two shields, there being four small shields in a transverse series; frontal longer than its distance from the end of the snout, slightly longer than the parietals; loreal deeper than long; præocular single, with a small subocular below it, between the third and fourth labials; two or three postoculars; temporals 2+3; eight upper labials, fourth and fifth entering the eye; four lower labials in contact with the anterior chin-shields; posterior chin-shields about two thirds the length of the anterior. Scales in 21 rows. Ventrals 193; anal entire; subcaudals 41. Cream-colour, with sixteen large, brown, black-edged spots above, longer than the interspaces between them; these spots indented in front and behind, and with at least a trace of a yellowish median line; the anterior spot is produced angularly to the posterior border of the frontal; the rest of the upper surface of the head speckled with dark brown; tail with a yellow vertebral line; an irregular series of small blackish spots along each abdominal edge, the lower surfaces being otherwise immaculate.

Total length 560 millim.; tail 65.

Southern India.

a. ♀ (V. 193; C. 41). Wynad. Col. Beddome [C.]. (Type.)

2. Simotes purpurascens.

Xenodon purpurascens, *Schleg. Phys. Serp.* ii. p. 90, pl. iii. figs. 13 & 14 (1837), *and Abbild.* p. 47, pl. xiv. (1844); *Cantor, Cat. Mal. Rept.* p. 67 (1847).
Simotes trinotatus, *Dum. & Bibr.* vii. p. 631 (1854); *Günth. Rept. Brit. Ind.* p. 219 (1864).
—— albocinctus (*non Cant.*), *Dum. & Bibr.* p. 633, pl. lxxxii. fig. 1.
Calamaria brachyorrhos (*non Schleg.*), *Motl. & Dillw. Nat. Hist. Lab.* p. 49, pl. — (1855).
Simotes purpurascens, var. C, *Günth. Cat.* p. 25 (1858).
—— purpurascens, *Jan, Arch. Zool. Anat. Phys.* ii. 1862, p. 235, *and Icon. Gén.* 12, pl. v. fig. 2 (1865); *Bouleng. Proc. Zool. Soc.* 1890, p. 34; *v. Lidth de Jeude, Notes Leyd. Mus.* xii. 1890, p. 255.
—— labuanensis, *Günth. Rept. Brit. Ind.* p. 217.
—— catenifer, *Stoliczka, Journ. As. Soc. Beng.* xlii. 1873, p. 121, pl. xi. fig. 3.
—— dennysi, *Blanf. Proc. Zool. Soc.* 1881, p. 218, pl. xxi. fig. 1.
—— affinis, *Fischer, Abh. naturw. Ver. Hamb.* ix. 1885, p. 4, pl. i. fig. 1.

Nasal divided; portion of rostral seen from above as long as its distance from the frontal; suture between the internasals as long as or longer than that between the præfrontals; frontal large, considerably longer than its distance from the end of the snout, as long as or longer than the parietals; loreal as long as deep; one or two præoculars, with one or two suboculars below; two or three postoculars; temporals 2+3 or 2+2; usually eight upper labials, fourth and fifth entering the eye, or fourth excluded by the second subocular; four or five lower labials in contact with the anterior chin-shields, which are about twice as long as the posterior. Scales in 19 or 21 rows. Ventrals 160–210, angulate laterally; anal entire; subcaudals 40–60. Brown above, with darker, black-edged sinuous transverse bands, or with yellowish black-edged cross bars; a large arrow-headed dark-brown marking on the occiput and nape, with the apex on the frontal and usually confluent with a dark-brown chevron-shaped transverse band passing through the eyes; a dark oblique temporal streak; belly yellowish, with square blackish spots, which may be confined to the sides.

Total length 680 millim.; tail 110.

S. China, Cochinchina, Siam, Malay Peninsula, Sumatra, Borneo, Java.

A. Pale transverse bands narrower than the interspaces between them; scales in 19 rows. (*S. purpurascens*, Schleg.)

a. Yg. (V. 182; C. 43). Java.
b. Hgr. (V. 169; C. 51). Matang, Borneo.

B. Narrow dark transverse bands; scales in 19 rows.
(*S. labuanensis*, Gthr.)

a. Hgr. (V. 186; C. 54). Singapore. Governor of Singapore [P.].
b. Yg. (V. 166; C. 53). Deli, Sumatra. Mr. Iversen [C.].
c. Yg. (V. 169; C. 59). Sumatra. Dr. Bleeker.
d–e. ♂ (V. 170; C. ?) & Nias. Hr. Sundermann [C.].
♀ (V. 174; C. 53).
f. ♂ (V. 171; C. ?). Labuan, Borneo. L. L. Dillwyn, Esq. [P.]. Types of *S. labuanensis*.
g. ♂ (V. 187; C. 60). —— ?

C. Narrow dark transverse bands; scales in 21 rows.
(*S. trinotatus*, D. & B.)

a. ♀ (V. 190; C. 53). Pinang. Dr. Cantor.
b. Hgr. (V. 165; C. 53). Deli, Sumatra. Prof. Moesch [C.].

3. Simotes cyclurus.

Coronella cyclura, *Cantor, Proc. Zool. Soc.* 1839, p. 50.
Simotes bicatenatus, *Günth. Rept. Brit. Ind.* p. 217 (1864); *Theob. Journ. Linn. Soc.* x. 1868, p. 40; *Stoliczka, Journ. As. Soc. Beng.* xl. 1871, p. 430; *Anders. Proc. Zool. Soc.* 1871, p. 170.
—— fasciolatus, *Günth. l. c.* p. 218, pl. xx. fig. B.
—— cochinchinensis, *Günth. l. c.* p. 219, pl. xx. fig. C.
—— brevicauda, *Steind. Novara, Rept.* p. 61, pl. iii. figs. 13 & 14 (1867).
—— obscurus, *Theob. Cat. Rept. As. Soc. Mus.* p. 48 (1868).
—— crassus, *Theob. l. c.*
—— cyclurus, *Bouleng. Faun. Ind., Rept.* p. 311 (1890); *W. L. Sclater, Journ. As. Soc. Beng.* lx. 1890, p. 235.

Nasal divided; portion of rostral seen from above nearly as long as its distance from the frontal; suture between the internasals as long as or shorter than that between the præfrontals; frontal as long as its distance from the end of the snout or a little longer, and as long as the parietals; loreal as long as deep, or a little deeper than long; præocular single, usually with a small subocular below it, between the third and fourth labials; two postoculars; temporals 2+2; normally eight upper labials, fourth and fifth entering the eye; four, rarely three, lower labials in contact with the anterior chin-shields; posterior chin-shields one half or two thirds the length of the anterior. Scales in 19 or 21 rows. Ventrals 156–210, angulate laterally; anal undivided; subcaudals 37–58. Pale brown or greyish above, with or without four darker brown stripes, the median pair separated by the vertebral series of scales; antero-lateral border of some of the scales frequently black; markings on the head as in *S. albocinctus*; lower surfaces yellowish, uniform, or with squarish brown spots on each side below the angle of the ventrals; these blotches may be subconfluent or form two chain-like series; subcaudals unspotted or with a few small scattered dots.

Total length 700 millim.; tail 100.

Bengal and Assam, Burma, Siam, Cochinchina, and Southern China.

A. Above with very ill-defined dark longitudinal streaks; belly with a lateral series of square brown spots. Scales in 19 rows. (*S. cyclurus, S. bicatenatus.*)

a.	♂ (V. 175; C. 37).	Toungoo.	E. W. Oates, Esq. [P.].
b.	Yg. (V. 170; C. 37).	Pegu.	W. Theobald, Esq. [C.].
c.	Yg. (V. 156; C. 43).	S. China.	R. Swinhoe, Esq. [C.].
d.	♂ (V. 166; C. 43).	——?	College of Surgeons. (Type of *S. bicatenatus.*)

B. Above with four more or less distinct dark longitudinal streaks and a light vertebral line; belly with a lateral series of square brown spots. Scales in 19 rows.

a. ♀ (V. 175; C. 40). Pegu. W. Theobald, Esq. [C.].
b. ♀ (V. 167; C. 44). Tenasserim. Col. Beddome [C.].

C. Like the preceding, but ventral spots absent or reduced to a few small dots.

a–b. Hgr. (V. 171, 173; Pegu. W. Theobald, Esq. [C.].
C. 46, 43).

D. No longitudinal streaks, no ventral spots. Scales in 19 rows.

a, b. ♂ (V. 167; C. 46) Pegu. W. Theobald, Esq. [C.].
& yg. (V. 150; C. 49).

E. No longitudinal streaks, no ventral spots. Scales in 21 rows. (*S. fasciolatus, S. cochinchinensis.*)

a, b. ♂ (V. 160; C. 42) Pachebone, Siam. M. Mouhot [C.]. (Types
& ♀ (V. 165; C. 45). of *S. fasciolatus.*)
c. Hgr. (V. 173; C. 41). Siam. M. H. Newman, Esq. [P.].
d. Yg. (V. 210; C. 47). Lao Mountains. M. Mouhot [C.]. (Type of *S. cochinchinensis*).

F. Above with four dark stripes as in B, which are crossed by black bars; belly immaculate. Scales in 21 rows.

a. Hgr. (V. 180; C. 53). Bia-po, Karin Hills. M. L. Fea [C.].

4. Simotes albocinctus.

Coronella albocincta, *Cantor, Proc. Zool. Soc.* 1839, p. 50.
Xenodon purpurascens, var., *Cantor, Cat. Mal. Rept.* p. 67 (1847); *Blyth, Journ. As. Soc. Beng.* xxiii. 1854, p. 289.
Coronella punctulatus, *Gray, Ann. & Mag. N. H.* (2) xii. 1853, p. 389.
Simotes purpurascens, vars. D *and* E, *Günth. Cat.* p. 25 (1858).
—— punctulatus, *Günth. Rept. Brit. Ind.* p. 217 (1864); *Anders. Proc. Zool. Soc.* 1871, p. 169.

Simotes albocinctus, part., *Günth. l. c.* p. 218.
—— amabilis, *Günth. Ann. & Mag. N. H.* (4) i. 1868, p. 416, pl. xvii. fig. A.
—— albocinctus, *Bouleng. Faun. Ind., Rept.* p. 312 (1890).

Nasal divided; portion of rostral seen from above shorter than its distance from the frontal; suture between the internasals shorter than that between the præfrontals; frontal longer than its distance from the end of the snout, as long as the parietals; loreal as long as deep or a little longer, seldom united with the præfrontal; præocular single or (rarely) divided into two; two postoculars; temporals 1+2; normally seven upper labials, third and fourth entering the eye; four (rarely five) lower labials in contact with the anterior chin-shields; posterior chin-shields about two thirds the length of the anterior. Scales in 19 rows. Ventrals 177-205; anal undivided; subcaudals 51-69. Brown or brick-red above; head yellowish, with a dark brown or black-edged red crescentic band across the forehead to below each eye, sometimes interrupted in the middle, an oblique one on each side from the parietal to behind the angle of the mouth, and a chevron-shaped one from the frontal to the nape, sometimes with a detached round spot in front; lower surfaces yellowish or coralline-red, more or less spotted or marbled with black; the black spots may be small, squarish, far apart, and confined to the sides, or the black may predominate on the belly; or the ventrals may be alternately black and yellow.

Total length 800 millim.; tail 130.

Eastern and Central Himalayas (to 4000 feet), Khasi and Arakan Hills.

A. Upper parts unspotted, with 27 to 34 light, black-edged cross bands two scales wide, separated by 5 to 8 series of scales. (*S. albocinctus*, Cant.)

a. ♀ (V. 203; C. 53).   Khasi Hills.   Sir J. Hooker [P.]. (Type of *C. punctulatus*.)
b-e. ♂ (V. 183; C. 64), Himalayas.   T. C. Jerdon, Esq. [P.].
  hgr. (V. 184, 194; C. 59, 54), & yg. (V. 195; C. 54).
f. ♂ (V. 195; C. 65).   Himalayas.   Col. Beddome [C.].
g-h. ♀ (V. 203; C. 51)   Darjeeling.   W. T. Blanford, Esq. [P.].
  & yg. (V. 188; C. 66).
i. Skull.   Darjeeling.   W. T. Blanford, Esq. [P.].

B. Upper parts unspotted, with 55 light, black-edged cross bars, separated by 3 or 4 series of scales. (*S. amabilis*, Gthr.)

a. Yg. (V. 178; C. 64).   Arakan Hills.   W. Theobald, Esq. [C.]. (Type of *S. amabilis*.)

C. Upper parts with dark spots, without cross bands.

a-b. ♀ (V. 202; C. 55)   Khasi Hills.   Sir J. Hooker [P.].
  & yg. (V. 193; C. 62).
c. ♂ (V. 193; C. 65).   Sikkim.   Messrs. v. Schlagintweit [C.].
d. ♀ (V. 205; C. ?).   Darjeeling.   T. C. Jerdon, Esq. [P.].

e–f. ♂ (V. 194; C. 63) & hgr. (V. 194; C. 69).	Darjeeling.	W. T. Blanford, Esq. [P.].
g. ♀ (V. 203; C. 56).	Nepaul.	B. H. Hodgson, Esq. [P.].
h–i. ♂ (V. 182, 177; C. 62, 68).	——?	Dr. Griffith.

5. Simotes formosanus. (PLATE VIII. fig. 2.)

Simotes formosanus, *Günth. Ann. & Mag. N. H.* (4) ix. 1872, p. 20; Fischer, *Abh. naturw. Ver. Hamb.* ix. 1886, p. 12.

Nasal divided ; portion of rostral seen from above as long as its distance from the frontal ; suture between the internasals shorter than that between the præfrontals ; frontal longer than its distance from the end of the snout, as long as the parietals ; loreal square ; præocular single or divided ; two postoculars ; temporals 1+2 ; seven or eight upper labials, third and fourth or fourth and fifth entering the eye ; three or four labials in contact with the anterior chin-shields, which measure about once and a half the size of the posterior. Scales in 19 rows. Ventrals obtusely angulate, 162–173 ; anal entire ; subcaudals 46–55. Pale brown above, with a very indistinct lighter vertebral line and ill-defined dark cross bands produced by the black borders of some of the scales ; ventrals yellowish, brown on the sides, or with a lateral series of small blackish spots.

Total length 600 millim. ; tail 95.

S. China.

| a. ♂ (V. 163; C. 55). | Takao, Formosa. | R. Swinhoe, Esq. [C.]. (Type.) |
| b. ♀ (V. 173; C. 46). | Swatow. | Christiania Museum. |

6. Simotes violaceus.

Coronella violacea, *Cantor, Proc. Zool. Soc.* 1839, p. 50.
Simotes cinereus, *Günth. Rept. Brit. Ind.* p. 215 (1864).
—— swinhonis, *Günth. l. c.* pl. xx. fig. E.
—— multifasciatus, *Jan, Icon. Gén.* 12, pl. iv. fig. 2 (1865).
—— semifasciatus, *Anders. Journ. As. Soc. Beng.* xl. 1871, p. 16.
—— violaceus, *Bouleng. Faun. Ind., Rept.* p. 312 (1890).

Nasal divided ; portion of rostral seen from above as long as its distance from the frontal or a little shorter ; suture between the internasals usually shorter than that between the præfrontals ; frontal as long as its distance from the end of the snout, as long as the parietals ; loreal usually longer than deep ; præocular single, usually with a small subocular below, between the third and fourth labials ; one or two postoculars ; temporals 1+2 ; eight upper labials, fourth and fifth entering the eye ; three or four lower labials in contact with the anterior chin-shields ; posterior chin-shields one half or less than one half the size of the anterior. Scales in 17 rows. Ventrals 160–196 ; anal undivided ; subcaudals 33–41. Pale brown, purplish or reddish above ; markings on the head very indistinct.

Total length 760 millim.; tail 75.
Bengal, Assam, Burma, Camboja, Southern China.

A. Brown or reddish above, without dark markings; belly unspotted. (*S. violaceus, S. cinereus.*)

a. ♂ (V. 164; C.?).	Camboja.		M. Mouhot [C.].
			(Type of *S. cinereus.*)
b. ♂ (V. 169; C. 35).	Toungyi, S. Shan States, 5000 ft.		Lieut. Blakeway [C.].

B. The black edges of some of the scales forming more or less distinct dark cross bands; belly unspotted. (*S. swinhonis.*)

a. ♂ (V. 161; C. 37).	Amoy.		R. Swinhoe, Esq. [C.].
b-c. ♂ (V. 165; C. 39)	——?		Haslar Hospital.
& ♀ (V. 174; C. 34).			(Types of *S. swinhonis.*)
d. ♂ (V. 160; C. 38).	Hainan.		R. Swinhoe, Esq. [C.].
e. ♀ (V. 166; C. 33).	Island of Hong Kong.		H.M.S. 'Challenger.'

C. Intermediate between A and D.

a. ♂ (V. 172; C. 38). Tenasserim. Col. Beddome [C.].

D. Above with more or less distinct dark cross bands; belly with quadrangular brown spots. (*S. multifasciatus, S. semifasciatus.*)

a. ♂ (V. 175; C. 38).	Assam.		W. T. Blanford, Esq. [P.].
b. ♂ (V. 171; C. 34).	Toungyi, S. Shan States, 5000 ft.		Lieut. Blakeway [C.].

7. Simotes woodmasoni.

Simotes wood-masoni, *W. L. Sclater, Journ. As. Soc. Beng.* lx. 1891, p. 235, pl. vi. fig. 2.

Nasal divided; portion of the rostral seen from above a little shorter than its distance from the frontal; suture between the internasals shorter than that between the præfrontals; frontal longer than its distance from the end of the snout, as long as the parietals; loreal small, longer than deep; one præocular; one subocular, separating the third labial from the eye; two postoculars; temporals 1+2; six upper labials, the fourth alone entering the eye; four lower labials in contact with the anterior chin-shields; posterior chin-shields small. Scales in 17 rows. Ventrals strongly angulate, 180-186; anal undivided; subcaudals 57. Brick-reddish above, with traces of seven lighter stripes; head with a dark median longitudinal mark extending from the frontal to the nape, where it bifurcates, an oblique dark streak across the anterior nasal and the three anterior labials, a dark streak across the forehead through the eyes to the fourth and fifth labials, and an oblique dark streak across the parietals and the sides of the nape; dusky reddish beneath.

Andaman and Nicobar Islands.

8. Simotes octolineatus.

Russell, Ind. Serp. ii. pl. xxxviii. (1801).
Elaps octolineatus, *Schneid. Hist. Amph.* ii. p. 299 (1801).
Coluber octolineatus, *Shaw, Zool.* iii. p. 540 (1802); *Daud. Rept.* vii. p. 17 (1803).
Coronella octolineata, *Schleg. Phys. Serp.* ii. p. 77 (1837).
Simotes octolineatus, *Dum. & Bibr.* vii. p. 634, pl. lxxxii. fig. 3 (1854); *Günth. Cat.* p. 24 (1858); *Jan, Arch. Zool. Anat. Phys.* ii. 1863, p. 235, *and Icon. Gén.* 12, pl. v. fig. 1 (1865); *Bouleng. Faun. Ind., Rept.* p. 313 (1890), *and Ann. & Mag. N. H.* (6) ix. 1892, p. 74.
—— meyerinkii, *Steindachn. Sitzb. Ak. Wien,* c. 1891, p. 294.

Nasal divided; portion of rostral seen from above a little shorter than its distance from the frontal; suture between the internasals nearly as long as that between the præfrontals; frontal longer than its distance from the end of the snout, as long as the parietals; loreal as long as deep; præocular single; two postoculars; temporals 2+2; six upper labials, third and fourth entering the eye; four lower labials in contact with the anterior chin-shields; posterior chin-shields about two thirds the length of the anterior. Scales in 17 rows. Ventrals obtusely angulate, 156–197; anal undivided; subcaudals 43–61. Yellow with six black longitudinal stripes, or blackish with yellow longitudinal lines, which are much narrower than the interspaces between them; head yellow, the two median dorsal bands meeting on the frontal; a black crescentic cross band anteriorly, passing through the eyes, and an oblique band on each side from the parietal shield to below the angle of the mouth; lower surfaces uniform or with a series of black spots on each side of the ventrals, sometimes confluent into a line.

Total length 680 millim.; tail 100.

Java, Borneo, Sooloo Islands, Sumatra, Malay Peninsula; Anamallay Hills, S. India.

A. Yellowish above, with six black longitudinal stripes.

a. Yg. (V. 168; C. 54).	Anamallays.	Col. Beddome [C.].
b. ♂ (V. 165; C. 61).	Singapore.	Dr. Dennys [P.].
c–d. ♀ (V. 178, 183; C. 54, 51).	Sumatra.	Mrs. J. Crosley [P.].
e. ♀ (V. 175; C. 54) & hgr. (V. 183; C. 54).	Bandjermassing, Borneo.	L. L. Dillwyn, Esq. [P.].
f. Yg. (V. 168; C. 61).	Java.	J. Bowring, Esq. [P.].

B. Black above, with five narrow yellow longitudinal lines.

a. Yg. (V. 197; C. 61).	Sumatra.	Sir S. Raffles.
b–c. ♂ (V. 171; C. 59) & yg. (V. 171; C. 61).	Nias.	Hr. Sundermann [C.].
d. ♀ (V. 175; C. 53).	Borneo.	Sir H. Low [C.].

C. Markings disappearing in the adult, which is dark brown above, with a somewhat ill-defined light vertebral stripe. (*S. meyerinkii,* Steind.)

a. ♀ (V. 158; C. 48).   Tawi-Tawi, Sooloo Ids.   A. Everett, Esq. [C.].

9. Simotes phænochalinus.

? Xenodon ancorus, *Girard, Proc. Ac. Philad.* 1857, p. 182, *and U. S. Explor. Exped., Herp.* p. 167 (1858).
Simotes purpurascens, var. C, part., *Günth. Cat.* p. 25 (1858).
—— phænochalinus, *Cope, Proc. Acad. Philad.* 1860, p. 244.
—— aphanospilus, *Cope, l. c.* p. 245.
—— ancoralis, *Jan, Arch. Zool. Anat. Phys.* ii. 1863, p. 233, *and Icon. Gén.* 11, pl. vi. fig. 2 (1865); *Steind. Novara, Rept.* p. 61 (1867).

Nasal divided; portion of rostral seen from above as long as or a little shorter than its distance from the frontal; suture between the internasals usually shorter than that between the præfrontals; frontal a little longer than its distance from the end of the snout, as long as the parietals; loreal longer than deep; one præ- and two postoculars; temporals 1+2 (in one specimen 2+2); seven upper labials, third and fourth entering the eye; four lower labials in contact with the anterior chin-shields, which are about once and a half as long as the posterior. Scales in 17 rows. Ventrals angulate laterally, 156-173; anal entire; subcaudals 36-45. Pale brown above, the vertebral region sometimes more reddish, with large transverse rhomboidal or escutcheon-shaped, purplish, black-edged spots; a large arrow-headed, purplish, black-edged marking on the head and nape, often confluent with a transverse band passing through the eyes; an oblique temporal streak; lower parts yellowish, uniform or with small brown spots.

Total length 670 millim.; tail 95.

Philippine Islands.

a-b. ♂ (V. 156; C. 42) & ♀ (V. 172; C. 39).	Philippines.	H. Cuming, Esq. [C.].
c. ♂ (V. 157; C. 45).	Luzon.	Dr. A. B. Meyer [C.].
d. ♀ (V. 170; C. 36).	Java (?).	
e. ♀ (V. 173; C. 36).	——?	Dr. Günther [P.].

10. Simotes forbesii.

Simotes forbesii, *Bouleng. Proc. Zool. Soc.* 1883, p. 387, pl. xlii.

Nasal divided; portion of rostral seen from above as long as its distance from the frontal; suture between the internasals shorter than that between the præfrontals; frontal longer than its distance from the end of the snout, as long as the parietals; loreal slightly deeper than long; one præ- and two postoculars; temporals 1+2; seven upper labials, third and fourth entering the eye; four lower labials in contact with the anterior chin-shields, which are nearly twice as long as the posterior. Scales in 17 rows. Ventrals slightly angulate laterally, 155-165; anal entire; subcaudals 45. Greyish brown above, the borders of the scales darker; head with symmetrical dark markings; two fine dark brown lines along the back, separated by three longitudinal series of scales; belly yellowish, with a lateral series of brown spots, more or less confluent into a stripe.

Total length 305 millim.; tail 58.

Timor Laut.

a-b. ♀ (V. 155, 165; C. 45, 45).	Timor Laut.	H. O. Forbes, Esq. [C.]. (Types.)

11. Simotes signatus.

Simotes purpurascens, var. F, *Günth. Cat.* p. 26 (1858).
—— signatus, *Günth. Rept. Brit. Ind.* p. 215, pl. xx. fig. F (1864).

Nasal divided; portion of rostral seen from above shorter than its distance from the frontal; suture between the internasals as long as or shorter than that between the præfrontals; frontal longer than its distance from the end of the snout, as long as the parietals; a small square loreal; one præ- and one postocular; temporals 1+2; seven upper labials, third and fourth entering the eye; three or four lower labials in contact with the anterior chin-shields, which are a little longer than the posterior. Scales in 17 rows. Ventrals 141–157; anal entire; subcaudals 47–59. Dark brown above, with yellowish cross bands or transverse rhomboidal spots; the first two cross bands chevron-shaped; head with dark brown symmetrical markings separated by yellowish interspaces; belly yellowish, with a lateral series of brown spots.

Total length 520 millim.; tail 130.
Malay Peninsula, Sumatra, Java.

a. ♂ (V. 149; C. 59).	Singapore.	Gen. Hardwicke [P.].
b. ♀ (V. 157; C. 47).	Singapore.	Haslar Collection.
		(Types.)
c. ♀ (V. 151; C. 50).	Deli, Sumatra.	Prof. Moesch [C.].
d. ♂ (V. 141; C. 52).	Java.	Dr. Ploem [C.].

12. Simotes subcarinatus.

Simotes subcarinatus, *Günth. Proc. Zool. Soc.* 1872, p. 595, pl. xxxix. fig. B.

Nasal divided; portion of rostral seen from above shorter than its distance from the frontal; suture between the internasals shorter than that between the præfrontals; frontal nearly as broad as long, longer than its distance from the end of the snout, as long as the parietals; a small square loreal; one præ- and one postocular; temporals 1+1+2; seven upper labials, third and fourth entering the eye; four lower labials in contact with the anterior chin-shields, which are nearly twice as large as the posterior. Scales feebly keeled, in 17 rows. Ventrals 155; anal entire; subcaudals 50. Dark brown above, with yellowish, black-edged cross streaks, the anterior of which are chevron-shaped; head yellowish, with symmetrical dark brown markings; belly yellowish, with a few brown spots on the sides.

Total length 395 millim.; tail 80.
Borneo.

a. ♀ (V. 155; C. 50).	Sarawak.	(Type.)

13. Simotes annulifer. (Plate VIII. fig. 3.)

Simotes annulifer, *Bouleng. Proc. Zool. Soc.* 1893, p. 524.

Nasal divided; portion of rostral seen from above slightly shorter than its distance from the frontal; suture between the internasals slightly shorter than that between the præfrontals; frontal as broad as long, longer than its distance from the end of the snout, shorter than the parietals; a small loreal; one præ- and two postoculars; temporals 1+2; seven upper labials, third and fourth entering the eye; four lower labials in contact with the anterior chin-shields, which are longer than the posterior. Scales in 15 rows. Ventrals 153; anal entire; subcaudals 49. Brown above, with 26 black annuli on the back, enclosing large and oval yellowish-brown spots; sides black-spotted, with vertical and oblique yellowish lines; head yellowish brown above, with a dark brown bar across the forehead, passing through the eye; a large λ-shaped marking from the frontal shield to the nape, and an oblique bar on the temple; labials, chin, and throat black-spotted; lower parts white, with a series of small black spots on each side.

Total length 160 millim.; tail 30.

North Borneo.

a. Yg. (V. 153; C. 49).   N. Borneo.   A. Everett, Esq. [C.].
(Type.)

14. Simotes tæniatus.

Simotes tæniatus, *Günth. Proc. Zool. Soc.* 1861, p. 189, *and Rept. Brit. Ind.* p. 216, pl. xx. fig. A (1864); *Steind. Novara, Rept.* p. 60 (1867); *F. Müll. Verh. nat. Ges. Basel,* vii. 1882, p. 144.
—— quadrilineatus, *Jan, Nouv. Arch. Mus.* ii. 1866, *Bull.* p. 7, *and Icon. Gén.* 12, pl. iv. fig. 3 (1865).

Nasal divided; portion of rostral seen from above about half as long as its distance from the frontal; suture between the internasals shorter than that between the præfrontals; frontal slightly longer than its distance from the end of the snout, a little shorter than the parietals; loreal longer than deep; one præocular, usually with a small subocular below it, between the third and fourth labials; two postoculars; temporals 1+1+2, anterior very small; eight upper labials, fourth and fifth entering the eye; four lower labials in contact with the anterior chin-shields, which are about once and a half as long as the posterior. Scales in 17 or 19 rows. Ventrals 146–169; anal entire; subcaudals 30–47. Brown above, with a darker dorsal band enclosing a light vertebral line; a dark lateral streak or series of spots; a dark interocular band, passing through the eyes and extending to the mouth, or a large triangular spot covering the upper surface of the snout; an oblique dark temporal band, sometimes descending to or even extending across the throat, confluent or not with a large nuchal spot; usually a blackish spot at the base, and another at the end of the tail; lower parts whitish, with squarish black spots which may be confined to the hinder portion of the body.

Total length 320 millim.; tail 60.

Siam, Camboja, Cochinchina.

a, b, c–e. ♂ (Sc. 19, 17, 17, 17; V. 159, 146, 149, 149; C. 43, 42, 41, 40) & ♀ (Sc. 19; V. 161; C. 34).	Camboja.	M. Mouhot [C.]. (Types.)
f. ♀ (Sc. 17; V. 169; C. 30).	Bangkok.	Dr. Günther [P.].
g. ♂ (Sc. 19; V. 153; C. 44).	Siam.	Sir R. Schomburgk [P.].
h–i. ♂ (Sc. 17, 19; V. 150, 152; C. 41, 43).	Siam.	M. H. Newman, Esq. [P.].

15. Simotes chinensis. (PLATE IX. fig. 1.)

Simotes chinensis, *Günth. Ann. & Mag. N. H.* (6) i. 1888, p. 16.

Nasal divided; portion of rostral seen from above one half to three-fifths as long as its distance from the frontal; suture between the internasals as long as or longer than that between the præfrontals; frontal longer than its distance from the end of the snout, as long as the parietals; loreal a little longer than deep; one præ- and two postoculars; a subocular may be present below the præocular; temporals $1+2$ or $2+2$; eight upper labials, fourth and fifth, or fifth, entering the eye; four lower labials in contact with the anterior chin-shields, which are about once and a half the length of the posterior. Scales in 17 rows. Ventrals angulate laterally, 170–191; anal entire; subcaudals 55–60. Pale grey-brown above, with narrow blackish-brown transverse bands, 15 to 17 in number; a dark brown arrow-headed marking on the occiput and nape, with the apex on the frontal shield; a crescentic dark brown band anteriorly, passing through the eyes; yellowish inferiorly, with quadrangular black spots.

Total length 600 millim.; tail 120.

China.

a. Yg. (V. 191; C. 55).	Mountains north of Kiu Kiang.	A. E. Pratt, Esq. [C.]. (Type.)
b–d. ♂ (V. 176; C. 59) & yg. (V. 172, 170; C. 60, 58).	Hoi-How, Hainan.	J. Neumann, Esq. [P.].

16. Simotes vaillanti.

Simotes vaillanti, *Sauvage, Bull. Soc. Philom.* (7) i. 1877, p. 107.

Nasal divided; one præ- and two postoculars; temporals $1+2$; eight upper labials, fourth and fifth entering the eye. Scales in 19 rows. Anal divided. Olive-brown above, with or without two dorsal black lines commencing from the eyes, and with black transverse spots; belly yellowish.

China.

17. Simotes beddomii. (PLATE IX. fig. 2.)

Simotes beddomii, *Bouleng. Faun. Ind., Rept.* p. 314 (1890).

Nasal undivided; rostral comparatively small, the portion seen from above much shorter than its distance from the frontal; suture between the internasals much shorter than that between the præfrontals; frontal longer than its distance from the end of the snout, a little shorter than the parietals; loreal longer than deep; præocular single; two postoculars; temporals 1+2; eight upper labials, fourth and fifth entering the eye; four or five lower labials in contact with the anterior chin-shields; posterior chin-shields about two thirds the length of the anterior. Scales in 17 rows. Ventrals 167-168; anal divided; subcaudals 43-46. Brown above, with three light longitudinal lines, intersected by dark brown transverse spots or lines; head with well-marked dark brown markings, viz. a crescentic band anteriorly, passing through the eyes, and a chevron-shaped band behind, the apex on the frontal; each of the four anterior labials with a brown spot; a dark brown nuchal spot, angular anteriorly, divided behind by the light vertebral line; lower surfaces yellowish, with very few, scattered, small brown spots; tail immaculate.

Total length 330 millim.; tail 50.

Southern India.

a-b. ♂ (V. 167; C. 43) & yg. (V. 168; C. 46).	Wynad.	Col. Beddome [C.]. (Types.)

18. Simotes arnensis.

Seba, Thes. ii. pl. lxii. fig. 4 (1735); *Russell, Ind. Serp.* i. pls. xxxv. & xxxviii. (1796).
Coluber arnensis, *Shaw, Zool.* iii. p. 526 (1802).
—— russelius, *Daud. Rept.* vi. p. 395, pl. lxxvi. fig. 2 (1803).
Coronella russelii, *Schleg. Phys. Serp.* ii. p. 78 (1837).
Coluber monticolus, *Cantor, Proc. Zool. Soc.* 1839, p. 52.
Simotes russellii, *Dum. & Bibr.* vii. p. 628 (1854); *Günth. Cat.* p. 24 (1858); *Jan, Arch. Zool. Anat. Phys.* ii. 1863, p. 233, and *Icon. Gén.* 11, pl. vi. fig. 1 (1865); *Günth. Rept. Brit. Ind.* p. 213 (1864).
—— albiventer, *Günth. Rept. Brit. Ind.* p. 213.
—— arnensis, *Bouleng. Faun. Ind., Rept.* p. 314 (1890).

Nasal divided; portion of rostral seen from above as long as its distance from the frontal or a little shorter; suture between the internasals usually nearly as long as that between the præfrontals; frontal as long as its distance from the end of the snout or a little shorter, and a little shorter than the parietals; loreal, if distinct, longer than deep, frequently united with the præfrontal; præocular single; two postoculars; temporals 1+2; seven upper labials, third and fourth entering the eye; four lower labials in contact with the anterior chin-shields; posterior chin-shields one half or two thirds the length of the anterior. Scales in 17 rows. Ventrals angulate

230	COLUBRIDÆ.

laterally, 170-202; anal divided; subcaudals 41-59. Pale brown or orange above, with well-defined black cross bands, which vary in number and in width according to individuals, and may be edged with white; an angular or transverse black band between the eyes, another behind, with the apex on the frontal, and a third on the nape; lower surfaces uniform yellowish, rarely spotted with brown or with a brown posterior border to the ventrals.

Total length 600 millim.; tail 90.

India and Ceylon, northwards to Nepal and the Himalayas.

A. Belly uniform yellowish.

a–d. ♂ (V. 178, 172; C. 59, 50), ♀ (V. 192; C. 45), & yg. (V.186,188; C. 54, 50).	India.	
e. ♀ (V. 183; C. 45).	Nepal.	B. H. Hodgson, Esq. [P.].
f. Yg. (V. 187; C. 53).	Sikkim.	Messrs. v. Schlagintweit [C.].
g. ♀ (V. 190; C. 47).	Deccan.	Col. Sykes [C.].
h. ♂ (V. 186; C. 54).	Bombay.	Dr. Leith [P.].
i. Yg. (V. 202; C. 48).	Deesa.	Dr. Leith [P.].
k. ♂ (V. 175; C. 53).	Madras.	Rev. G. Smith [P.].
l, m. ♂ (V. 170; C. 50) & ♀ (V. 185; C. 46).	Anamallays.	Col. Beddome [C.].
n. ♀ (V. 190; C. 47).	Ceylon.	Sir J. Banks [P.].
o. ♂ (V. 179; C. ?).	Near Kandy.	Capt. Gascoigne [P.]. (Type of *S. albiventer*.)
p. ♀ (V. 199; C. 47).	——?	

B. Belly spotted with brown.

a. Yg. (V. 185; C. 46).	Nepal.	B. H. Hodgson, Esq. [P.].
b. ♀ (V. 180; C. 41).	S. India.	T. C. Jerdon, Esq. [P.].

19. Simotes theobaldi. (PLATE IX. fig. 3.)

Simotes theobaldi, *Günth. Ann. & Mag. N. H.* (4) i. 1868, p. 417; *Bouleng. Faun. Ind., Rept.* p. 315 (1890).

Nasal divided or semi-divided; portion of rostral seen from above shorter than its distance from the frontal; suture between the internasals much shorter than that between the præfrontals; frontal a little longer than its distance from the end of the snout, a little shorter than the parietals; loreal twice as long as deep, longer than the posterior nasal; præocular single; two postoculars; temporals 1+2; eight upper labials, fourth and fifth entering the eye; four lower labials in contact with the anterior chin-shields; posterior chin-shields about two thirds the length of the anterior. Scales in 17 rows. Ventrals 171-180; anal divided; subcaudals 34-42. Brown above, with three light longitudinal lines, between which are transverse blackish bars; head with well-marked dark markings, viz. a crescentic band anteriorly, passing through the eyes, and a chevron-shaped band behind, the apex on the frontal; a large blackish nuchal spot, partly divided by the light vertebral line; lower surfaces yellowish, with or without square black spots; tail immaculate.

Total length 380 millim.; tail 30.

Burma.

A. Belly with square black spots.

a. ♀ (V. 180; C. 35).	Pegu.	W. Theobald, Esq. [C.]. (Type.)

B. Belly unspotted.

a. ♀ (V. 174; C 35).	Minhla.	M. L. Fea [C.].
b. ♀ (V. 174; C. 34).	Near Toungoo.	E. W. Oates, Esq. [P.].
c. ♂ (V. 171; C. 42).	Thayetmyo.	E. Y. Watson, Esq. [P.].

20. Simotes cruentatus. (PLATE X. fig. 1.)

Simotes cruentatus, *Günth. Ann. & Mag. N. H.* (4) i. 1868, p. 417; *Theob. Journ. Linn. Soc.* x. 1868, p. 41; *Stoliczka, Proc. As. Soc. Beng.* 1872, p. 145; *Bouleng. Faun. Ind., Rept.* p. 315 (1890).

Nasal divided; portion of rostral seen from above shorter than its distance from the frontal; suture between the internasals much shorter than that between the præfrontals; frontal as long as its distance from the end of the snout, slightly shorter than the parietals; loreal longer than deep, but usually shorter than in the preceding species; præocular single; two postoculars; temporals 1+2; eight (rarely seven) upper labials, fourth and fifth (or third and fourth) entering the eye; four or five lower labials in contact with the anterior chin-shields; posterior chin-shields about two thirds the length of the anterior. Scales in 17 rows. Ventrals 148–173; anal divided; subcaudals 27–37. Brown above, with four more or less distinct darker longitudinal lines; symmetrical dark markings on head broken up, rather indistinct in the adult, an oblique spot below the eye and a large patch on the occiput being distinguishable; lower surfaces yellowish (coral-red in life) with square black spots, at least on the posterior third of the belly; a large black spot at the base of the tail, and sometimes a second near the tip.

Total length 350 millim.; tail 45.

Pegu.

a–d. ♂ (V. 164, 165, 163, 164; C. 33, 37, 37, 36) & ♀ (V. 167; C. 30).	Pegu.	W. Theobald, Esq. [C.]. (Types.)
e. Yg. (V. 148; C. 36).	Pegu.	W. Theobald, Esq. [C.].

f–g. ♂ (V. 169, 160; C. 31, 35).	Toungoo.	E. W. Oates, Esq. [P.].
h–i. ♂ (V. 162, 165; C. 37, 37).	Burma.	Col. Beddome [C.].

21. Simotes torquatus.

Simotes torquatus, *Bouleng. Ann. Mus. Genova*, (2) vi. 1888, p. 597, pl. v. fig. 1, *and Faun. Ind., Rept.* p. 316 (1890).

Nasal undivided; portion of rostral seen from above shorter than its distance from the frontal; suture between the internasals shorter than that between the præfrontals; frontal longer than its distance from the end of the snout and a little shorter than the parietals; loreal usually a little longer than deep; one præ- and two postoculars; temporals 1+2; seven upper labials, third and fourth entering the eye; four lower labials in contact with the anterior chin-shields; posterior chin-shields about two thirds the size of the anterior. Scales in 15 rows. Ventrals 144–159; anal divided; subcaudals 27–34. Grey-brown above, with four rather indistinct darker longitudinal streaks, replaced in the young by series of small blackish spots; upper surface of head with sometimes very indistinct symmetrical markings; a large dark brown (black in the young) spot below the eye, and a broad band of the same colour across the occiput, behind the parietal shields; lower parts white, usually with some black quadrangular spots on the posterior ventrals.

Total length 290 millim.; tail 38.

Upper Burma.

a–b. ♀ (V. 150; C. 31) & yg. (V. 144; C. 30).	Bhamo.	M. L. Fea [C.].	(Types.)

22. Simotes planiceps.

Simotes planiceps, *Bouleng. Ann. Mus. Genova*, (2) vi. 1888, p. 597, pl. v. fig. 2, *and Faun. Ind., Rept.* p. 316 (1890).

Head much depressed; nasal undivided; rostral very much produced above, entirely separating the internasals and wedged in between the præfrontals; frontal elongate, longer than its distance from the end of the snout and than the parietals; loreal longer than deep; one præ- and two postoculars; temporals 1+1; five upper labials, third entering the eye; four lower labials in contact with the anterior chin-shields; posterior chin-shields hardly half as large as the anterior. Scales in 13 rows. Ventrals 132; anal divided; subcaudals 27. Pale brown above, with small oblique black markings occupying the anterior outer border of some of the scales; a black cross band on the occiput, behind the parietals; upper lip yellow, with an oblique black streak below the eye, crossing the suture between the third and fourth upper labials;

lower parts pinkish, with square black spots mostly arranged in pairs.

Total length 130 millim.; tail 15.

Minhla, Burma.

93. OLIGODON.

Oligodon, *Boie, Isis,* 1827, p. 519; *Wagler, Syst. Amph.* p. 191 (1830); *Dum. & Bibr. Erp. Gén.* vii. p. 54 (1854); *Günth. Cat. Col. Sn.* p. 20 (1858); *Jan, Arch. Zool. Anat. Phys.* ii. 1862, p. 36; *Günth. Rept. Brit. Ind.* p. 205 (1864); *Bouleng. Faun. Ind., Rept.* p. 317 (1890).
Calamaria, part., *Schleg. Phys. Serp.* ii. p. 25 (1837).
Homalosoma, part., *Jan, l. c.* p. 33.
Rhynchocalamus, *Günth. Proc. Zool. Soc.* 1864, p. 491.
Tripeltis, *Cope, Proc. Amer. Philos. Soc.* xxiii. 1886, p. 487.

Characters of *Simotes*, but maxillary teeth fewer still (6 to 8), and no pterygoid teeth, the palate being entirely edentulous or with two or three teeth on each palatine. Scales in 15 or 17 rows.

Southern Asia; Lower Egypt.

This genus may have to be united with the preceding, *S. venustus*, being, as pointed out by Dr. Günther (Ann. & Mag. N. H. (4) i. 1868, p. 416), intermediate between *Simotes* and *Oligodon*.

Synopsis of the Species.

I. Scales in 17 rows.

  A. Anal divided; no loreal shield; two postoculars.

Portion of rostral seen from above as long as, or a little shorter than, its distance from the frontal; latter shield as long as the parietals	1. *venustus*, p. 235.
Portion of rostral seen from above as long as or a little shorter than its distance from the frontal; latter shield shorter than the parietals	2. *travancoricus*, p. 236.
Portion of rostral seen from above much shorter than its distance from the frontal	3. *affinis*, p. 236.

  B. Anal entire; loreal usually present.

Two postoculars; subcaudals 35–46	4. *bitorquatus*, p. 237.
One postocular; subcaudals 52–62	5. *trilineatus*, p. 238.

II. Scales in 15 rows.

  A. Anal entire.

    1. Nasal divided.

One postocular; no loreal; only the third upper labial entering the eye	6. *modestus*, p. 238.

One postocular; loreal present; third
 and fourth labials entering the
 eye .... 7. *notospilus*, p. 239.
Two postoculars; loreal present...... 8. *everetti*, p. 239.

  2. Nasal undivided; two post-
 oculars .... 9. *propinquus*, p. 240.

B. Anal divided.

  1. Nasal divided.

    a. No internasals.............. 10. *brevicauda*, p. 240.

    b. A pair of internasals.

      α. A single postocular...... 11. *dorsalis*, p. 241.

      β. Two postoculars.

        * Ventrals not more than 160, subcaudals not more
 than 35.

Portion of rostral seen from above about
 half as long as its distance from the
 frontal....................... 12. *templetonii*, p. 241.
Portion of rostral seen from above mea-
 suring about two thirds its distance
 from the frontal .... 13. *sublineatus*, p. 242.
Portion of rostral seen from above as
 long as its distance from the frontal. 14. *ellioti*, p. 242.

        ** Ventrals 154 or more, subcaudals more than 35.

Portion of rostral seen from above as
 long as or a little shorter than its
 distance from the frontal.......... 15. *subgriseus*, p. 243.
Portion of rostral seen from above con-
 siderably shorter than its distance
 from the frontal .... 16. *vertebralis*, p. 245.

  2. Nasal undivided.

Two postoculars; six upper labials;
 ventrals 145–151 .... 17. *waandersii*, p. 245.
One postocular; seven upper labials;
 ventrals 181–229 .... 18. *melanocephalus*, p. 246.

Oligodon dorsale, Berthold, Götting. Nachr. 1859, p. 179, stated to be from Bengal, perhaps belongs to a different genus. It is described as:—" Grey above, with a white median line, and a curved brown interocular band; white beneath; rostral shield large; scales rhomboidal, smooth, in 13 rows; ventrals 183, anal single, subcaudals 46; tail one fifth of total length."

93. OLIGODON.

Table showing Numbers of Scales and Shields.

	Sc.	V.	A.	C.	Lab.
venustus	17	143–162	2	28–36	7 (6)
travancoricus	17	145–151	2	35–37	7
affinis	17	129–142	2	25–36	7
bitorquatus	17	140–155	1	35–46	7
trilineatus	17	145–157	1	52–62	7
modestus	15	158–170	1	41	6
notospilus	15	142	1	32	7
everetti	15	154	1	46	7
propinquus	15	140	1	27	7
brevicauda	15	164–173	2	25–29	7
dorsalis	15	174–188	2	37–51	7
templetonii	15	135–152	2	28–31	7
sublineatus	15	136–160	2	26–35	7
ellioti	15	152	2	29	7
subgriseus	15	158–218	2	38–56	7
vertebralis	15	154	2	54	7
waandersii	15	145–151	2	22–28	6
melanocephalus	15	181–229	2	53–68	7

1. Oligodon venustus.

Oligodon venustus, *Jerdon, Journ. As. Soc. Beng.* xxii. 1853, p. 528; *Bouleng. Faun. Ind., Rept.* p. 317 (1890).
Simotes binotatus (non *D. & B.*), *Günth. Cat.* p. 24 (1858).
—— venustus, *Günth. Rept. Brit. Ind.* p. 213 (1864), and *Ann. & Mag. N. H.* (4) i. 1868, p. 416.

Nasal divided; portion of rostral seen from above as long as its distance from the frontal or a little shorter; suture between the internasals shorter than that between the præfrontals; frontal longer than its distance from the end of the snout, as long as the parietals; no loreal, the posterior nasal sometimes forming a suture with the præocular; præocular single; two postoculars; temporals 1+2; usually seven (rarely six) upper labials, third and fourth entering the eye, sixth frequently excluded from the labial border and taking the position of a lower anterior temporal; four lower labials in contact with the anterior chin-shields; posterior chin-shields one half to two thirds the length of the anterior. Scales in 17 rows. Ventrals 143–162; anal divided; subcaudals 28–36. Pale brown or greyish above, with a series of paired large oval or rhomboidal blackish spots edged with yellowish, which are usually united mesially; a large, more irregular spot below each dorsal spot; head-markings consisting of a crescentic anterior cross band, passing through the eye, an oblique band on each side, from the parietal to below the angle of the mouth, and a broad angular band on the nape; top of head frequently with black vermiculations. Lower surfaces black and yellow, the two colours in nearly equal proportions, except under the tail, where the yellow predominates, or where sometimes black spots are entirely wanting.

Total length 480 millim.; tail 65.
South-western India.

a–b. ♂ (V. 144; C. 34) & ♀ (V. 152; C. 30).	Madras Presidency.	T. C. Jerdon, Esq. [P.].
c. Hgr. ♂ (V. 146; C. 34).	Madras Presidency.	Col. Beddome [P.].
d. ♀ (V. 151; C. 33).	Nilgherries, 7000 ft.	Col. Beddome [C.].
e. Hgr. ♀ (V. 154; C. 29).	Coonoor, Nilgherries, 5900 ft.	W. Davison, Esq. [P.].
f–g. Hgr. (V. 162; C. 28) & yg. (V. 147; C. 32).	Nilgherries.	E. A. Minchin, Esq. [P.].

2. Oligodon travancoricus. (PLATE X. fig. 2.)

Oligodon travancoricus, Beddome, Proc. Zool. Soc. 1877, p. 685; *Bouleng. Faun. Ind., Rept.* p. 318 (1890).

Nasal divided; portion of rostral seen from above as long as or a little shorter than its distance from the frontal; suture between the internasals shorter than that between the præfrontals; frontal as long as or longer than its distance from the end of the snout, shorter than the parietals; no loreal, the posterior nasal forming a suture with the præocular, or the præfrontal forming a suture with the second labial: præocular single; two postoculars; temporals 1+2 or 2+3; upper labials seven, third and fourth entering the eye, sixth excluded from the labial margin; three lower labials in contact with the anterior chin-shields; posterior chin-shields two thirds the length of the anterior. Scales in 17 rows. Ventrals 145–151; anal divided; subcaudals 35–37. Greyish or pale brown above, with 25 to 33 black or dark brown, light-edged cross bands on the body, and five or six pairs of spots on the tail; three broad black, light-edged transverse bands on the head, viz., a frontal, an occipital, and a nuchal, connected or not longitudinally on the median line; a small white spot may be present in the middle between the parietals. Lower surfaces white, with large square black spots.

Total length 450 millim.; tail 65. I have examined a specimen belonging to the Trevandrum Museum, which measures 465 millim.; tail 65.

Travancore hills.

a. Hgr. ♂ (V. 145; C. 37).	Tinnevelly hills.	Col. Beddome [C.]. (Type.)
b–c. ♂ (V. 147, 146; C. 37, 35).	High Range, Travancore.	H. S. Ferguson, Esq. [P.].

3. Oligodon affinis.

Oligodon affinis, Günth. Ann. & Mag. N. H. (3) ix. 1862, p. 58, *and Rept. Brit. Ind.* p. 209, pl. xix. fig. B (1864); *Bouleng. Faun. Ind., Rept.* p. 318 (1890).

Nasal divided; portion of rostral seen from above half as long as its distance from the frontal; suture between the internasals as long as that between the præfrontals, or a little shorter; frontal much longer than its distance from the end of the snout, as long as the parietals; no loreal, the posterior nasal sometimes forming a suture with the præocular; præocular single; two postoculars; temporals 1+2, or 1+1+2; upper labials seven, third and fourth entering the eye; four lower labials in contact with the anterior chin-shields; posterior chin-shields about two thirds the length of the anterior. Scales in 17 rows. Ventrals 129–142; anal divided; subcaudals 25–36. Brown above, with more or less distinct darker cross lines; head with dark symmetrical transverse markings, which are usually connected by a median longitudinal streak. Lower surfaces white (in spirit) with square black spots, both colours being distributed in nearly equal proportion.

Total length 330 millim.; tail 50.

South-western India.

a. ♀ (V. 132; C. 27).	Anamallays.	Col. Beddome [C.]. (Type.)
b–c, d–f. ♂ (V. 137, 129; C. 33, 36), ♀ (V. 136, 139; C. 25, 28), & yg. (V. 142; C. 26).	Wynad, 3000 feet.	Col. Beddome [C.].

4. Oligodon bitorquatus.

Russell, Ind. Serp. ii. p. 39, pl. xxxiv.
Oligodon bitorquatus, *Boie, Isis,* 1827, p. 519.
Calamaria oligodon, *Schleg. Phys. Serp.* ii. p. 40, pl. i. figs. 27–29 (1837), *and Abbild.* p. 96, pl. xxv.
Oligodon subquadratus, *Dum. & Bibr.* vii. p. 55 (1854); *Günth. Cat.* p. 21 (1858); *Jan, Arch. Zool. Anat. Phys.* ii. 1862, p. 37, *and Icon. Gén.* 13, pl. iv. figs. 5 & 6 (1865).
Rabdosoma amboinense, *Bleek. Nat. Tijdschr. Nederl. Ind.* xxii. 1860, p. 42.

Nasal divided; portion of rostral seen from above shorter than its distance from the frontal; suture between the internasals shorter than that between the præfrontals; frontal longer than its distance from the end of the snout, as long as or a little shorter than the parietals; loreal usually present, a little longer than deep; one præ- and two postoculars; temporals 1+2 or 2+2; seven upper labials, third and fourth entering the eye; three or four lower labials in contact with the anterior chin-shields; posterior chin-shields one half to two thirds the length of the anterior. Scales in 17 rows. Ventrals 140–155; anal entire; subcaudals 35–46. Purplish brown or blackish above, with yellow (or red) dots and usually a vertebral series of larger spots; head with symmetrical blackish markings and one or two chevron-shaped yellowish bands, the second on the occiput and nape; belly red, with quadrangular or transverse black spots.

Total length 370 millim.; tail 67.

Java, Amboyna.

a-b. ♂ (V. 144, 144; C. 37, 45).		Java.	Leyden Museum.
c. ♀ (V. 142; C. ?).		Java.	Hr. Frühstorfer [C.].
d. ♀ (V. 155; C. 46).		Salak, Java.	R. Kirkpatrick, Esq. [P.].
e. ♂ (V. 148; C. 41).		Batavia.	
f. ♂ (V. 150; C. 37).		Willis Mts., Kediri, Java, 5000 feet.	Baron v. Huegel [C.].
g. Yg. (V. 155; C. 44).		Amboyna.	Dr. Bleeker. (Type of *Rabdosoma amboinense*.)

5. Oligodon trilineatus.

Simotes trilineatus, *Dum. & Bibr.* vii. p. 636 (1854); *Jan, Arch. Zool. Anat. Phys.* ii. 1862, p. 324, *and Icon. Gén.* 12, pl. iv. fig. 1 (1865).
Oligodon trilineatus, *Fischer, Abh. naturw. Ver. Hamb.* ix. 1885, p. 7.

Nasal divided; portion of rostral seen from above nearly as long as its distance from the frontal; suture between the internasals nearly as long as that between the præfrontals; frontal longer than its distance from the end of the snout, as long as the parietals; loreal small, longer than deep; one præ- and one postocular; temporals 1+2 or 2+2; seven upper labials, third and fourth entering the eye; four lower labials in contact with the anterior chin-shields, which are nearly twice as long as the posterior. Scales in 17 rows. Ventrals 145–157; anal entire; subcaudals 52–62. Dark brown or blackish above and below; head yellowish brown, with blackish oblique bands interrupted on the crown; a yellow vertebral stripe, a fine yellowish line along each side of the back, and a white line along each side of the belly.

Total length 330 millim.; tail 80.

Pulo Nias.

a-b. ♀ (V. 147; C. 60) & hgr. (V. 157; C. 54).		Nias.	Herr Sundermann [C.].

6. Oligodon modestus. (PLATE X. fig. 3.)

Oligodon modestus, *Günth. Rept. Brit. Ind.* p. 210 (1864), *and Proc. Zool. Soc.* 1879, p. 77.

Nasal divided; portion of rostral seen from above as long as its distance from the frontal; suture between the internasals a little shorter than that between the præfrontals; frontal longer than its distance from the end of the snout, as long as the parietals; no loreal; one præ- and one postocular; temporals 1+2 or 1+3; six upper labials, third largest and entering the eye; three or four lower labials in contact with the anterior chin-shields, which are longer than the posterior. Scales in 15 rows. Ventrals 158–170; anal entire; subcaudals 41. Dark brown above, with a yellowish vertebral streak; a yellowish chevron-shaped band on the occiput; lower parts yellowish, with quadrangular black spots.

Total length 350 millim.; tail 55.
Philippine Islands.

a. ♂ (V. 158; C. 41).   Philippine Islands.   H. Cuming, Esq. [C.].
(Type.)
b. ♂ (V. 170; C. 41).   S. Negros.   A. Everett, Esq. [C.].

7. Oligodon notospilus.

Oligodon notospilus, *Günth*. Proc. Zool. Soc. 1873, p. 169, pl. xviii. fig. A.

Nasal divided; portion of rostral seen from above shorter than its distance from the frontal; suture between the internasals shorter than that between the præfrontals; frontal longer than its distance from the end of the snout, as long as the parietals; loreal very small, longer than deep; one præ- and one postocular; temporals 1 + 2; seven upper labials, third and fourth entering the eye; three lower labials in contact with the anterior chin-shields, which are a little longer than the posterior. Scales in 15 rows. Ventrals 142; anal entire; subcaudals 32. Dark purplish brown above, with yellow dots and a series of large transverse, rhomboidal, yellow, black-edged spots; head yellow, with two chevron-shaped black bands, the anterior passing through the eyes, the posterior with the point on the frontal shield; uniform yellowish inferiorly.

Total length 260 millim.; tail 40.
Philippine Islands.

a. ♀ (V. 142; C. 32).   Mindanao.   Dr. A. B. Meyer [C.].
(Type.)

8. Oligodon everetti. (Plate XI. fig. 1.)

Oligodon everetti, *Bouleng*. Proc. Zool. Soc. 1893, p. 524.

Nasal divided; portion of rostral seen from above slightly shorter than its distance from the frontal; suture between the internasals shorter than that between the præfrontals; frontal longer than its distance from the end of the snout, slightly shorter than the parietals; loreal very small, longer than deep; one præ- and two postoculars; temporals 1 + 2; seven upper labials, third and fourth entering the eye; four lower labials in contact with the anterior chin-shields, which are longer than the posterior. Scales in 15 rows. Ventrals 154; anal entire; subcaudals 46. Slaty grey above, with three blackish-brown stripes, the middle one three scales wide and enclosing a series of small yellowish-brown rhomboidal spots; head brown above, with two chevron-shaped black bands, the anterior passing through the eyes, the posterior with the point on the frontal shield; uniform coral-red beneath, the outer ends of the ventral shields black.

Total length 370 millim.; tail 70.
North Borneo.

a. ♀ (V. 154; C. 46).   Mt. Kina Baloo.   A. Everett, Esq. [C.].
(Type.)

9. Oligodon propinquus.

Oligodon propinquus, *Jan, Arch. Zool. Anat. Phys.* ii. 1862, p. 38, and *Icon. Gén.* 48, pl. i. fig. 1 (1876).

Nasal undivided; portion of rostral seen from above a little shorter than its distance from the frontal; suture between the internasals longer than that between the præfrontals; frontal longer than its distance from the end of the snout, as long as the parietals; loreal small, a little longer than deep; one præ- and two postoculars; temporals 1+2; seven upper labials, third and fourth entering the eye; four lower labials in contact with the anterior chin-shields, which are a little longer than the posterior. Scales in 15 rows. Ventrals 140; anal entire; subcaudals 27. Black above, with yellowish dots; a series of small yellowish spots along the back; belly whitish.

Total length 285 millim.; tail 40.

Java.

10. Oligodon brevicauda.

Oligodon brevicauda, *Günth. Ann. & Mag. N. H.* (3) ix. 1862, p. 58, and *Rept. Brit. Ind.* p. 211, pl. xix. fig. A (1864); *Bouleng. Faun. Ind., Rept.* p. 319 (1890).

Nasal divided; portion of rostral seen from above longer than its distance from the frontal; no internasals; frontal longer than its distance from the end of the snout, as long as the parietals; no loreal, the posterior nasal forming a suture with the præocular; præocular single; two postoculars; temporals 1+2; seven upper labials, third and fourth entering the eye; four lower labials in contact with the anterior chin-shields; posterior chin-shields about two thirds the length of the anterior. Scales in 15 rows. Ventrals 164–173; anal divided; subcaudals 25–29. Brown above, with a light vertebral band, most marked posteriorly, bordered on each side by a dark brown or black band; a black narrow streak along each side; a rhomboidal dark spot on the frontal, confluent with a broad crescentic transverse band anteriorly, which passes through the eyes; a dark band from behind the eye to the angle of the mouth; a large dark nuchal spot. Lower surface red, with large quadrangular or transverse black spots; tail without, or with only a few black spots. Spec. *d* very dark, almost black above and below, the black markings being just distinguishable.

Total length 480 millim.; tail 50.

South-western India.

a.	♀ (V. 173; C. 28).	Anamallays.	Col. Beddome [C.]. (Type.)
b.	♀ (V. 165; C. 25).	Anamallays.	Col. Beddome [C.].
c.	♀ (V. 172; C. 29).	West slope of Nilgherries, 3000 ft.	Col. Beddome [C.].
d.	♀ (V. 164; C. 29).	Peermad, Travancore, 3300 ft.	H. S. Ferguson, Esq. [P.].

11. Oligodon dorsalis.

Elaps dorsalis, *Gray, Ill. Ind. Zool.* ii. pl. lxxxv. fig. 1 (1834).
Oligodon dorsalis, *Günth. Cat.* p. 22 (1858), *and Rept. Brit. Ind.* p. 210 (1864); *Anders. Proc. Zool. Soc.* 1871, p. 168; *Bouleng. Faun. Ind., Rept.* p. 319 (1890).

Nasal divided; portion of rostral seen from above shorter than its distance from the frontal; suture between the internasals as long as that between the præfrontals or shorter; frontal longer than its distance from the end of the snout, as long as the parietals; loreal as long as deep; præocular single; postocular single; temporals 1+2; seven upper labials, third and fourth entering the eye; four lower labials in contact with the anterior chin-shields; posterior chin-shields about two thirds the length of the anterior. Scales in 15 rows. Ventrals 174–188; anal divided; subcaudals 37–51. Brown above, with a yellowish vertebral streak, on each side of which is a series of small black spots; a black lateral streak; a large subtriangular blackish spot on the forehead, connected with a very large occipital spot by a longitudinal streak on the frontal; lower surfaces black and yellow (in spirit), the black predominating on the belly, the yellow on the tail.

Total length 360 millim.; tail 65.

Khasi, Naga, and Chittagong hills.

a.	♂ (V. 177; C. 47).	——?	Gen. Hardwicke [P.].	(Type.)
b.	♂ (V. 174; C. 51).	——?	Dr. Griffith.	

12. Oligodon templetonii.

Oligodon templetonii, *Günth. Ann. & Mag. N. H.* (3) ix. 1862, p. 57, *and Rept. Brit. Ind.* p. 209, pl. xix. fig. C (1864); *Bouleng. Faun. Ind., Rept.* p. 320 (1890).

Nasal divided; portion of rostral seen from above about half as long as its distance from the frontal; suture between the internasals longer or shorter than that between the præfrontals; frontal much longer than its distance from the end of the snout, as long as the parietals; loreal longer than deep, sometimes entering the eye; præocular single; two or three postoculars; temporals 1+2: seven upper labials, third and fourth entering the eye, sixth usually excluded from the labial border; three or four lower labials in contact with the anterior chin-shields; posterior chin-shields about two thirds the length of the anterior. Scales in 15 rows. Ventrals 135–152; anal divided; subcaudals 28–31. Brown above, with a yellowish vertebral streak, which becomes more distinct on the tail, and is crossed by about eighteen narrow dark brown bands; head-markings very indistinct except an oblique band below the eye; lower surfaces white with square black spots, both colours being distributed in nearly equal proportion.

Total length 270 millim.; tail 40.

Ceylon.

a. ♂ (V. 135; C. 31).		Ceylon.	R. Templeton, Esq. [P.]. (Type.)
b–c. ♂ (V. 139; C. 30) & yg. (V. 152; C. 28).		Udagama.	E. E. Green, Esq. [P.].

13. Oligodon sublineatus.

Oligodon sublineatus, *Dum. & Bibr.* vii. p. 57 (1854); *Günth. Cat.* p. 21 (1858); *Jan, Arch. Zool. Anat. Phys.* ii. 1862, p. 38; *Günth. Rept. Brit. Ind.* p. 209 (1864); *Jan, Icon. Gén.* 48, pl. i. fig. 2 (1876); *Bouleng. Faun. Ind., Rept.* p. 320 (1890); *W. L. Scluter, Journ. As. Soc. Beng.* lx. 1891, p. 237.

Nasal divided; portion of rostral seen from above shorter than its distance from the frontal; suture between the internasals nearly as long as that between the præfrontals; frontal longer than its distance from the end of the snout, as long as the parietals, or slightly longer; loreal as long as deep, or a little longer, rarely absent; præocular single; two postoculars; temporals 1+2; seven upper labials, third and fourth entering the eye; three or four lower labials in contact with the anterior chin-shields; posterior chin-shields about two thirds the length of the anterior. Scales in 15 rows. Ventrals 136–160; anal divided; subcaudals 26–35. Pale brown above, some of the scales edged with dark brown; frequently a more or less regular series of paired dark brown dorsal spots; head with dark brown markings; an angular transverse band anteriorly, passing through the eyes; a longitudinal band from the middle of the frontal to a little beyond the parietals, and a large spot on each side of the nape; the longitudinal band and the nuchal spots sometimes confluent; lower surfaces yellowish, with three (rarely two) longitudinal series of small dark brown spots, the lateral series often confluent into a line.

Total length 315 millim.; tail 35.

Ceylon, Nicobars.

a–b. ♀ (V. 151, 160; C. 27, 28).	Punduloya, 4000 ft.	E. E. Green, Esq. [P.].
c. ♂ (V. 138; C. 33).	Kandy.	Capt. Gascoigne [P.].
d. Yg. (V. 159; C. 26).	Ceylon.	G. H. K. Thwaites, Esq. [P.].
e, f, g–o. ♂ (V. 138,136; C. 35, 34), ♀ (V. 149, 146, 152; C. 28, 27, 26, 26), & hgr. (V. 150, 147, 137, 146; C. 29, 29, 32, 31).	Ceylon.	

14. Oligodon ellioti.

Oligodon ellioti, *Günth. Rept. Brit. Ind.* p. 207, pl. xix. fig. G (1864); *Bouleng. Faun. Ind., Rept.* p. 321 (1890).

Nasal divided; portion of rostral seen from above as long as its distance from the frontal; suture between the internasals as long as that between the præfrontals; frontal longer than its distance from the end of the snout, as long as the parietals; loreal as long as deep; præocular single; two postoculars; temporals 1+2; seven upper labials, third and fourth entering the eye; three lower labials in contact with the anterior chin-shields; posterior chin-shields about two thirds the length of the anterior. Scales in 15 rows. Ventrals 152; anal divided; subcaudals 29. Brown above, with a median series of large rhombic black spots, on each side of which is a small spot separated by a whitish border; head with black markings, viz. an angular band across the forehead, passing through the eyes, and a second behind, with the apex on the frontal, descending to below the angle of the mouth, and confluent with a large nuchal spot; an angular interrupted, narrow black band across the throat; lower surfaces whitish, unspotted.

Total length 270 millim.; tail 30.

Madras Presidency.

a. ♀ (V. 152; C. 29).   Madras Pres.   Sir W. Elliot [P.]. (Type.)

15. Oligodon subgriseus.

Oligodon subgriseus, *Dum. & Bibr.* vii. p. 59 (1854); *Günth. Cat.* p. 21 (1858); *Jan, Arch. Zool. Anat. Phys.* ii. 1862, p. 39; *Günth. Rept. Brit. Ind.* p. 207, pl. xix. fig. F (1864); *Jan, Icon. Gén.* 48, pl. i. fig. 3 (1876); *Blanf. Journ. As. Soc. Beng.* xliii. 1879, p. 114; *Bouleng. Faun. Ind., Rept.* p. 321 (1890); *W. L. Sclater, Journ. As. Soc. Beng.* lx. 1891, p. 237.

Simotes binotatus, *Dum. & Bibr. t. c.* p. 630; *Jan, l. cc. Icon.* 11, pl. vi. fig. 3 (1865); *Günth. Rept. Brit. Ind.* p. 214.

Xenodon dubium, *Jerdon, Journ. As. Soc. Beng.* xxii. 1853, p. 528 (1854).

Oligodon spilonotus, *Günth. l. c.* p. 207, pl. xix. fig. E.

—— fasciatus, *Günth. l. c.* p. 208, pl. xix. fig. D.

Nasal divided; portion of rostral seen from above as long as, or a little shorter than, its distance from the frontal; suture between the internasals usually shorter than that between the præfrontals; frontal longer than its distance from the end of snout, as long as the parietals; loreal about as long as deep; præocular single; two postoculars; temporals 1+2; seven upper labials, third and fourth entering the eye; three or four lower labials in contact with the anterior chin-shields; posterior chin-shields one half to two thirds the length of the anterior. Scales in 15 rows. Ventrals angulate laterally, 158-218; anal usually divided; subcaudals 38-56. Brown above, with a series of large rhomboidal dark spots or transverse bands, or pairs of spots, with or without a more or less distinct light vertebral line; head with dark markings, usually consisting of a crescentic band across the forehead, through the eyes, to the fourth and fifth labials, a band, widening posteriorly, from the parietals or from the frontal to the angle of the mouth, and a large spot, bifid posteriorly, from the frontal to the nape; the upper surface of the head sometimes almost entirely blackish brown, with small yellowish markings; lower surfaces immaculate or with small brown spots or dots on each side.

Total length 480 millim.; tail 65.

From Baluchistan, Sind, and Bengal to Southern India and Ceylon.

A. Belly unspotted.

a. ♂ (V. 186; C. 55).	Kurrachee.	Dr. Leith [P.].
b, c. ♀ (V. 218; C. ?) & yg. (V. 200; C. 48).	Sind.	Dr. Leith [P.].
d. ♀ (V. 204; C. 44).	Ajmere, Rajputana.	W. T. Blanford, Esq. [P.].
e. ♂ (V. 169 C. 45).	Aska, Ganjam.	E. A. Minchin, Esq. [P.].
f–h. Hgr. (V. 179, 182, 190; C. 43, 51, 42).	Bangalore.	E. A. Minchin, Esq. [P.].
i. Yg. (V. 203; C. 45).	Madras.	J. E. Boileau, Esq. [P.].
k–l. ♂ (V. 166, 160; C. 47, 45).	Madras.	— Bevan, Esq. [P.]. (Types of *O. spilonotus*).
m. Hgr. (V. 167; C. 48).	Malabar.	(Type of *O. spilonotus*).
n, o, p. ♀ (V. 178, 171, 168; C. 40, 43, 41).	Malabar.	Col. Beddome [C.].
q, r–s. ♂ (V. 173; C. 53), ♀ (V. 187; C. 43), & hgr. (V. 174; C. 48).	Wynad.	Col. Beddome [C.].
t–v, w. ♂ (V. 175, 184; C. 51, 53) & hgr. (V. 182, 178; C. 46, 52).	Nilgherries.	Hon. J. Dormer [P.].
x–y, z–β. ♀ (V. 185, 185; C. 50, 41), hgr. (V. 173, 177; C. 39, 46) & yg. (V. 164; C. 42).	Nilgherries.	Col. Beddome [C.].
γ–ζ. ♀ (V. 176, 182, 182; C. 51, 49, 52), & yg. (V. 160; C. 55).	Anamallays.	Col. Beddome [C.].
η. ♀ (V. 169; C. 44).	Merchiston, Travancore.	H. S. Ferguson, Esq. [P.].
θ. ♀ (V. 158; C. 47).	Poumadi, Travancore.	H. S. Ferguson, Esq. [P.].
ι. ♂ (V. 171; C. 48).	Trincomalee.	Major Barrett [P.].
κ. ♂ (V. 168; C. 48).	Ceylon.	W. Ferguson, Esq. [P.].
λ. ♀ (V. 189; C. 42).	Ceylon.	E. W. H. Holdsworth, Esq. [C.].
μ. ♀ (V. 187; C. 45).	Ceylon.	

B. Belly with small brown spots or dots.

a–b. ♂ (V. 159; C. 45) & ♀ (V. 174; C. 40).	Deccan.	Col. Sykes [P.]. } Types of *O. fasciatus*.
c–d. ♂ (V. 158, 158; C. 46, 40).	Deccan.	E. India Comp.
e–o. ♂ (V. 164, 165, 166; C. 44, 42, 41), ♀ (V. 178, 175; C. 40, 38), & yg. (V. 178, 169, 163, 175, 168; C. 39, 42, 42, 38, 42).	Matheran.	Dr. Leith [P.].

16. Oligodon vertebralis. (Plate XI. fig. 2.)

Simotes vertebralis, *Günth. Ann. & Mag. N. H.* (3) xv. 1865, p. 91.

Nasal divided; portion of rostral seen from above considerably shorter than its distance from the frontal; suture between the internasals longer than that between the præfrontals: frontal longer than its distance from the end of the snout, a little shorter than the parietals; a small square loreal; one præ- and two postoculars; temporals 1+2: seven upper labials, third and fourth entering the eye; four lower labials in contact with the anterior chin-shields, which are nearly twice as large as the posterior. Scales in 15 rows. Ventrals 154; anal divided; subcaudals 54. Brown above, with small yellow, black-edged spots, the largest of which form a vertebral series; head yellowish, with two chevron-shaped, brown, black-edged bands, the anterior passing through the eyes, the posterior with the apex on the frontal; lower parts yellowish, throat spotted with brown.

Total length 345 millim.; tail 75.

Borneo.

a. ♂ (V. 154; C. 54).   Bandjermassing.   L. L. Dillwyn, Esq. [P.].
                                               (Type.)

17. Oligodon waandersii. (Plate XI. fig. 3.)

Rabdion waandersii, *Bleek. Nat. Tijdschr. Nederl. Ind.* xxii. 1860, p. 83.

——— cruciatum, *Bleek. l. c.* p. 82.

Oligodon waandersii, *Günth. Ann. & Mag. N. H.* (3) xv. 1865, p. 91.

Nasal undivided; portion of rostral seen from above shorter than its distance from the frontal; suture between the internasals as long as that between the præfrontals; frontal longer than its distance from the end of the snout, as long as the parietals; no loreal, nasal in contact with præocular; one præ- and two postoculars; temporals 1+2; six upper labials, third and fourth entering the eye; three or four lower labials in contact with the anterior chin-shields, which are about twice as long as the posterior. Scales in 15 rows. Ventrals 145–151; anal divided; subcaudals 22-28. Reddish brown above, with a few widely separated yellow brown-edged small spots, mostly disposed in pairs; a yellowish collar; lower parts uniform yellowish.

Total length 220 millim.; tail 23.

Boni Id., Celebes.

a. ♂ (V. 151; C. 28).   Boni.   Dr. Bleeker. (Type of *Rabdion waandersii.*)
b. Yg. (V. 145; C. 22).   Boni.   Dr. Bleeker. (Type of *Rabdion cruciatum.*)

18. Oligodon melanocephalus.

Homalosoma melanocephalum, *Jan, Arch. Zool. Anat. Phys.* ii. 1862,
p. 34, *and Icon. Gén.* 13, pl. iii. fig. 4 (1865); *Boettg. Ber.
Senckenb. Ges.* 1878-79, p. 60; *F. Müll. Verh. nat. Ges. Basel,*
vii. 1885, p. 678.
Rhynchocalamus melanocephalus, *Günth. Proc. Zool. Soc.* 1864,
p. 491, *and Zool. Rec.* 1865, p. 152; *Boettg. Ber. Senckenb. Ges.*
1879-80, p. 139; *Tristram, Faun. Palest.* pl. xvi. fig. 1 (1884).
Coronella melanocephala, *Peters, Mon. Berl. Ac.* 1869, p. 439.
Oligodon melanocephalus, *Bouleng. Faun. Ind., Rept.* p. 317 (1890).

Nasal undivided; portion of rostral seen from above as long as
or a little longer than its distance from the frontal; suture between
the internasals nearly as long as that between the præfrontals;
frontal longer than its distance from the end of the snout, a little
shorter than the parietals; a small square loreal present or absent;
one præ- and one postocular; temporals 1+1 or 1+2; seven upper
labials, third and fourth entering the eye; three or four lower
labials in contact with the anterior chin-shields; posterior chin-
shields very small. Scales in 15 rows. Ventrals 181-229; anal
divided; subcaudals 53-68. Yellow or reddish above; upper
surface of head and nape bluish black; lower parts yellowish white.

Total length 460 millim.; tail 75.

Syria, Sinaitic Peninsula, Lower Egypt.

a. ♀ (V. 218; C. 54).   Merom.   Canon Tristram [C.].
   (Type of *Rhynchocalamus melanocephalus*.)
b. ♀ (V. 211; C. 56).   Haifa.   A. Smith Woodward,
    Esq. [P.].
c. Hgr. (V. 229; C. 59).   Sinaitic Peninsula.   H. C. Hart, Esq. [C.].

94. PROSYMNA.

Prosymna, *Gray, Cat. Sn.* p. 80 (1849); *Jan, Arch. Zool. Anat. Phys.*
ii. 1862, p. 55; *Peters, Reise n. Mossamb.* iii. p. 106 (1882).
Temnorhynchus (*non Hope*), *Smith, Ill. Zool. S. Afr., Rept., App.*
p. 17 (1849); *Peters, Mon. Berl. Ac.* 1867, p. 235.
Rhinostoma, part., *Günth. Cat. Col. Sn.* p. 8 (1858).
Ligonirostra, *Cope, Am. Journ. Sc. & Arts,* (2) xxxv. 1863, p. 457.

Maxillary bone short, with seven or eight teeth increasing in size
posteriorly, the first tooth minute, falling below the centre of the

Fig. 18.

Maxillary of *Prosymna sundevallii*.

eye, the hindermost very large, strongly compressed, blade-like;
palate toothless, or with a few minute teeth on the pterygoids;

mandibular teeth few, very small, equal. Head not distinct from neck; snout much depressed, projecting, with angular horizontal edge; eye rather small, with vertically subelliptic pupil; nostril in a semidivided nasal, a horizontal suture extending from the nostril to the loreal; præfrontal usually single. Body cylindrical, short; scales smooth or keeled, with apical pits, in 15 or 17 rows. Tail short, ending in a horny spine; subcaudals in two rows.

Tropical and South Africa.

Synopsis of the Species.

I. Scales smooth.

  A. Two internasals; two superposed anterior temporals.
                              1. *sundevallii*, p. 247.

  B. A single internasal; a single anterior temporal.

    1. Third and fourth or fourth and fifth upper labials entering the eye; two or three postoculars.

Width of the frontal not half the width of
  the head: subcaudals 50 .... 2. *frontalis*, p. 248.
Width of the frontal more than half the
  width of the head; subcaudals 19-34 .. 3. *ambigua*, p. 248.

    2. Second and third upper labials entering the eye; one postocular .... 4. *meleagris*, p. 249.

II. Scales keeled .... 5. *jani*, p. 249.

1. Prosymna sundevallii.

Temnorhynchus sundevalli, *Smith, Ill. Zool. S. Afr., Rept., App.* p. 17 (1849); *Peters, Mon. Berl. Ac.* 1867, p. 235.
Rhinostoma cupreum, *Günth. Cat.* p. 9 (1858).
Temnorhynchus frontalis, part., *Peters, l. c.* p. 236, pl. —. fig. 2.

Rostral very large and broad, with sharp horizontal edge, usually in contact with the præfrontal; a pair of internasals, separated from each other or in contact with their inner angles; præfrontal single (rarely divided); frontal large, more than half the width of the head, considerably longer than the parietals; loreal square or rather deeper than broad; one præ- and two or three postoculars; temporals 2+2 or 2+3; usually seven upper labials, third and fourth entering the eye (rarely eight, fourth and fifth entering the eye); only one pair of well-developed chin-shields. Scales smooth, in 15 rows. Ventrals 125-169; anal entire; subcaudals 23-38. Pale brown above, each scale edged with darker; head yellowish, with a more or less distinct dark brown band between the eyes and a large

rows of small brown spots along the back; uniform whitish inferiorly.
Total length 300 millim.; tail 43.
South Africa.

a. ♀ (V. 164; C. 24).   Durban, Natal.   Capt. Munn [P.].
b. Hgr. (V. 155; C. 25).   Orange R.   Dr. Kannemeyer [P.].
c. ♂ (V. 154; C. 38).   S. Africa.   (Type of *Rhinostoma cupreum*.)
d-e, f-h. ♂ (V. 144; C.   S. Africa.
28), ♀ (V. 161; C. 23),
& hgr. (V. 130, 125,
135; C. 28, 27, 28).

2. Prosymna frontalis.

Temnorhynchus frontalis, part., *Peters, Mon. Berl. Ac.* 1867, p. 236, pl. —. fig. 1.
Prosymna frontalis, *Bocage, Jorn. Sc. Lisb.* iv. 1873, p. 217, and viii. 1882, p. 288.

Closely allied to *P. sundevallii*, but frontal smaller, not half the width of the head, rostral smaller and separated from the præfrontal by a single band-like internasal, temporals 1+2, and tail more elongate. Ventrals 167; subcaudals 50.
Total length 135 millim.; tail 57.
Otjimbingue and Mossamedes, S.W. Africa.

3. Prosymna ambigua.

Prosymna ambiguus, *Bocage, Jorn. Sc. Lisb.* iv. 1873, p. 218.
Ligonirostra stuhlmanni, *Pfeffer, Jahrb. Hamb. Wiss. Anst.* x. 1893, p. 78, pl. i. figs. 8-10.

Rostral very large and broad, with angular horizontal edge; a single internasal and a single præfrontal; frontal large, more than half the width of the head, a little longer than the parietals; loreal longer than deep; one præ- and two postoculars; temporals 1+2; six or seven upper labials, third and fourth entering the eye; only one pair of well-developed chin-shields. Scales smooth, in 15 or 17 rows. Ventrals 133-152; anal entire; subcaudals 19-34. Blackish above, each scale usually greyish in the centre; lower parts whitish or brown.
Total length 255 millim.; tail 27.
East Africa, Angola.

a-k. ♂ (V. 134, 137, 133;   Zanzibar.   Sir J. Kirk [C.].
C. 32, 31, 34), ♀ (V.
150, 141, 151; C. 24,
19, 21), hgr. (V. 137,
140, 140; C. 24, 21, 20),
& yg. (V. 152; C. 22).
l. Ad., bad state.   Shiré Valley, Zambezi.

4. Prosymna meleagris.

Calamaria meleagris, *Reinh. Vid. Selsk. Afhandl.* x. 1843, p. 238, pl. i. figs. 4-6.
Prosymna meleagris, *Gray, Cat.* p. 80 (1849); *Jan, Arch. Zool. Anat. Phys.* ii. 1862, p. 55.
Temnorhynchus meleagris, *Peters, Mon. Berl. Ac.* 1875, p. 198.

Rostral very large, with angular horizontal edge; a single internasal and a single præfrontal; frontal at least half the width of the head, longer than the parietals; loreal much longer than deep; one præ- and one postocular; temporals 1+2; five upper labials, second and third entering the eye; only one pair of well-developed chin-shields. Scales smooth, in 15 rows. Ventrals 140-170; anal entire; subcaudals 22-34. Black above each scale with a whitish terminal dot; uniform yellowish white inferiorly.

Total length 270 millim.; tail 40.

West Africa.

a. ♂ (V. 140; C. 33).	Lagos.	Dr. J. G. Fischer.
b. ♂ (V. 154; C. 33).	W. Africa.	

5. Prosymna jani.

Prosymna janii, *Bianconi, Mem. Acc. Bologna*, (2) i. 1862, p. 470, pl. i.; *Jan, Arch. Zool. Anat. Phys.* ii. 1862, p. 56, *and Icon. Gén.* 48, pl. ii. fig. 1 (1876); *Peters, Reise n. Mossamb.* iii. p. 106 (1882).

Upper head-shields as in *P. meleagris*; two præ- and two or three postoculars; temporals 1+2; six upper labials, third and fourth entering the eye. Scales keeled, in 17 rows. Ventrals 117; anal entire; subcaudals 32. Pale reddish brown above; the greater part of the upper surface of the head and nape black, leaving a pale spot on each supraocular and a cordiform one on the middle of the nape a black cross bar on the neck, followed by a double series of black spots along the anterior two thirds of the back; yellowish wh[ck] inferiorly.

Total length 180 millim.; tail 32.

Inhambane, Mozambique.

95. LEPTOCALAMUS.

Leptocalamus, *Günth. Ann. & Mag. N. H.* (4) ix. 1872, p. 16.
Enulius (*non Cope*), *Bocourt, Miss. Sc. Mex., Rept.* p. 536 (1883).

Maxillary very short, not extending forwards beyond the anterior extremity of the palatines, with 4 or 5 small teeth, followed, after an interspace, by a very large, compressed, blade-like fang; mandibular teeth small, equal: teeth on palatines and pterygoids few. Head not or but slightly distinct from neck; snout depressed, strongly projecting; eye small, with round pupil; nostril between two nasals; loreal (or loreal and præfrontal) entering the eye. Body cylindrical, elongate; scales smooth, with apical pits, in 15 or 17 rows; ventrals rounded. Tail long, ending obtusely; subcaudals in two rows. Tropical America.

Synopsis of the Species.

Scales in 17 rows; portion of rostral visible from above one half or three fifths its distance from the frontal frontal as long as its distance from the end of the snout; præfrontal entering the eye .... 1. *torquatus*, p. 250.

Scales in 17 rows; upper portion of rostral as long as its distance from the frontal; frontal longer than its distance from the end of the snout; a small præocular .... 2. *sumichrasti*, p. 250.

Scales in 15 rows; upper portion of rostral about one third its distance from the frontal; frontal longer than its distance from the end of the snout; præfrontal entering the eye .... 3. *sclateri*, p. 251.

1. Leptocalamus torquatus.

Leptocalamus torquatus, *Günth. Ann. & Mag. N. H.* (4) ix. 1872, p. 17, pl. iii. fig. A.
Geophis unicolor, *Fischer, Abh. naturw. Ver. Bremen,* vii. 1880, p. 227, pl. xv. figs. 1-3.
Enulius murinus (non Cope), *Bocourt, Miss. Sc. Mex., Rept.* p. 537, pl. xxxv. fig. 9 (1883).
Leptocalamus unicolor, *Cope, Proc. Amer. Philos. Soc.* xxii. 1885, p. 178; *Günth. Biol. C.-Am., Rept.* p. 100 (1893).
Geagras longicaudatus, *Cope, Am. Nat.* 1884, p. 162.

Rostral large, broad, the portion visible from above one half or three fifths its distance from the frontal; internasals shorter than the præfrontals, which largely enter the eyes; frontal large, as long as broad, as long as its distance from the end of the snout, a little shorter than the parietals; supraocular very small; loreal as long as the nasals; no præocular; two postoculars; temporals 1+2; seven upper labials, third and fourth entering the eye; a single pair of small chin-shields. Scales in 17 rows. Ventrals 177-203; anal divided; subcaudals 90-102. Pale reddish brown above, uniform or with a yellow band across the occiput; lower parts uniform yellowish.

Total length 305 millim.

Mexico.

a. ♂ (V. 183; C. ?). S. America (?). (Type.)

2. Leptocalamus sumichrasti.

Enulius sumichrasti, *Bocourt, Miss. Sc. Mex., Rept.* p. 538, pl. xxxi. fig. 6 (1883).
Geagras sumichrasti, *Cope, Am. Nat.* 1884, p. 162.

Snout shorter than in *L. torquatus*; upper portion of rostral as long as its distance from the frontal; præfrontal separated from the eye by a small præocular. Scales in 17 rows. Ventrals 191; subcaudals 93. Uniform grey-brown above, yellowish white beneath.

Total length 346 millim.; tail 105.

Isthmus of Tehuantepec.

3. Leptocalamus sclateri. (PLATE XII. fig. 1.)

Rostral smaller than in *L. torquatus*, nearly twice as broad as deep, the portion visible from above measuring about one third its distance from the frontal; internasals shorter than the præfrontals, which largely enter the eyes; frontal large, kite-shaped, slightly longer than broad, much longer than its distance from the end of the snout, as long as the parietals; supraocular very small; loreal longer than the nasals; no præocular; two postoculars; temporals 1+2; seven upper labials, third and fourth entering the eye; five small equal chin-shields, the anterior a pair in contact with three labials. Scales in 15 rows. Ventrals 144; anal divided; subcaudals 98. Uniform dark brown above; head yellow, with a black blotch on the snout and another round each eye; lower parts uniform yellowish.

Total length 380 millim.; tail 150.

S. America.

a. ♀ (V. 144; C. 98).   S. America.   P. L. Sclater, Esq [P.].

96. ARRHYTON.

Arrhyton, *Günth. Cat. Col. Sn.* p. 244 (1858).
Cryptodacus, *Gundl. & Peters, Mon. Berl. Ac.* 1861, p. 1002; *Bocourt, Miss. Sc. Mex., Rept.* p. 560 (1883).
Colorhogia, *Cope, Proc. Ac. Philad.* 1862, p. 81.

Maxillary bone short, with 8 small teeth, followed, after a considerable interspace, by a strongly enlarged fang; palatine and pterygoid teeth few; mandibular teeth small, equal. Head slightly distinct from neck; eye rather small, with round pupil; nasal entire or semidivided. Body cylindrical, moderately elongate; scales smooth, without apical pits, in 15 or 17 rows; ventrals rounded. Tail moderate; subcaudals in two rows.

Cuba.

Synopsis of the Species.

Nasal semidivided; præfrontals distinct, in
  contact with labials .... 1. *tæniatum*, p. 252.
Nasal entire; præfrontals and loreal distinct. 2. *vittatum*, p. 252.
Nasal entire; præfrontals coalesced; loreal
  distinct .... 3. *redimitum*, p. 252.

1. Arrhyton tæniatum. (Plate XII. fig. 2.)

Arrhyton tæniatum, *Günth. Cat.* p. 244 (1858); *Cope, Proc. Ac. Philad.* 1860, p. 421; *Gundl. Repert. Fis.-Nat. Cuba*, ii. 1866, p. 116, *and Erp. Cubana*, p. 75 (1880).
Arrhyton fulvum, *Cope, Proc. Acad. Philad.* 1862, p. 82.

Rostral twice as broad as deep, visible from above; nasal semidivided; suture between the internasals hardly half as long as that between the præfrontals; frontal slightly longer than broad, thrice as broad as the supraocular, as long as its distance from the end of the snout, shorter than the parietals; no loreal, præfrontal in contact with second labial; one præ- and two postoculars; temporals 1+2; seven upper labials, third and fourth entering the eye; four lower labials in contact with the anterior chin-shields, which are longer than the posterior. Scales in 17 rows. Ventrals 171; anal divided; subcaudals 75. Yellowish; upper surface of head and three longitudinal streaks brown; lower parts white.

Total length 230 millim.; tail 50.

Cuba.

a. Hgr. (V. 171; C. 75).   Cuba.   Zoological Society.
(Type.)

2. Arrhyton vittatum.

Cryptodacus vittatus, *Gundl. & Peters, Mon. Berl. Ac.* 1861, p. 1002; *Cope, Proc. Ac. Philad.* 1862, p. 339; *Gundl. Repert. Fis.-Nat. Cuba*, ii. 1866, p. 116; *Bocourt, Miss. Sc. Mex., Rept.* p. 561, pl. xxxv. fig. 7 (1883).
Arrhyton bivittatum, *Cope, Proc. Acad. Philad.* 1862, p. 82.

Rostral twice as broad as deep, visible from above; nasal undivided; suture between the internasals shorter than that between the præfrontals; frontal as long as broad or a little longer, much broader than the supraocular, as long as its distance from the end of the snout, a little shorter than the parietals; loreal square or longer than deep; one præ- and two postoculars; temporals 1+2; seven upper labials, third and fourth entering the eye; four lower labials in contact with the anterior chin-shields, which are as long as or longer than the posterior. Scales in 15 or 17 rows. Ventrals 115-116; anal divided; subcaudals 64-73. Yellowish above, with three brown longitudinal lines, the median sometimes indistinct; yellowish white inferiorly.

Total length 297 millim.; tail 105.

Cuba.

3. Arrhyton redimitum.

Colorhogia redimita, *Cope, Proc. Ac. Philad.* 1862, p. 81.
Cryptodacus redimitus, *Bocourt, Miss. Sc. Mex., Rept.* p. 561, pl. xxxv. fig. 8 (1883).

Nasal undivided; præfrontals fused to a single shield; frontal broad, as broad as long, shorter than the parietals; loreal longer than deep; one præ- and one postocular; temporals 1+2; seven upper labials, third and fourth entering the eye; anterior chin-shields longer than the posterior. Scales in 17 rows. Ventrals 141; anal divided; subcaudals 120. Pale grey-brown above, with three darker longitudinal streaks; lips and lower parts yellowish white.

Total length 390 millim.; tail 90.

Cuba.

97. SIMOPHIS.

Rhinostoma, part., *Fitzing. N. Class. Rept.* p. 29 (1826); *Günth. Cat. Col. Sn.* p. 8 (1858).
Heterodon, part., *Schleg. Phys. Serp.* ii. p. 96 (1837).
Simophis, *Peters, Mon. Berl. Ac.* 1860, p. 521.
Rhinaspis (*Fitz.*), *Jan, Arch. Zool. Anat. Phys.* ii. 1863, p. 215.

Maxillary teeth 20 to 22, equal; mandibular teeth equal. Head slightly distinct from neck, with prominent, cuneiform snout; eye moderate, pupil round; rostral large, with obtuse transverse keel; nostril between two nasals. Body elongate, cylindrical; scales smooth, with apical pits, in 15 rows; ventrals angulate laterally. Tail moderate; subcaudals in two rows.

Brazil.

1. Simophis rhinostoma.

Heterodon rhinostoma, *Schleg. Phys. Serp.* ii. p. 100, pl. iii. figs. 17-19 (1837).
Rhinostoma schlegelii, *Günth. Cat.* p. 8 (1858):
Simophis rhinostoma, *Peters, Mon. Berl. Ac.* 1860, p. 521.
Rhinaspis proboscideus, *Jan, Arch. Zool. Anat. Phys.* ii. 1863. p. 215.

Portion of rostral visible from above measuring about two thirds its distance from the frontal; suture between the internasals as long as that between the præfrontals; frontal a little shorter than its distance from the end of the snout, as long as the parietals; loreal square; one præ- and two postoculars; temporals 2+2; seven upper labials, third and fourth entering the eye; four lower labials in contact with the anterior chin-shields, which are as long as the posterior. Scales in 15 rows. Ventrals 170-190; anal divided; subcaudals 64-66. Yellowish above (red in life?) with black rings arranged in threes, the narrow interspaces between the three rings black-spotted; snout black and yellow; vertex and occiput black, with a yellow bar across the frontal and a round yellow spot in the middle of the suture between the parietals; belly with black transverse spots in addition to the rings.

Total length 510 millim.; tail 105.

Brazil.

a. ♀ (V. 172: C. 64). Brazil. Prof. C. Machado [P.].

2. Simophis rohdii.

Rhinaspis rohdei, *Boettg. Zeitschr. f. Naturw.* (4) iv. 1885, p. 231.

Distinguished from the preceding in having the rostral more pointed behind, forming a right angle; eight upper labials, fourth and fifth entering the eye; and scales in 17 rows. Ventrals 171; anal divided; subcaudals 67. Coloration same as in *S. rhinostoma*.
Total length 730 millim.; tail 140.
Paraguay.

98. SCAPHIOPHIS.

Scaphiophis, *Peters, Mon. Berl. Ac.* 1870, p. 644.

Maxillary teeth very small, subequal, 12 or 13 on each side; mandibular teeth small, subequal. Head short, rarely distinct from neck, with broad, beak-shaped, hooked snout; rostral very large, with sharp horizontal edge, concave inferiorly; nostril a curved slit between two nasals; a series of suboculars; eye moderate, with round pupil. Body elongate; scales smooth, with apical pits, in 21 series or more; ventrals rounded. Tail moderate or rather short; subcaudals in two rows, or the anterior single.

Tropical Africa.

1. Scaphiophis albopunctatus.

Scaphiophis albopunctatus, *Peters, Mon. Berl. Ac.* 1870, p. 645, pl. i. fig. 4; *Bocourt, Ann. Sc. Nat.* (6) ii. 1875, art. 3; *Fischer, Jahrb. Hamb. Wiss. Anst.* ii. 1885, p. 100, pl. iii. fig. 6; *Bocage, Jorn. Sc. Lisb.* xi. 1887, p. 195; *Mocquard, Bull. Soc. Philom.* (7) xi. 1887, p. 77.
—— raffreyi, *Bocourt, l. c.*

The upper surface of the rostral once and a half to twice as long as its distance from the frontal; internasals and præfrontals broad, the former as long as or a little shorter than the latter; frontal a little longer than broad, nearly twice as broad as the supraocular; parietals broken up into four shields or more; usually two small superposed loreals; one præ- and two postoculars; three suboculars, separating the eye from the labials; temporals small, scale-like; five upper labials, fifth very long; a single pair of large chin-shields, in contact with three lower labials on each side. Scales in 25 to 31 rows on the neck, 21 to 25 on the middle of the body. Ventrals 210–240; anal divided; subcaudals 51–65. Pale brown above, uniform or spotted with black, or with whitish dots; lower parts white.

Total length 960 millim.; tail 160.

Tropical Africa, from the coast of Guinea and the Congo to the Upper Nile and Abyssinia.

a. ♂ (Sc. 23; V. 214; C. 15).	Semmio, E. Central Africa.	Herr F. Bohndorff [C.].

99. CONTIA.

Leptophis, part., *Bell, Zool. Journ.* ii. 1826, p. 328; *Baird & Gir. Cat. N. Am. Rept.* p. 106 (1853).
Calamaria, part., *Schleg. Phys. Serp.* ii. p. 25 (1837).
Herpetodryas, part., *Schleg. l. c.* p. 173; *Dum. & Bibr. Erp. Gén.* vii. p. 203 (1854).
Chlorosoma, part., *Baird & Gir. l. c.* p. 108.
Contia, *Baird & Gir. l. c.* p. 110; *Cope, Proc. Ac. Philad.* 1860, p. 251; *Bocourt, Miss. Sc. Mex., Rept.* p. 556 (1883); *Cope, Proc. U.S. Nat. Mus.* xiv. 1892, p. 599.
Sonora, *Baird & Gir. l. c.* p. 117.
Ablabes, part., *Dum. & Bibr. t. c.* p. 304; *Günth. Cat. Col. Sn.* p. 27 (1858).
Lamprosoma (*non Kirby*), *Hallow. Proc. Ac. Philad.* 1856, p. 311; *Bocourt, l. c.* p. 558.
Conopsis, *Günth. l. c.* p. 6; *Bocourt, l. c.* p. 563.
Cyclophis, part., *Günth. l. c.* p. 119.
Toluca, *Kennicott, U.S. Mex. Bound. Surv.* ii. *Rept.* p. 23 (1859); *Cope, Proc. Ac. Philad.* 1860, p. 241.
Chionactis, *Cope, l. c. and Proc. U.S. Nat. Mus.* xiv. 1892, p. 604.
Liopeltis, part., *Cope, Proc. Ac. Philad.* 1860, p. 559; *Jan, Elenco sist. Ofid.* p. 81 (1863).
Homalosoma, part., *Jan, Arch. Zool. Anat. Phys.* ii. 1862, p. 33.
Oxyrhina (*non Meig*), *Jan, l. c.* p. 59.
Achirhina, *Jan, l. c.* p. 61.
Exorhina, *Jan, l. c.*
Psilosoma, part., *Jan, l. c.*
Eirenis, *Jan, t. c.* 1863, p. 256.
Pseudoficimia, *Bocourt, l. c.* p. 572.
Phyllophilophis, *Garman, N. Am. Rept.* p. 40 (1883).
Pseudocyclophis, *Boettg. Zool. Anz.* 1888, p. 262.
Pseudocyclophis, part., *Bouleng. Faun. Ind., Rept.* p. 299 (1890).

Maxillary teeth small, equal or subequal, 12 to 20; mandibular teeth subequal. Head small, not or but slightly distinct from neck; eye small or moderate, with round or vertically subelliptic pupil; nasal often single or semidivided; loreal sometimes absent; internasals sometimes fused with the præfrontals. Body cylindrical; scales smooth or keeled, with apical pits, in 13 to 19 rows; ventrals rounded. Tail short, moderate or long; subcaudals in two rows.

South-western Asia and Sind; America.

Synopsis of the Species.

I. Rostral shield moderate.

   A. Scales keeled, in 15 rows; ventrals 148–166; subcaudals 111–148 .... 1. *æstiva*, p. 258.

   B. Scales smooth.

      1. Nasal entire (exceptionally semidivided).

a. Seven upper labials; portion of rostral visible from above not more than half as long as its distance from the frontal.

    α. Ventrals 103–148.

Scales in 15 rows; subcaudals 65–94 ....   2. *vernalis*, p. 258.
Scales in 13 rows; subcaudals 51–64 ....   3. *agassizii*, p. 259.
Scales in 15 to 19 rows; subcaudals 24–52.   11. *coronella*, p. 264.

    β. Ventrals 150–235; subcaudals 48–85; scales in 15 or 17 rows.

      * Rostral as deep as broad or a little broader than deep; scales in 15 or 17 rows; ventrals 150–191.

Frontal at least twice as long as broad ..   4. *decemlineata*, p. 260.
Frontal not twice as long as broad ......   5. *collaris*, p. 260.

      ** Rostral considerably broader than deep; scales in 15 rows.

        † Frontal as long as the parietals; ventrals 158–171.
                                        6. *fasciata*, p. 262.

        †† Frontal much shorter than the parietals.

Frontal once and two thirds as long as broad; ventrals 186 ....   7. *angusticeps*, p. 262.
Frontal not once and a half as long as broad; ventrals 159–187 ....   8. *rothi*, p. 262.
Frontal once and a half as long as broad, about once and a half as broad as the supraocular; ventrals 194–216 ......   9 *persica*, p. 263.
Frontal once and a half as long as broad, nearly twice as broad as the supraocular; ventrals 211–235 ....   10. *walteri*, p. 263.

b. Seven upper labials; portion of rostral visible from above measuring one half to two thirds its distance from the frontal.

    α. Scales in 13 rows; ventrals 126–137; subcaudals 37–46.................   12. *taylori*, p. 265.

    β. Scales in 15 rows.

      * Body without black cross bars.

Ventrals 145–163 ....   13. *episcopa*, p. 265.
Ventrals 183 ....   14. *torquata*, p. 266.

      ** Body with black cross bars.

Ventrals 158–167; subcaudals 50–52 ....   15. *isozona*, p. 266.
Ventrals 147–158; subcaudals 34–44 ....   16. *occipitalis*, p. 266.

c. Eight upper labials; ventrals 133; subcaudals 50; scales in 17 rows .... 17. *pachyura*, p. 267.

2. Nasal divided or semidivided; scales in 15 rows; ventrals 147–167; subcaudals 29–39.

Two postoculars..................... 18. *mitis*, p. 267.
Three postoculars .... 19. *semiannulata*, [p. 268.

II. Rostral large, the portion seen from above as long as or a little shorter than its distance from the frontal; scales in 17 rows.

Nasal entire or semidivided; ventrals 110–137; subcaudals 23–40 .... 20. *nasus*, p. 268.
Nasal divided; ventrals 141–160; subcaudals 38–48 .... 21. *frontalis*, p. 270.

TABLE SHOWING NUMBERS OF SCALES AND SHIELDS.

I. Old-World species.

	Scales.	Ventrals.	Caudals.	Postoculars.	Labials.
decemlineata ...	17	152–175	64–85	2	7
collaris	15–17	150–191	50–78	1–2	7
fasciata	15	158–171	48–62	2	7
angusticeps ...	15	186	86	1	7
rothi	15	159–187	40–60	2	7
persica	15	194–216	63–77	1	7
walteri	15	211–235	73–82	1	7
coronella	15–19	103–148	24–52	1–2	7

II. American species.

	Scales.	Ventrals.	Caudals.	Postoculars.	Labials.
æstiva	15	148–166	111–148	2	7
vernalis	15	119–148	65–94	1–2	7
agassizii	13	128–138	51–64	2	7
taylori	13	126–137	37–46	2	7
episcopa	15	145–163	35–57	2	7
torquata	15	183	38	2	7
isozona	15	158–167	50–52	2	7
occipitalis	15	147–158	34–44	2	7
pachyura	17	133	50	2	8
mitis	15	147–167	29–39	2	7
semiannulata ...	15	149	39	3	7
nasus	17	110–137	23–40	2	7
frontalis	17	141–160	38–48	2	7

1. Contia æstiva.

Coluber æstivus, *Linn. S. N.* i. p. 387 (1766); *Daud. Rept.* vii. p. 101 (1803); *Harl. Journ. Ac. Philad.* v. 1827, p. 357, *and Phys. Med. Res.* p. 121 (1835).

Leptophis æstivus, *Bell, Zool. Journ.* ii. 1826, p. 329; *Holbr. N. Am. Herp.* iv. p. 17, pl. iii. (1842); *Baird & Gir. Cat. N. Am. Rept.* p. 106 (1853), *and Rep. U.S. Explor. R. R.* x. pt. iii. pl. xxxii. fig. 79 (1859).

Herpetodryas æstivus, *Schleg. Phys. Serp.* ii. p. 186, pl. vii. figs. 12 & 13 (1837); *Dum. & Bibr.* vii. p. 209 (1854).

Leptophis majalis, *Baird & Gir. ll. cc.* p. 107, fig. 80, *and Marcy's Explor. Red Riv.* p. 232, pl. ix. (1853).

Cyclophis æstivus, *Günth. Cat.* p. 119 (1858); *Garm. N. Am. Rept.* p. 40 (1883).

Opheodrys æstivus, *Cope, Proc. Ac. Philad.* 1860, p. 560.

Liopeltis æstivus, *Jan, Elenco,* p. 81 (1863), *and Icon. Gén.* 31, pl. v. fig. 1 (1869).

Philophyllophis majalis, *Garm. Bull. Essex Inst.* xxiv. 1892, p. 108.

Phyllophilophis æstivus, *H. Garm. Bull. Illin. Lab.* iii. 1892, p. 283.

Rostral broader than deep, just visible from above; nasal undivided; suture between the internasals shorter than that between the præfrontals; frontal once and a half as long as broad, as long as or a little longer than its distance from the end of the snout, shorter than the parietals; loreal longer than deep; one præ- and two postoculars; temporals 1+2; seven upper labials, third and fourth entering the eye; four lower labials in contact with the anterior chin-shields; posterior chin-shields longer than the anterior, and in contact anteriorly. Scales striated and keeled, in 15 rows. Ventrals 148–166; anal divided; subcaudals 111–148. Uniform green above, upper lip and lower parts yellowish or greenish white.

Total length 670 millim.; tail 260.

United States, east of the Rocky Mountains.

a. ♀ (V. 159; C. 114).	Michigan.	
b. ♂ (V. 148; C. 130).	Bloomington, Indiana.	C. H. Bollman, Esq. [C.].
c. ♀ (V. 151; C. 120).	Delaware.	H. Doubleday, Esq. [P.].
d. Yg. (V. 152; C. 148).	Florida.	Dr. Guillemard [P.].
e. ♀ (V. 153; C. 127).	New Orleans.	M. Sallé [C.].
f. ♀ (V. 162; C. 114).	Duval Co., Texas.	W. Taylor, Esq. [C.].
g. ♂ (V. 159; C. 130).	Texas.	
h. Skull.	United States.	

2. Contia vernalis.

Coluber vernalis, *Harlan, Journ. Ac. Philad.* v. 1827, p. 361, *and Phys. Med. Res.* p. 124 (1835); *Storer, Rep. Fish. & Rept. Mass.* p. 224 (1839); *Holbr. N. Am. Herp.* iii. p. 79, pl. xvii. (1842); *De Kay, N. Y. Faun.* iii. p. 40, pl. xi. fig. 22 (1842).

Chlorosoma vernalis, *Baird & Gir. Cat. N. Am. Rept.* p. 108 (1853), *and Rep. U.S. Explor. R. R.* x. pt. iii. pl. xxxii. fig. 81 (1859).

Herpetodryas vernalis, *Hallow. Proc. Ac. Philad.* 1856, p. 243.

Cyclophis vernalis, *Günth. Cat.* p. 119 (1858); *Garm. N. Am. Rept.*

p. 39, pl. iii. fig. 4 (1883), and *Bull. Essex Inst.* xxiv. 1892, p. 108;
H. Garm. *Bull. Illin. Lab.* iii. 1892, p. 282.
Liopeltis vernalis, *Cope, Proc. Ac. Philad.* 1860, p. 560; *Jan, Icon. Gén.* 31, pl. v. fig. 3 (1869).

Rostral broader than deep, the portion visible from above nearly half as long as its distance from the end of the snout; nasal undivided; suture between the internasals shorter than that between the præfrontals; frontal nearly once and a half as long as broad, longer than its distance from the end of the snout, shorter than the parietals; loreal small, square, or a little deeper than long, or a little longer than deep, rarely united with the nasal; one (rarely two) præ- and two postoculars; temporals 1+2 (rarely 1+1); seven upper labials, third and fourth entering the eye; four lower labials in contact with the anterior chin-shields; posterior chinshields longer than the anterior, and in contact anteriorly. Scales in 15 rows. Ventrals 119–148; anal divided; subcaudals 65–94. Uniform green above, upper lip and lower parts yellowish or greenish white.

Total length 450 millim.; tail 160.

North America, east of the Rocky Mountains.

a. ♀ (V. 127; C. 65).	Halifax.	
b. ♂ (V. 129; C. 80).	Illinois.	Smithsonian Instit. [P.].
c. ♀ (V. 135; C. 77).	Florida.	
d–e. ♂ (V. 119; C. 74) & ♀ (V. 131; C. 75).	N. America.	
f. ♂ (V. 126; C. 89).	—?	(Named by Dr. Shaw *C. cyaneus*.)

3. Contia agassizii.

Eirenis agassizii, *Jan, Arch. Zool. Anat. Phys.* ii. 1863, p. 260, *and Icon. Gén.* 15, pl. v. fig. 3 (1866).
Liopeltis brevicauda, *Jan, Elenco,* p. 82 (1863), *and Icon.* 31, pl. vi. fig. 1 (1869).
Ablabes agassizii, *Bouleng. Ann. & Mag. N. H.* (5) xvi. 1885, p. 87.

Rostral as deep as broad, visible from above; nasal entire or semidivided; suture between the internasals as long as or shorter than that between the præfrontals; frontal once and a half to once and two thirds as long as broad, longer than its distance from the end of the snout, as long as the parietals; loreal small, square, rarely united with the nasal; one præ- and two postoculars; temporals 1+1 (rarely 1+2); seven upper labials, third and fourth entering the eye; four lower labials in contact with the anterior chin-shields, which are longer than the posterior; latter in contact with each other. Scales in 13 rows. Ventrals 128–138; anal divided; subcaudals 51–64. Olive above, uniform or with two light, black-edged dorsal streaks; sometimes a reddish-brown vertebral band; labials whitish, edged with black; lower parts uniform greenish white.

Total length 380 millim.; tail 85.

Southern Brazil, Uruguay.

a-d. ♂ (V. 132, 135; C. 63, Rio Grande do Sul. Dr. H. v. Ihering [C.].
62), ♀ (V. 138; C. 51), &
hgr. (V. 134; C. 63).
e. Hgr. ♂ (V. 128; C. 61). Uruguay.

4. Contia decemlineata.

Ablabes decemlineatus, *Dum. & Bibr.* vii. p. 327 (1854); *Günth. Proc. Zool. Soc.* 1864, p. 489.

Eirenis collaris, vars. inornata, decemlineata, *et* quadrilineata, *Jan, Arch. Zool. Anat. Phys.* ii. 1863, p. 256, *and Icon. Gén.* 15, pl. iv. figs. 2–4 (1866).

—— decemlineatus, *F. Müll. Verh. nat. Ges. Basel*, vi. 1878, p. 595.

—— inornatus, *F. Müll. l. c.*

Ablabes modestus, vars., *Boettg. Ber. Senck. Ges.* 1879–80, p. 144.

Homalosoma modestus, part., *Peracca, Boll. Mus. Torino*, ix. 1894, no. 167, p. 14.

Rostral as deep as broad or a little broader than deep, just visible from above; nasal undivided; suture between the internasals as long as or a little shorter than that between the præfrontals; frontal narrow, at least twice as long as broad, not broader than the supraocular, as long as or a little longer than its distance from the end of the snout, shorter than the parietals; loreal considerably longer than deep; one (rarely two) præ- and two postoculars; temporals 1+2 or 1+3; seven upper labials, third and fourth entering the eye; four, rarely five, lower labials in contact with the anterior chin-shields; posterior chin-shields as long as or a little shorter than the anterior and in contact with each other. Scales in 17 rows. Ventrals 152–175; anal divided; subcaudals 64–85. Pale brown above, uniform or with black longitudinal lines arranged in pairs; no dark cross band on the nape; lower parts uniform white.

Total length 730 millim.; tail 160.

Syria.

a. ♂ (V. 164; C. 73).	Lebanon.	Canon Tristram [C.].
b-e. ♂ (V. 168, 170; C. 76, 75), ♀ (V. 175; C. 66), & yg. (V. 167; C. 73).	Merom.	Canon Tristram [C.].
f-h, i, k. ♂ (V. 166, 167, 168, 168; C. 78, 81, 76, ?) & ♀ (V. 159; C. ?).	Galilee.	Canon Tristram [C.].
l. Hgr. (V. 162; C. 85).	Mt. Hermon.	Canon Tristram [C.].

5. Contia collaris.

Coluber collaris, *Ménétr. Cat. Rais.* p. 67 (1832).

—— nigricollis, *Dwigubsky, Nat. Hist. Russ. Emp., Amph. (Russian)*, p. 26 (1832).

—— reticulatus (*non Ménétr.*), *Krynicky, Bull. Soc. Nat. Mosc.* 1837, p. 60.

Coronella modesta, *Martin, Proc. Zool. Soc.* 1838, p. 82.

Tyria argonauta, *Eichw. Bull. Soc. Nat. Mosc.* 1839, p. 306, *and Faun. Casp.-Cauc.* p. 114, pl. xxvi. figs. 1 & 2 (1841).

Psammophis moniliger, *Nordm. in Demid. Voy. Russ. Mér.* iii. p. 342, pl. iv. fig. 1 (1840).
Coronella collaris, *Berth. in Wagn. Reise n. Kolchis*, p. 332 (1850).
Ablabes modestus, *Günth. Cat.* p. 27 (1858); *Strauch, Schl. Russ. R.* p. 36, pl. i. fig. 1 (1873); *Boettg. Ber. Senck. Ges.* 1890, p. 294.
Contia modesta, *Cope, Proc. Ac. Philad.* 1862, p. 339.
Eirenis collaris, *Jan, Arch. Zool. Anat. Phys.* ii. 1863, p. 257, *and Icon. Gén.* 15, pl. iv. fig. 1 (1866).
Ablabes collaris, *Strauch, l. c.* p. 41, pl. i. fig. 2.
Cyclophis modestus, *Blanf. Zool. E. Pers.* p. 403 (1876); *Boettg. Sitzb. Ac. Berl.* 1888, p. 171, *and Ber. Senck. Ges.* 1892, p. 147; *Méhely, Zool. Anz.* 1894, p. 85.
—— collaris, *Blanf. l. c.* p. 405.
Ablabes modestus, var. semimaculata, *Boettg. Ber. Offenb. Ver. Nat.* 1876, p. 58, pl. —.
Cyclophis modestus, var. punctatolineata, *Boettg. Ber. Senck. Ges.* 1892, p. 147.

Rostral a little broader than deep, just visible from above; nasal undivided; suture between the internasals as long as or a little shorter than that between the præfrontals; frontal once and a half to once and two thirds as long as broad, longer than its distance from the end of the snout, shorter than the parietals; loreal square or a little longer than deep; one (rarely two) præ- and two (rarely one) postoculars; temporals 1+2; seven upper labials, third and fourth entering the eye; four (rarely five) lower labials in contact with the anterior chin-shields; posterior chin-shields smaller than the anterior, and usually separated from each other by scales. Scales in 15 or 17 rows. Ventrals 150–191; anal divided; subcaudals 50–78. Pale olive above, each scale lighter in the centre: head and nape in the young black, with yellow spots or cross bands, the black nuchal collar edged with yellow; these markings becoming indistinct in the adult; lower parts uniform white. Some varieties have longitudinal series of dark dots, which may form lines on the tail.

Total length 480 millim.; tail 110.

From the Archipelago, Cyprus, and Asia Minor to the Caucasus and Persia.

a–b. ♂ (Sc. 15; V. 151; C. 62) & ♀ (Sc. 15; V. 164; C. 53).	Areshsky, Gov. Elizabethpol.	St. Petersburg Mus. [E.].
c–d. ♂ (Sc. 17; V. 176; C. 78) & yg. (Sc. 17; V. 165; C. 75).	Constantinople.	
e. Many specs. (Sc. 17), ♂ (V. 164, 162, 161, 164, 168, 163, 156; C. 73, 64, 65, 66, 66, 68, 62), ♀ (V. 172, 172; C. 56, 55), & yg. (V. 159, 177, 170, 150, 174; C. 67, 62, 55, 69, 63).	Xanthus.	Sir C. Fellows [P.].
f–g. ♀ & yg. (Sc. 17; V. 185, 154; C. 59, 78).	Valley of the Meinder.	R. MacAndrew, Esq. [P.].
h. Yg. (Sc. 15; V. 158; C. 60).	Ruins of Nineveh.	W. K. Loftus, Esq. [P.].
i–k. Yg. (Sc. 15, 17; V. 163, 173; C. ?, 61).		Euphrates Exped.

6. Contia fasciata.

Eirenis fasciatus, *Jan, Arch. Zool. Anat. Phys.* ii. 1863, p. 260, *and Icon. Gén.* 15, pl. v. fig. 2 (1866).
Cyclophis fasciatus, *Blanf. Zool. E. Pers.* p. 406 (1876); *Boettg. Zool. Jahrb.* iii. 1888, p. 920.
Ablabes fasciatus, *Boettg. Ber. Senck. Ges.* 1879-80, p. 142.

Rostral broader than deep; nasal undivided; frontal as long as its distance from the end of the snout, as long as the parietals; loreal very small, as long as deep; one præ- and two postoculars; temporals 1+2 or 1+1; seven upper labials, third and fourth entering the eye; chin-shields small, posterior separated by a scale. Scales in 15 rows. Ventrals 158-171; anal divided; subcaudals 48-62. Sandy grey above, with numerous narrow cross bands of brownish olive, breaking up into spots on the posterior third of the body and on the tail; belly salmon-colour in life, the anterior portion of each ventral shield brown near the sides.

Total length 340 millim.; tail 82.

Syria, Persia, Transcaspia.

7. Contia angusticeps.

Head narrow and elongate. Rostral twice as broad as deep, just visible from above; nasal undivided; suture between the internasals longer than that between the præfrontals; frontal once and two thirds as long as broad, hardly once and a half as broad as the supraocular, longer than its distance from the end of the snout, much shorter than the parietals; a very small loreal, longer than deep; one præ- and one postocular; temporals 1+1; seven upper labials, third and fourth entering the eye; five lower labials in contact with the anterior chin-shields, which are a little longer than the posterior. Scales in 15 rows. Ventrals 186; anal divided; subcaudals 86. Pale greyish brown above, each scale darker in the centre; a darker bar between the eye, another across the parietals and temples, and a third across the nape; lower parts uniform white.

Total length 340 millim.; tail 85.

Cherat, Baluchistan.

This species is described from a single specimen in the Indian Museum, Calcutta, submitted to me by Mr. W. L. Sclater.

8. Contia rothi.

Eirenis rothi, *Jan, Arch. Zool. Anat. Phys.* ii. 1863, p. 259, *and Icon. Gén.* 15, pl. v. fig. 1 (1866); *F. Müll. Verh. nat. Ges. Basel,* vi. 1878, p. 659.
Ablabes modestus (*non Mart.*), *Günth. Proc. Zool. Soc.* 1864, p. 489.
—— rothi, *Boettg. Ber. Senck. Ges.* 1879-80, p. 143.
Homalosoma collaris, *Peracca, Boll. Mus. Torino,* ix. 1894, no. 167, p. 15.

Rostral much broader than deep, just visible from above; nasal undivided; suture between the internasals as long as or shorter than that between the præfrontals; frontal not once and a half as long as broad, as long as or longer than its distance from the end of the snout, shorter than the parietals; loreal very small, longer than deep; one præ- and two postoculars; temporals 1+1; seven upper labials, third and fourth entering the eye; four lower labials in contact with the anterior chin-shields; posterior chin-shields smaller and in contact with each other. Scales in 15 rows. Ventrals 159–183; anal divided; subcaudals 40–53. Brownish yellow above; head and nape black, with three or four yellow transverse lines; the black of the nape descending to the sides of the throat; lower parts uniform white.

Total length 300 millim.; tail 55.

Syria.

a–b. ♂ (V. 174; C. 52) & hgr. (V. 161; C. 53).	Sarona, near Jaffa.	Senckenberg Mus. [E.].
c. ♂ (V. 159; C. 47).	Lebanon.	Canon Tristram [C.].
d–e. Yg. (V. 159, 160; C. 47, 47).	Galilee.	Canon Tristram [C.].

9. Contia persica.

Cyclophis persicus, *Anders. Proc. Zool. Soc.* 1872, p. 392, fig.; *Blanf. Zool. E. Persia*, p. 408, pl. xxviii. fig. 1 (1876).
Pseudocyclophis persicus, *Boettg. Zool. Jahrb.* iii. 1888, p. 922.

Rostral nearly twice as broad as deep, just visible from above; nasal undivided; suture between the internasals nearly as long as that between the præfrontals; frontal once and a half as long as broad, about once and a half as broad as the supraocular, as long as its distance from the end of the snout, much shorter than the parietals; no loreal, nasal touching or nearly reaching the præocular; one præ- and one postocular; temporals 1+1; seven upper labials, third and fourth entering the eye; four lower labials in contact with the anterior chin-shields, which are at least twice as large as the posterior. Scales in 15 rows. Ventrals 194–216; anal divided; subcaudals 63–77. Yellowish or pale olive above, unspotted; upper surface of head and nape black, the black descending in front of the eye to the anterior lower labials; the rest of the labial region, the rostral, and the lower parts uniform yellowish white.

Total length 400 millim.; tail 90. The specimen in the collection measures 230; tail 45.

Persia.

a. Hgr. (V. 196; C. 63).   Persia.

10. Contia walteri.

Pseudocyclophis walteri, *Boettg. Zool. Anz.* 1888, p. 262, *and Zool. Jahrb.* iii. 1888, p. 922, pl. xxxiv. fig. 1; *Bouleng. Faun. Ind., Rept.* p. 300 (1890), *and Proc. Zool. Soc.* 1891, p. 632.

Rostral once and a half as broad as deep, visible from above; nasal undivided; suture between the internasals longer than that between the præfrontals; frontal once and a half as long as broad, nearly twice as broad as the supraocular, as long as its distance from the end of the snout, and much shorter than the parietals; loreal, if present, small, longer than deep; one præ- and one postocular; temporals 1+1; seven upper labials, third and fourth entering the eye; four or five lower labials in contact with the anterior chin-shields, which are longer than the posterior. Scales in 15 rows. Ventrals 211-235; anal divided; subcaudals 73-82. Pale olive or reddish above, anteriorly with more or less distinct black transverse lines, posteriorly uniform or with small black spots; head sometimes black above; lower parts white.

Total length 450 millim.; tail 110.

Transcaspia and Sind.

a. Hgr. (V. 235; C. 73).	Ashkabad, Transcaspia.	M. P. A. Warentzoff [C.].
b. ♀. (V. 211; C. 78).	Kolustan, Sind.	W. T. Blanford, Esq. [P.].

11. Contia coronella.

Calamaria coronella, *Schleg. Phys. Serp.* ii. p. 48 (1837).
Homalosoma coronelloides, *Jan, Arch. Zool. Anat. Phys.* ii. 1862, p. 34, *and Icon. Gén.* 13, pl. iii. fig. 5 (1865).
—— coronella, *Jan, ll. cc.* p. 36, pl. iv. fig. 3; *Lortet, Arch. Mus. Lyon,* iii. 1883, p. 184, pl. xix. fig. 3; *Peracca, Boll. Mus. Torino,* ix. 1894, no. 167, p. 13.
Ablabes coronella, *Günth. Proc. Zool. Soc.* 1864, p. 489; *Boettg. Ber. Senck. Ges.* 1879-80, p. 140.

Rostral much broader than deep, well visible from above; suture between the internasals as long as or shorter than that between the præfrontals; nasal undivided; frontal longer than its distance from the end of the snout, nearly as long as or longer than the parietals, not more than once and a half as broad as the supraocular; a small square loreal present or absent; one præocular; one postocular (rarely two); temporals 1+1 or 1+2; seven upper labials, third and fourth entering the eye; anterior chin-shields short and in contact with three or four labials, second pair very small and separated by a scale. Scales in 17 rows (rarely 15 or 19). Ventrals 103-148; anal divided; subcaudals 24-52. Pale brown above, with brown spots which may be confluent into transverse bands; a broad crescentic dark brown collar, descending to and sometimes extending across the throat; lips yellowish, with a blackish vertical streak below the eye and another between the two last labials; yellowish white inferiorly; belly with round brown spots.

Total length 295 millim.; tail 45.

Syria.

a. ♀ (Sc.17; V.128; C.29).	Merom.	Canon Tristram [C.].
b-c. ♂ (Sc. 17; V. 103, 103; C. 28, 29).	Lake of Galilee.	Canon Tristram [C.].
d. ♀ (Sc.19; V.123; C. 24).	Lebanon.	Canon Tristram [C.].
e-g. ♀ (Sc. 17; V. 120; C. 31) & yg. (Sc. 17; V. 127, 123; C. 32, 27).	Haifa.	Senckenberg Museum [E.].

12. Contia taylori. (Plate XII. fig. 3.)

Rostral broader than deep, the portion visible from above one half to two thirds as long as its distance from the frontal; nasal undivided; suture between the internasals as long as or shorter than that between the præfrontals; frontal about once and a half as long as broad, broader than the supraocular, longer than its distance from the end of the snout, a little shorter than the parietals; loreal small, longer than deep; one præ- and two postoculars; temporals 1+1 or 1+2; seven upper labials, third and fourth entering the eye; three or four lower labials in contact with the anterior chin-shields; posterior chin-shields very small and separated from each other. Scales in 13 rows. Ventrals 126–137; anal divided; subcaudals 37–46. Pale brown above, each scale darker along the centre; upper lip and lower parts white.

Total length 270 millim.; tail 55.

Texas, North Mexico.

a–c. ♂ (V. 126, 133, 137; C. 46, 46, 46).		Duval Co., Texas.	W. Taylor, Esq. [P.].
d. ♀ (V. 136; C. 37).		Nuevo Leon.	W. Taylor, Esq. [P.]

13. Contia episcopa.

Lamprosoma episcopum, *Kennicott, U. S. Mex. Bound. Surv., Rept.* p. 22, pl. viii. fig. 2 (1859); *Bocourt, Miss. Sc. Mex., Rept.* p. 559, pl. xxxiv. fig. 4 (1883).

Contia episcopa, *Cope, Proc. Ac. Philad.* 1860, p. 251, *and Bull. U. S. Nat. Mus.* no. 17, 1880, p. 20; *Garm. N. Am. Rept.* p. 94, pl. vi. fig. 2 (1883); *Cope, Proc. U. S. Nat. Mus.* xiv. 1892, p. 600.

Homalosoma episcopum, *Jan, Arch. Zool. An. Phys.* ii. 1862, p. 35, *and Icon. Gén.* 13, pl. iv. fig. 2 (1865).

Rostral a little broader than deep, the portion visible from above about half as long as its distance from the frontal; nasal undivided; suture between the internasals shorter than that between the præfrontals; frontal about once and a half as long as broad, broader than the supraocular, longer than its distance from the end of the snout, a little shorter than the parietals; loreal small, longer than deep; one præ- and two postoculars; temporals 1+2 or 1+1; seven upper labials, third and fourth entering the eye; four lower labials in contact with the anterior chin-shields; posterior chin-shields very small. Scales in 15 rows. Ventrals 145–163; anal divided; subcaudals 35–57. Uniform yellowish or greenish brown, each scale with a lighter margin; lower parts uniform yellowish white.

Total length 250 millim.; tail 60.

Texas, North Mexico.

a–b. ♂ (V. 145; C. 57) & hgr. (V. 148; C. 48).   Nuevo Leon.   W. Taylor, Esq. [P.].

14. Contia torquata.

Contia episcopa torquata, *Cope, Bull. U.S. Nat. Mus.* no. 17, 1880, p. 21.

Like *C. episcopa*, but ventrals 183. Subcaudals 38. Ground-colour light yellow tinged with brown above; three median dorsal rows of scales orange; top of head, from anterior border of frontal to end of parietals, black; a black cross band on the nape; sometimes a dark stripe along each side of the body; lips and below immaculate.

N.W. Texas.

15. Contia isozona.

Contia isozona, *Cope, Proc. Acad. Philad.* 1866, p. 304; *Yarrow, Wheeler's Rep. Explor. W.* 100*th Mer.* v. p. 537, pl. xviii. fig. 1 (1875); *Garm. N. Am. Rept.* p. 92 (1883).
—— episcopa isozona, *Cope, Bull. U.S. Nat. Mus.* no. 17, p. 21 (1880).

Rostral a little broader than deep, the portion visible from above measuring about half its distance from the frontal; nasal undivided; suture between the internasals shorter than that between the præfrontals; frontal longer than its distance from the end of the snout, shorter than the parietals; a small loreal, longer than deep; one præ- and two postoculars; temporals 1+2; seven upper labials, third and fourth entering the eye; three lower labials in contact with the anterior chin-shields; posterior chin-shields very small and separated by a scale. Scales in 15 rows. Ventrals 158–167; anal divided; subcaudals 50–52. Orange above, with black cross bands; tail with complete annuli; loreal and postocular region, posterior half of frontal and supraoculars, and anterior half of parietals black; sides and lower surface whitish.

Total length 250 millim; tail 50.

Utah, Nevada, Arizona, Texas.

a. ♀ (V. 159; C. 50).   Fort Yuma, Arizona.   Hr. A. Forrer [C.].

16. Contia occipitalis.

Rhinostoma occipitale, *Hallow. Proc. Ac. Philad.* 1854, p. 95.
Lamprosoma occipitale, *Hallow. Proc. Ac. Philad.* 1856, p. 110; *Baird, U.S. Mex. Bound. Surv.* ii. *Rept.* p. 21, pl. xxi. fig. 1 (1859); *Bocourt, Miss. Sc. Mex., Rept.* p. 558, pl. xxxiv. fig. 6 (1883).
Chionactis occipitalis, *Cope, Check-List N. Am. Rept.* p. 35 (1875), and *Proc. U.S. Nat. Mus.* xiv. 1892, p. 605.
Contia occipitalis, *Garm. N. Am. Rept.* p. 91 (1883).

Rostral a little broader than deep, the portion visible from above about half as long as its distance from the frontal; nasal undivided; suture between the internasals shorter than that between the præfrontals; frontal not once and a half as long as broad, broader than the supraoculars, longer than its distance from the end of the snout, shorter than the parietals; loreal very small, longer than deep; one præ- and two postoculars; temporals 1+2; seven upper labials, third and fourth entering the eye; four lower labials in contact with the anterior chin-shields; posterior chin-shields very small and separated from each other. Scales in 15 rows. Ventrals 147–158 (172?); anal divided; subcaudals 34–44. Yellowish above, reddish in the middle, with black transverse bands forming complete annuli round the tail; white inferiorly.

Total length 285 millim.; tail 56.

Arizona and Texas.

17. Contia pachyura.

Contia pachyura, *Cope, Journ. Ac. Philad.* viii. 1876, p. 145.

Frontal rather elongate, but shorter than the parietals; loreal large, deeper than long; one præ- and two postoculars; temporals 1+1; eight upper labials, fourth and fifth entering the eye; chin-shields equal, rather elongate. Scales in 17 rows. Ventrals 133; subcaudals 50. Black, the lower lateral rows of scales with a rufous shade; scales of the first row with grey tips; head blackish brown, with a black line from eye above labials; latter yellowish, unspotted; belly yellowish, each shield with a black extremity.

Total length 335 millim.

Sipurio, Costa Rica.

18. Contia mitis.

Contia mitis, *Baird & Gir. Cat. N. Am. Rept.* p. 110 (1853); *Girard, U.S. Explor. Exped., Herp.* p. 125, pl. x. figs. 6–12 (1858), *and Rep. U.S. Explor. R. R.* x. pt. iii. pl. xxxvi. fig. 7 (1859); *Cope, Proc. Ac. Philad.* 1861, p. 74; *Garm. N. Am. Rept.* p. 93 (1883); *Bocourt, Miss. Sc. Mex., Rept.* p. 557, pl. xxxiv. fig. 3 (1883).
Ablabes purpureocauda, *Günth. Cat.* p. 245 (1858).
Homalosoma mite, *Jan, Arch. Zool. Anat. Phys.* ii. 1862, p. 35, *and Icon. Gén.* 13, pl. iv. fig. 1 (1865).

Rostral broader than deep, just visible from above; nasal divided or semidivided; suture between the internasals as long as or shorter than that between the præfrontals; frontal not once and a half as long as broad, longer than its distance from the end of the snout, shorter than the parietals; loreal small, longer than deep; one præ- and two postoculars; temporals 1+2; seven upper labials, third and fourth entering the eye; four lower labials in contact with the anterior chin-shields; posterior chin-shields very small and in contact with each other. Scales in 15 rows. Ventrals 147–167; anal divided; subcaudals 29–39. Brown above, sides darker; a yellowish line along each side of the back; a blackish streak on each side of the head, passing through the eye, and a black spot on the rostral; ventrals whitish, with a broad black anterior border.

Total length 310 millim.; tail 40.

California.

a. ♂ (V. 147; C. 36).		California.	Smithsonian Institution. (As typical of *C. mitis*.)
b-c. ♂ (V. 160; C. 35) & ♀ (V. 166; C. 32).		California.	Mr. Bridges [C.]. (Types of *A. purpureocauda*.)
d. Yg. (V. 151; C. 34).		Murphy, California.	Christiania Mus.
e. ♀ (V. 161; C. 29).		Vancouver Id.	H.M.S. 'Plumper.'

19. Contia semiannulata.

Sonora semiannulata, *Baird & Gir. Cat. N. Am. Rept.* p. 117 (1853), and *Rep. U. S. Explor. R. R.* x. pt. iii. pl. xxxiii. fig. 88 (1859), and *U.S. Mex. Bound. Surv.* ii. *Rept.* pl. xix. fig. 3 (1859).
Contia semiannulata, *Garm. N. Am. Rept.* p. 90 (1883).

Nasal divided; frontal broader than the supraocular; loreal longer than deep; one præocular; three postoculars, upper largest and slightly touching the frontal; temporals 1+2; seven upper labials, third and fourth entering the eye. Scales in 15 rows. Ventrals 149; anal divided; subcaudals 39. Orange above, with black cross bands, which form annuli round the tail; sides and belly dull green.

Total length 240 millim.; tail 50.

Sonora.

According to Cope (Proc. U.S. Nat. Mus. xiv. 1892, p. 601) this species may prove to be the same as *C. isozona*, the bad state and abnormal condition of Baird and Girard's specimens possibly accounting for the differences on which the two have been separated.

20. Contia nasus.

Conopsis nasus, *Günth. Cat.* p. 6 (1858); *Peters, Mon. Berl. Ac.* 1869, p. 875; *Bocourt, Miss. Sc. Mex., Rept.* p. 563, pl. xxxv. fig. 2 (1883); *Günth. Biol. C.-Am., Rept.* p. 97, pl. xxxiv. fig. B (1893)
Oxyrhina varians, *Jan, Arch. Zool. Anat. Phys.* ii. 1862, p. 60.
—— (Exorhina) maculata, *Jan, l. c.* p. 61, *and Icon. Gén.* 48, pl. ii. figs. 2-4 (1876).
Conopsis maculatus, *Bocourt, l. c.* p. 564, pl. xxxv. fig. 3.
—— lineatus, *Bocourt, l. c.* p. 565, pl. xxxv. fig. 4, *and Bull. Soc. Zool. Fr.* 1892, p. 41.
—— varians, *Bocourt, l. c.* p. 566, pl. xxxv. fig. 5; *Dugès, La Naturaleza*, (2) i. 1888, p. 123.
Ficimia nasus, *Garm. N. Am. Rept.* p. 83 (1883).
—— maculata, *Garm. l. c.* p. 84.
Chionactis diasii, *Cope, Proc. U.S. Nat. Mus.* ix. 1886, p. 188.

Snout prominent, more or less pointed. Rostral large, the portion visible from above as long as its distance from the frontal; internasals often fused with the præfrontals; nasal entire or semidivided; frontal longer than its distance from the end of the snout, as long as the parietals, once and a half to once and two thirds as broad as the supraocular; a small square loreal present or absent; one præ- and two (rarely one) postoculars; temporals 1+2; seven upper labials, third and fourth entering the eye; anterior chin-shields short and in contact with three or four labials; second pair very small. Scales in 17 rows. Ventrals 110-137; anal divided; subcaudals 23-40. Grey or pale brown above, with dark brown (sometimes light-edged) spots or cross bands, or with a series of small dark spots forming a vertebral stripe, which may be accompanied by another more or less distinct stripe along each side; upper surface of head speckled with dark brown, sometimes with symmetrical markings; belly yellowish or red, uniform or with blackish dots or square black spots.

Total length 290 millim.; tail 50.

Mexico.

A. Internasals distinct.

a-e. ♂ (V. 123; C. 36) & ♀ (V. 137, 133, 135, 128; C. 29, 31, 28, 28).	La Cumbre de los Arrastrados, Talpa Mascota, Jalisco.	Dr. A. C. Buller [C.].
f-g. Yg. (V. 135, 122; C. 30, 26).	Puebla.	Paris Museum [E.].
h-l. ♀ (V. 129, 133, 131; C. 30, 30, 31) & yg. (V. 125; C. 29).	S. Mexico.	Mr. H. H. Smith [C.]. F. D. Godman, Esq. [P.].
m. ♂ (V. 131; C. 35).	Mexico.	

B. Internasal distinct on one side, fused with the præfrontal on the other.

n-o. Yg. (V. 128, 122; C. 27, 39).	La Cumbre de los Arrastrados.	Dr. A. C. Buller [C.].

C. Internasals fused with the præfrontals.

p-r. ♂ (V. 128, 125; C. 38, 37) & ♀ (V. 133; C. 29).	La Cumbre de los Arrastrados.	Dr. A. C. Buller [C.].
s-w. ♂ (V. 115, 119; C. 33, 36), ♀ (V. 130, 129; C. 23, 26), & yg. (V. 129; C. 28).	Omilteme, Guerrero.	Mr. H. H. Smith [C.]. F. D. Godman, Esq. [P.].
x. ♂ (V. 132; C. 36).	Milpas, Durango.	Hr. A. Forrer.
y-δ. ♂ (V. 131, 128; C. 40, 39) & ♀ (V. 129, 128, 130, 132; C. 31, ?, 33, 33).	Mexico.	M. Sallé [C.]. (Types.)
ε. ♂ (V. 131; C. 35).	California (?).	Mr. Bridges [C.].

Toluca lineata, Kennicott, U.S. Mex. Bound. Surv. ii. Rept. p. 23, pl. xxi. fig. 2 (1859), and Rep. U.S. Explor. R. R. x. pt. iii. pl. xxxv. fig. 8 (1859), = *Oxyrhina (Achirhina) defilippii*, Jan, Arch. Zool. Anat. Phys. ii. 1862, p. 61, are based on what are, in all probability, anomalous specimens of *C. nasus*, in which the frontal shield is so produced anteriorly as to separate the præfrontals.

21. Contia frontalis.

Toluca frontalis, *Cope, Proc. Ac. Philad.* 1864, p. 167.
Ficimia olivacea, part., *Peters, Mon. Berl. Ac.* 1869, p. 875.
Pseudoficimia pulchra, *Bocourt, Miss. Sc. Mex., Rept.* p. 572, pl. xxxv. fig. 12 (1883).
Ficimia frontalis, *Garm. N. Am. Rept.* p. 82 (1883).
Geagras frontalis, *Cope, Am. Nat.* 1884, p. 163.
Pseudoficimia frontalis, *Günth. Biol. C.-Am., Rept.* p. 96 (1893).

Snout short, projecting, pointed, slightly turned up at the end, concave above. Rostral large, its upper portion as long as or a little shorter than its distance from the frontal; nasal divided; internasals shorter than the præfrontals; frontal nearly as broad as long, as long as or a little shorter than its distance from the end of the snout, a little shorter than the parietals; supraocular about one third the width of the frontal; no loreal; one præ- and two postoculars; temporals 1+2; seven upper labials, second or second and third in contact with the præfrontal, third and fourth entering the eye; three or four lower labials in contact with the anterior chin-shields; posterior chin-shields very small. Scales in 17 rows. Ventrals 141–160; anal divided; subcaudals 38–48. Pale brown above, yellowish along the middle of the back, with three longitudinal series of brown, black-edged spots, those of the middle series largest, broader than long; a brown black-edged cross bar between the eyes, and vermiculations of the same colour on the back of the head; upper lip whitish, with some vertical brown spots; lower parts uniform white.

Total length 245 millim.; tail 47.

Mexico.

a. ♂ (V. 150; C. 48).   Ventanas, Durango.   Hr. A. Forrer [C.].
b. Hgr. (V. 153; C. 47).   Presidio, nr. Mazatlan.   Hr. A. Forrer [C.].

100. **FICIMIA.**

Ficimia, *Gray, Cat. Sn.* p. 80 (1849); *Jan, Arch. Zool. Anat. Phys.* ii. 1862, p. 58; *Bocourt, Miss. Sc. Mex., Rept.* p. 569 (1883); *Günth. Biol. C.-Am., Rept.* p. 98 (1893).
Amblymetopon, *Günth. Cat. Col. Sn.* p. 7 (1858).
Gyalopion, *Cope, Proc. Ac. Philad.* 1861, p. 243, *and Proc. U.S. Nat. Mus.* xiv. 1892, p. 603.

Maxillary teeth 13 to 15, small, equal; mandibular teeth small, equal. Head not distinct from neck; snout projecting, pointed; eye small, with round pupil; rostral very large, in contact with or approaching the frontal; internasals small or absent; nostril pierced in a divided or semidivided nasal, the anterior portion of which is fused with the first labial; no loreal, præfrontal in contact with the labials. Body cylindrical, short; scales smooth, with apical pits, in 17 rows; ventrals rounded. Tail short; subcaudals in two rows.

Arizona, Texas, Mexico.

Synopsis of the Species.

Rostral shield forming a suture with the
  frontal .... 1. *olivacea*, p. 271.
Rostral shield separated from the frontal;
  seven well-developed upper labials .. 2. *cana*, p. 272.
Rostral shield separated from the frontal; [p. 272.
  seventh upper labial very small .... 3. *quadrangularis*,

1. Ficimia olivacea.

Ficimia olivacea, *Gray, Cat.* p. 80 (1849); *Steind. Sitzb. Ak. Wien,* lxii. i. 1870, p. 344, pl. vi.; *Bocourt, Miss. Sc. Mex., Rept.* p. 570, pl. xxxv. fig. 11 (1883); *Günth. Biol. C.-Am., Rept.* p. 98, pl. xxxv. figs. B & C (1893).
Amblymetopon variegatum, *Günth. Cat.* p. 7 (1858).
Ficimia elaiocroma, *Jan, Arch. Zool. Anat. Phys.* ii. 1862, p. 58.
—— publia, *Cope, Proc. Ac. Philad.* 1866, p. 126, and *Am. Nat.* 1884, p. 163; *Günth. Biol. C.-Am., Rept.* p. 98.
—— variegata, *Cope, l. c.*
—— olivacea, part., *Peters, Mon. Berl. Ac.* 1869, p. 875.
—— ornata, *Bocourt, l. c.* p. 571, pl. xxxv. fig. 10.

Snout short, concave above, and strongly turned up at the end. Rostral forming a broad suture with the frontal, separating the præfrontals; internasals small or absent; frontal as broad as long, as long as its distance from the end of the snout, nearly as long as the parietals; supraocular about one third the width of the frontal; one præ- and two postoculars; temporals 1+1 or 1+2; seven upper labials, second in contact with the præfrontal, third and fourth entering the eye; three lower labials in contact with the anterior chin-shields; posterior chin-shields very small, scale-like. Scales in 17 rows. Ventrals 138–160; anal divided; subcaudals 34–42.

Total length 390 millim.; tail 65.

Mexico.

A. Uniform dark brown above, white below. (*F. olivacea*, Gray.)

a–b. ♂ (V. 152, 150;   Mexico.   Mr. Hugo Finck [C.].
C. 42, 41).   (Types.)

In one of the specimens the internasals are fused with the præfrontals, whilst in the other the internasal is distinct on one side.

B. Yellowish or pale brown above, with dark brown transverse spots or cross bars; sides with dark brown spots; head with symmetrical dark markings; lower parts uniform white.

  a. Internasals distinct. (*F. publia*, Cope.)

a. Yg. (V. 140; C. 36).   Cuernavaca.   M. Boucard [C.].
b. Yg. (V. 142; C. 36).   Yucatan.

b. Internasals fused with præfrontals. (*A. variegatum,* Gthr.)

c. Yg. (V. 160; C. 37).	Mexico.	M. Sallé [C.].
d. ♀ (V. 149; C. 36).	Mexico.	(Types of *A. variegatum.*)
e. Hgr. ♀ (V. 142; C. 35).	Mexico.	

2. Ficimia cana.

Gyalopion canum, *Cope, Proc. Ac. Philad.* 1860, p. 243; *Yarrow, Wheeler's Rep. U.S. Explor. Surv. W.* 100*th Mer.* v. p. 624, pl. xviii. fig. 2 (1875).

Ficimia cana, *Garm. N. Am. Rept.* p. 83 (1883); *Günth. Biol. C.-Am., Rept.* p. 99 (1893).

Closely allied to the preceding, but rostral pointed behind and wedged in between the præfrontals; internasals present, small; seven upper labials, as in the preceding. Ventrals 130–131; subcaudals 28. Pale yellowish or reddish brown above, with brown, dark-edged cross bars, which are broken up into spots on the sides; a large triangular dark spot on the nape, with the point between the parietals; a dark band across the head, to the lower labials, passing through the eyes across the interocular region; lower parts uniform white.

Total length 205 millim.; tail 28.

Arizona and Western Texas.

a. ♀ (V. 131; C. 28). El Paso, Texas. Hr. A. Forrer [C.].

3. Ficimia quadrangularis.

Ficimia quadrangularis, *Günth. Biol. C.-Am., Rept.* p. 99, pl. xxxv. fig. A (1893).

Scaling as in *F. cana,* but seventh upper labial very small, scale-like. Ventrals 129; subcaudals 23. Ground-colour nearly white, each scale with a narrow brown margin; a series of large sub-quadrangular blackish-brown spots along the back; no spots on the sides; a large blackish band on the nape, confluent with a large blackish blotch covering the frontal and supraocular shields and the greater part of the parietals; lower parts white.

Total length 320 millim.; tail 37.

Western Mexico.

a. ♀ (V. 129; C. 23). Presidio, near Mazatlan. Hr. A. Forrer [C.]. (Type.)

101. CHILOMENISCUS.

Chilomeniscus, *Cope, Proc. Ac. Philad.* 1860, p. 339, and 1861, p. 302, and *Proc. U.S. Nat. Mus.* xiv. 1892, p. 593.

Maxillary teeth equal or the posterior a little stouter. Head not distinct from neck; snout rounded, very prominent, and much depressed; eye small, with round pupil; rostral large; no loreal;

nasal single, confluent with the internasal. Body cylindrical, short; scales smooth, with apical pits, in 13 rows; ventrals rounded. Tail short; subcaudals in two rows.

Lower California, Sonora, Arizona.

1. Chilomeniscus stramineus.

Chilomeniscus stramineus, *Cope, Proc. Ac. Philad.* 1860, p. 339, and *Proc. U.S. Nat. Mus.* xiv. 1892, p. 594.
—— cinctus, *Cope, Proc. Ac. Philad.* 1861, p. 303.
Carphophis straminea, *Garm. N. Am. Rept.* p. 99 (1883).
—— cincta, *Garm. l. c.* p. 100.
Chilomeniscus stramineus fasciatus, *Cope, Proc. U.S. Nat. Mus.* xiv. 1892, p. 595.

Internasals separated by the rostral or forming a short suture; frontal a little longer than broad, nearly as long as the parietals; supraocular very small; nasal not reaching the single præocular; two postoculars; temporals 1+1; seven upper labials, third and fourth entering the eye; anterior chin-shields in contact with the symphysial, twice as long as the posterior. Scales in 13 rows. Ventrals 108–117; anal divided; subcaudals 21–26. Reddish brown above, dotted with dark brown, or with black cross bars which may form complete annuli.

Total length 235 millim.; tail 33.

Lower California and Sonora.

2. Chilomeniscus ephippicus.

Chilomeniscus ephippicus, *Cope, Proc. Ac. Philad.* 1875, p. 85; *Coues, Wheeler's Rep. Explor. Surv. W. 100th Mer.* v. p. 625, pl. xviii. fig. 3 (1875).

Rostral entirely separating the internasals; frontal a little longer than broad, nearly as long as the parietals; nasal in contact with the præocular; two postoculars; temporals 1+1; seven upper labials, third and fourth entering the eye; anterior chin-shields separated from the symphysial, posterior very small. Scales in 13 rows. Ventrals 113; anal divided; subcaudals 28. Reddish or yellow above, with black cross bars which do not quite reach the ventrals; tail with nearly complete black rings; snout red, rest of head black; belly white.

Total length 140 millim.; tail 20.

Nevada and Arizona.

102. HOMALOSOMA.

Duberria, part., *Fitzing. N. Class. Rept.* p. 29 (1826).
Homalosoma, *Wagl. Syst. Amph.* p. 190 (1830); *Dum. & Bibr. Erp. Gén.* vii. p. 109 (1854); *Günth. Cat. Col. Sn.* p. 19 (1858); *Peters, Reise n. Mossamb.* iii. p. 107 (1882).
Calamaria, part., *Schleg. Phys. Serp.* ii. p. 25 (1837).
Homalosoma, part., *Jan, Arch. Zool. Anat. Phys.* ii. 1862, p. 33.

Maxillary short, with 10 to 12 teeth; maxillary and mandibular teeth small, anterior longest. Head small, not distinct from neck; eye small, with round pupil; nostril in a single nasal; loreal small or absent. Body cylindrical, short; scales smooth, with apical pits, in 15 rows; ventrals rounded. Tail short, subcaudals in two rows.

Africa.

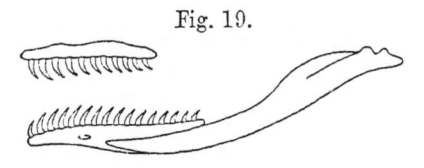

Fig. 19.

Maxillary and mandible of *Homalosoma lutrix*.

Synopsis of the Species.

I. Ventrals 115–144.

    A. Nasal not in contact with præocular; usually two postoculars; frontal as long as parietals .... 1. *lutrix*, p. 274.

    B. Nasal in contact with præocular; one postocular; frontal shorter than parietals.

Frontal much longer than broad; suture between the internasals longer than that between the præfrontals........ 2. *shiranum*, p. 276.

Frontal scarcely longer than broad; suture between the internasals shorter than that between the præfrontals ...... 3. *abyssinicum*, p. 276.

II. Ventrals 97–110; nasal not in contact with præocular; two postoculars; frontal as long as parietals.

                                          4. *variegatum*, p. 276.

1. Homalosoma lutrix.

Coluber lutrix, *Linn. S. N.* i. p. 375 (1766).
—— duberria (*Klein*), *Merr. Beitr.* p. 7, pl. i. (1790).
Elaps duberria, *Schneid. Hist. Amph.* ii. p. 297 (1801).
Coluber tetragonus, *Latr. Rept.* iv. p. 97 (1802); *Daud. Rept.* vii. p. 207 (1803).
—— arctiventris, *Daud. t. c.* p. 221; *Kuhl, Beitr.* p. 82 (1820).
Duberria arctiventris, *Fitzing. N. Class. Rept.* p. 55 (1826).
Calamaria arctiventris, *Schleg. Phys. Serp.* ii. p. 36, pl. i. figs. 24–26 (1837).
Homalosoma arctiventris, *Smith, Ill. Zool. S. Afr., Rept., App.* p. 16 (1849).
—— lutrix, *Dum. & Bibr.* vii. p. 110 (1854); *Günth. Cat.* p. 20 (1858); *Jan, Arch. Zool. Anat. Phys.* ii. 1862, p. 33, and *Icon. Gén.* 13, pl. iii. fig. 3 (1865).
Cyclophis catenatus, *Theob. Cat. Rept. As. Soc. Mus.* 1868, p. 49.

Rostral broader than deep, visible from above; suture between the internasals a little shorter than or nearly as long as that between the præfrontals; frontal once and a half to twice as long as broad, longer than its distance from the end of the snout, as long as the parietals, once and a half to once and two thirds as broad as the supraocular; nostril in the anterior half of the nasal, which is never in contact with the præocular; a small square loreal usually present; one præ- and two (rarely one) postoculars; temporals 1+2; six upper labials, third and fourth entering the eye; two pairs of chin-shields in contact with each other, the anterior in contact with three or four labials. Scales in 15 rows. Ventrals 115-144; anal entire; subcaudals 21-46. Coloration very variable; belly and lower surface of tail uniform yellowish in the middle and greyish on the sides, usually with a regular lateral series of black dots.

Total length 390 millim.; tail 65.

South Africa.

A. Pale brown or yellowish above, with a vertebral series of black dots; greyish olive or plumbeous on the sides, which are limited above by a series of black dots; lateral ventral dots large.

a-e. ♂ (V. 122, 127; C. 35, 40), hgr. (V. 127, 127; C. 34, 33), & yg. (V. 119; C. 31).   S. Africa.   Sir J. Macgrigor [P.].

f-g. ♀ (V. 134, 126; C. 28, 33).   S. Africa.

h. ♂, skeleton.   S. Africa.

B. Reddish brown above, blackish olive on the sides, greyish towards the belly; a fine black vertebral line; lateral ventral dots large.

i. ♀ (V. 126; C. 27).   S. Africa.

C. Like the preceding, but the vertebral line and the ventral dots indistinct.

k. ♀ (V. 118; C. 37).   Port Natal.   Mr. T. Ayres [C.].

D. Uniform olive-brown above, passing to greyish on the sides; the lateral ventral dots very small.

l. ♂ (V. 115; C. 45).   S. Africa.

E. Reddish brown or brick-red above, grey on the sides; lateral ventral dots well marked.

m. ♂ (V. 118; C. 42).   S. Africa.
n. ♀ (V. 128; C. 37).   Port Elizabeth.   Mr. J. L. Drege [P.].

F. Uniform reddish brown above and on the sides; lateral ventral dots well marked.

o. ♂ (V. 125; C. 46). Port Elizabeth. J. M. Leslie, Esq. [P.].

2. Homalosoma shiranum. (Plate XIII. fig. 1.)

Homalosoma lutrix (*non* L.), *Günth. Proc. Zool. Soc.* 1892, p. 555.

Rostral broader than deep, visible from above; suture between the internasals thrice as long as that between the præfrontals; frontal once and two thirds as long as broad, longer than its distance from the end of the snout, shorter than the parietals, nearly twice as broad as the supraocular; nostril in the anterior half of the nasal; no loreal, nasal forming a suture with the præocular; one præ- and one postocular; temporals 1+2; six upper labials, third and fourth entering the eye; two pairs of short chin-shields in contact with each other, the anterior in contact with three labials. Scales in 15 rows. Ventrals 142; anal entire; subcaudals 33. Uniform olive-brown above; belly yellowish white in the middle; outer ends of ventrals and lower rows of scales grey.

Total length 310 millim.; tail 47.

Nyassaland.

a. ♀ (V. 142; C. 33). Shiré Highlands. H. H. Johnston, Esq. [P.].

3. Homalosoma abyssinicum. (Plate XIII. fig. 2.)

Homolosoma lutrix (*non* L.), *Blanf. Zool. Abyss.* p. 458 (1870).

Rostral broader than deep, visible from above; suture between the internasals hardly half as long as that between the præfrontals; frontal scarcely longer than broad, as long as its distance from the end of the snout, shorter than the parietals, a little broader than the supraocular; nostril in the anterior half of the nasal; no loreal, nasal forming a suture with the præocular; one præ- and one postocular; temporals 1+2; six upper labials, third and fourth entering the eye; two pairs of chin-shields in contact with each other, the anterior in contact with three labials. Scales in 15 rows. Ventrals 120; anal entire; subcaudals 32. Olive-brown above, with a fine, interrupted, black vertebral line; sides black, dotted with whitish; belly greyish olive, spotted and freckled with black on the sides.

Total length 235 millim.; tail 43.

Abyssinia.

a. ♂ (V. 120; C. 32). Ashangi. W. T. Blanford, Esq. [P.].

4. Homalosoma variegatum.

Homalosoma variegatum, *Peters, Mon. Berl. Ac.* 1854, p. 622, *and Reise n. Mossamb.* iii. p. 107, pl. xvi. fig. 1 (1882); *Bouleng. Ann. & Mag. N. H.* (6) ii. 1888, p. 140.

Eye rather larger than in *H. lutrix*. Rostral broader than long, visible from above; suture between the internasals a little longer than that between the præfrontals; frontal once and a half as long as broad, longer than its distance from the end of the snout, as long as the parietals, nearly twice as broad as the supraocular; nostril just anterior to the middle of the nasal; a small loreal, nearly as deep as long; one præ- and two postoculars; temporals 1+2; six or seven upper labials, third and fourth or fourth and fifth entering the eye; two pairs of chin-shields in contact with each other, the anterior in contact with three labials. Scales in 15 rows. Ventrals 97–110; anal entire; subcaudals 25–36. Belly reticulated black and whitish.

Total length 250 millim.; tail 33.

South-east Africa (Inhambane and Delagoa Bay).

A. Dark brown above, with irregular lichen-like brownish-white variegation.

a. ♀ (V. 110; C. 25).　　Delagoa Bay.　　S.-African Museum [P.].

B. Olive-brown above, with three series of dark brown spots.

b. ♂ (V. 97; C. 36).　　Delagoa Bay.　　S.-African Museum [P.].

103. ABLABES.

Coronella, part., *Schleg. Phys. Serp.* ii. p. 50 (1837).
Herpetodryas, part., *Schleg. l. c.* p. 173.
Ablabes, part., *Dum. & Bibr. Erp. Gén.* vii. p. 304 (1854); *Günth. Cat. Col. Sn.* p. 27 (1858), *and Rept. Brit. Ind.* p. 223 (1864).
Cyclophis, part., *Günth. ll. cc.* pp. 119, 229.
Eurypholis, *Hallow. Proc. Ac. Philad.* 1860, p. 493.
Liopeltis, part., *Cope, Proc. Ac. Philad.* 1860, p. 559; *Jan, Elenco sist. Ofid.* p. 81 (1863).
Homalosoma, part., *Jan, Arch. Zool. Anat. Phys.* ii. 1862, p. 33.
Diadophis, part., *Jan, t. c.* 1863, p. 261.
Cyclophiops, *Bouleng. Ann. Mus. Genova*, (2) vi. 1888, p. 599.
Ablabes, *Bouleng. Faun. Ind., Rept.* p. 304 (1890).

Maxillary teeth small, equal, 15 to 30; mandibular teeth subequal. Head not or scarcely distinct from neck; eye rather small or moderate, with round pupil; loreal present or absent; nasal entire or divided. Body cylindrical, usually slender; scales smooth or feebly keeled, without apical pits, in 13 to 17 rows; ventrals not angulate laterally. Tail moderate or long; subcaudals in two rows.

South-eastern Asia and Japan.

Synopsis of the Species.

I. Scales in 15 rows.

   A. Scales feebly keeled .... 1. *semicarinatus*, p. 278.

   B. Scales smooth.

     1. Seven or eight upper labials.

       α. Loreal present; nasal divided or semidivided.

Rostral as deep as broad or a little broader than deep; anal divided .... 2. *major*, p. 279.
Rostral as deep as broad; anal entire .. 3. *doriæ*, p. 279.
Rostral much broader than deep ...... 4. *frenatus*, p. 280.

       β. Loreal present; nasal single . 5. *stoliczkæ*, p. 281.

       γ. No loreal; nasal single.

Snout twice as long as the eye .... 6. *tricolor*, p. 281.
Snout not twice as long as the eye .... 7. *calamaria*, p. 282.

     2. Six upper labials; nasal divided; loreal present.
                                                   8. *rappii*, p. 282.

II. Scales in 13 rows.

   A. Tail not more than one third of the total length.

Seven upper labials, third and fourth entering the eye .... 9. *baliodirus*, p. 283.
Eight upper labials, fourth and fifth entering the eye .... 10. *scriptus*, p. 284.

   B. Tail more than one third of the total length; subcaudals 77–103...................... 11. *longicauda*, p. 284.

III. Scales in 17 rows .... 12. *nicobariensis*, p. 285.

1. Ablabes semicarinatus.

Eurypholis semicarinatus, *Hallow. Proc. Ac. Philad.* 1860, p. 493.
Cyclophis nebulosus, *Günth. Ann. & Mag. N. H.* (4) i. 1868, p. 418, pl. xix. fig. C.
Ablabes semicarinatus, *Bouleng. Proc. Zool. Soc.* 1887, p. 148.

Rostral a little broader than deep, just visible from above; nasal divided; eye a little more than half the length of the snout; suture between the internasals much shorter than that between the præfrontals; frontal as long as or a little longer than its distance from the end of the snout, shorter than the parietals; loreal about twice as long as deep; one præ- and two postoculars; temporals 1+2; eight upper labials, fourth and fifth entering the eye; four lower labials in contact with the anterior chin-shields, which are a little shorter than the posterior. Scales in 15 rows, with a very feeble keel along their anterior half. Ventrals 174-192; anal divided; subcaudals 70-82. Olive above, the scales lighter in the centre; the lower scale on each side yellowish, margined with olive or black; young with dark spots anteriorly and four dark longitudinal bands on the hinder half of the body, which disappear or become rather indistinct in the adult; upper lip and lower parts uniform yellowish.

Total length 770 millim.; tail 160.

Loo Choo Islands and Japan.

a-e. ♂ (V. 190, 190, 190; C. 82, 76, ?) & ♀ (V. 192, 187; C. 70, 72).	Loo Choo Ids.	H. Pryer, Esq. [P.].
f. Yg. (V. 185; C. 76).	Great Loo Choo Island.	Mr. Holst [C.].
g. Yg. (V. 174; C. 74).	Nagasaki.	Mr. Whitely [C.]. (Type of *C. nebulosus.*)

2. Ablabes major.

Cyclophis major, *Günth. Cat.* p. 120 (1858), *and Rept. Brit. Ind.* p. 230, pl. xvii. fig. I (1864).
Herpetodryas chloris, *Hallow. Proc. Ac. Philad.* 1860, p. 503.
Ablabes major, *Boettg. Ber. Senckenb. Ges.* 1894, p. 140.

Rostral as deep as broad or a little broader than deep, just visible from above; nasal divided; eye about two thirds the length of the snout; suture between the internasals shorter than that between the præfrontals; frontal as long as or a little longer than its distance from the end of the snout, shorter than the parietals; loreal longer than deep; one præ- and two postoculars; temporals 1+2; eight upper labials, fourth and fifth entering the eye; four lower labials in contact with the anterior chin-shields, which are broader and a little longer than the posterior. Scales in 15 rows. Ventrals 163-177; anal divided; subcaudals 70-87. Uniform green above, yellow beneath.

Total length 960 millim.; tail 250.

China.

a. ♀ (V. 163; C. 71).	Near Ningpo.	(Type)
b. ♀ (V. 170; C. 82).	Chusan Archipelago.	J. J. Walker, Esq. [P.].
c. ♂ (V. 164; C. 79).	Shanghai.	R. Swinhoe, Esq. [C.].
d-e. ♀ (V. 170; C. 87) & yg. (V. 166; C. 82).	Mountains North of Kiu Kiang.	A. E. Pratt, Esq. [C.].
f-i. ♀ (V. 170, 168, 176, 171; C. 86, 86, 85, 83).	Island of Formosa.	R. Swinhoe, Esq. [C.].
k-m. ♀ (V. 168, 171, 169; C. 70, 76, 77).	China.	

3. Ablabes doriæ.

Cyclophiops doriæ, *Bouleng. Ann. Mus. Genova,* (2) vi. 1888. p. 599, pl. vi. fig. 1.

Ablabes doriæ, *Bouleng. Faun. Ind., Rept.* p. 306 (1890); *W. L. Sclater, Journ. As. Soc. Beng.* lx. 1891, p. 235.

Snout convex, profile curved from the frontal region to the lip; rostral as deep as broad, visible from above; nostril between two nasals; internasals a little shorter than the præfrontals; frontal longer than its distance from the end of the snout, shorter than the parietals; loreal small, a little longer than deep; a large præocular, usually with a second, very small, below; two or three postoculars; temporals 1+2 or 2+2; eight upper labials, fourth and fifth entering the eye; four lower labials in contact with the anterior chin-shields; posterior chin-shields about half as large as the anterior. Scales in 15 rows. Ventrals 173–187; anal undivided; subcaudals 77–80. Uniform green above, white below, the green colour extending on to the ends of the ventrals.

Total length 910 millim.; tail 210.

Kakhyen Hills and Assam.

a. Yg. (V. 173; C. 77).　　　Kakhyen Hills.　　　M. L. Fea [C.].
　　　　　　　　　　　　　　　　　　　　　　　(One of the types.)

4. Ablabes frenatus.

Dipsas monticola (*non Cant.*), *Blyth, Journ. As. Soc. Beng.* xxiii. 1854, p. 294.
Cyclophis frenatus, *Günth. Cat.* p. 120 (1858), *and Rept. Brit. Ind.* p. 230, pl. xix. fig. 1 (1864); *Jerdon, Proc. As. Soc. Beng.* 1870, p. 80; *Anders. Proc. Zool. Soc.* 1871, p. 188; *Theob. Cat. Rept. Brit. Ind.* p. 157 (1876); *Blanf. Zool. E. Persia*, p. 408 (1876).
Ablabes frenatus, *Bouleng. Faun. Ind., Rept.* p. 306 (1890).

Snout short, convex; eye rather large, three fourths the length of the snout; rostral broader than deep, visible from above; nasal divided or semidivided, suture between the internasals shorter than that between the præfrontals; frontal longer than its distance from the end of the snout, about two thirds the length of the parietals; loreal small, as long as deep or a little longer; one præocular; two postoculars, only the upper in contact with the parietal; temporals 1+2, rarely 2+2; upper labials seven, third and fourth entering the eye; four lower labials in contact with the anterior chin-shields, which are a little shorter than the posterior. Scales in 15 rows. Ventrals 151–163; anal divided; subcaudals 87–96. Olive above, the scales on the anterior part of the body black-edged; a broad black stripe from the eye to the nape, gradually narrowing and disappearing on the anterior fourth of the body; four light longitudinal narrow lines more or less distinct on the front half of the body; lower surface uniform yellow.

Total length 690 millim.; tail 230.

Khasi Hills; Assam.

a. ♂ (V. 161; C. 95).　["Mountains near Af-　Dr. Griffith.　E. India
　　　　　　　　　　　　ghanistan."] Khasi　　　　　　　Co. [P.].　(Type.)
　　　　　　　　　　　　Hills.
b. Hgr. (V. 151; C. 87).　Khasi Hills.　　　　T. C. Jerdon, Esq. [P.].
c. ♂ (V. 163; C. 96).　[Mesopotamia.]

5. Ablabes stoliczkæ.

Ablabes stoliczkæ, *W. L. Sclater, Journ. As. Soc. Beng.* lx. 1891, p. 234, pl. vi. fig. 1; *Bouleng. Ann. Mus. Genova*, (2) xiii. 1893, p. 325.

Snout rather long and much depressed, twice as long as the diameter of the eye; rostral much broader than deep, visible from above; nasal undivided, thrice as long as deep; internasals a little shorter than the præfrontals; frontal not broader than the supraocular, nearly as long as its distance from the end of the snout, shorter than the parietals; a small loreal; one præ- and two postoculars; temporals 1+2; eight upper labials, fourth and fifth entering the eye; four lower labials in contact with the anterior chin-shields, which are as long as the posterior. Scales in 15 rows. Ventrals 150-154; anal divided; subcaudals 116-134. Pale olive-brown above; upper lip and lower parts yellowish white; a black streak from the eye along the temple and the side of the neck, separating the brown colour of the upper parts from the white of the lower.

Total length 600 millim.; tail 255.

Mountains of Burma and Assam.

6. Ablabes tricolor.

Herpetodryas tricolor, *Schleg. Phys. Serp.* ii. p. 187, pl. vi. figs. 16-18 (1837).
Cyclophis tricolor, *Günth. Cat.* p. 121 (1858), *and Proc. Zool. Soc.* 1872, p. 590; *Stoliczka, Journ. As. Soc. Beng.* xlii. 1873, p. 122.
Liopeltis tricolor, *Cope, Proc. Ac. Philad.* 1860, p. 559; *Jan, Icon. Gén.* 31, pl. vi. fig. 2 (1869).

Snout rather long and depressed, twice as long as the diameter of the eye; rostral broader than deep, visible from above; nasal undivided; suture between the internasals as long as or a little shorter than that between the præfrontals; frontal not broader than the supraocular, as long as its distance from the end of the snout, shorter than the parietals; no loreal, the præfrontal in contact with the second and third labials; one præ- and two postoculars; temporals 1+2; eight upper labials, fourth and fifth entering the eye: four or five lower labials in contact with the anterior chin-shields, which are a little shorter than the posterior. Scales in 15 rows. Ventrals 140-187; anal divided; subcaudals 103-130. Olive or greenish above; a black streak on each side of the head and anterior part of the body, passing through the eye; upper lip and lower parts yellowish white; a pale olive streak along each side of the belly.

Total length 560 millim.; tail 220.

Java, Sumatra, Borneo.

a. ♂ (V. 140; C. 118).	Java.	Leyden Museum. (One of the types.)
b. ♀ (V. 187; C. 108).	Java.	A. Scott, Esq. [P.].

c–e. ♂ (V. 153, 160; C. 125, 124) & ♀ (V. 171; C. 127).	Borneo.	L. L. Dillwyn, Esq. [P.].
f. ♂ (V. 158; C. 130).	Borneo.	A. Everett, Esq. [C.].
g. ♀ (V. 148; C. 103).	——?	Dr. Bleeker. (*Ablabes schlegelii*, Blkr.)

7. Ablabes calamaria.

Cyclophis calamaria, *Günth. Cat.* p. 250 (1858), *and Rept. Brit. Ind.* p. 231, pl. xvii. fig. K (1864).
Homalosoma baliolum, *Jan, Arch. Zool. Anat. Phys.* ii. 1862, p. 36, *and Icon. Gén.* 13, pl. iv. fig. 4 (1865).
Cyclophis nasalis, *Günth. Rept. Brit. Ind.* p. 231, pl. xvii. fig. M.
Ablabes calamaria, *Bouleng. Faun. Ind., Rept.* p. 305 (1890).

Rostral broader than deep, visible from above; nasal single and united with the loreal; suture between the internasals as long as that between the præfrontals or shorter; frontal longer than its distance from the end of the snout, slightly shorter than the parietals; one præocular (rarely two); two postoculars, only the upper in contact with the parietal; temporals 1+2; seven upper labials, third and fourth entering the eye; four lower labials in contact with the anterior chin-shields, which are larger than the posterior. Scales in 15 rows. Ventrals 130–154; anal divided; subcaudals 64–76. Pale brown above, with or without two more or less distinct black longitudinal lines; lower parts uniform yellowish.

Total length 410 millim.; tail 110.

Ceylon, Madras Presidency, Bombay.

a, b. ♂ (V. 130, 134; C. 75, 76).	Ceylon.	(Types.)
c. ♂ (V. 134; C. 68).	Ceylon.	R. Templeton, Esq. [P.].
d. ♀ (V. 136; C. 64).	Kotagiri, Nilgherries.	Dr. J. R. Henderson [P.].
e. Hgr. (V. 142; C. 64).	Sevagherry Ghat.	Col. Beddome [C.].
f. ♀ (V. 130; C. 69).	Madras Presidency.	Col. Beddome [C.].
g–i. ♀ (V. 154, 143, 154; C. 70, 68, 64).	Matheran.	Dr. Leith [P.].
k. ♀ (V. 149; C. 72).	——?	Dr. Günther [P.]. (Type of *C. nasalis*.)

8. Ablabes rappii.

Ablabes rappii, *Günth. Proc. Zool. Soc.* 1860, p. 154, pl. xxvi. fig. B, *and Rept. Brit. Ind.* p. 225 (1864); *Anders. Proc. Zool. Soc.* 1871, p. 171; *Bouleng. Faun. Ind., Rept.* p. 307 (1890).
—— owenii, *Günth. Proc. Zool. Soc.* 1860, p. 155, pl. xxvi. fig. A.

Rostral twice as broad as deep, just visible from above; nostril between two nasals; suture between the internasals a little shorter than that between the præfrontals; frontal slightly shorter than its distance from the end of the snout, a little shorter than the parietals; loreal as long as deep or a little longer than deep; one præocular; two postoculars, only the upper in contact with the parietal; temporals 1+1; six upper labials, third and fourth entering the eye; four lower labials in contact with the anterior chin-shields, which equal or a little exceed the posterior in length. Scales in 15 rows. Ventrals 178-195; anal divided; subcaudals 60-75. Brown above, with a broad dark collar and a double series of transverse dark spots on the anterior part of the body; these markings most distinct in the young; lower parts uniform yellowish.

Total length 460 millim.; tail 110.

Himalayas from Simla to Darjeeling.

a. ♂ (V. 185; C. 63).	Sikkim.	Messrs. v. Schlagintweit [C.]. (Type.)
b. Yg. (V. 184; C. 60).	Sikkim.	Messrs. v. Schlagintweit [C.]. (Type of *A. owenii*.)
c-d. ♂ (V. 189; C. 74) & hgr. (V. 178; C. 68).	Darjeeling.	W. T. Blanford, Esq. [P.].
e. ♂ (V. 180; C. 71).	Darjeeling.	T. C. Jerdon, Esq. [P.].
f. ♂ (V. 182; C. 69) & hgr. (V. 184; C. 75).	Darjeeling.	
g. ♀ (V. 195; C. ?).	Nepal.	B. H. Hodgson, Esq. [P.].

9. Ablabes baliodirus.

Coronella baliodeira, *Boie, Isis,* 1827, p. 539; *Schleg. Phys. Serp.* ii. p. 64, pl. ii. figs. 9 & 10 (1837); *Cantor, Cat. Mal. Rept.* p. 66 (1847).

Ablabes baliodeirus, *Dum. & Bibr.* vii. p. 313 (1854); *Günth. Cat.* p. 29 (1858), *and Rept. Brit. Ind.* p. 224 (1864).

Diadophis baliodeirus, *Jan, Arch. Zool. Anat. Phys.* ii. 1863, p. 263, *and Icon. Gén.* 15, pl. v. fig. 4 (1866).

Ablabes baliodirus, var. cinctus, *Fischer, Abh. naturw. Ver. Hamb.* ix. 1886, p. 8, pl. i. fig. 2.

Rostral broader than deep, just visible from above; nasal divided; eye three fourths the length of the snout; suture between the internasals shorter than that between the præfrontals; frontal longer than its distance from the end of the snout, shorter than the parietals; a small loreal, as long as deep or deeper than long; one or two præ- and two postoculars; temporals 1+2; seven upper labials, third and fourth entering the eye; four or five lower labials in contact with the anterior chin-shields, which are as long as or shorter than the posterior. Scales in 13 rows. Ventrals 118-137; anal divided; subcaudals 58-75. Brown above, usually with small black-edged yellow spots, sometimes with blackish cross bands or two series of large alternating black blotches on the anterior part of the body, or with two light, dark-edged streaks along the body and a third on the nape; head sometimes yellowish; lips yellowish white, with the labial sutures black; lower parts yellowish or coral-red, the brown of the upper parts usually descending to the sides of the ventrals, each of which may bear a black dot at the end; sometimes a black line along the middle of the lower surface of the tail.

Total length 400 millim.; tail 125.

Java, Borneo, Sumatra, Pinang.

a. ♂ (V. 127; C. 66).	Java.	Leyden Museum. (One of the types.)
b-c. ♂ (V. 123; C. 63) & ♀ (V. 134; C. 58).	Java.	Hr. Frühstorfer [C.].
d. ♂ (V. 127; C. 69).	Java.	
e. ♀ (V. 137; C. 64).	Salak, Java.	R. Kirkpatrick, Esq. [C.].
f. ♀ (V. 127; C. 59).	Willis Mts., Kediri, Java, 5000 ft.	Baron v. Huegel [C.].
g. ♂ (V. 124; C. ?).	Borneo.	Sir E. Belcher [P.].
h. ♀ (V. 134; C. 60).	Sarawak.	
i. ♂ (V. 130; C. 75).	Sarawak.	Rajah Brooke [P.].
k. ♀ (V. 135; C. 70).	Baram, Sarawak.	A. Everett, Esq. [P.].
l. ♂ (V. 118; C. 66).	Bunguran Id., Natuna Ids.	A. Everett, Esq. [C.].
m. ♀ (V. 137; C. ?).	Padang, Sumatra.	
n-o. ♂ (V. 120, 124; C. 68, 74).	Deli, Sumatra.	Prof. Moesch [C.].
p-q. ♂ (V. 121, 119; C. 71, 70).	Pinang.	Dr. Cantor.

10. Ablabes scriptus.

Ablabes scriptus, *Theob. Journ. Linn. Soc.* x. 1868, p. 42, *and Cat. Rept. As. Soc. Mus.* 1868, p. 49, *and Cat. Rept. Brit. Ind.* p. 154 (1876); *Bouleng. Faun. Ind., Rept.* p. 305 (1890).

Closely allied to *A. baliodirus*. Scales in 13 rows. Præocular one; postoculars two, small; loreal very small, much smaller than postocular; præfrontals broader than long; upper labials eight, the third, fourth, and fifth entering the orbit; a long narrow temporal, forming a suture with both postoculars and the sixth and seventh labials; seventh labial largest, more than twice as broad as the temporal; two pairs of chin-shields, the hinder rather larger than the other, first in contact with four labials; lower labials seven, fifth largest. Colour above brown; a few black dots on either side of spine on the front part of the trunk; a black mark under the eye, followed by a white upright border involving the postoculars; a black-bordered white patch on the last upper labial, and a white collar on the nape. Beneath white.

Martaban, Burma.

11. Ablabes longicauda.

Ablabes longicaudus, *Peters, Mon. Berl. Ac.* 1871, p. 574, *and Ann. Mus. Genova*, iii. 1872, p. 35, pl. v. fig. 1.

—— quinquestriatus, *F. Müller, Verh. nat. Ges. Basel*, vi. 1878, p. 657, pl. ii. fig. B, and vii. 1882, p. 143.

Rostral broader than deep, just visible from above; nasal divided; eye three fourths the length of the snout; suture between the internasals as long as or a little shorter than that between the præfrontals; frontal longer than its distance from the end of the snout, shorter than the parietals; a small loreal, as long as deep or a little deeper than long, sometimes confluent with the posterior nasal; one or two præ- and two postoculars; temporals 1+2;

eight (or seven) upper labials, third, fourth, and fifth (or third and fourth) entering the eye; four lower labials in contact with the anterior chin-shields, which are slightly longer than the posterior. Scales in 13 rows. Ventrals 114-122; anal divided; subcaudals 77-103. Lustrous dark brown or black above, with three yellowish longitudinal lines lost on the hinder half of the body; the vertebral line beginning behind the nape, the laterals widening anteriorly and descending obliquely with their black borders towards the throat; a yellow chevron-shaped collar, pointing backwards; the black of the upper surface of the head descending as a triangular patch on the temple; eye bordered with black; upper labials yellowish, with a few small black spots: lower parts uniform yellowish white.

Total length 430 millim.; tail 170.

Borneo, Sumatra, Pinang.

a-b. ♂ (V. 118, 116; C. 84, 77).	Rejang River, Sarawak.	BrookeLow, Esq. [P.].
c. ♂ (V. 117; C. 93).	Rejang River.	Rajah Brooke [P.].
d. ♂ (V. 122; C. ?).	Mt. Batu-Song, Sarawak, 2000 ft.	C. Hose, Esq. [C.].

The following snake is perhaps to be regarded as a variety of *A. longicauda*; the description is, unfortunately, insufficient, and the figure represents the animal with a vertical pupil:—

DIADOPHIS BIPUNCTATUS, *v. Lidth de Jeude, in M. Weber, Zool. Ergebn.* p. 184, pl. xvi. fig. 9 (1890).

One præ- and two postoculars; temporals 1+2; seven upper labials, third and fourth entering the eye. Scales in 13 rows. Ventrals 117; subcaudals 90. Upper parts dark brown, lower parts yellow, without stripes or spots; head and neck black; two small white spots on the parietal shields; three yellow cross bands on each side of the neck, the first pair meeting in the middle and forming a collar, the third not clearly marked; a yellow spot behind and under the orbit; the three anterior upper labials partly yellow, partly black; a triangular black spot at each end of the ventral and caudal shields, close to the outer row of scales, the spots forming a continuous black line along each side of the belly.

Kaju Tanam, Sumatra.

12. Ablabes nicobariensis.

Ablabes nicobariensis, *Stoliczka, Journ. As. Soc. Beng.* xxxix. 1870, p. 184, pl. xi. fig. 1; *Bouleng. Faun. Ind., Rept.* p. 307 (1890).

Rostral low, wide, not reaching the top of the head; nostril between two nasals; internasals about half the size of the præfrontals; frontal somewhat larger than the supraoculars; parietals about one fourth larger than the frontal, in contact with both postoculars; loreal united with the postnasal; one præ- and two postoculars; temporals 1+2: upper labials seven, third and fourth entering the eye; both pairs of chin-shields subequal in size. Scales in 17 rows. Ventrals 189; anal divided; subcaudals 87. Anterior half of the body reddish brown above, posterior blackish grey; head above blackish, the first three labials with yellow spots; a short broad yellow streak from behind and below the eye posteriorly to the angle of the mouth; a black collar, margined on both sides with an interrupted yellow band, of which the anterior is the most distinct; an indistinct series of blackish-grey dorsal spots, almost forming a dark undulating band; sides marbled and freckled blackish grey, this colour being separated from the upper brown one by a series of closely-set black spots, which are partially conspicuous on the posterior part of the body; chin dusky; lower parts yellow with a vermilion tinge, each ventral with a large black spot near its outer extremity.

Total length 440 millim.; tail 110.

Camorta, Nicobars.

104. GRAYIA.

Heteronotus (*non Lap.*), *Hallow. Proc. Ac. Philad.* 1857, p. 67.
Grayia, *Günth. Cat. Col. Sn.* p. 50 (1858).
Glaniolestes, *Cope, Proc. Ac. Philad.* 1862, p. 191.
Leionotus (*non Bibr.*), *Jan, Elenco sist. Ofid.* p. 68 (1863), *and Arch. Zool. Anat. Phys.* iii. 1865, p. 240.
Macrophis, *Bocage, Jorn. Sc. Lisb.* i. 1866, p. 67.

Maxillary teeth equal, 22 to 25; mandibular teeth equal. Head distinct from neck; eye moderate, with round pupil; nostril directed upwards, in a semidivided nasal shield. Body cylindrical, stout; scales short and smooth, without apical pits, in 17 or 19 rows; ventrals rounded. Tail rather long; subcaudals in two rows.

Tropical Africa.

Synopsis of the Species.

I. Fourth upper labial entering the eye; two postoculars; two superposed anterior temporals.

Seven or eight upper labials; ventrals 145–161.   1. *smythii*, p. 286.
Nine upper labials; ventrals 219 . . . .   2. *furcata*, p. 287.

II. Fourth and fifth upper labials entering the eye; three postoculars; a single anterior temporal . . . .   3. *giardi*, p. 288.

1. Grayia smythii. (PLATE XIII. fig. 3.)

Coluber smythii, *Leach, in Tuckey's Explor. R. Zaire, App.* p. 409 (1818).
—— lævis (*non Lacép.*), *Hallow. Proc. Ac. Philad.* 1844, p. 118.
Coronella triangularis, *Hallow. Proc. Ac. Philad.* 1854, p. 100.
Heteronotus triangularis, *Hallow. Proc. Ac. Philad.* 1857, p. 68.
Grayia silurophaga, *Günth. Cat.* p. 51 (1858); *F. Müll. Verh. nat. Ges. Basel,* vii. 1885, p. 683.
Leionotus schlegeli, *Jan, Elenco,* p. 68 (1863), *and Arch. Zool. Anat. Phys.* iii. 1865, p. 241.

Grayia triangularis, *Bocage, Jorn. Sc. Lisb.* i. 1866, p. 47; *Günth. Ann. & Mag. N. H.* (6) i. 1888, p. 325; *Boetty. Ber. Senck. Ges.* 1887–88, p. 51.
Macrophis ornatus, *Bocage, t. c.* p. 67.

Rostral broader than deep, just visible from above; internasals longer than broad, as long as or longer than the præfrontals; frontal once and two thirds to twice as long as broad, longer than its distance from the end of the snout, as long as or a little shorter than the parietals; loreal longer than deep; one præ- and two postoculars; temporals 2+3; seven or eight upper labials, fourth entering the eye; four or five lower labials in contact with the anterior chin-shields, which are a little shorter than the posterior. Scales in 17 (rarely 19) rows. Ventrals 145–161; anal divided; subcaudals 74–103. Brown or olive above, with black cross bands tapering to a point towards the ventrals, or with light black-edged cross bands widening towards the belly; or dark olive with the scales edged with black; or uniform blackish olive; belly yellow, uniform or spotted with black, or entirely blackish.

Total length 1250 millim.; tail 450.

West Africa, from Sierra Leone to Angola.

A. Belly yellow, with or without black spots.

a.	♀ (V. 147; C. 93).	Boma, Congo.	Dr. Leach [P.]. (Type of *C. smythii* and of *G. silurophaga*.)
b.	Yg. (V. 150; C. ?).	Mouth of the Loango.	Mr. H. I. Duggan [C.].
c.	♂ (Sc. 19; V. 147; C. ?).	Gaboon.	
d.	♀ (V. 161; C. ?).	Opposite Fernando Po.	Mrs. Burton [P.].
e–f.	♂ (V. 154; C. 99) & ♀ (V. 161; C. 89).	Oil River.	H. H. Johnston, Esq. [P.].
g.	♂ (V. 148; C. 97).	Sierra Leone.	Sir A. Kennedy [P.].
h.	Hgr. ♂ (V. 151; C. 97).	W. Africa.	Mr. Rich [C.].
i.	Hgr. ♂ (V. 149; C. 101).	Africa.	(Types of *G. silurophaga*.)

B. Belly black.

k.	Yg. (V. 147; C. 82).	Gaboon.
l.	Yg. (V. 145; C. ?).	Cette Cama, Gaboon.
m.	Yg. (V. 148; C. 74).	Congo.

2. Grayia furcata.

Grayia furcata, *Mocquard, Bull. Soc. Philom.* (7) xi. 1887, p. 71.

Distinguished from *G. smythii* by the smaller and more irregular temporals, nine upper labials instead of seven or eight, and the more numerous ventral shields (219). Subcaudals 88. In colour resembling the barred form or *G. smythii*; posterior ventral and subcaudal regions blackish.

Total length 1012 millim.; tail 292.

Brazzaville, Congo.

3. Grayia giardi.

Grayia giardi, *Dollo, Bull. Mus. Belg.* iv. 1886, p. 158, fig.

One præ- and three postoculars; temporals 1+2; eight upper labials, fourth and fifth entering the eye. Scales in 19 rows.
Lake Tanganyika.

105. XENUROPHIS.

Xenurophis, *Günth. Ann. & Mag. N. H.* (3) xii. 1863, p. 357.

Maxillary teeth 35, closely set, equal; mandibular teeth equal. Head distinct from neck; eye large, with round pupil. Body cylindrical; scales smooth, without pits, in 15 rows; ventrals rounded. Tail long, with two dorsal series of very large, shield-like scales; subcaudals in two rows.
West Africa.

1. Xenurophis cæsar.

Xenurophis cæsar, *Günth. l. c.* pl. vi. fig. C; *Bouleng. Zool. Rec.* 1891, Rept. p. 11.
Grayia longicaudata, *Mocquard, Bull. Soc. Philom.* (8) iii. 1891, C. R. p. 9.

Eye as long as the snout. Rostral twice as broad as deep, just visible from above; internasals as long as or a little longer than broad, a little shorter than the præfrontals; frontal not broader than the supraocular, once and two thirds to twice as long as broad, slightly longer than its distance from the end of the snout, as long as the parietals; loreal as long as deep, or deeper than long; one præ- and two postoculars; temporals 2+3; eight upper labials, fourth and fifth entering the eye; five lower labials in contact with the anterior chin-shields, which are as long as or a little shorter than the posterior. Scales in 15 rows (in 4 rows on the tail). Ventrals 125-149; anal divided; subcaudals 143-161. Dark olive above, with regular narrow, whitish, black-edged cross bands; a few light spots on the head and two whitish black-edged oblique bands on each side, the anterior behind the eye, the second on the temple; lower parts uniform yellowish white.

Total length 600 millim. (tail injured) When intact the tail is nearly as long as the body.
West Africa.

a. ♀ (V. 145; C. ?).   Fernando Po.   (Type.)
b. Hgr. (V. 149; C. 143).   Mouth of the Loango.   Mr. H. I. Duggan [C.].

106. VIRGINIA.

Virginia, *Baird & Gir. Cat. N. Am. Rept.* p. 127 (1853); *Jan, Arch. Zool. Anat. Phys.* ii. 1862, p. 23; *Bocourt, Miss. Sc. Mex., Rept.* p. 542 (1883); *Cope, Proc. U.S. Nat. Mus.* xiv. 1892, p. 599.
Carphophis, part., *Dum. & Bibr. Erp. Gén.* vii. p. 131 (1854).

Maxillary teeth 22, small, subequal; mandibular teeth small, equal. Head small, not distinct from neck; eye small, with round pupil; no præocular; loreal and præfrontal entering the eye. Body cylindrical; scales smooth or feebly keeled, in 15 or 17 rows, without apical pits; ventrals rounded. Tail short; subcaudals in two rows. North America.

1. Virginia valeriæ.

Virginia valeriæ, *Baird & Gir. Cat. N. Am. Rept.* p. 127 (1853); *Jan, Arch. Zool. Anat. Phys.* ii. 1862, p. 24, *and Icon. Gén.* 12, pl. ii. fig. 5 (1865); *Garm. N. Am. Rept.* p. 98, pl. vii. fig. 3 (1883); *H. Garm. Bull. Illin. Lab.* iii. 1892, p. 307.
Carphophis harpertii, *Dum. & Bibr.* vii. p. 135 (1854).
Virginia harperti, *Cope, Check-list N. Am. Rept.* p. 35 (1875); *Garm. l. c.* p. 99.
—— harpertii, part., *Bocourt, Miss. Sc. Mex., Rept.* p. 543, pl. xxxii. fig. 3 (1883).

Rostral nearly as deep as broad, visible from above; internasals much shorter than the præfrontals; frontal longer than broad, shorter than the parietals; loreal once and a half to twice and a half as long as deep; two or three postoculars; temporals 1+2; six upper labials, third and fourth entering the eye; four lower labials in contact with the anterior chin-shields, which are as long as or shorter than the posterior. Scales in 15 rows, smooth or feebly keeled. Ventrals 111–128; anal divided; subcaudals 24–37. Yellowish or greyish brown above, uniform or with minute black dots which may form two or four longitudinal series; dull yellow beneath.
Total length 220 millim.; tail 40.

United States, from Maryland and Illinois to Georgia and Texas.

a–c. ♂ (V. 112, 113; C. 34, Raleigh, N. Carolina. Messrs. Brimley [C.]. 36) & ♀ (V. 115; C. 24).

2. Virginia elegans.

Virginia elegans, *Kennicott, Proc. Ac. Philad.* 1859, p. 99; *Jan, Arch. Zool. Anat. Phys.* ii. 1862, p. 24, *and Icon. Gén.* 12, pl. ii. fig. 6 (1865); *Cope, Proc. U.S. Nat. Mus.* xiv. 1892, p. 599.
—— harpertii, part., *Bocourt, Miss. Sc. Mex., Rept.* p. 543, pl. xxxii. fig. 4 (1883).

Differs from the preceding in the narrower scales (which may be feebly keeled) in 17 rows. Ventrals 135; subcaudals 31. Uniform light olivaceous brown above, dull yellowish white beneath.
Southern Illinois, Arkansas, Texas.

107. ABASTOR.

Pseudoeryx, part., *Fitziny. N. Class. Rept.* p. 29 (1826).
Helicops, part., *Wagler, Syst. Amph.* p. 170 (1830).
Homalopsis, part., *Schleg. Phys. Serp.* ii. p. 297 (1837).
Abastor, *Gray, Cat. Sn.* p. 78 (1849); *Baird & Gir. Cat. N. Am. Rept.* p. 125 (1853); *Cope, Proc. U.S. Nat. Mus.* xiv. 1892, p. 603.
Calopisma, part., *Dum. & Bibr. Erp. Gén.* vii. p. 336 (1854); *Jan, Arch. Zool. Anat. Phys.* iii. 1865, p. 241.

Maxillary teeth subequal, 15 to 17; mandibular teeth subequal.

Head small, not distinct from neck; eye small, with round pupil; nostril directed upwards, in a semidivided nasal; no præocular, loreal and præfrontal entering the eye. Body cylindrical; scales smooth, without apical pits, in 19 rows; ventrals rounded. Tail moderate or rather short; subcaudals in two rows.

North America.

1. Abastor erythrogrammus.

Coluber erythrogrammus, *Daud.* Rept. vii. p. 93, pl. lxxxiii. fig. 2 (1803).
—— seriatus, *Hermann*, Obs. Zool. p. 273 (1804).
Homalopsis erythrogrammus, *Boie*, Isis, 1827, p. 551.
—— plicatilis, var., *Schleg.* Phys. Serp. ii. p. 355 (1837).
Helicops erythrogrammus, *Holbr.* N. Am. Herp. iii. p. 107, pl. xxv. (1842).
Abastor erythrogrammus, *Gray*, Cat. p. 78 (1849); *Baird & Gir.* Cat. N. Am. Rept. p. 125 (1853).
Calopisma erythrogrammum, *Dum. & Bibr.* vii. p. 337 (1854); *Jan*, Arch. Zool. Anat. Phys. iii. 1865, p. 243, *and Icon. Gén.* 29, pl. iv. fig. 2, & pl. v. fig. 1 (1868).
Hydrops erythrogrammus, *Garm.* N. Am. Rept. p. 35 (1883); *H. Garm.* Bull. Illin. Lab. iii. 1892, p. 280.

Rostral large, broader than deep, the portion visible from above nearly as long as its distance from the frontal; internasals broader than long, much shorter than the præfrontals; frontal once and two thirds to twice as long as broad, longer than its distance from the end of the snout, a little shorter than the parietals; loreal at least twice as long as deep; two postoculars; temporals 1+2; seven (rarely eight) upper labials, third and fourth (or fourth and fifth) entering the eye; four lower labials in contact with the anterior chin-shields, which are a little longer than the posterior. Scales in 19 rows. Ventrals 157–185; anal divided; subcaudals 37–55. Black above, with three red stripes; the two or three lower rows of scales yellow or reddish; lower parts yellowish or red, with two or three regular series of roundish black spots.

Total length 900 millim.; tail 120.

Mississippi Valley and eastwards.

a–e. ♂ (V. 166; C. 46), ♀ (V. 178; C. 40), hgr. (V. 165; C. 47), & yg. (V. 166, 157; C. 47, 51). N. America. Lord Ampthill [P.].
f. Yg. (V. 166; C. 49). N. America. Mrs. Phillips [P.].
g. Yg. (V. 182; C. 37). N. America.
h. Yg. (V. 167; C. 50). Marion Co., Florida. A. Erwin Brown, Esq. [P.].

108. FARANCIA.

Homalopsis, part., *Schleg.* Phys. Serp. ii. p. 297 (1837).
Farancia, *Gray*, Zool. Misc. p. 68 (1842), *and* Cat. Sn. p. 74 (1849); *Baird & Gir.* Cat. N. Am. Rept. p. 123 (1853); *Cope*, Proc. U.S. Nat. Mus. xiv. 1892, p. 605.
Calopisma, part., *Dum. & Bibr.* Erp. Gén. vii. p. 336 (1854); *Jan*, Arch. Zool. Anat. Phys. iii. 1865, p. 241.

Maxillary teeth subequal, 17 or 18; mandibular teeth subequal. Head small, not distinct from neck; eye small, with round pupil; nostril directed upwards, in a semidivided nasal; a single internasal; no præocular, loreal and præfrontal entering the eye. Body cylindrical; scales smooth, without apical pits, in 19 rows; ventrals rounded. Tail short; subcaudals in two rows.

North America.

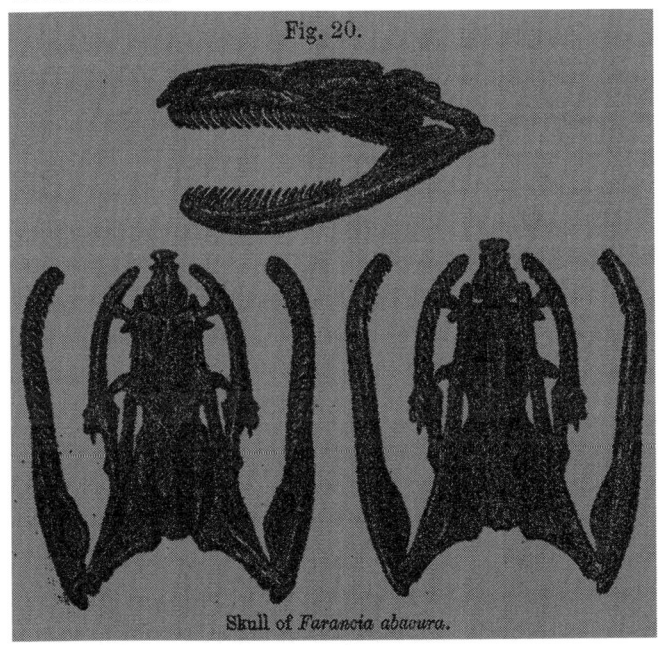

Fig. 20.

Skull of *Farancia abacura*.

1. Farancia abacura.

Coluber ovivorus, part., *Daud. Rept.* vi. p. 341 (1803).
—— abacurus, *Holbr. N. Am. Herp.* p. 119, pl. xxiii. (1836).
Homalopsis reinwardtii, *Schleg. Phys. Serp.* ii. p. 357 (1837).
Helicops abacurus, *Holbr. op. cit.* ed. 2, iii. p. 107, pl. xxvi. (1842).
Hydrops reinwardtii, *Gray, Zool. Misc.* p. 68 (1842).
Farancia drummondi, *Gray, l. c.*
—— fasciata, *Gray, Cat.* p. 74 (1849).
—— abacurus, *Baird & Gir. Cat. N. Am. Rept.* p. 123 (1853).
Calopisma abacurum, *Dum. & Bibr.* vii. p. 342, pl. lxv. (1854).
—— reinwardti, *Jan, Arch. Zool. Anat. Phys.* iii. 1865, p. 243, *and Icon. Gén.* 29, pl. vi. (1868).
Hydrops abacurus, *Garm. N. Am. Rept.* p. 36, pl. i. fig. 5 (1883); *H. Garm. Bull. Illin. Lab.* iii. 1892, p. 281.

Rostral broader than deep, just visible from above, forming a suture with the internasal, which is much broader than long; frontal about once and a half as long as broad, longer than its distance from the end of the snout, shorter than the parietals; loreal much longer than deep; two postoculars; temporals 1+2; seven upper labials, third and fourth entering the eye; four lower labials in contact with the anterior chin-shields, which are longer than the posterior. Scales in 19 rows, more or less distinctly keeled in the ischiadic region. Ventrals 168-206; anal divided; subcaudals 34-49. Black above; sides and belly red, with black cross bars or alternating spots.

Total length 1020 millim.; tail 150.

Mississippi Valley and eastwards.

a, b, c. ♂ (V. 170; C. 46) & ♀ (V. 189, 191; C. 36, 36).	New Orleans.	M. Sallé [C.].
d. ♀ (V. 199; C. 37), skin & skeleton.	N. America.	
e. Yg. (V. 172; C. 34).	N. America.	Mrs. Drummond [P.]. (Type of *F. drummondi*.)
f. Yg. (V. 168; C. ?).	N. America.	Sir R. Murchison [P.].

109. PETALOGNATHUS.

Dipsas, part., *Schleg. Phys. Serp.* ii. p. 257 (1837).
Petalognathus, *Dum. & Bibr. Mém. Ac. Sc.* xxiii. 1853, p. 466, *and Erp. Gén.* vii. p. 463 (1854).
Leptognathus, part., *Günth. Cat. Col. Sn.* p. 177 (1858); *Jan, Elenco sist. Ofid.* p. 100 (1863); *Cope, Proc. Ac. Philad.* 1868, p. 107.

Fig. 21.

Skull of *Petalognathus nebulatus*.

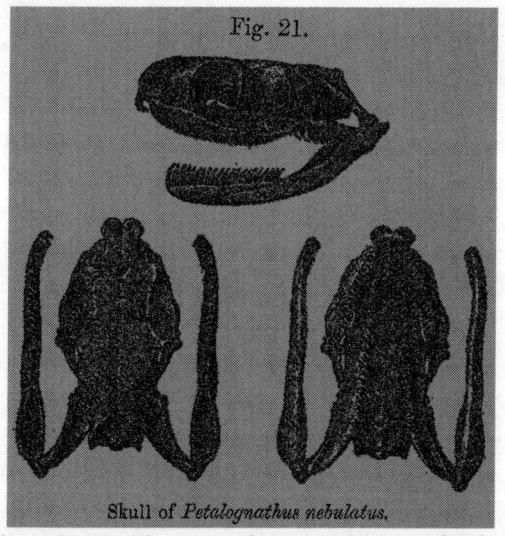

Maxillary short, with 15 or 16 teeth, which gradually decrease in size; posterior mandibular teeth gradually decreasing in size. Head short and thick, very distinct from neck; eye large, with vertically elliptic pupil; no præocular, loreal and præfrontal entering the eye. Body elongate, compressed; scales smooth, without pits, in 15 rows, the middle row formed of enlarged scales; ventrals rounded. Tail long; subcaudals in two rows.

Tropical America.

1. Petalognathus nebulatus.

Coluber nebulatus, *Linn. Mus. Ad. Frid.* p. 32, pl. xxiv. fig. 1 (1754), and *S. N.* i. p. 383 (1766); *Daud. Rept.* vi. p. 413 (1803).
—— sibon, *Linn. S. N.* i. p. 383; *Daud. t. c.* p. 435.
Cerastes nebulatus, *Laur. Syn. Rept.* p. 83 (1768).
Dipsas nebulatus, *Boie, Isis,* 1827, p. 560; *Schleg. Phys. Serp.* ii. p. 275, pl. xi. figs. 14 & 15 (1837).
Petalognathus nebulatus, *Dum. & Bibr.* vii. p. 464 (1854).
Leptognathus nebulatus, *Günth. Cat.* p. 177 (1858); *Cope, Proc. Ac. Philad.* 1868, p. 108; *Jan, Icon. Gén.* 37, pl. v. fig. 3 (1870); *Garm. N. Am. Rept.* p. 13 (1883).

Snout short, as long as or a little longer than the diameter of the eye. Rostral as broad as deep or a little broader than deep, not or scarcely visible from above; internasals about half the length of the præfrontals; frontal once and one fourth to once and one third as long as broad, as long as its distance from the end of the snout, shorter than the parietals; loreal a little longer than deep; two (rarely three) postoculars; temporals $1+2$; seven (rarely eight) upper labials, fourth and fifth (or fourth, fifth, and sixth) entering the eye; five (rarely six) lower labials in contact with the anterior chin-shields, which are longer than the posterior. Scales smooth, in 17 rows. Ventrals 170–197; anal entire; subcaudals 73–102. Grey or grey-brown above, with blackish spots, or mottled black and white, with more or less regular cross bars which may have a light edge; belly yellowish, with black dots and more or less regular black cross bars or large alternating spots.

Total length 790 millim.; tail 210.

Central and South America, from Mexico to Northern Brazil and Ecuador; Trinidad and Tobago.

a–b. ♂ (V. 177, 176; C. 85, 83).	Mexico.	Mr. Hugo Finck [C.].
c. ♀ (V. 174; C. 81).	City of Mexico.	Mr. Doorman [C.].
d–g. ♂ (V. 178, 176; C. 82, 92) & ♀ (V. 174, 175; C. 81, 77).	Teapa, Tabasco.	F. D. Godman, Esq. [P.].
h. ♀ (V. 170; C. 90).	N. of Coban, Guatemala.	F. C. Sarg, Esq. [C.].
i. ♀ (V. 178; C. 84).	Hacienda Rosa de Jericho, Nicaragua, 3250 ft.	Dr. E. Rothschuh [C.].
k. ♂ (V. 194; C. 93).	W. Ecuador.	Mr. Fraser [C.].
l–m. ♀ (V. 175; C. 86) & yg. (V. 179; C. 98).	Venezuela.	Mr. Dyson [C.].
n. ♀ (V. 183; C. 91).	Tobago.	A. Ludlam, Esq. [P.].

o-u. ♂ (V. 181, 180; C. 89, 88) & ♀ (V. 173, 174, 174, 182, 182; C. 82, 84, 76, 82, 102). Demerara. Mr. Snellgrove [C.].

v-η. ♂ (V. 181, 181, 174; C. 90, 95, 86), ♀ (V. 176, 173, 174, 177, 172, 180; C. 84, 85, 74, 79, 75, ?), & yg. (V. 173, 181; C. 73, 90). Berbice.

θ. ♂ (V. 192; C. 92). Pernambuco. J. P. G. Smith, Esq. [P.].
ι. Skeleton. Berbice.

110. TROPIDODIPSAS.

Tropidodipsas, *Günth. Cat. Sn.* p. 180 (1858), and *Zool. Rec.* 1872, p. 75.
Galedon, *Jan, Elenco sist. Ofid.* p. 95 (1863).
Leptognathus, part., *Jan, l. c.* p. 100; *Cope, Proc. Ac. Philad.* 1868, p. 107.
Tropidogeophis, *F. Müll. Verh. nat. Ges. Basel*, vii. 1878, p. 411.
Tropidoclonium, part., *Bocourt, Miss. Sc. Mex., Rept.* p. 738 (1893).

Maxillary short, with 12 to 16 teeth, which decrease in size in front and behind; posterior mandibular teeth gradually decreasing in size. Head distinct from neck; eye moderate or large, with vertically elliptic pupil; præoculars may be absent. Body elongate, cylindrical or a little compressed; scales smooth or feebly keeled, without pits, in 13 to 17 rows; ventrals rounded. Tail moderate; subcaudals in two rows.

Central America.

Synopis of the Species.

I. Eye rather large; loreal shield not entering the eye; two præoculars; frontal a little longer than broad.

Scales in 17 rows .... 1. *fasciata*, p. 295.
Scales in 15 rows .... 2. *philippii*, p. 295.

II. Eye moderate.

   A. Scales feebly keeled, in 17 rows; one or two præoculars.

Loreal at least once and two thirds as long as deep, entering the eye; frontal not broader than long; six upper labials .... 3. *fischeri*, p. 296.
Loreal short, entering the eye or separated from it by the præoculars; frontal a little broader than long .... 4. *sartorii*, p. 296.

   B. Scales smooth, in 15 or 13 rows; no præocular.

Scales in 15 rows; frontal as long as broad. 5. *annulifera*, p. 297.
Scales in 13 rows; frontal longer than broad. 6. *anthracops*, p. 297.

1. Tropidodipsas fasciata.

Tropidodipsas fasciata, part., *Günth. Cat.* p. 181 (1858).
Leptognathus fasciata, *Cope, Proc. Ac. Philad.* 1868, pp. 109 & 137.
—— (Tropidodipsas) subannulatus, *F. Müll. Verh. nat. Ges. Basel,* viii. 1887, p. 274, pl. i. fig. 5.

Eye rather large. Rostral a little broader than deep, just visible from above ; internasals two fifths the length of the præfrontals ; frontal a little longer than broad, as long as its distance from the end of the snout, shorter than the parietals ; loreal as long as deep, not entering the eye ; two præoculars ; two postoculars ; temporals 1+2 or 2+2 ; seven or eight upper labials, fourth and fifth entering the eye, sixth or seventh largest; four or five lower labials in contact with the anterior chin-shields, which are a little longer than the posterior. Scales in 17 rows, dorsals feebly keeled. Ventrals 171–176 ; anal entire ; subcaudals 71–78. Dark brown or black, with 20 to 38 white (red ?) or pale brown annuli, some of which are complete, others broken up above and below and forming alternating cross bars ; throat and a bar across occiput white or pale brown.

Total length 600 millim. ; tail 140.

Mexico.

a–b. ♀ (V. 176 ; C. 71) & yg. (V. 171 ; C. 78).	—— ?	Zoological Society.	
c. Hgr., bad state.		Mexico.	Mr. Hugo Finck [C.].

(Types.)

2. Tropidodipsas philippii.

Tropidodipsas fasciata, part., *Günth. Cat.* p. 181 (1858).
Leptognathus philippii, *Jan, Elenco,* p. 101 (1863), *and Icon. Gén.* 37, pl. v. fig. 1 (1870).
—— albocinctus, *Fischer, Jahrb. Hamb. Wiss. Anst.* ii. 1885, p. 107, pl. iv. fig. 9.

Eye rather large. Rostral a little broader than deep, just visible from above; internasals two fifths the length of the præfrontals ; frontal a little longer than broad, as long as its distance from the end of the snout, shorter than the parietals ; loreal as long as or slightly longer than deep, not entering the eye ; two præoculars ; two postoculars ; temporals 1+2 ; seven upper labials, fourth and fifth entering the eye, sixth largest; six lower labials in contact with the anterior chin-shields, which are more than twice as long as the posterior. Scales in 15 rows, dorsals faintly keeled. Ventrals 178–180 ; anal entire ; subcaudals 67–71. A series of large roundish blackish-brown spots, separated by narrow whitish interspaces, occupies the upper parts, the spots descending to the sides of the belly ; upper lip white ; belly white, with a few dark brown spots ; the dark brown predominates under the tail.

Total length 330 millim. ; tail 80.

Western Mexico, California (?).

a. ♀ (V. 180 ; C. 71). Mexico.   (One of the types of *T. fasciata.*)
b. ♀ (V. 178 ; C. 67). San Diego, California (?).   Dr. J. G. Fischer.
(Type of *L. albocinctus.*)

3. Tropidodipsas fischeri.

Virginia fasciata, *Fischer, Jahrb. Hamb. Wiss. Anst.* ii. 1885, p. 95.
Elapoides fasciatus, *Cope, Bull. U.S. Nat. Mus.* no. 32, 1887, p. 85.
Tropidoclonium annulatum, *Bocourt, Le Natur.* 1892, p. 132, *and Miss. Sc. Mex., Rept.* p. 738, pl. liv. fig. 3 (1893).
Geophis fasciata, *Günth. Biol. C.-Amer., Rept.* p. 93, pl. xxxiv. fig. A (1893).

Eye moderate. Rostral a little broader than deep, just visible from above; internasals two fifths the length of the præfrontals; frontal as long as or a little longer than broad, as long as its distance from the end of the snout, shorter than the parietals; loreal once and two thirds or twice as long as deep, entering the eye, with a præocular above it; two postoculars; temporals 1+2; six upper labials, third and fourth entering the eye, fifth largest; four lower labials in contact with the anterior chin-shields, which are twice as long as the posterior. Scales in 17 rows, dorsals feebly keeled. Ventrals 174–196; anal entire; subcaudals 51–59. Brownish grey above, with numerous black cross bars, which expand into rhomboidal blotches on the back, and descend as narrow bars to the sides of the belly, some of the anterior forming more or less complete annuli; head black above, lips white; lower parts white, with a few black spots on the belly and numerous ones under the tail.

Total length 540 millim.; tail 110.
Guatemala.

a. ♂ (V. 180; C. 57).　　　Guatemala.　　　Stuttgart Mus. [E.].
　　　　　　　　　　　　　　　　　　　　　　(One of the types.)

4. Tropidodipsas sartorii.

Tropidodipsas fasciata, part., *Günth. Cat.* p. 181 (1858).
—— sartorii, *Cope, Proc. Ac. Philad.* 1863, p. 100.
Leptognathus dumerilii, *Jan, Elenco,* p. 101 (1863), *and Icon. Gén.* 37, pl. v. fig. 2 (1870).
—— sartorii, *Cope, Proc. Ac. Philad.* 1868, pp. 109 & 137.
Geophis annulatus, *Peters, Mon. Berl. Ac.* 1870, p. 643, pl. i. f. 2; *F. Müll. Verh. nat. Ges. Basel,* vi. 1878, p. 409.
Galedon annularis, *Jan, Icon. Gén.* 36, pl. v. fig. 1 (1870).
Leptognathus sexscutatus, *Bocourt, Bull. Soc. Philom.* (7) viii. 1884, p. 137.
—— leucostomus, *Bocourt, l. c.* p. 138.
—— (Tropidodipsas) bernouillii, *F. Müll. Verh. nat. Ges. Basel,* viii. 1887, p. 272, pl. i. fig. 3.
—— (Tropidodipsas) cuculliceps, *F. Müll. l. c.* p. 273, pl. i. fig. 4.

Eye moderate. Rostral a little broader than deep, just visible from above; internasals half the length of the præfrontals; frontal a little broader than long, a little shorter than its distance from the end of the snout, much shorter than the parietals; loreal short and separated from the eye by the præoculars, or fused with the lower præocular; two postoculars; temporals 1+2; seven or eight upper labials (exceptionally six), fourth and fifth or fifth and sixth entering the eye, sixth largest; five lower labials in contact with the anterior chin-shields, which are twice as long as the posterior. Scales in 17 rows, dorsals feebly keeled. Ventrals 170–190; anal entire; subcaudals 51–74. Black, with 16 to 30 white (red), usually black-dotted annuli, which may be interrupted on the belly; snout and chin black; throat, and a bar across occiput or nape, white (red).

Total length 620 millim.; tail 140.

Mexico and Guatemala.

A. Most of the annuli interrupted on the belly.

a. ♀ (V. 170; C. 51).　　Mexico.　　(One of the types of *T. fasciata*.)

Loreal shield entering the eye, between two præoculars, on one side, separated from the eye by the præoculars on the other side.

b. ♀ (V. 190; C. 56).　　Yucatan.

Loreal bordering the eye on both sides.

c. ♀ (V. 182; C. 56).　　——?　　　　　　Zoological Society.

Loreal shield entering the eye, between two præoculars, on one side, separated from the eye by the præoculars on the other side.

B. Annuli all complete.

d. ♂ (V. 181; C. 68).　　Hacienda Cubilgiitz, north　F. C. Sarg, Esq. of Coban, Vera Paz.　　　[P.].

Loreal bordering the eye on both sides.

e. ♂ (V. 179; C. 66).　　Orizaba.

Loreal bordering the eye, longitudinally cleft.

5. Tropidodipsas annulifera. (PLATE XIV. fig. 1.)

Eye moderate. Rostral a little broader than deep, just visible from above; internasals half the length of the præfrontals; frontal as broad as long, slightly shorter than its distance from the end of the snout, much shorter than the parietals; loreal once and a half as long as deep, entering the eye; præfrontal entering the eye; no præocular; two postoculars; temporals 1+2 or 3; six or seven upper labials, third or fourth or fourth and fifth entering the eye; four or five lower labials in contact with the anterior chin-shields, which are once and a half as long as the posterior. Scales smooth, in 15 rows. Ventrals 156; anal entire; subcaudals 52. Black, with 20 white (red?) complete annuli, which are broader on the belly than on the back; throat, and a band across the occiput, white.

Total length 430 millim.; tail 90.

Habitat unknown.

a. ♂ (V. 156; C. 52).　　　　——?　　　　Zoological Society.

6. Tropidodipsas anthracops.

Leptognathus anthracops, *Cope, Proc. Ac. Philad.* 1868, pp. 108 & 136.

Physiognomy of *T. sartorii*. Frontal longer than broad; loreal longer than deep, entering the eye; no præocular; two postoculars; temporals 1+2; seven upper labials, fourth and fifth entering the eye, sixth largest. Scales smooth, in 13 rows. Ventrals 177; anal entire; subcaudals 76. Black, with yellow rings continuous on the belly, but not on the anterior parts above; yellow scales black-edged; temples and nape yellow.

Total length 480 millim.; tail 130.

Nicaragua.

111. DIROSEMA.

Maxillary short, with 10 to 16 teeth, the anterior and posterior of which are the shortest; mandibular teeth small, decreasing in length posteriorly. Head distinct from neck; eye small, with vertically elliptic pupil; præocular absent, loreal and præfrontal entering the eye. Body moderately elongate, cylindrical; scales without pits, smooth or feebly keeled, in 15 or 17 rows; ventrals rounded. Tail short; subcaudals in two rows.

Central America.

Connects *Tropidodipsas* with *Geophis*.

Synopsis of the Species.

I. Scales in 17 rows.

Nostril between two nasals; fifth upper labial largest; scales smooth........ 1. *bicolor*, p. 298.
Nostril between two nasals; sixth upper labial largest; scales faintly keeled .. 2. *omiltemanum*, p. 299.
Nasal undivided; sixth upper labial largest; scales keeled............. 3. *psephotum*, p. 299.

II. Scales in 15 rows................ 4. *brachycephalum*, [p. 299.

1. Dirosema bicolor. (Plate XIV. fig. 2.)

Geophis bicolor, *Günth. Ann. & Mag. N. H.* (4) i. 1868, p. 416, and *Biol. C.-Am., Rept.* p. 91 (1893).
Rhabdosoma guttulatum, *Cope, Proc. Amer. Philos. Soc.* xxii. 1885, p. 385.

Snout broad, rounded. Rostral a little broader than deep, visible from above; internasals much broader than long, two fifths to one half the length of the præfrontals; frontal as long as broad or a little broader than long, as long as its distance from the end of the snout; nostril between two rather large nasals; loreal twice as long as deep; two postoculars (rarely one); temporals 1+2 or 1+1+2; six upper labials, third and fourth entering the eye, fifth largest and usually in contact with the parietal; two pairs of chin-shields, anterior longest and in contact with three or four lower labials. Scales smooth, in 17 rows. Ventrals 151–168; anal entire; subcaudals 38–50. Black above, with a more or less distinct lateral series of whitish dots; the scales of the two or three outer series white in the centre; upper lip white; lower parts white, uniform or spotted with black.

Total length 425 millim.; tail 95.

Mexico.

a–d. ♂ (V.159,151; C.47, 50) & ♀ (V. 167, 168; C. 38, 41).	City of Mexico.	Mr. Doorman [C.]. (Types.)
e–f, g. ♂ (V. 149; C. 46) & ♀ (V. 163,161; C. 42, 46).	La Cumbre de los Arrastrados, Jalisco, 8500 ft.	Dr. A. C. Buller [C.].

2. Dirosema omiltemanum.

Geophis omiltemana, *Günth. Biol. C.-Am., Rept.* p. 92, pl. xxxiii. fig. A (1893).

Snout broad, rounded. Rostral a little broader than deep, visible from above; internasals much broader than long, about two fifths the length of the præfrontals; frontal broader than long, as long as or slightly longer than its distance from the end of the snout, much shorter than the parietals; nostril between two rather large nasals; loreal at least twice as long as deep; a small subocular may be present between the loreal and the third and fourth labials; two postoculars; temporals 1+2; six upper labials, fourth or third and fourth entering the eye, sixth very long; two pairs of chin-shields, the anterior a little the longer and in contact with four lower labials. Scales in 17 rows, dorsals faintly keeled. Ventrals 150–166; anal entire; subcaudals 38–51. Black, with whitish cross bars widening towards the belly, which is white; these bars less distinct and closer together on the posterior part of the body; upper lip white; subcaudals brown or speckled with brown.

Total length 305 millim.; tail 65.

Mexico.

a–c. ♂ (V. 150; C. 51) & hgr. (V. 166, 156; C. 38, 50).		Omilteme, Guerrero.	Mr. H. H. Smith [C.]; F. D. Godman, Esq. [P.]. (Types.)

3. Dirosema psephotum.

Catostoma psephotum, *Cope, Journ. Ac. Philad.* (2) viii. 1876, p. 146.
Elapoidis psephotus, *Cope, Proc. Amer. Philos. Soc.* xxii. 1885, p. 386.
Geophis psephota, *Günth. Biol. C.-Am., Rept.* p. 94 (1893).

Nasal undivided; two postoculars; temporals 1+2; six upper labials, third and fourth entering the eye, sixth longest; two pairs of chin-shields, posterior short. Scales keeled, in 17 rows. Ventrals 162; anal divided; subcaudals 73. Above uniform black; below black with the half or less of an occasional ventral shield red, forming a tessellated pattern.

Total length 480 millim.; tail 128.

Costa Rica.

4. Dirosema brachycephalum.

Colobognathus brachycephalus, *Cope, Proc. Ac. Philad.* 1871, p. 211.
Catostoma brachycephalum, *Cope, Journ. Ac. Philad.* (2) viii. 1876, p. 147.
Elapoidis brachycephalus, *Cope, Bull. U.S. Nat. Mus.* no. 32, 1887, p. 85.

One postocular; six upper labials, third and fourth entering the eye, fifth in contact with the parietal. Scales in 15 rows, with a faint trace of carination near the posterior part of the body. Ventrals 124; anal entire; subcaudals 38. Black above, with a yellowish collar and a yellowish lateral streak; ventrals reddish, brown-margined.

Total length 200 millim.

Costa Rica.

112. ATRACTUS.

Atractus, *Wagler, Isis,* 1828, p. 741.
Brachyorrhos, part., *Wagler, Syst. Amph.* p. 190 (1830).
Calamaria, part., *Schleg. Phys. Serp.* ii. p. 25 (1837).
Rabdosoma, part., *Dum. & Bibr. Mém. Ac. Sc.* xxiii. 1853, p. 440, and *Erp. Gén.* vii. p. 91 (1854); *Günth. Cat. Col. Sn.* p. 10 (1858); *Jan, Arch. Zool. Anat. Phys.* ii. 1862, p. 10.
Isoscelis, *Günth. l. c.* p. 204.
Adelphicos, *Jan, l. c.* p. 18; *Bocourt, Miss. Sc. Mex., Rept.* p. 553 (1883).
Rhegnops, *Cope, Proc. Ac. Philad.* 1866, p. 128.
Rabdosoma, *Bocourt, l. c.* p. 538.

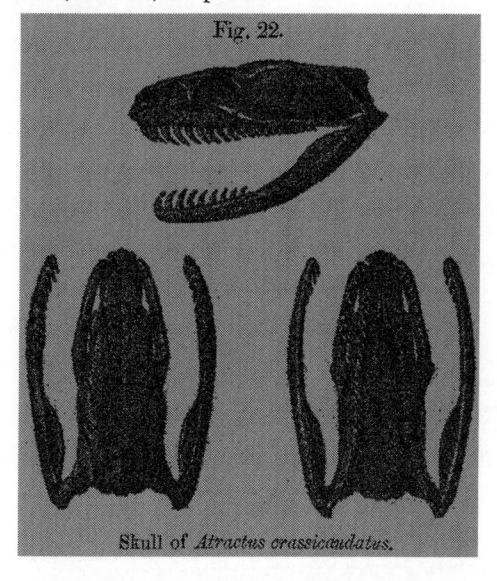

Fig. 22.

Skull of *Atractus crassicaudatus.*

Maxillary short, with 8 to 12 teeth; maxillary and mandibular teeth decreasing in size posteriorly. Head not distinct from neck; eye small, with round or subelliptic pupil; nostril between two nasals; præocular usually absent, loreal and præfrontal entering the eye. Body cylindrical; scales smooth, without pits, in 15 or 17 rows; ventrals rounded. Tail short or rather long; subcaudals in two rows.
Central and South America.

Synopsis of the Species.

I. Symphysial shield separated from the chin-shields.

A. Loreal not more than once and a half as long as deep.

Scales in 15 rows; one postocular .... 1. *elaps,* p. 302.
Scales in 17 rows; one postocular .... 2. *latifrons,* p. 303.
Scales in 17 rows; two postoculars.... 3. *modestus,* p. 304.

B. Loreal at least twice as long as deep.
  1. Two pairs of chin-shields; usually a single postocular.

Scales in 15 rows; præfrontals as long
  as broad; eight upper labials ...... 4. *vittatus*, p. 304.
Scales in 17 rows; præfrontals as long
  as broad; six upper labials .... 5. *latifrontalis*, p. 304.
Scales in 17 rows; præfrontals much
  longer than broad; six upper labials. 6. *longiceps*, p. 305.

  2. A single pair of moderately broad chin-shields; anal entire.
    a. Scales in 17 rows.
      α. Two postoculars.
        * Præfrontals a little broader than long.

Ventrals 140; rostral broader than deep;
  six upper labials .... 7. *peruvianus*, p. 305.
Ventrals 143-158; rostral broader than
  deep; seven upper labials .... 8. *guentheri*, p. 305.
Ventrals 175-191; rostral nearly as
  deep as broad; seven upper labials .. 9. *bocourti*, p. 306.

        ** Præfrontals as long as broad; ventrals 146-161.

Rostral once and a half as broad as deep;
  frontal slightly broader than long .. 10. *maculatus*, p. 306.
Rostral nearly as deep as broad; frontal
  as long as broad or a little longer.... 15. *crassicaudatus*, p. 310.

        *** Præfrontals a little longer than broad.

Frontal as long as broad or slightly
  longer than broad; loreal twice and
  a half to thrice and a half as long as
  broad; ventrals 163-182 .... 11. *major*, p. 307.
Frontal a little longer than broad; loreal
  twice as long as broad; ventrals 160. 12. *isthmicus*, p. 307.
Frontal as long as broad or a little
  broader than long; loreal twice and
  a half to thrice and a half as long as
  deep; ventrals 143-160 .... 13. *badius*, p. 308.

      β. A single postocular; frontal longer than broad; eight
          upper labials .... 14. *torquatus*, p. 309.

    b. Scales in 15 rows.
      α. Snout obtuse; ventrals 150-172; subcaudals 16-32.
        * Eight upper labials.

Præfrontals longer than broad; both
  postoculars in contact with temporal. 16. *duboisi*, p. 310.
Præfrontals as long as broad; lower post-
  ocular not in contact with temporal.. 17. *occipitoalbus*, p. 310.

** Seven upper labials.

Frontal a little broader than long .... 18. *reticulatus*, p. 311.
Frontal once and a half as broad as
long .... 19. *emmeli*, p. 311.

    β. Snout more or less pointed; ventrals 125–150; subcaudals 11–19 .... 20. *trilineatus*, p. 312.

3. A single pair of very large chin-shields; anal divided.
    21. *quadrivirgatus*, p. 312.

II. Symphysial shield in contact with the chin-shields; tail long and thick .... 22. *favæ*, p. 313.

TABLE SHOWING NUMBERS OF SCALES AND SHIELDS.

	Scales.	Ventrals.	Anal.	Subcaudals.	Labials.	Postoculars.
elaps	15	144–167	1	23–37	6	1
latifrons	17	142–150	1	35–40	6	1
modestus	17	173	1	38	6	2
vittatus	15	148	1	18	8	1–2
latifrontalis	17	172–179	1	24–32	6	1–2
longiceps	17	173	1	28	6	1
peruvianus	17	140	1	31	6	2
quentheri	17	143–158	1	19–33	7	2
bocourti	17	175–191	1	25–39	7	2
maculatus	17	148–159	1	21–26	7	2
major	17	163–182	1	30–45	7	2
isthmicus	17	160	1	34	7	2
badius	17	143–160	1	20–47	7–8	2
torquatus	17	140–165	1	35–47	8	1
crassicaudatus	17	146–161	1	19–27	6–7	2
duboisi	15	150–172	1	16–32	8	2
occipitoalbus	15	150	1	22	8	2
reticulatus	15	156	1	26	7	2
emmeli	15	167–170	1	28–30	7	2
trilineatus	15	125–150	1	11–19	7–8	1–2
quadrivirgatus	15	126–144	2	25–43	7	2
favæ	17	171–185	1	58–66	7	2

1. Atractus elaps.

Rhabdosoma elaps, *Günth. Cat.* p. 241 (1858), *and Zool. Rec.* 1865, p. 151.
Rabdosoma brevifrenum, *Jan, Arch. Zool. Anat. Phys.* ii. 1862, p. 12, *and Icon. Gén.* 10, pl. iv. fig. 1 (1865).
Geophis elaps, *Günth. Ann. & Mag. N. H.* (4) i. 1868, p. 415.

Snout rounded. Rostral rather large, broader than deep, the portion visible from above at least half as long as its distance from the frontal; internasals nearly twice as broad as long; frontal as long as broad or a little broader than long, as long as its distance from the end of the snout, shorter than the parietals; loreal short, not or but slightly longer than deep; one postocular; temporals 1+2; six upper labials, third and fourth entering the eye; three or four lower labials in contact with the single pair of moderately large chin-shields, which are separated from the symphysial. Scales in 15 rows. Ventrals 144–167; anal entire; subcaudals 23–37. Black above, with light annuli which widen towards the belly; or reddish with black annuli disposed in pairs; belly barred black and yellow.

Total length 610 millim.; tail 60.

Ecuador and Brazil.

a. ♂ (V. 144; C. 29).	W. Ecuador.	Mr. Fraser [C.]. (Type.)
b. ♀ (V. 167; C. 26).	Canelos, Ecuador.	Mr. Buckley [C.].
c. ♂ (V. 163; C. 37).	Pebas.	Mr. Hauxwell [C.].
d. ♀ (V. 155; C. 28).	Upper Amazon.	Mr. E. Bartlett [C.].

2. Atractus latifrons.

Geophis latifrons, *Günth. Ann. & Mag. N. H.* (4) i. 1868, p. 415, pl. xix. fig. B, and ix. 1872, p. 15.

Snout rounded. Rostral rather large, a little broader than deep, the portion visible from above at least half as long as its distance from the frontal; internasals once and a half to twice as broad as long; frontal as long as broad or a little broader than long, as long as its distance from the end of the snout, shorter than the parietals; loreal short, not or but slightly longer than deep; one postocular; temporals 1+2; six upper labials, third and fourth entering the eye; three or four lower labials in contact with the single pair of moderately large chin-shields, which are separated from the symphysial. Scales in 17 rows. Ventrals 142–150; anal entire; subcaudals 35–40. Body with black and yellow or red rings, which may be broken up and irregular on the belly, the scales on the red rings black-edged; head black above, with a yellow vertical bar in front of the eye, and a triangular yellow blotch on the temple, extending or not across the occiput.

Total length 410 millim.; tail 70.

Upper Amazon.

A. Black rings disposed in 12 to 17 pairs, each enclosing a narrow yellow interspace and separated by broad red rings; a yellow bar across the occiput.

a. Hgr. (V. 150; C. 36).	Pebas.	Mr. Hauxwell [C.]. (Type.)
b. Hgr. (V. 142; C. 35).	Moyobamba.	Mr. A. H. Roff [C.].
c. ♂ (V. 143; C. ?).	Upper Amazon.	Mr. E. Bartlett [C.].

B. Black rings 65, separated by narrow yellow rings; no yellow across the occiput.

d. ♂ (V. 146; C. 40).	Upper Amazon.	H. W. Bates, Esq. [C.].

3. Atractus modestus. (Plate XV. fig. 1.)

Rhabdosoma crassicaudatum (*non* Dum. & Bibr.), *Günth. Proc. Zool. Soc.* 1859, p. 411.

Snout rounded. Rostral rather large, broader than deep, the portion visible from above one third as long as its distance from the frontal; internasals moderately large, a little broader than long; frontal as long as broad, as long as its distance from the rostral, shorter than the parietals; loreal once and a half as long as deep; two postoculars; temporals 1+2; six upper labials, third and fourth entering the eye; four lower labials in contact with the single pair of moderately large chin-shields, which are separated from the symphysial. Scales in 17 rows. Ventrals 173; anal entire; subcaudals 38. Uniform plumbeous above, on the sides of the ventrals, and on the subcaudals; ventrals yellowish in the middle.

Total length 380 millim.; tail 50.

Ecuador.

a. ♂ (V. 173; C. 38).   W. Ecuador.   Mr. Fraser [C.].

4. Atractus vittatus. (Plate XV. fig. 2.)

Rhabdosoma crassicaudatum, part., *Günth. Cat.* p. 11 (1858).

Snout obtusely pointed. Rostral small, as deep as broad, visible from above; internasals very small; præfrontals as long as broad; frontal as long as broad, as long as its distance from the rostral, much shorter than the parietals; loreal thrice as long as deep; one (or two) postoculars; temporals 1+1; eight upper labials, fourth, fifth, and sixth entering the eye; two pairs of chin-shields, the anterior rather elongate, in contact with four lower labials, and separated from the symphysial, the posterior two thirds the length of the anterior and in contact along their inner borders. Scales in 15 rows. Ventrals 148; anal entire; subcaudals 18. Dark brown above, with four black longitudinal lines, the space between the inner and outer pair lighter, reddish brown; belly yellow, much spotted with black; two black stripes along the lower surface of the tail.

Total length 405 millim.; tail 35.

Venezuela.

a. ♀ (V. 148; C. 18).   Caracas.

5. Atractus latifrontalis.

Geophis latifrontalis, *Garman, N. Am. Rept.* p. 103 (1883), *and Bull. Essex Inst.* xix. 1888, p. 127.

Rhabdosoma mutitorques, *Cope, Proc. Amer. Philos. Soc.* xxii. 1885, p. 384.

Geophis mutitorques, *Günth. Biol. C.-Am., Rept.* p. 93 (1893).

Snout broad, rounded. Rostral broader than deep; internasals small; præfrontals as long as broad; frontal broader than long;

one postocular (exceptionally two); temporals 1+2; six upper labials, third and fourth entering the eye; two pairs of chin-shields, posterior half as long as anterior. Scales in 17 rows. Ventrals 172-179; anal entire; subcaudals 24-32. Uniform dark plumbeous above, margins of scales lighter; a yellow occipital bar may be present, crossing the parietal shields; ventrals white, mottled with leaden, or uniform leaden.

Total length 455 millim.; tail 45.

Mexico (San Luis Potosi and Hidalgo).

6. Atractus longiceps.

Rhabdosoma longiceps, *Cope, Proc. U.S. Nat. Mus.* ix. 1886, p. 189.

Agrees in most respects with *A. latifrontalis*, but snout more elongate and præfrontals much longer than broad. Ventrals 173; subcaudals 28. Extremity of tail with a compressed horny cap. Colour everywhere blackish; some brownish shades on the sides near the head; free edges of ventral and subcaudal shields, and of lateral scales, lighter.

Total length 445 millim.; tail 44.

Vera Cruz, Mexico.

7. Atractus peruvianus.

Rabdosoma peruvianum, *Jan, Arch. Zool. Anat. Phys.* ii. 1862, p. 12, and *Icon. Gén.* 10, pl. iv. fig. 2 (1865).

Snout obtuse. Rostral much broader than deep, just visible from above; internasals small; præfrontals not quite as long as broad; frontal small, as long as broad, as long as the præfrontals, much shorter than the parietals; loreal at least twice as long as deep; two postoculars; temporals 1+2; six upper labials, third and fourth entering the eye; four lower labials in contact with the single pair of chin-shields, which are moderately large and separated from the symphysial. Scales in 17 rows. Ventrals 140; anal entire; subcaudals 31. Brown above, with five longitudinal series of black spots; ventrals black, with a white spot at the outer end; lower surface of tail white, spotted with black.

Total length 280 millim.; tail 40.

Peru.

8. Atractus guentheri.

Rhabdosoma maculatum, part., *Günth. Cat.* p. 241 (1858).
Geophis guentheri, *Wucherer, Proc. Zool. Soc.* 1861, p. 115, pl. xix. fig. 1; *Günth. Zool. Rec.* 1865, p. 151.
Rabdosoma univittatum, *Jan, Arch. Zool. Anat. Phys.* ii. 1862, p. 15, and *Icon. Gén.* 11, pl. ii. fig. 2 (1865).

Snout rounded. Rostral small, broader than deep, just visible from above; internasals very small; præfrontals broader than long; frontal a little broader than long (or as long as broad), as long as its distance from the end of the snout, much shorter than the parietals; loreal twice as long as deep; two postoculars; temporals 1+2 (or 2+2); seven upper labials, third and fourth entering the eye; four (or three) lower labials in contact with the single pair of chin-shields, which are moderately large and separated from the symphysial. Scales in 17 rows. Ventrals 143–158; anal entire; subcaudals 19–33. Pale yellowish brown or orange above, with or without dark edges to the scales; uniform whitish beneath.

Total length 330 millim.; tail 45.

Brazil; Venezuela.

A. A broad black vertebral stripe and a black line along each side, replaced by a series of spots on the anterior part of the body.

a. ♂ (V. 143; C. 30).   Caunavieras, south   Dr. Wucherer [C.].
                         of Bahia.   (Type.)

B. Two or four longitudinal series of black, light-edged, alternating spots or narrow cross bars.

b, c. ♀ (V. 152; C. 19)   Rio Janeiro.   A. Fry, Esq. [P.].
& hgr. (V. 158; C. 21).
d, e. Yg. (V. 145, 143; C.   Brazil?   G. Lennox Conyngham,
26, 27).   Esq. [P.].

9. Atractus bocourti.

Rabdosoma maculatum, part., *Bocourt, Miss. Sc. Mex., Rept.* p. 540, pl. xxxv. fig. 1 (1883).

Snout obtuse. Rostral small, nearly as deep as broad, just visible from above; internasals very small; præfrontals slightly broader than long; frontal a little broader than long, as long as its distance from the rostral, much shorter than the parietals; loreal twice and a half as long as deep; two postoculars; temporals 1+2; seven upper labials, third and fourth entering the eye; four lower labials in contact with the single pair of chin-shields, which are moderately large and separated from the symphysial. Scales in 17 rows. Ventrals 175–191; anal entire; subcaudals 25–39. Pale brown above, with five longitudinal series of small black spots; a black streak on each side of the head, passing through the eye, another along the nape, and a fifth along the tail; yellowish beneath, with some irregular black blotches and on each side a series of partly confluent small spots or short streaks, forming two lines along the belly.

Total length 390 millim.; tail 40.

Northern Peru, Ecuador.

a. ♀ (V. 175; C. 25).   Acomayo, N. Peru.   Messrs. Veitch [P.].

10. Atractus maculatus. (Plate XIV. fig. 3.)

Isoscelis maculata, *Günth. Cat.* p. 204 (1858).
Rhabdosoma maculatum, part., *Günth. l. c.* p. 241.
Rabdosoma zebrinum, *Jan, Arch. Zool. Anat. Phys.* ii. 1862, p. 15.

Snout rounded. Rostral small, once and a half as broad as deep, just visible from above; internasals very small, a little broader than long; præfrontals as long as broad; frontal slightly broader than long, as long as its distance from the end of the snout, much shorter than the parietals; loreal nearly thrice as long as deep; two postoculars; temporals 1+2; seven upper labials, third and fourth entering the eye; four lower labials in contact with the single pair of chin-shields, which are moderately large and separated from the symphysial. Scales in 17 rows. Ventrals 148-159; anal entire; subcaudals 21-26. Pale yellowish or reddish brown above, some of the scales edged with black, with numerous regular black transverse spots or bars; upper lip and lower parts uniform whitish.

Total length 335 millim.; tail 40.

Brazil.

a. ♂ (V. 148; C. 26).   Brazil?   (Type.)
b. ♀ (V. 159; C. 21).   Pernambuco.   W. A. Forbes, Esq. [P.].

11. Atractus major.

Rhabdosoma maculatum (*non Günth.*, 1858), *Günth. Proc. Zool. Soc.* 1859, p. 411.
Rabdosoma badium, part., *Jan, Arch. Zool. Anat. Phys.* ii. 1862, p. 13, *and Icon. Gén.* 11, pl. i. fig. 1 (1865).

Snout obtuse. Rostral small, nearly as deep as broad, just visible from above; internasals very small; præfrontals longer than broad; frontal as long as broad, or slightly longer than broad, as long as its distance from the internasals or from the rostral, much shorter than the parietals; loreal twice and a half to thrice and a half as long as deep; two postoculars; temporals 1+2; seven upper labials, third and fourth entering the eye; three lower labials in contact with the single pair of chin-shields, which are moderately large, and separated from the symphysial. Scales in 17 rows. Ventrals 163-182; anal entire; subcaudals 30-45. Brown above, with darker spots, which may be large and light-edged, forming a handsome pattern; labials yellowish, with brown sutures; yellowish beneath, dotted or spotted with brown or black.

Total length 700 millim.; tail 80.

Ecuador, Brazil.

a-b, c. ♀ (V. 174, 168; C. 43, 30) & yg. (V. 163; C. 43).   W. Ecuador.   Mr. Fraser [C.].
d. ♀ (V. 170; C. 31).   Intac, Ecuador.   Mr. Buckley [C.].
e. Yg. (V. 182; C. 36).   Pallatanga, Ecuador. Mr. Buckley [C.].
f. Yg. (V. 169; C. 45).   Canelos, Ecuador.   Mr. Buckley [C.].

12. Atractus isthmicus.

Rabdosoma zebrinum (*non Jan*), *Bocourt, Miss. Sc. Mex., Rept.* p. 539, pl. xxxiv. fig. 1 (1883).

Rhegnops zebrinus, *Cope, Proc. Amer. Philos. Soc.* xxii. 1885, p. 178.
Geophis zebrina, *Günth. Biol. C.-Am., Rept.* p. 94 (1893).

Snout rounded. Rostral small, nearly as deep as broad; internasals very small, broader than long; præfrontals a little longer than broad; frontal a little longer than broad, as long as its distance from the end of the snout, much shorter than the parietals; loreal twice as long as deep; two postoculars; temporals 1+2; seven upper labials, third and fourth entering the eye; four lower labials in contact with the single pair of chin-shields, which are moderately large and separated from the symphysial. Scales in 17 rows. Ventrals 160; anal entire; subcaudals 34. Pinkish yellow above, with about forty irregular transverse brown spots; lower parts dotted all over with reddish brown.

Total length 155 millim.; tail 18.

Tehuantepec.

13. Atractus badius.

Brachyorrhos badius, *Boie, Isis*, 1827, p. 540.
—— flammigerus, *Boie, l. c.*
Calamaria badia, part., *Schleg. Phys. Serp.* ii. p. 35 (1837).
Rabdosoma badium, *Dum. & Bibr.* vii. p. 95 (1854); *Günth. Cat.* p. 11 (1858).
—— badium, part., *Jan, Arch. Zool. Anat. Phys.* ii. 1862, p. 13, *and Icon. Gén.* 10, pl. iv. fig. 3 (1865).
—— badium, var. multicinctum, *Jan, ll. cc.* p. 14, pl. iv. fig. 5.
? Rabdosoma badium, var. rubinianum, *Jan, ll. cc.* fig. 4.
? Rabdosoma dubium, *Jan, ll. cc.* p. 18, *Icon.* 11, pl. iii. fig. 4.
Rhabdosoma microrhynchum, *Cope, Proc. Ac. Philad.* 1868, p. 102.
Rabdosoma maculatum, part., *Bocourt, Miss. Sc. Mex., Rept.* p. 539, pl. xxxiv. fig. 2 (1883).
Geophis badius, *Boettg. Ber. Senckenb. Ges.* 1888, p. 192.

Snout obtuse. Rostral small, a little broader than deep, just visible from above; internasals very small; frontal as long as broad or a little broader than long, as long as or shorter than its distance from the end of the snout, much shorter than the parietals; loreal twice and a half to thrice and a half as long as deep; two postoculars; temporals 1+2; seven (exceptionally eight) upper labials, third and fourth (or fourth and fifth) entering the eye; three or four lower labials in contact with the single pair of chin-shields, which are rather elongate, moderately broad, and separated from the symphysial. Scales in 17 rows. Ventrals 143–160; anal entire; subcaudals 20–47. Coloration very variable.

Total length 430 millim.; tail 60.

Guianas, Northern Brazil, Peru, Ecuador.

A. Dark brown above; anterior part of body paler brown or reddish, with pairs of black annuli; belly uniform whitish or with a median series of black spots.

a. ♀ (V. 155; C. 38).   Surinam.
b–c. ♀ (V. 150, 153;   Surinam.   Hr. Kappler [C.].
   C. 38, 38).

d. ♀ (V. 154; C. 35). Surinam. C. W. Ellascombe, Esq. [P.].
e. ♂ (V. 146; C. 47). Cayenne.

B. Pale reddish brown above, with more or less regular black cross bars; belly spotted with brown.

f. ♀ (V. 154; C. ?). Surinam. Hr. Kappler [C.].

C. Black above, with pale brown cross bars; belly with a median series of black spots.

g. ♂ (V. 158; C. 38). Yurimaguas, Hual- Dr. Hahnel [C.]. laga R., N. Peru.

D. Pale reddish brown above; snout black, back of head yellow, followed by a black collar; back with two alternating series of black spots, connected on the anterior part of the body by a black vertebral line; belly dotted all over with brown.

h. ♂ (V. 147; C. 30). Manáos, Brazil. Sr. A. Peixoto [C.].

E. Brown above, with black-edged yellow cross bands; back of head yellow; belly dotted with brown, with two series of brown spots along the middle line.

i–k. ♀ (V. 145, 154; Para.
C. 24, 24).

14. Atractus torquatus.

Calamaria badia, part., *Schleg. Phys. Serp.* ii. p. 35 (1837).
Rabdosoma torquatum, *Dum. & Bibr.* vii. p. 101 (1854).
—— varium, *Jan, Arch. Zool. Anat. Phys.* ii. 1862, pp. 18, 125, and *Icon. Gén.* 11, pl. iii. fig. 3 (1865).

Snout obtusely pointed. Rostral small, nearly as deep as broad, just visible from above; internasals very small; præfrontals at least as long as broad; frontal once and one third to once and a half as long as broad, as long as its distance from the end of the snout, shorter than the parietals; loreal three times as long as deep; one postocular; temporals 1+1 or 1+2; eight upper labials, fourth and fifth entering the eye; four (or three) lower labials in contact with the single pair of chin-shields, which are rather elongate, moderately broad, and separated from the symphysial. Scales in 17 rows. Ventrals 140–165; anal entire; subcaudals 35–47. Pale brown above, with small darker spots or narrow cross bars; yellowish beneath, with small brown spots.
Total length 600 millim.; tail 75.
Guianas.

a, b. ♂ (V. 155; C. 47) & Demerara Falls.
♀ (V. 165; C. 43).

15. **Atractus crassicaudatus.**

Rabdosoma crassicaudatum, *Dum. & Bibr.* vii. p. 103 (1854); *Jan, Arch. Zool. Anat. Phys.* ii. 1862, p. 15, *and Icon. Gén.* 11, pl. ii. fig. 1 (1865).
Rhabdosoma crassicaudatum, part., *Günth. Cat.* p. 11 (1858).

Snout obtuse. Rostral small, nearly as deep as broad, just visible from above; internasals very small; præfrontals as long as broad; frontal as broad as long or a little longer than broad, as long as its distance from the end of the snout, much shorter than the parietals; loreal at least twice as long as deep; two postoculars; temporals 1+2; seven upper labials (rarely six), third and fourth (or third) entering the eye; three lower labials in contact with the single pair of chin-shields, which are moderately large and separated from the symphysial. Scales in 17 rows. Ventrals 146–161; anal entire; subcaudals 19–27. Dark purplish brown or blackish above and below, with small yellowish spots on the back and larger ones on the belly; outer row of scales yellowish; a yellowish blotch on each temple.

Total length 230 millim.; tail 25. Reaches to 420 millim.

Colombia and Venezuela.

a. ♀ (V. 150; C. 22).   Colombia.
b. Yg. (V. 146; C. 23).   Bogota.
c. ♂ (V. 147; C. 25).   Caracas.
d, e. Skulls.   Caracas.

16. **Atractus duboisi.**

Rabdosoma duboisi, *Bouleng. Bull. Soc. Zool. France,* 1880, p. 44.

Snout obtuse. Rostral small, nearly as deep as broad; internasals very small; præfrontals longer than broad; frontal slightly broader than long, as long as its distance from the end of the snout, shorter than the parietals; loreal thrice as long as deep; two postoculars; temporals 1+2; eight upper labials, fourth and fifth entering the eye; four or five lower labials in contact with the single pair of chin-shields, which are rather elongate, moderately broad, and separated from the symphysial. Scales in 15 rows. Ventrals 150–172; anal entire; subcaudals 16–32. Dark brown or blackish above, with two longitudinal series of small round yellow spots; ventrals dark brown or black in the middle, yellow on the sides; subcaudals black.

Total length 120 millim.; tail 9. Reaches to 335 millim.; tail 25.

Andes of Ecuador.

a. Yg. (V. 172; C. 16).   Intac.   Mr. Buckley [C.].

17. **Atractus occipitoalbus.**

Rabdosoma occipitoalbum, *Jan, Arch. Zool. Anat. Phys.* ii. 1862, p. 16, *and Icon. Gén.* 11, pl. ii. fig. 4 (1865).

Snout obtuse. Rostral small, as deep as broad, just visible from above; internasals small; præfrontals as long as broad; frontal as long as broad, as long as its distance from the end of the snout, shorter than the parietals; loreal nearly three times as long as deep; two postoculars, lower very small and not in contact with the temporal; temporals 1+2; eight upper labials, fourth and fifth entering the eye; four lower labials in contact with the single pair of chin-shields, which are rather elongate, moderately broad, and separated from the symphysial. Scales in 15 rows. Ventrals 150; anal entire; subcaudals 22. Black above and beneath; parietal shields, temporal and gular regions white.

Total length 205 millim.; tail 24.

Western Andes of Ecuador.

18. Atractus reticulatus. (PLATE XV. fig. 3.)

Geophis reticulatus, *Bouleng. Ann. & Mag. N. H.* (5) xvi. 1885, p. 87.

Snout rounded. Rostral a little broader than deep, just visible from above; internasals very small; præfrontals as long as broad; frontal a little broader than long, as long as its distance from the end of the snout, shorter than the parietals; loreal twice as long as deep; two postoculars; temporals 1+2; seven upper labials, third and fourth entering the eye; four lower labials in contact with the single pair of chin-shields, which are moderately large and separated from the symphysial. Scales in 15 rows. Ventrals 156; anal entire; subcaudals 26. Pale brownish above, each scale edged with dark brown; an ill-defined dark brown collar; lower parts uniform whitish.

Total length 315 millim.; tail 40.

Southern Brazil.

a. ♀ (V. 156, C. 26). S. Lorenzo, Rio Grande do Sul. Dr. H. von Ihering [C.]. (Type.)

19. Atractus emmeli.

Geophis emmeli, *Boetty. Ber. Senckenb. Ges.* 1888, p. 192, fig.

Snout rounded. Rostral somewhat broader than deep, just visible from above; internasals very small; præfrontals as long as broad; frontal nearly once and a half as broad as long, a little shorter than the præfrontals, much shorter than the parietals; loreal more than twice as long as deep; two postoculars; temporals 1+2; seven upper labials, third and fourth entering the eye; four lower labials in contact with the single pair of chin-shields, which are rather elongate, moderately broad, and separated from the symphysial. Scales in 15 rows. Ventrals 167–170; anal entire; subcaudals 28–30. Dark brown or olive-grey above, somewhat paler across the back of the head; yellowish, greenish, or greyish beneath, spotted with black; posterior ventrals and subcaudals black, with light edges.

Total length 303 millim.; tail 34.

River Mapiri, Bolivia.

20. Atractus trilineatus.

Atractus trilineatus, *Wagl. Isis*, 1828, p. 742, pl. x. figs. 1-4.
Rabdosoma lineatum, *Dum. & Bibr.* vii. p. 105 (1854); *Jan, Arch. Zool. Anat. Phys.* ii. 1862, p. 17, *and Icon. Gén.* 11, pl. ii. fig. 5 (1865); *Garm. Proc. Amer. Philos. Soc.* xxiv. 1887, p. 280.
Rhabdosoma lineatum, part., *Günth. Cat.* p. 11 (1858).
Rabdosoma trivirgatum, *Jan, ll. cc.* p. 17, pl. iii. fig. 1.
—— punctatovittatum, *Jan, ll. cc.* p. 17, pl iii. fig. 2.
Geophis lineatus, *Günth. Ann. & Mag. N. H.* (4) ix. 1872, p. 15.

Snout obtusely pointed. Rostral small, as deep as broad, visible from above; internasals small; præfrontals at least as long as broad; frontal as broad as long, or a little broader than long, about as long as its distance from the rostral, shorter than the parietals; loreal at least twice as long as deep; two postoculars (rarely one); temporals 1+2, rarely 2+2; eight (rarely seven) upper labials, fourth and fifth (or third and fourth) entering the eye; four or five lower labials in contact with the single pair of chin-shields, which are rather elongate, moderately broad, and separated from the symphysial. Scales in 15 rows. Ventrals 125-150; anal entire; subcaudals 11-19. Brown above, with three or four darker longitudinal lines; white beneath, uniform or with a few brown dots.

Total length 225 millim.; tail 15.

Guianas, Trinidad.

a-c. ♂ (V. 125; C. 18) & ♀ (V. 137, 139; C. 12, 14).	Berbice.	
d. ♀ (V. 131; C. 11).	Demerara River.	S. S. Lawrence, Esq. [P.].
e-f. ♂ (V. 131; C. 19) & ♀ (V. 140; C. 12).	Maccasseema, Brit. Guiana.	W. L. Sclater, Esq. [P.].
g. ♀ (V. 143; C. 13).	Brit. Guiana.	
h-l. ♀ (V. 147, 140; C. 13, 13) & yg. (V. 140, 136; C. 14, 19).	Trinidad.	H. B. Guppy, Esq. [P.].
m-q. ♂ (V. 129; C. 17) & yg. (V. 130, 140, 135, 137; C. 17, 12, 12, 19).	S. America.	Mr. Mather [C.].
r. ♀ (V. 133; C. 12).	S. America.	C. Baker, Esq. [P.].

21. Atractus quadrivirgatus.

Rhabdosoma lineatum, part., *Günth. Cat.* p. 11 (1858).
Adelphicos quadrivirgatum, *Jan, Arch. Zool. Anat. Phys.* ii. 1862, p. 19, *and Icon. Gén.* 11, pl. iii. fig. 5 (1865); *F. Müll. Verh. nat. Ges. Basel*, vi. 1878, p. 654; *Bocourt, Miss. Sc. Mex., Rept.* p. 554, pl. xxxii. figs. 11 & 12 (1883); *Günth. Biol. C.-Am., Rept.* p. 94 (1893).

Rhegnops visoninus, *Cope, Proc. Ac. Philad.* 1866, p. 128.
—— sargii, *Fischer, Jahrb. Hamb. Wiss. Anst.* ii. 1885, p. 92.
Adelphicus sargii, *Cope, Bull. U.S. Nat. Mus.* no. 32, 1887, p. 85.
—— visoninus, *Cope, l. c.*

Snout obtuse. Rostral small, a little deeper than broad, visible from above; præfrontals nearly as long as broad; frontal nearly as broad as long, as long as its distance from the rostral or from the tip of the snout, shorter than the parietals; loreal at least twice as long as deep; two postoculars; temporals 1+1, rarely 1+2; seven upper labials, third and fourth entering the eye; a single pair of large chin-shields, separated from the symphysial; in consequence of the great width of the chin-shields, the three anterior lower lalials are much reduced, very narrow. Scales in 15 rows. Ventrals 126–144; anal divided; subcaudals 25–43. Pale reddish brown above, with four or five dark brown or blackish longitudinal lines or stripes; white beneath, with a brown line along the middle of the tail, and sometimes also along the belly.

Total length 365 millim.; tail 55.

Central America.

a. ♀ (V. 133; C. 35).	Mexico.	M. Sallé [C.].
b. Yg. (V. 126; C. 43).	Honduras.	Mr. Dyson [C.].
c. Hgr. ♀ (V. 141; C. 28).	Guatemala.	Stuttgart Mus. [E.].
		(One of the types of *R. sargii.*)
d. ♀ (V. 144; C. 28).	—— ?	

22. Atractus favæ.

Calamaria favæ, *Filippi, Cat. Rag. Serp. Mus. Pavia*, p. 16 (1840).
Rabdosoma longicaudatum, *Dum. & Bibr.* vii. p. 106 (1854); *Günth. Cat.* p. 12 (1858); *Jan, Arch. Zool. Anat. Phys.* ii. 1862, p. 15, and *Icon. Gén.* 11, pl. i. fig. 2 (1865).
—— favæ, *Jan, ll. cc.* p. 16, pl. ii. fig. 3.

Snout rounded. Rostral moderately large, a little broader than deep, just visible from above; internasals small, broader than long; præfrontals broader than long; frontal as long as broad, as long as its distance from the frontal, much shorter than the parietals; loreal nearly twice as long as deep; a small præocular sometimes present, above the loreal; two postoculars; temporals 1+2; seven upper labials, third and fourth entering the eye; a single pair of moderately large chin-shields, in contact with the symphysial and, on each side, with three lower labials. Scales in 17 rows. Ventrals 171–185; anal entire; subcaudals 58–66. Tail long and thick, ending in an obtuse point. Brown above, with darker spots and a rather indistinct darker vertebral line; yellow (red?) beneath, with large black blotches, which mostly alternate.

Total length 420 millim.; tail 110.

Brazil?

a. ♂ (V. 173; C. 58).	—— ?	Dr. Günther [P.].
b. ♀ (V. 172; C. 66).	—— ?	

113. GEOPHIS.

Catostoma (*non Lesueur*), *Wagler, Syst. Amph.* p. 194 (1830).
Geophis, *Wagler, l. c.* p. 342; *Bocourt, Miss. Sc. Mex., Rept.* p. 528 (1883).
Rabdosoma, part., *Dum. & Bibr. Erp. Gén.* vii. p. 91 (1854); *Günth. Cat. Col. Sn.* p. 10 (1858); *Jan, Arch. Zool. Anat. Phys.* ii. 1862, p. 10.
Colobognathus, *Peters, Mon. Berl. Ac.* 1859, p. 275.
Geophidium, *Peters, Mon. Berl. Ac.* 1861, p. 923.
Elapoides, part., *Jan, l. c.* p. 20.
Colophrys, *Cope, Proc. Ac. Philad.* 1868, p. 130.
Parageophis, *Bocourt, l. c.* p. 534.

Maxillary short, with 7 to 12 small, equal teeth; mandibular teeth subequal. Head small, not distinct from neck; eye small, with round or vertically subelliptic pupil; nostril between two nasals; no præocular, loreal and præfrontal entering the eye; internasals and supraoculars present or absent; parietals in contact with labials. Body cylindrical; scales smooth or keeled, without apical pits, in 15 or 17 rows; ventrals rounded. Tail moderate or short; subcaudals in two rows.

Central and South America.

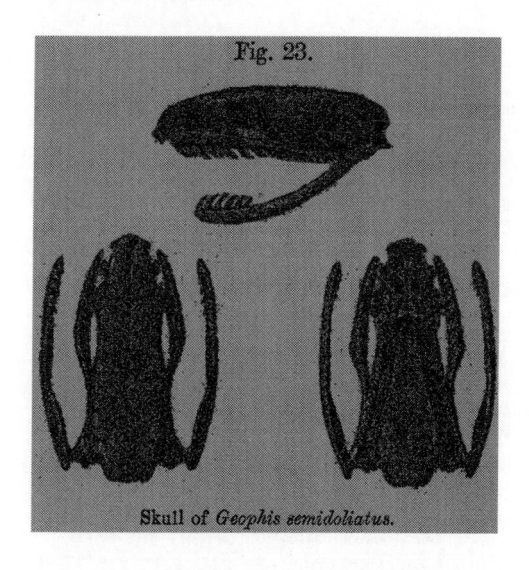

Fig. 23.

Skull of *Geophis semidoliatus*.

Synopsis of the Species.

I. A single pair of chin-shields; scales in 15 rows; ventrals 170.
   1. *poeppigii*, p. 316.

113. GEOPHIS.

II. Two pairs of chin-shields.

   A. Maxillary bone extending in front much beyond the palatines; scales perfectly smooth.

A supraocular; third upper labial entering the eye; scales in 15 rows; ventrals 147–174 .... 2. *semidoliatus*, p. 316.
No supraocular; third and fourth upper labials entering the eye; scales in 17 rows; ventrals 135–144 .... 3. *rhodogaster*, p. 317.

   B. Maxillary bone extending in front a little beyond the palatines.

Scales in 15 rows, perfectly smooth; ventrals 176 .... 4. *dugesii*, p. 317.
Scales in 15 rows, striated and keeled; ventrals 129–133.................. 5. *sallæi*, p. 318.
Scales in 17 rows, striated and keeled on the posterior half of the body; ventrals 131–154 .... 6. *chalybæus*, p. 318.

   C. Maxillary bone not extending in front beyond the palatines.

     1. Upper portion of rostral not measuring more than one third its distance from the frontal; scales in 15 rows.

Scales perfectly smooth on the anterior part of the body, striated and very feebly keeled further back .... 7. *hoffmanni*, p. 319.
Scales very distinctly striated and keeled, the keel rather strong on the posterior part of the body and on the tail ...... 8. *dolichocephalus*, [p. 320.

     2. Upper portion of rostral measuring at least two thirds its distance from the frontal; scales all smooth, or those at the base of the tail very faintly keeled.

       a. Scales in 15 rows.

A supraocular; præfrontals as broad as long .... 9. *petersii*, p. 321.
A supraocular; præfrontals longer than broad........................... 10. *championi*, p. 321.
No supraocular..................... 11. *godmani*, p. 322.

       b. Scales in 17 rows.

First lower labial in contact with its fellow behind the symphysial............. 12. *dubius*, p. 322.
Symphysial in contact with the anterior chin-shields .... 13. *rostralis*, p. 323.

1. Geophis poeppigii.

Rabdosoma pöppigi, *Jan, Arch. Zool. Anat. Phys.* ii. 1862, p. 11, *and Icon. Gén.* 10, pl. iii. fig. 4 (1865).

Eye very small, much shorter than its distance from the mouth. Snout rounded. Rostral rather large, once and a half as broad as deep, the portion visible from above about half as long as its distance from the frontal; internasals and præfrontals much broader than long; frontal much broader than long, as long as its distance from the rostral, much shorter than the parietals; supraocular small; loreal nearly twice as long as deep, shorter than the two nasals together; one postocular; six upper labials, third and fourth entering the eye, fifth largest; a single pair of chin-shields, in contact with four lower labials and separated from the symphysial. Scales smooth, in 15 rows. Ventrals 170; anal entire; subcaudals ——? Brown above; barred brown and white beneath; on the anterior part of the body the white bars of the ventral surface extend some way up the sides; upper lip and temporal region white.

Total length 570 millim.

Brazil.

2. Geophis semidoliatus.

Rabdosoma semidoliatum, *Dum. & Bibr.* vii. p. 93 (1854).
Rhabdosoma semidoliatum, part., *Günth. Cat.* p. 10 (1858).
Geophis semidoliatus, *Peters, Mon. Berl. Ac.* 1859, p. 276; *Bocourt, Miss. Sc. Mex., Rept.* p. 534, pl. xxxi. fig. 7 (1883); *Günth. Biol. C.-Am., Rept.* p. 90 (1893).
Catostoma semidoliatum, *Cope, Proc. Ac. Philad.* 1860, p. 339.
Elapoides semidoliatus, *Jan, Arch. Zool. Anat. Phys.* ii. 1862, p. 22, *and Icon. Gén.* 12, pl. ii. fig. 1 (1865).

Maxillary extending considerably beyond the palatine in front, the first tooth below the nostril. Eye extremely small, its length hardly half its distance from the mouth. Snout rounded, projecting. Rostral rather large, a little broader than deep, the portion visible from above measuring one third to two fifths its distance from the frontal; internasals a little broader than long, about two fifths the length of the præfrontals, which are nearly as long as broad; frontal as long as broad, or slightly broader than long; as long as its distance from the rostral, much shorter than the parietals; supraocular small; loreal once and two thirds to twice as long as broad; a minute postocular, sometimes fused with the supraocular; five upper labials, third entering the eye; four lower labials in contact with the anterior chin-shields, which are longer than the posterior and separated from the symphysial. Scales in 15 rows, everywhere perfectly smooth. Ventrals 147–174; anal entire; subcaudals 24–28. More or less regularly barred black and yellow or orange above, uniform yellowish beneath; head and nape black, with a yellowish bar across the temples and the parietal shields.

Total length 330 millim.; tail 30.

Mexico.

a-c. ♂ (V. 155; C. 28)   Jalapa.
 & ♀ (V. 173, 172;
 C. 25, ?).
d-e. ♂ (V. 147; C. 26)   Jalapa.   Mr. Hoege [C.].
 & ♀ (V. 174; C. 24).
f. Yg. (V. 162; C. 27).   Orizaba.
g. ♂ (V. 147; C. 26).   Huatuzco.   F. D. Godman, Esq. [P.].
h. ♀ (V. 160; C. 25).   Mexico.   Mr. Hugo Finck [C.].
i. Skull.   Jalapa.

3. Geophis rhodogaster.

Colophrys rhodogaster, *Cope, Proc. Ac. Philad.* 1868, p. 130, fig.
Geophis rhodogaster, *Bocourt, Miss. Sc. Mex., Rept.* p. 531, pl. xxxi.
 fig. 12 (1883).
—— chalybæa, part., *Günth. Biol. C.-Am., Rept.* p. 87 (1893).

Maxillary extending considerably beyond the palatine in front, the first tooth below the suture between the first and second labial shields. Eye very small, measuring two thirds to three fourths its distance from the mouth. Snout rounded, feebly projecting. Rostral moderate, slightly broader than deep, the portion visible from above measuring about one third its distance from the frontal; internasals nearly twice as broad as long, one half to two fifths the length of the præfrontals; præfrontals a little broader than long; frontal broader than long, as long as or a little longer than its distance from the end of the snout, a little shorter than the parietals; no supraocular, frontal bordering the eye; loreal nearly twice as long as deep; one postocular; six upper labials, third and fourth entering the eye; three lower labials in contact with the anterior chin-shields, which are as long as or a little longer than the posterior, and separated from the symphysial. Scales in 17 rows, everywhere perfectly smooth. Ventrals 135–144; anal entire; subcaudals 30–43. Uniform dark brown above; upper lip, outer row of scales, and lower parts uniform orange.

Total length 345 millim.; tail 55.

Yucatan, Guatemala, Costa Rica.

a-b. ♂ (V. 141; C. 32) &   Rio Chisoy, below the   O. Salvin, Esq. [C.].
 ♀ (V. 136; C. 33).   town of Cubulco,
          Guatemala.

4. Geophis dugesii.

Geophis dugesii, *Bocourt, Miss. Sc. Mex., Rept.* p. 573, pl. xxxvii.
 fig. 1 (1883); *Dugès, La Naturaleza*, vi. 1884, p. 359, pl. ix. fig. 1.
Elapoides dugesi, *Cope, Proc. Amer. Philos. Soc.* xxii. 1885, p. 386.

Eye quite as long as its distance from the mouth. Snout rounded, feebly prominent. Rostral broader than deep, the portion visible from above measuring hardly one third its distance from the frontal; internasals nearly twice as broad as long, two fifths the length of the præfrontals, which are as broad as long; frontal slightly longer than broad, longer than its distance from the end of the snout, shorter than the parietals; a narrow supraocular; loreal nearly twice as long as deep; one postocular; six upper labials, third and fourth entering the eye; four lower labials in contact with the anterior chin-shields, which are longer than the posterior and separated from the symphysial. Scales smooth, in 15 rows. Ventrals 176; anal entire; subcaudals 44. Black above, with widely separated yellowish cross bars; lower parts uniform yellowish.

Total length 220 millim.; tail 40.

Tangancienaro, Mexico.

5. Geophis sallæi. (Plate XVI. fig. 1.)

Geophis chalybæa, part., *Günth. Biol. C.-Am., Rept.* p. 87 (1893).

Maxillary extending a little beyond the palatine in front, the first tooth corresponding to the suture between the second and third labial shields. Eye very small, measuring two thirds to three fourths its distance from the mouth. Snout rounded, feebly projecting. Rostral moderate, a little broader than deep, the portion visible from above measuring one fourth to one third its distance from the frontal; internasals much broader than long, one fifth to one fourth the length of the præfrontals, which are longer than broad; frontal as long as broad or slightly broader than long, as long as its distance from the end of the snout, shorter than the parietals; a very small supraocular; loreal twice to twice and a half as long as deep; one postocular; six upper labials, third and fourth entering the eye; three or four lower labials in contact with the anterior chin-shields, which are as long as or a little longer than the posterior and separated from the symphysial. Scales in 15 rows, smooth on the nape, finely striated and feebly but very distinctly keeled on the body and base of tail. Ventrals 129-133; anal entire; subcaudals 30-41. Blackish brown above, scales of outer row lighter in the centre; belly uniform yellowish white; subcaudals edged with brown.

Total length 300 millim.; tail 55.

Mexico.

a, b-c. ♂ (V. 129; C. 41)   Mexico.   M. Sallé [C.].
& ♀ (V. 133, 131; C. 36, 30).

6. Geophis chalybæus.

Catostoma chalybæum, *Wagl. Syst. Amph.* p. 194 (1830).
Geophis chalybeus, *Peters, Mon. Berl. Ac.* 1859, p. 275; *Bocourt, Miss. Sc. Mex., Rept.* p. 530, pl. xxxi. fig. 11 (1883).
Elapoides sieboldi, *Jan, Arch. Zool. Anat. Phys.* ii. 1862, p. 21, *and Icon. Gén.* 12, pl. i. fig. 4 (1865).
Catostoma nasale, *Cope, Proc. Ac. Philad.* 1868, p. 131, fig.
Ninia sieboldii, *Garm. N. Am. Rept.* p. 96 (1883).
Rhabdosoma nasale, *Cope, Proc. Amer. Philos. Soc.* xxii. 1885, p. 385.

Elapoides chalybæus, *Cope, l. c.* p. 386.
Geophis chalybæa, part., *Günth. Biol. C.-Am., Rept.* p. 87 (1893).
—— mœsta, part., *Günth. l. c.* p. 90, pl. xxxiii. fig. C.

Maxillary extending a little beyond the palatine in front, the first tooth corresponding to the suture between the second and third labial shields. Eye very small, measuring two thirds to three fourths its distance from the mouth. Snout rounded, feebly projecting. Rostral moderate, as deep as broad or a little broader than deep, the portion visible from above measuring one third to two fifths its distance from the frontal; internasals much broader than long, one fourth to one third the length of the præfrontals, which are longer than broad; frontal as long as broad or slightly broader than long, as long as its distance from the end of the snout, a little shorter than the parietals; a very small supraocular; loreal twice to twice and a half as long as deep; one postocular; six upper labials, third and fourth entering the eye; three or four lower labials in contact with the anterior chin-shields, which are longer than the posterior and separated from the symphysial. Scales in 17 rows, smooth on the anterior part of the body, finely striated and feebly keeled on the posterior part and on the base of the tail. Ventrals 131–154; anal entire; subcaudals 25–40. Blackish brown above, scales of outer row lighter in the centre; young with a yellowish blotch across temples and occiput; belly uniform yellowish white; subcaudals edged with brown.

Total length 330 millim.; tail 45.

Mexico and Guatemala.

a. Yg. (V. 132; C. 40). Amula, Guerrero. Mr. H. H. Smith [C.]; F. D. Godman, Esq. [P.].
b–c. ♀ (V. 138, 140; C. 30, ?). Dueñas, Guatemala. O. Salvin, Esq. [C.].
d. Hgr., very bad state. Pacific coast of Guatemala. O. Salvin, Esq. [C.].

7. Geophis hoffmanni.

Colobognathus hoffmanni, *Peters, Mon. Berl. Ac.* 1859, p. 276, pl. — fig. 2.
Elapoides hoffmanni, *Jan, Arch. Zool. Anat. Phys.* ii. 1862, p. 22, and *Icon. Gén.* 12, pl. ii. fig. 3 (1865).
Geophis mœstus, *Günth. Ann. & Mag. N. H.* (4) ix. 1872, p. 15.
—— hoffmanni, *Bocourt, Miss. Sc. Mex., Rept.* p. 529, pl. xxxi. fig. 8 (1883).
Rhabdosoma mœstum, *Cope, Bull. U.S. Nat. Mus.* no. 32, 1887, p. 85.
Geophis chalybæa, part., *Günth. Biol. C.-Am., Rept.* p. 87 (1893).
—— mœsta, part., *Günth. l. c.* p. 90.

Maxillary not extending beyond palatine in front, the first tooth corresponding to the suture between the second and third labial shields. Eye small, as long as or a little shorter than its distance from the mouth. Snout rounded, feebly projecting. Rostral moderate, as deep as broad or slightly broader than deep, the portion visible from above measuring one fourth to one third its distance from the frontal; internasals much broader than long, one fourth to one third the length of the præfrontals, which are longer than broad; frontal as long as broad or slightly broader than long, as long as or a little longer than its distance from the end of the snout, shorter than the parietals; a very small supraocular; loreal twice to twice and a half as long as deep; one postocular (rarely two); six upper labials, third and fourth entering the eye; three lower labials in contact with the anterior chin-shields, which are a little longer than the posterior and separated from the symphysial. Scales in 15 rows, smooth on the anterior part of the body, finely striated and feebly keeled on the posterior part and on the base of the tail. Ventrals 124-145; anal entire; subcaudals 28-36. Blackish brown above, young with a more or less distinct yellowish blotch across temples and occiput; ventrals and subcaudals brown edged with whitish, or whitish edged with brown.

Total length 450 millim.; tail 70.
Costa Rica.

a. Hgr. ♂ (V. 130; C. 34).	Costa Rica.	Prof. Peters [E.]. (One of the types.)
b-e, f. ♀ (V. 141, 141, 145, 138; C. 34, 34, 36, 34) & hgr. (V. 135; C. 36).	Irazu, Costa Rica.	O. Salvin & F. D. Godman, Esqrs. [P.].
g. Yg. (V. 145; C. 36).	Cartago, Costa Rica.	(Type of *G. mœstus*.)

8. Geophis dolichocephalus.

Colobognathus dolichocephalus, *Cope, Proc. Ac. Philad.* 1871, p. 211 *; *Günth. Zool. Rec.* 1872, p. 72.
Geophis chalybæus, var., *Günth. Ann. & Mag. N. H.* (4) ix. 1872, p. 16.
Catostoma dolichocephalum, *Cope, Journ. Ac. Philad.* (2) viii. 1876, p. 147.
Elapoides dolichocephalus, *Cope, Proc. Amer. Philos. Soc.* xxii. 1885, p. 386.
Geophis dolichocephalus, *Günth. Biol. C.-Am., Rept.* p. 87 (1893).
—— chalybæa, var. quadrangularis, *Günth. l. c.* p. 89, pl. xxxiii. fig. B.

Maxillary not extending beyond palatine in front, the first tooth corresponding to the suture between the second and third labial shields. Eye small, nearly as long as its distance from the mouth. Rostral moderate, as deep as broad, the portion visible from above measuring about one third its distance from the frontal; internasals broader than long, one fourth to one third the length of the præfrontals, which are longer than broad; frontal as long as broad, as

---

* It appeared to me probable that "thirteen longitudinal series of scales," in the description, was a slip of the pen or a misprint for "fifteen longitudinal series." Having communicated with Prof. Cope on this matter, he has kindly examined the type specimen, and informs me that the scales are in 15 rows.

long as its distance from the end of the snout, much shorter than the parietals; a small supraocular; loreal twice and a half as long as deep; one postocular; six upper labials, third and fourth entering the eye; four lower labials in contact with the anterior chin-shields, which are longer than the posterior and separated from the symphysial. Scales in 15 rows, smooth on the nape, finely striated and keeled on the body and tail, the keels quite strong on the posterior part of the back and on the base of the tail. Ventrals 131–138; anal entire; subcaudals 39–46. Black above, with two alternating series of large yellowish spots, some of which are confluent and form cross bars; belly uniform white, tail blackish.

Total length 290 millim.; tail 35.

Costa Rica.

a–b. ♀ (V. 135, 138; C. ?, 46). Cartago, Costa Rica.

9. Geophis petersii. (PLATE XVI. fig. 2.)

Geophis chalybæa, part., *Günth. Biol. C.-Am., Rept.* p. 87 (1893).

Maxillary not extending beyond palatine in front, the first tooth corresponding to the suture between the second and third labial shields. Eye very small, measuring two thirds to three fourths its distance from the mouth. Snout obtusely pointed, prominent. Rostral slightly deeper than broad, the portion visible from above measuring three fourths its distance from the frontal; internasals twice as broad as long, one third the length of the præfrontals, which are as long as broad; frontal as long as broad, a little longer than its distance from the end of the snout, as long as the parietals; a small supraocular; loreal twice and a half as long as broad; one postocular; six upper labials, third and fourth entering the eye; three lower labials in contact with the anterior chin-shields, which are not longer than the posterior and separated from the symphysial. Scales in 15 rows, smooth, a few at the base of the tail faintly keeled. Ventrals 144–145; anal entire; subcaudals 35–36. Uniform dark brown above, white beneath.

Total length 200 millim.; tail 28.

Mexico.

a–b. ♂ (V. 144; C. 36) City of Mexico. Mr. Doorman [C.].
& ♀ (V. 145; C. 35).

10. Geophis championi. (PLATE XVI. fig. 3.)

Geophis chalybæa, part., *Günth. Biol. C.-Am., Rept.* p. 87 (1893).

Maxillary not extending beyond palatine in front, the first tooth corresponding to the suture between the second and third labial shields. Eye very small, measuring two thirds its distance from the mouth. Snout obtusely pointed, prominent. Rostral a little deeper than broad, the portion visible from above measuring three fourths its distance from the frontal; internasals broader than long, half as long as the præfrontals, which are a little longer than broad; frontal nearly as long as broad, as long as its distance from the end of the snout, a little shorter than the parietals; a small supraocular; loreal twice as long as deep; one postocular; six upper labials, third and fourth entering the eye; three lower labials in contact with the anterior chin-shields, which are not longer than the posterior and separated from the symphysial. Scales in 15 rows, all perfectly smooth. Ventrals 125; anal entire; subcaudals 33. Uniform iridescent black above; ventrals and subcaudals whitish edged with blackish.

Total length 250 millim.; tail 40.

Panama.

a. ♂ (V. 125; C. 33).   Chiriqui.   J. G. Champion, Esq. [C.]; F. D. Godman, Esq. [P.].

11. Geophis godmani. (PLATE XVI. fig. 4.)

Geophis chalybæa, part., *Günth. Biol. C.-Am., Rept.* p. 87 (1893).

Maxillary not extending beyond palatine in front, the first tooth corresponding to the suture between the second and third labial shields. Eye small, a little shorter than its distance from the mouth. Snout obtusely pointed, very prominent. Rostral rather large, slightly deeper than broad, the portion visible from above measuring two thirds or three fourths its distance from the frontal; internasals a little broader than long, half as long as the præfrontals, which are a little longer than broad; frontal as long as broad, as long as or a little shorter than its distance from the end of the snout, shorter than the parietals; no supraocular; præfrontal forming a narrow suture with the parietal between the frontal and the eye; loreal once and a half as long as broad; a minute postocular; six upper labials, third and fourth entering the eye; three lower labials in contact with the anterior chin-shields, which are not longer than the posterior and separated from the symphysial. Scales in 15 rows, all perfectly smooth. Ventrals 144–145; anal entire; subcaudals 27–28. Uniform blackish brown above, yellowish beneath; a few transverse brown blotches under the tail, sometimes also on the belly.

Total length 400 millim.; tail 55.

Costa Rica.

a–b. ♀ (V. 145; C. 28) & hgr. (V. 144; C. 27).   Irazu, Costa Rica.   O. Salvin & F. D. Godman, Esqrs. [P.].

12. Geophis dubius.

Geophidium dubium, *Peters, Mon. Berl. Ac.* 1861, p. 923.
Geophis fuscus, *Fischer, Abh. naturw. Ver. Hamb.* ix. 1886, p. 11, pl. ii. fig. 5.
—— chalybæa, part., *Günth. Biol. C.-Am., Rept.* p. 87 (1893).

Maxillary not extending beyond palatine in front, the first tooth corresponding to the suture between the second and third labial shields. Eye very small, its length about two thirds its distance from the mouth. Snout obtusely pointed, very prominent. Rostral rather large, as deep as broad, or a little deeper than broad, the portion visible from above measuring at least two thirds its distance from the frontal; internasals often fused with the præfrontals *; frontal as long as broad or slightly broader than long, as long as or a little longer than its distance from the end of the snout, a little shorter than the parietals; a very small supraocular; loreal once and a half to twice as long as broad; one postocular; six upper labials, third and fourth entering the eye; three or four lower labials in contact with the anterior chin-shields, which are as long as or a little longer than the posterior and separated from the symphysial. Scales in 17 rows, perfectly smooth, a few at the base of the tail with faint traces of keels. Ventrals 134-146; anal entire; subcaudals 43-48. Uniform dark brown or blackish above, yellowish white beneath.

Total length 340 millim.; tail 60.

Mexico.

a-b. ♂ (V. 134; C. 48) & ♀ (V. 146; C. 43).	Mexico.	M. Sallé [C.].
c. Hgr. ♂ (V. 141; C. 48).	Jalapa.	Dr. J. G. Fischer. (Type of *G. fuscus*.)

13. Geophis rostralis.

Elapoides rostralis, *Jan, Icon. Gén.* 12, pl. ii. fig. 2 (1865).
Geophis dubius, *Bocourt, Miss. Sc. Mex., Rept.* p. 532, pl. xxxi. fig. 9 (1883).
—— rostralis, *Bocourt, l. c.* p. 533, pl. xxxi. fig. 10; *Günth. Biol. C.-Am., Rept.* p. 89 (1893).
Rhabdosoma rostrale, *Cope, Bull. U.S. Nat. Mus.* no. 32, 1887, p. 85.

Maxillary not extending beyond palatine in front, the first tooth corresponding to the suture between the second and third labial shields. Eye very small, its length about two thirds its distance from the mouth. Snout obtusely pointed, very prominent. Rostral rather large, as deep as broad, the portion visible from above measuring at least two thirds its distance from the frontal; the internasals may be fused with the præfrontals †; frontal slightly broader than long, as long as its distance from the end of the snout, a little shorter than the parietals; a very small supraocular; loreal once and a half to twice as long as deep; one postocular, sometimes fused with the supraocular; six upper labials, third and fourth entering the eye; anterior chin-shields longer than the posterior, in contact with the

---

* In spec. *a* the internasals are distinct; in *b* the left internasal is fused with the præfrontal; in *c* the internasals are no longer distinguishable.

† As in the specimen in the Collection.

symphysial and with three labials. Scales in 17 rows, perfectly smooth, a few at the base of the tail with faint traces of keels. Ventrals 136-151; anal entire; subcaudals 36-48. Uniform blackish brown above, yellowish white beneath.

Total length 310 millim.; tail 50.

Mexico.

a. ♀ (V.146; C. 40). Mexico. M. Sallé [C.].

114. CARPHOPHIS.

Calamaria, part., *Schleg. Phys. Serp.* ii. p. 25 (1837).
Carphophis, *Gervais, in D'Orbigny, Dict. d'Hist. Nat.* iii. p. 191 (1849); *Günth. Cat. Col. Sn.* p. 17 (1858); *Jan, Arch. Zool. Anat. Phys.* ii. 1862, p. 23; *Bocourt, Miss. Sc. Mex., Rept.* p. 535 (1883).
Carphophiops, *Gerv. l. c.*; *Cope, Proc. Ac. Philad.* 1860, p. 78, *and Proc. U.S. Nat. Mus.* xiv. 1892, p. 596.
Carpophis, part., *Dum. & Bibr. Mém. Ac. Sc.* xxiii. 1853, p. 442.
Celuta, *Baird & Gir. Cat. N. Am. Rept.* p. 129 (1853); *Cope, Proc. U.S. Nat. Mus.* xiv. 1892, p. 599.
Carphophis, part., *Dum. & Bibr. Erp. Gén.* vii. p. 131 (1854).

Maxillary teeth very small, subequal, about 10; mandibular teeth subequal. Head small, not distinct from neck; eye small, with round pupil; internasals present or absent; supraocular very small; nostril in a single nasal; no præocular, loreal and præfrontal entering the eye. Body cylindrical; scales smooth, without apical pits, in 13 rows; ventrals rounded. Tail short; subcaudals in two rows.

North America.

1. Carphophis amœnus.

Coluber amœnus, *Say, Journ. Ac. Philad.* iv. 1825, p. 237; *Harl. Journ. Ac. Philad.* v. 1827, p. 355, *and Med. Phys. Res.* p. 118 (1835); *Storer, Rep. Fish. & Rept. Mass.* p. 226 (1839).
Calamaria amœna, *Schleg. Phys. Serp.* ii. p. 31, pl. i. figs. 19 & 20 (1837).
Brachyorrhos amœnus, *Holbr. N. Am. Herp.* iii. p. 115, pl. xxvii. (1842).
Carphophis amœna, *Gerv. in D'Orb. Dict. d'Hist. Nat.* iii. p. 191 (1849); *Dum. & Bibr.* vii. p.131 (1854); *Günth. Cat.* p. 18 (1858); *Jan, Arch. Zool. Anat. Phys.* ii. 1862, p. 23, *and Icon. Gén.* 12, pl. ii. fig. 4 (1865); *Garm. N. Am. Rept.* p. 100, pl. vii. fig. 1 (1883); *Bocourt, Miss. Sc. Mex., Rept.* p. 535, pl. xxxii. figs. 1 & 2 (1883); *H. Garm. Bull. Illin. Lab.* iii. 1892, p. 309; *Hay, Batr. & Rept. Indiana*, p. 78 (1893).
Carphophiops vermiformis, *Gerv. l. c.*
Celuta amœna, *Baird & Gir. Cat. N. Am. Rept.* p. 129 (1853).
—— vermis, *Kennicott, Proc. Ac. Philad.* 1859, p. 99.
—— helenæ, *Kennicott, l. c.* p. 100.
Carphophiops amœna, *Cope, Proc. Ac. Philad.* 1860, p.79, *and Proc. U.S. Nat. Mus.* xiv. 1892, p. 596.

Carphophiops vermis, *Cope, Check-List N. Am. Rept.* p. 34 (1876), and *Proc. U.S. Nat. Mus.* xiv. 1892, p. 597.
—— helenæ, *Cope, l. c.*
Carphophis helenæ, *Garm. l. c.* p. 100; *H. Garm. l. c.* p. 308.

Snout rounded, somewhat projecting. Rostral as deep as broad, the portion visible from above one half to two thirds as long as its distance from the frontal; nasal entire; internasals distinct, or fused with the præfrontals; supraocular very small; frontal as long as broad or a little longer, longer than its distance from the end of the snout, shorter than the parietals; loreal about twice as long as deep; an elongate temporal; five upper labials, third and fourth entering the eye, fifth largest; four lower labials in contact with the anterior chin-shields, which are longer than the posterior. Scales in 13 rows. Ventrals 112-138; anal divided; subcaudals 24-37. Brown above, yellowish (red in life) beneath.

Total length 300 millim.; tail 50.

Mississippi Valley and eastwards as far north as Massachusetts.

A. Internasals distinct.

a. ♀ (V. 131; C. 28).	Delaware.	E. Doubleday, Esq. [P.].
b. ♂ (V. 123; C. 37).	Pennsylvania.	Christiania Museum.
c. ♀ (V. 121; C. 28).	Pennsylvania.	
d, e. ♀ (V. 138, 131; C. 27, 27).	N. America.	Lord Ampthill [P.].

B. No internasals.

f. ♂ (V. 126; C. 36).	Illinois.	
g. ♀ (V. 131; C. 25).	Illinois.	Smithsonian Institution.
h-k. ♀ (V. 120, 125, 127; C. 24, 25, 28).	Bloomington, Indiana.	C. Bollman, Esq. [C.].

115. STILOSOMA.*

Stilosoma, *A. E. Brown, Proc. Ac. Philad.* 1890, p. 199; *Cope, Proc. U.S. Nat. Mus.* xiv. 1892, p. 595.

Head not distinct from neck; eye moderately small; one nasal; internasals distinct, or fused with the præfrontals; no loreal; præocular present or absent; parietals in contact with the labials. Body very slender, cylindrical and rigid; scales smooth, without pits, in 19 rows; ventrals rounded. Tail short; subcaudals in two rows.

Florida.

1. Stilosoma extenuatum.

Stilosoma extenuata, *A. E. Brown, l. c.*

Snout rounded. Rostral prominent but not recurved; supra-

---

* Mr. A. Erwin Brown informs me (12 Aug., 1893) that, since describing this remarkable new genus, he has received three more specimens from Lake Kerr, Florida, no two of which agree in the lepidosis of the head. I have therefore modified the diagnosis according to the notes he has kindly communicated to me.

ocular short and broad; frontal hexagonal; parietal large, in contact with the fifth labial; two small postoculars; three temporals, the first between the fifth and sixth labials and the parietal; six upper labials, third and fourth entering the eye, fifth largest; three pairs of chin-shields. Scales in 19 rows. Ventrals 235; anal entire; subcaudals 44. Silvery grey above, with a series of dark brown black-edged spots; an elongate triangular dark patch on the parietals, pointing backwards; a dark bar running back from the eye to the upper margin of the labials; fore part of the head maculated with black; lower surface silvery grey, much blotched with black.

Total length 532 millim.; tail 50.

Florida.

116. GEAGRAS.

Geagras, *Cope, Journ. Ac. Philad.* (2) viii. 1876, p. 141.
Sphenocalamus, *Fischer, Oster-Progr. Akad. Gymn. Hamb.* 1883, p. 5.

Maxillary teeth 12, very small, equal; mandibular teeth very small, equal. Head not distinct from neck; snout much depressed, strongly projecting, with angular horizontal edge; eye very small, with round pupil; nostril between two large nasals and the internasal; no loreal, posterior nasal in contact with the præocular. Body cylindrical, short; scales smooth, without apical pits, in 15 rows; ventrals rounded. Tail short; subcaudals in two rows.

Mexico.

1. Geagras redimitus.

Geagras redimitus, *Cope, l. c.,* and *Proc. Amer. Philos. Soc.* xxii. 1885, p. 177.
Sphenocalamus lineolatus, *Fischer, l. c.* pl. —. figs. 3–5.
Geophis redimita, *Günth. Biol. C.-Am., Rept.* p. 92 (1893).

Rostral very large, flat beneath, its upper portion convex and as long as its distance from the frontal; internasals broad, shorter than the præfrontals; frontal large, a little longer than broad, longer than its distance from the end of the snout, a little shorter than the parietals; supraocular narrow; one præ- and one postocular; temporals 1+2; six upper labials, fourth bordering the eye, fifth in contact with the parietal; two pairs of chin-shields, anterior in contact with the symphysial, posterior small. Scales in 15 rows. Ventrals 117–119; anal divided; subcaudals 25–30. Pale reddish above, each scale on the anterior half of the body with a brown median streak; head brown above, with two yellow longitudinal streaks meeting on the rostral; lower parts uniform yellowish.

Total length 185 millim.; tail 25.

Western Mexico.

a. ♀ (V. 117; C. 29).   Mexico.   L. Greening, Esq. [P.].

117. MACROCALAMUS.

Macrocalamus, *Günth. Rept. Brit. Ind.* p. 198 (1864).

Maxillary teeth about 10, subequal; mandibular teeth subequal. Head not distinct from neck; eye small, with round pupil; nostril pierced between a nasal and the first upper labial; no internasals; no loreal; a præocular. Body cylindrical; scales smooth, without apical pits, in 15 rows; ventrals rounded. Tail short; subcaudals in two rows.

Malay Peninsula?

1. Macrocalamus lateralis.

Macrocalamus lateralis, *Günth. l. c.* p. 199, pl. xviii. fig. D.

Rostral deeper than broad, extending on to the upper surface of the head, separating the nasals; frontal longer than broad, a little longer than its distance from the end of the snout, a little shorter than the parietals; one præ- and one postocular: temporals 1+2; eight upper labials, fourth and fifth entering the eye; first lower labial in contact with its fellow behind the symphysial; four lower labials in contact with the anterior chin-shields; posterior chin-shields small and separated from each other. Scales in 15 rows. Ventrals 118; anal entire; subcaudals 20. Reddish brown above, with a few small blackish spots; a dark brown lateral stripe, enclosing a series of small whitish spots; whitish inferiorly, with a few brown dots.

Total length 300 millim.; tail 32.

Malay Peninsula?

a. ♂ (V. 118; C. 20). ———? Gen. Hardwicke [P.]. (Type.)

118. IDIOPHOLIS.

Idiopholis, *Mocquard, Le Natur.* 1892, p. 35, *and Mém. Soc. Zool. Fr.* v. 1892, p. 191.

Maxillary teeth numerous, very small and equal. Head not distinct from neck; eye very small, with round pupil; nostril pierced between two small nasals; a pair of small internasals separated by a small azygous shield; no supraocular; no loreal; no præocular; no anterior temporal. Body cylindrical; scales smooth, without apical pits, in 15 rows; ventrals rounded. Tail short; subcaudals in two rows.

Borneo.

1. Idiopholis collaris.

Idiopholis collaris, *Mocquard, Le Natur.* 1892, p. 35, *and Mém. Soc. Zool. Fr.* v. 1892, p. 191, pl. vii. fig. 1.

Snout short, obtuse. Rostral narrow, deeper than broad; frontal broader than long, as long as the præfrontals, about half as long as the parietals; six upper labials, third and fourth entering the eye, fifth largest and forming a suture with the parietal; anterior chin-shields longer than the posterior. Scales in 15 rows. Ventrals 127; anal entire; subcaudals 28. Dark brown above, somewhat lighter beneath; a broad yellowish-white collar behind the parietals.

Total length 190 millim.

Sebrocang Valley, N.E. Borneo.

119. RHABDOPHIDIUM.

Rabdion, part., *Dum. & Bibr. Mém. Ac. Sci.* xxiii. 1853, p. 441, *and Erp. Gén.* vii. p. 115 (1854)*.
Rhabdion, *Günth. Cat. Col. Sn.* p. 243 (1858); *Jan, Arch. Zool. Anat. Phys.* ii. 1862, p. 28.

Maxillary teeth 10 to 12, subequal; mandibular teeth subequal. Head not distinct from neck; eye small, with round pupil; nostril pierced in a small nasal, which is in contact with or narrowly separated from the præocular; a pair of internasals; no temporals, the parietals in contact with the labials. Body cylindrical; scales smooth, without apical pits, in 15 rows; ventrals rounded. Tail short; subcaudals in two rows.

Celebes.

1. Rhabdophidium forsteni.

Rabdion forsteni, *Dum. & Bibr.* vii. p. 116 (1854); *Günth. Cat.* p. 243 (1858); *Jan, Arch. Zool. Anat. Phys.* ii. 1862, p. 29, *and Icon. Gén.* 13, pl. i. fig. 4 (1865).

Snout rather pointed. Rostral broader than deep, well visible from above; suture between the internasals a little shorter than that between the præfrontals; frontal longer than broad, longer than its distance from the end of the snout, a little shorter than the parietals, about twice as broad as the supraocular; a rather large præocular, reaching or nearly reaching the nasal; one postocular; five (or six) upper labials, third and fourth entering the eye; first lower labial in contact with its fellow behind the symphysial; three lower labials in contact with the anterior chin-shields, which are longer than the posterior. Scales in 15 rows. Ventrals 137-158; anal entire; subcaudals 22-31. Uniform blackish above, brown beneath.

Total length 370 millim.; tail 45.

Celebes.

a, b. ♀ (V. 153, 158; C. 26, 22).   Celebes.

120. PSEUDORHABDIUM.

Rabdion, part., *Dum. & Bibr. Mém. Ac. Sc.* xxiii. 1853, p. 441, *and Erp. Gén.* vii. p. 115 (1854).
Pseudorabdion, *Jan, Arch. Zool. Anat. Phys.* ii. 1862, p. 10.
Oxycalamus, *Günth. Rept. Brit. Ind.* p. 199 (1864).

Maxillary teeth 10 to 12, subequal; anterior mandibular teeth slightly longer than the posterior. Head not distinct from neck;

---

* Name preoccupied (*Rhabdium*, Wallr., 1833).

eye small, with round pupil; nostril pierced in a minute nasal; internasals small; no loreal; præocular small or absent; no temporals, the parietals in contact with the labials. Body cylindrical; scales smooth, without apical pits, in 15 rows; ventrals rounded. Tail short; subcaudals in two rows.

Malay Peninsula and Archipelago.

1. Pseudorhabdium longiceps.

Calamaria longiceps, *Cantor, Cat. Mal. Rept.* p. 63, pl. —. fig. 1 (1847).
Rabdion torquatum, *Dum. & Bibr.* vii. p. 119 (1854); *Peters, Mon. Berl. Ac.* 1861, p. 684.
Pseudorabdion torquatum, *Jan, Arch. Zool. Anat. Phys.* ii. 1862, p. 10, *and Icon. Gén.* 10, pl. iii. fig. 3 (1865).
Oxycalamus longiceps, *Günth. Rept. Brit. Ind.* p. 199 (1864); *Stoliczka, Journ. As. Soc. Beng.* xlii. 1873, p. 120.
Pseudorhabdion longiceps, *Bouleng. Ann. & Mag. N. H.* (5) xvi. 1885, p. 389.

Snout rather pointed. Rostral small, as deep as broad, well visible from above; suture between the internasals one third or one fourth the length of that between the præfrontals; frontal a little longer than broad, as long as or a little shorter than its distance from the end of the snout, shorter than the parietals, more than twice as broad as the supraocular; præocular small (rarely absent); one postocular; five upper labials, third and fourth entering the eye; symphysial in contact with the anterior chin-shields; three lower labials in contact with the anterior chin-shields, which are about twice as large as the posterior. Scales in 15 rows. Ventrals 129–146; anal entire; subcaudals 10–28. Tail pointed. Iridescent brown or black, with or without a yellowish collar; usually a yellowish vertical spot above the angle of the mouth.

Total length 230 millim.; tail 35 (♂).

Malay Peninsula, Sumatra, Borneo, Philippines, Celebes.

a. ♂ (V. 132; C. 25).		Pinang.	Dr. Cantor. (Type.)
b–c. ♂ (V. 132; C. 24) & ♀ (V. 141; C. 19).		Perak.	L. Wray, Esq. [P.].
d. ♀ (V. 143; C. 18).		Perak.	G. E. Dobson, Esq. [P.].
e. ♀ (V. 133; C. 20).		Singapore.	Dr. Dennys [P.].
f. Yg. (V. 130; C. 28).		Singapore.	H. N. Ridley, Esq. [P.].
g–h. ♂ (V. 132; C. 28) & ♀ (V. 146; C. 19).		Deli, Sumatra.	Prof. Moesch [C.].
i. ♀ (V. 137; C. 10).		Nias.	Hr. Sundermann [C.].
k. ♀ (V. 137; C. 21).		Pontiana, Borneo.	Prof. Peters [P.].
l. ♂ (V. 129; C. 28).		Borneo.	

2. Pseudorhabdium oxycephalum.

Rhabdosoma oxycephalum, *Günth. Cat.* p. 242 (1858).
Oxycalamus oxycephalus, *Günth. Proc. Zool. Soc.* 1873, p. 168, fig.

Closely allied to the preceding, but differing in the following

characters:—Frontal a little broader than long, about half as long as the parietals; supraocular smaller still and confluent with the postocular; no præocular. Uniform iridescent blackish brown.
Total length 280 millim.; tail 20.
Philippine Islands.

a. ♀ (V. 152; C. 16).   Philippines.   H. Cuming, Esq. [C.].   (Type.)
b. ♂ (V. 136; C. 23).   Negros.   Dr. A. B. Meyer [C.].

121. CALAMARIA.

Calamaria, *Boie, in Férussac, Bull. Sc. Nat.* ix. 1826, p. 236, *and Isis*, 1827, p. 519; *Dum. & Bibr. Erp. Gén.* vii. p. 60 (1854); *Günth. Cat. Col. Sn.* p. 3 (1858); *Jan, Arch. Zool. Anat. Phys.* ii. 1862, p. 4; *Günth. Rept. Brit. Ind.* p. 195 (1864); *Bouleng. Faun. Ind., Rept.* p. 281 (1890).
Calamaria, part., *Schleg. Phys. Serp.* ii. p. 25 (1837).
Typhlocalamus, *Günth. Proc. Zool. Soc.* 1872, p. 595.

Maxillary teeth 8 to 11, subequal; anterior mandibular teeth a little longer than the posterior. Head not distinct from neck; eye small, with round pupil; nostril pierced in a minute nasal; no loreal; no internasals; præocular present or absent; no temporals,

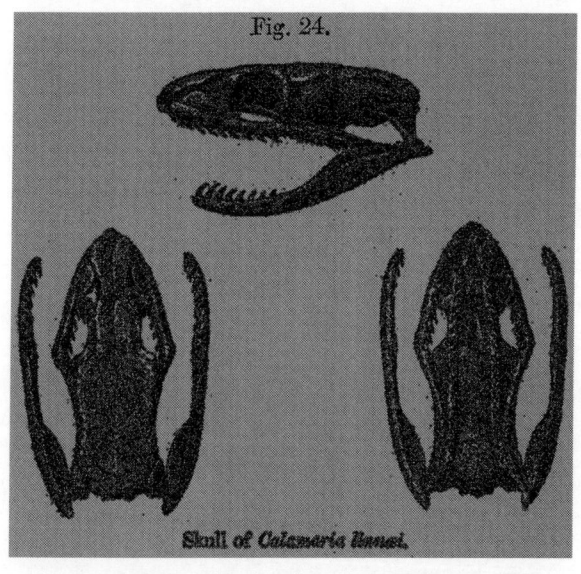

Fig. 24.

Skull of *Calamaria lumei*.

the parietals in contact with the labials. Body cylindrical; scales smooth, without apical pits, in 13 rows; ventrals rounded. Tail short; subcaudals in two rows.

From Assam, Burma, and S. China to the Malay Archipelago.

Synopsis of the Species.

I. 129 to 225 ventrals.
  A. Five or six upper labials, third and fourth or fourth and fifth entering the eye.
    1. Frontal longer than broad.
      a. Symphysial in contact with the chin-shields.
        α. Frontal not twice as broad as the supraocular.
          * Rostral broader than deep.

Diameter of the eye less than its distance from the mouth	1. *lumbricoidea*, p. 333.
Diameter of the eye equal to its distance from the mouth	2. *vermiformis*, p. 333.
Diameter of the eye more than its distance from the mouth; frontal shorter than the parietals	3. *stahlknechtii*, p. 335.
Diameter of the eye more than its distance from the mouth; frontal as long as the parietals	4. *baluensis*, p. 335.

          ** Rostral as deep as broad.
            † Frontal as long as the parietals.

Frontal once and two thirds as long as broad	5. *grabowskii*, p. 336.
Frontal once and a half as long as broad.	6. *albiventer*, p. 336.
Frontal once and one third as long as broad	7. *margaritifera*, p. 336.

            †† Frontal shorter than the parietals.

Ventrals 126–144	8. *prakkii*, p. 337.
Ventrals 175–195	9. *grayi*, p. 338.

        β. Frontal at least twice as broad as the supraocular.

Rostral as deep as broad; frontal as long the parietals	10. *bitorques*, p. 338.
Rostral as deep as broad; frontal shorter than the parietals	11. *gervaisii*, p. 338.
Rostral a little broader than deep; frontal a little shorter than the parietals	12. *sumatrana*, p. 339.

      b. First lower labial in contact with its fellow behind the symphysial.
        a. A præocular; a postocular.
          * Frontal not more than twice as broad as the supraocular.
            † Diameter of the eye much more than its distance from the mouth.. 13. *everetti*, p. 340.

†† Diameter of the eye not more than its distance
from the mouth.

Rostral as deep as broad ; first and second
labials subequal in size .... 14. *virgulata*, p. 340.
Rostral nearly as deep as. broad ; second
labial larger than first .... 15. *leucogaster*, p. 341.
Rostral much broader than deep ; anal
divided .... 16. *occipitalis*, p. 342.

** Frontal more than twice as broad as the supra-
ocular............ 17. *bicolor*, p. 342.

β. A præocular ; postocular fused with the supraocular.
18. *lateralis*, p. 342.

γ. A præocular ; two postoculars.
19. *beccarii*, p. 343.

δ. No præocular.......... 20. *rebentischii*, p. 343.

2. Frontal as broad as long ; rostral broader than deep.

a. First lower labial in contact with its fellow behind the
symphysial.

No præocular .... 21. *agamensis*, p. 343.
A præocular .... 22. *leucocephala*, p. 344.

b. Symphysial in contact with the chin-shields ; no præocular.
23. *schlegelii*, p. 345.

B. Four upper labials, second and third entering the eye.
1. Frontal longer than broad.
a. Symphysial in contact with the chin-shields.
α. A præocular.

Tail obtuse ; the portion of the rostral
visible from above shorter than its
distance from the frontal .... 24. *linnæi*, p. 345.
Tail obtuse ; the portion of the rostral
visible from above at least as long as
its distance from the frontal........ 25. *borneensis*, p. 347.
Tail ending in a sharp point.......... 26. *benjaminsii*, p. 347.

β. No præocular .... 27. *javanica*, p. 347.

b. First lower labial in contact with its fellow behind the
symphysial.

Frontal as long as the parietals, not twice
as broad as the supraocular .... 28. *brevis*, p. 348.
Frontal shorter than the parietals, at
least twice as broad as the supraocular. 29. *pavimentata*, p. 348.

2. Frontal as broad as long.

First lower labial in contact with its
fellow behind the symphysial ; tail
rounded at the end .... 30. *septentrionalis*, p. 349.

Symphysial in contact with the chin-
shields; tail pointed .... 31. *melanota*, p. 349.

C. Four upper labials, third entering
the eye .... 32. *lovii*, p. 350.

II. Ventrals 300-320; no distinct postocular....................... 33. *gracillima*, p. 350.

1. Calamaria lumbricoidea.

Calamaria lumbricoidea, *Boie, Isis*, 1827, p. 540; *Schleg. Phys. Serp.* ii.
p. 27, pl. i. figs. 14-16 (1837); *Dum. & Bibr.* vii. p. 89 (1854); *Jan,
Arch. Zool. Anat. Phys.* ii. 1862, p. 8, *and Icon. Gén.* 10, pl. ii.
fig. 2 (1865); *v. Lidth de Jeude, Notes Leyd. Mus.* xii. 1890, p. 254.
—— lumbricoidea, part., *Günth. Cat.* p. 5 (1858).
—— variabilis, *v. Lidth de Jeude, in M. Weber, Zool. Ergebn.*
p. 183, pl. xvi. fig. 8 (1890).

Rostral broader than deep, well visible from above; frontal longer than broad, slightly shorter than the parietals, not twice as broad as the supraocular; one præ- and one postocular; diameter of the eye less than its distance from the mouth; five upper labials, the four anterior subequal in size, third and fourth entering the eye; symphysial in contact with the anterior chin-shields; both pairs of chin-shields in contact with each other. 13 rows of scales. Ventrals 177–217; anal entire; subcaudals 16–23. Tail ending in a point. Dark brown above, whitish inferiorly; lower surface of the tail with a dark brown longitudinal streak.

Total length 420 millim.; tail 35.

Java; Celebes.

A. Head uniform dark brown above.

a. ♂ (V. 191; C. 22).   Java.   Leyden Museum.

B. A yellowish band across the parietal shields.

b. ♂ (V. 177; C. 19).   Java.

2. Calamaria vermiformis.

Calamaria vermiformis, *Dum. & Bibr.* vii. p. 85 (1854); *Jan, Arch.
Zool. Anat. Phys.* ii. 1862, p. 8, *and Icon. Gén.* 10, pl. ii. fig. 3
(1865).
—— temminckii, *Dum. & Bibr. t. c.* p. 87; *Günth. Cat.* p. 5 (1858).
—— schlegelii (*non D. & B.*), *Günth. l. c.*
—— dimidiata, *Bleek. Nat. Tijdschr. Nederl. Ind.* xxi. 1860, p. 295.
—— melanorhynchus, *Bleek. l. c.*
—— alkeni, *Bleek. l. c.*
—— flaviceps, *Günth. Ann. & Mag. N. H.* (3) xv. 1865, p. 90.
—— vermiformis, var. sumatranus, *v. Lidth de Jeude, Notes Leyd.
Mus.* xii. 1890, p. 18.

Rostral a little broader than deep, visible from above; frontal longer than broad, not twice as broad as the supraocular, shorter than the parietals; a præ- and a postocular; diameter of the eye equal to its distance from the mouth; five or six upper labials, third and fourth entering the eye; symphysial in contact with the anterior chin-shields; posterior chin-shields in contact or separated. 13 rows of scales. Ventrals 153–210; anal entire; subcaudals 15–26. Tail ending in a point. Dark brown or black above; beneath yellowish, with or without large black transverse spots or cross bands, or uniform blackish brown.

Total length 410 millim.; tail 33.

Sumatra, Borneo, Java, Moluccas.

A. Head yellow; body with transverse series of small yellowish spots; belly with large black spots or cross bands.

a. ♀ (V. 164; C. 20).   Sarawak.   Sir J. Brooke [P.]   (Types of
b. Hgr. (V. 153; C. 26).   Borneo.   A. R. Wallace, Esq. [P.]   *C. flaviceps*.)

B. Head yellow, with a black lateral streak, passing through the eye; above with yellowish cross bars disposed regularly, each of which in the middle expands into a longitudinally oval spot; belly barred black and yellow, the yellow bars confluent with the dorsals.

a. Hgr. (V. 184; C. 26).   Mt. Batu-Song, Sarawak, 1000 ft.   C. Hose, Esq. [C.]

C. Head yellow above, entirely or partially; body without spots; belly with black cross bands.

a. ♂ (V. 154; C. 25).   Borneo.   A. R. Wallace, Esq. [P.].
b. ♀ (V. 169; C. 18).   Borneo.
c. Yg. (V. 193; C. 22).   Sumatra.   Dr. Bleeker. (Type of *C. melanorhynchus*.)
d. Hgr. (V. 167; C. 20).   E. coast of Sumatra.   Mrs. Findlay [P.].

D. Head uniform dark brown or black above, like the body; belly with large black spots or cross bands.

a. ♀ (V. 161; C. 20).   Matang, Borneo.
b. ♂ (V. 188; C. 24).   Sumatra.   Dr. Bleeker. (Type of *C. alkeni*.)
c–d. ♂ (V. 163, 164; C. 22, 22).   Padang, Sumatra.
e. ♀ (V. 194; C. 18).   Java.   C. Bowring, Esq. [P.].
f–g. ♂ (V. 162; C. 23) & ♀ (V. 183; C. 17).   Ternate.   H.M.S. 'Challenger.'

E. As in the preceding, but lower parts without or with mere traces of spots.

a. Yg. (V. 202; C. 17).   Java.   Dr. Bleeker. (Type of *C. dimidiata*.)
b. Yg. (V. 210; C. 15).   Java.   Dr. Ploem [C.].

c, d. ♀ (V. 205, 177;   Java.
C. 15, 19).
e. (♂ V. 159; C. 22).   Padang, Sumatra.

F. Blackish brown above and below; labial and gular regions and a lateral streak running along the two outer rows of scales yellowish.

a. ♀ (V. 179; C. 17).   Deli, Sumatra.   Hr. Hartert [C.]; Fischer Collection.

3. Calamaria stahlknechtii.

Calamaria stahlknechti, *Stoliczka, Journ. As. Soc. Beng.* xlii. 1873, p. 119, pl. xi. fig. 2; *Bouleng. Ann. & Mag. N. H.* (5) xvi. 1885, p. 388.

Rostral a little broader than deep, visible from above; frontal longer than broad, not twice as broad as the supraocular, shorter than the parietals; a præ- and a postocular; diameter of the eye exceeding its distance from the mouth; five upper labials, third and fourth entering the eye; symphysial in contact with the anterior chin-shields; posterior chin-shields separated from each other. Scales in 13 rows. Ventrals 147–163; anal entire; subcaudals 22–23. Tail ending in a point. Uniform brown or olive above; yellowish (red) inferiorly, with or without more or less distinct black transverse spots or cross bands.

Total length 265 millim.; tail 24.

Sumatra and Nias.

a–c. ♂ (V. 147, 148, 150; C. 23,   Nias.   Hr. Sundermann [C.].
22, 22).

4. Calamaria baluensis. (PLATE XVII. fig. 1.)

Calamaria baluensis, *Bouleng. Proc. Zool. Soc.* 1893, p. 524.

Rostral a little broader than deep, visible from above; frontal nearly twice as long as broad, not twice as broad as the supraocular, as long as the parietals; eye rather large, its diameter much greater than its distance from the mouth; one præ- and one postocular; five upper labials, third and fourth entering the eye; symphysial in contact with the anterior chin-shields; both pairs of chin-shields in contact with each other. 13 rows of scales. Ventrals 175; anal entire; subcaudals 28. Tail ending in a point. Brown above, with small black spots; an interrupted black streak along each side of the head and neck, passing through the eye; upper lip and lower parts white; belly with three longitudinal series of small black spots; a black line along the lower surface of the tail.

Total length 340 millim.; tail 33.

North Borneo.

a. ♂ (V. 175; C. 28).   Mt. Kina Baloo.   A. Everett, Esq. [C.].
(Type.)

5. Calamaria grabowskii.

Calamaria grabowskyi, *Fischer, Arch. f. Nat.* 1885, p. 50, pl. iv. fig. 1.

Rostral nearly as deep as broad; frontal once and two thirds as long as broad, not twice as broad as the supraocular, as long as the parietals; a præ- and a postocular; diameter of the eye more than its distance from the mouth; five upper labials, third and fourth entering the eye; symphysial in contact with the anterior chin-shields; both pairs of chin-shields in contact with each other. 13 rows of scales. Ventrals 187; anal entire: subcaudals 20–22. Tail ending in a point. Brown above, with small black spots; ventral shields and outer row of scales black, edged with white.

Total length 430 millim.; tail 32.

Borneo.

a–b. ♀ (V. 187, 187; C. 20, 22).   S.E. Borneo.   Hr. Grabowsky [C.]; Dr. J. G. Fischer. (Types.)

5. Calamaria albiventer.

Changulia albiventer, *Gray, Ill. Ind. Zool.* ii. pl. lxxxvi. fig. 3 (1834).
Calamaria linnæi, var., *Cantor, Cat. Mal. Rept.* p. 62 (1847).
—— albiventer, *Günth. Cat.* p. 4 (1858), *and Rept. Brit. Ind.* p. 197 (1864).

Rostral nearly as deep as broad, well visible from above; frontal once and a half as long as broad, as long as the parietals, not twice as broad as the supraocular; one præ- and one postocular; diameter of the eye equal to its distance from the mouth; five upper labials, the four anterior subequal in size, third and fourth entering the eye; symphysial in contact with the anterior chin-shields; both pairs of chin-shields in contact with each other. 13 rows of scales. Ventrals 143–167; anal entire; subcaudals 16–21. Tail ending in a point. Brown above, with four yellow (vermilion) longitudinal streaks; lower parts uniform yellow (carmine), with a black streak running along the junction of the ventrals with the lower row of scales; a black line along the middle of the tail.

Total length 390 millim.; tail 20.

Pinang.

a, b. ♀ (V. 160, 157; C. 17, 17).   Pinang.   Gen. Hardwicke [P.]. (Types.)
c–d. ♀ (V. 167, 166; C. 17, 16).   Pinang.   Dr. Cantor.
e–f. ♂ (V. 143; C. 21) & ♀ ——?
(V. 159; C. 16).

7. Calamaria margaritifera. (Plate XVII. fig. 2.)

Calamaria margaritophora, *Bleeker, Nat. Tijdschr. Nederl. Ind.* xxi. 1860, p. 294.
—— maculolineata, *Peters, Mon. Berl. Ac.* 1863, p. 403.

Rostral as deep as broad, well visible from above; frontal longer than broad, nearly as long as the parietals, not twice as broad as the supraocular; one præocular; one postocular; diameter of the eye equal to its distance from the mouth; five upper labials, third and fourth entering the eye; symphysial in contact with the anterior chin-shields; posterior chin-shields separated (or in contact). Scales in 13 rows. Ventrals 165 (149); anal entire; subcaudals 12 (19). Tail ending in a point. Brown above, with longitudinal black lines or series of spots; head blackish above; scales of the outer series with a yellowish spot; lower parts uniform yellowish.

Total length 280 millim.; tail 13.

Sumatra, Java.

a. ♀ (V. 165; C. 12). ——? Dr. Bleeker. (One of the types.)

The following appears to be closely allied to, and may have to be regarded as a colour-variety of, *C. margaritifera* :—

CALAMARIA HOEVENII.

Calamaria hoevenii, *Edeling, Nat. Tijdschr. Nederl. Ind.* xxxi. 1870, p. 380.

Rostral well visible from above; frontal once and a half as long as broad; five upper labials, third and fourth entering the eye; symphysial in contact with the anterior chin-shields; a small scale between the two pairs of chin-shields. 13 rows of scales. Ventrals 169; anal entire; subcaudals 9. Brown above, head marbled with black; back with transverse pale black-edged bands; a series of whitish spots along the outer row of scales; lower parts yellowish with black spots.

Total length 305 millim.; tail 13.

Sumatra.

8. Calamaria prakkii.

Calamaria prakkei, *v. Lidth de Jeude, Notes Leyd. Mus.* xv. 1893, p. 252.

Rostral as deep as broad, well visible from above; frontal once and a half as long as broad, shorter than the parietals; one præ- and one postocular; five upper labials, third and fourth entering the eye; symphysial in contact with the anterior chin-shields; no azygous shield between the four chin-shields. Scales in 13 rows. Ventrals 126–144; anal entire; subcaudals 25–30. Tail obtuse, with a conical scale at the end. Dark brown above, with bluish gloss; scales of the outer row with a light spot; some specimens brown, with rather indistinct blackish longitudinal lines; an indication of a light collar may be present; belly yellow, with or without small dark spots; lower surface of tail usually with a median black line.

Total length 260 millim.; tail 40.

North Borneo.

9. Calamaria grayi.

Calamaria grayi, *Günth. Cat.* p. 6 (1858).
—— philippinica, *Steindachn. Verh. zool.-bot. Ges. Wien*, xvii. 1867, p. 514, pl. xiii. figs. 4-6.

Snout very short and broadly rounded. Rostral nearly as deep as broad, well visible from above; frontal a little longer than broad, shorter than the parietals, not twice as broad as the supraocular; one præ- and one postocular; diameter of the eye less than its distance from the mouth; five upper labials, the four anterior subequal in size, third and fourth entering the eye; symphysial in contact with the anterior chin-shields; two pairs of chin-shields, in contact with each other. Scales in 13 rows. Ventrals 175-195; anal entire; subcaudals 14-24. Tail ending in a rather obtuse point. Young reddish white, with black rings; adult uniform blackish above, alternately barred black and white below.

Total length 365 millim.; tail 35.

Philippine Islands.

a, b. ♂ (V. 175; C. 24) & yg. (V. 195; C. 16).	Philippines.	H. Cuming, Esq. [C.]. (Types.)

10. Calamaria bitorques.

Calamaria gervaisii, part., *Günth. Cat.* p. 4 (1858).
—— bitorques, *Peters, Mon. Berl. Ac.* 1872, p. 585.

Rostral as deep as broad, well visible from above; frontal longer than broad, as long as the parietals, about twice as broad as the supraocular; one præ- and one postocular; the diameter of the eye nearly equals its distance from the mouth; five upper labials, the four anterior subequal in size, third and fourth entering the eye; symphysial in contact with the anterior chin-shields; both pairs of chin-shields in contact with each other. 13 rows of scales. Ventrals 151-199; anal entire; subcaudals 13-21. Tail ending in a point. Brown above, the young with dark transverse bands separated by yellowish interspaces; only the two or three anterior of these cross bands persist in the adult; lower parts uniform yellowish.

Total length 265 millim.; tail 20.

Philippine Islands.

a. ♀ (V. 199; C. 13).	Luzon.	Dr. A. B. Meyer [C.].
b, c. ♀ (V. 183; C. 15) & yg. (V. 151; C. 18).	Philippines.	H. Cuming, Esq. [C.].
d, e. ♂ (V. 158; C. 20) & yg. (V. 158; C. 21).	Philippines.	Hr. Salmin [C.].

11. Calamaria gervaisii.

Calamaria virgulata (*non Boie*), *Eydoux & Gervais, in Guér. Mag. Zool.* Cl. iii. 1837, pl. xvi. figs. 7-10, *and Voy. Favorite,* v. *Zool.* pl. xxx. figs. 7-10 (1839).
—— gervaisii, *Dum. & Bibr.* vii. p. 76 (1854); *Jan, Arch. Zool. Anat. Phys.* ii. 1862, p. 8, *and Icon. Gén.* 10, pl. ii. fig. 1 (1865).
—— gervaisii, part., *Günth. Cat.* p. 4 (1858).

Rostral as deep as broad, well visible from above; frontal longer than broad, shorter than the parietals, about twice as broad as the supraocular; one præ- and one postocular; diameter of the eye less than its distance from the mouth; five upper labials, the four anterior subequal in size, third and fourth entering the eye; symphysial in contact with the anterior chin-shields; both pairs of chin-shields in contact with each other. 13 rows of scales. Ventrals 152-178; anal entire; subcaudals 12-18. Tail ending in an obtuse point. Brown or blackish above, with or without darker or lighter spots and streaks; a black spot at the outer extremity of each ventral shield, and a blackish line along the lower surface of the tail.

Total length 255 millim.; tail 20.

Philippine Islands.

A. Brown above, with three light dorsal streaks or series of spots bordered by series of black dots; belly uniform yellowish or with a few scattered black dots.

a, b. ♀ (V. 167; C. 13) &   Philippines.   H. Cuming, Esq. [C.].
    yg. (V. 170; C. 17).
c. ♀ (V. 171; C. 12).   Luzon.   Dr. A. B. Meyer [C.].

B. Brown above, with four more or less distinct blackish longitudinal lines; ventrals dotted with brown anteriorly.

d, e. ♂ (V. 162; C. 18) &   Philippines.   H. Cuming, Esq. [C.].
    ♀ (V. 178; C. 12).
f. ♂ (V. 163; C. 17).   Philippines.   Hr. Salmin [C.].

C. Blackish brown above, with minute whitish dots; ventrals black anteriorly.

g. ♀ (V. 169; C. 14).   S. Negros.   A. Everett, Esq. [C.].

12. Calamaria sumatrana.

Calamaria sumatrana, *Edeling, Nat. Tijdschr. Nederl. Ind.* xxxi. 1870, p. 379; *Bouleng. Proc. Zool. Soc.* 1890, p. 34; *W. L. Sclater, Journ. As. Soc. Beng.* lx. 1891, p. 233.

Rostral a little broader than deep; frontal longer than broad, a little shorter than the parietals, rather more than twice as broad as the supraocular; one præ- and one postocular; the diameter of the eye equals its distance from the mouth; five upper labials, third and fourth entering the eye; two pairs of chin-shields in contact with each other, the anterior in contact with the symphysial. Scales in 13 rows. Ventrals 129-176; anal entire; subcaudals 12-31*. Tail ending in a point. Reddish brown or dark brown above, with five black longitudinal lines or series of small black spots; each scale of the outer row with a white spot; a more or less distinct yellow collar on the nape, continuous or narrowly interrupted in the middle; sometimes a similar marking at the base of the tail; lower parts uniform yellowish, with a more or less distinct black line along the middle of the tail.

---

* The male specimen from Singapore in the Calcutta Museum has 129 ventrals and 31 subcaudals.

Total length 265 millim.; tail 12.
Sumatra, Singapore.

a. ♀ (V. 168; C. 13). District of Deli, Sumatra. Prof. Moesch [C.].
b-c. ♂ (V. 142; C. 24) Sumatra? Dr. Bleeker.
& ♀ (V. 153; C. 14).

13. Calamaria everetti. (PLATE XVIII. figs. 1 & 2.)

Calamaria everetti, *Bouleng. Proc. Zool. Soc.* 1893, p. 525, *and Ann. & Mag. N. H.* (6) xiv. 1894, p. 84.

Rostral broader than deep, well visible from above; frontal once and a half as long as broad, shorter than the parietals, not twice as broad as the supraocular; one præ- and one postocular; eye rather large, its diameter much more than its distance from the mouth; five upper labials, third and fourth entering the eye; two pairs of chin-shields in contact with each other; first lower labial in contact with its fellow behind the symphysial. Scales in 13 rows. Ventrals 144–184; anal entire; subcaudals 16–23. Tail ending in a point. Brown above, with longitudinal series of darker spots, forming two lines along each side; each scale of the outer row white in the centre, dark brown on the borders; upper surface of head brown, spotted with darker; lower parts yellowish, with a dark line along the middle of the tail.

Total length 330 millim.; tail 18.
Borneo, Palawan.

A. Nape dark brown, with a yellow collar; belly unspotted.

a. Yg. (V. 144; C. 23). Sarawak. A. Everett, Esq. [C.]. (Type.)

B. No collar; belly unspotted.

b. Yg. (V. 175; C. 18). Palawan. A. Everett, Esq. [C.].

C. No collar; a series of black dots along the middle of the belly.

c. ♀ (V. 184; C. 16). Palawan. A. Everett, Esq. [C.].

14. Calamaria virgulata.

Calamaria virgulata, *Boie, Isis,* 1827, p. 540; *v. Lidth de Jeude, Notes Leyd. Mus.* xii. 1890, p. 254.
—— modesta, *Dum. & Bibr.* vii. p. 74 (1854); *Jan, Arch. Zool. Anat. Phys.* ii. 1862, p. 8, *and Icon. Gén.* 10, pl. ii. fig. 5 (1865); *Günth. Proc. Zool. Soc.* 1873, p. 168.
—— monochrous, *Bleek. Nat. Tijdschr. Nederl. Ind.* xxi. 1860, p. 293.
—— bogorensis, *v. Lidth de Jeude, in M. Weber, Zool. Ergebn.* p. 182, pl. xvi. figs. 6 & 7 (1890).

Rostral nearly as deep as broad, well visible from above; frontal longer than broad, shorter than the parietals, about twice as broad as the supraocular; one præ- and one postocular; diameter of the eye equal to its distance from the mouth; five upper labials, third and fourth entering the eye; first pair of lower labials in contact behind the symphysial; two pairs of chin-shields, in contact with each other. Scales in 13 rows. Ventrals 151–199; anal entire; sub-

caudals 15–28. Tail obtuse. Dark brown or black above, uniform or with yellowish spots, or yellowish with brown spots; lower parts blackish, uniform or variegated with yellow, or yellow in the middle and black on the sides.
Total length 440 millim.; tail 30.
Sumatra, Java, Celebes.

A. Dark brown or black above, with yellow spots; lower parts black and yellow.

a–b. ♂ (V. 170; C. 27) Manado, Celebes. Dr. A. B. Meyer [C.].
& ♀ (V. 193; C. 19).

B. Dark brown with yellow spots above, uniform black beneath.

c. ♀ (V. 198; C. 15). Java. Dr. Ploem [C.].

C. Dark brown above, with small black spots or lines, without yellow spots or with only a few on the anterior part of the body; ventrals yellow in the middle, black on the sides.

d. ♀ (V. 199; C. 19). Java. Dr. Ploem [C.].
e–g, h–i. ♂ (V. 158, 156, 156, 163; C. 26, 28, 26, 25) & ♀ (V. 183; C. 15). Rarahan, Java. Dr. Beccari [C.]; Marquis G. Doria [P.].
k–l, m–o. ♂ (V. 164, 162; C. 24, 24) & ♀ (V. 178, 174, 179; C. 20, 18, 19). Mt. Singalang, Sumatra. Dr. Beccari [C.]; Marquis G. Doria [P.].
p. ♀ (V. 196; C. 18). Manado, Celebes. Dr. A. B. Meyer [C.].

D. Uniform black above and below.

q. ♂ (V. 151; C. 22). Fort de Kock, Sumatra. Dr. Bleeker. (Type of *C. monochrous*.)

15. Calamaria leucogaster.

Calamaria leucogaster, *Bleeker, Nat. Tijdschr. Nederl. Ind.* xxi. 1860, p. 293; *Günth. Ann. & Mag. N. H.* (3) xv. 1865, p. 89; *Edeling, Nat. Tijdschr. Nederl. Ind.* xxxi. 1870, p. 381.
—— arcticeps, *Günth. Ann. & Mag. N. H.* (3) xviii. 1866, p. 25, pl. vi. fig. C.

Snout rather pointed; rostral as deep as broad, well visible from above; frontal longer than broad, shorter than the parietals, about twice as broad as the supraocular; one præ- and one postocular; diameter of the eye equal to its distance from the mouth; five upper labials, second larger than first or third, third and fourth entering the eye, or (spec. *d*) six upper labials, fourth and fifth entering the eye; first pair of lower labials forming a suture behind the symphysial; two pairs of chin-shields, in contact with each other. Scales in 13 rows. Ventrals 133–151; anal entire; subcaudals 14–17. Tail ending in a point. Brown above; nape blackish, followed by a narrow yellow collar; tail with two blackish transverse spots; uniform yellowish inferiorly.

Total length 200 millim.; tail 13.
Sumatra, Borneo.

A. Back uniform purplish brown, with mere traces of darker lines.

a. Hgr. (V. 133; C. 17).　　Ampatlawang.　　Dr. Bleeker.
(One of the types.)

B. Pale brown above, with eight blackish longitudinal lines.

b. ♀ (V. 151; C. 16).　　Borneo.　　L. L. Dillwyn, Esq. [P.].
(Type of *C. arcticeps*.)
c. ♀ (V. 149; C. 14).　　Borneo.　　Dr. Collingwood [P.].
d. ♀ (V. 147; C. 14).　　Labuan.　　A. Everett, Esq. [C.].
e. Hgr. (V. 143; C. 14).　　Kina Baloo.　　A. Everett, Esq. [C.].

16. Calamaria occipitalis.

Calamaria occipitalis, *Jan, Arch. Zool. Anat. Phys.* ii. 1862, p. 9, *and Icon. Gén.* 10, pl. iii. fig. 1 (1865).

Rostral much broader than deep, just visible from above; frontal longer than broad, as long as the parietals, once and a half as broad as the supraocular; one præocular; one postocular; diameter of the eye less than its distance from the mouth; five upper labials, third and fourth entering the eye; first lower labial in contact with its fellow behind the symphysial; two pairs of chin-shields, in contact with each other. Scales in 13 rows. Ventrals 177–179; anal divided; subcaudals 20–21. Tail ending in a point. Blackish above; upper lip, a band across the occiput, two outer series of scales, and lower parts whitish; a blackish line along the lower surface of the tail.

Total length 500 millim.; tail 40.

Java.

17. Calamaria bicolor.

Calamaria bicolor, *Dum. & Bibr.* vii. p. 78 (1854); *Jan, Arch. Zool. Anat. Phys.* ii. 1862, p. 9, *and Icon. Gén.* 10, pl. ii. fig. 6 (1865).

Rostral as deep as broad, visible from above; frontal longer than broad, nearly as long as the parietals, about thrice as broad as the supraocular; a præocular and a postocular; diameter of the eye a little less than its distance from the mouth; five upper labials, third and fourth entering the eye; first pair of lower labials forming a suture behind the symphysial; two pairs of chin-shields, in contact with each other. Scales in 13 rows. Ventrals 143–152; anal entire; subcaudals 24–29. Tail ending in a point. Black above; upper lip, two outer series of scales, and lower parts whitish; lower surface of tail with or without a black longitudinal line.

Total length 375 millim.; tail 57.

Borneo.

a. ♂ (V. 143; C. 29).　　Sarawak.　　Rajah Brooke [P.].

18. Calamaria lateralis.

Calamaria lateralis, *Mocquard, Le Naturaliste*, 1890, p. 154, *and Nouv. Arch. Mus.* (3) ii. 1890, p. 136, pl. viii. fig. 4.

Rostral just visible from above; frontal longer than broad, as long as the parietals; one præocular; postocular fused with the supraocular; diameter of the eye less than its distance from the mouth; five upper labials, third and fourth entering the eye, fifth largest; first lower labial in contact with its fellow behind the symphysial; both pairs of chin-shields in contact with each other. 13 rows of scales. Ventrals 146; anal entire; subcaudals 21. Tail ending in a point. Blackish brown, a little lighter below; a white lateral stripe, running along the second and third series of scales.

Total length 245 millim.; tail 12.

Mt. Kina Baloo, N. Borneo.

The fusion of the postocular with the supraocular may prove not to be a constant character.

19. Calamaria beccarii.

Calamaria beccarii, *Peters, An. Mus. Genova,* iii. 1872, p. 34.

Rostral visible from above; frontal longer than broad; one præ- and two postoculars; five upper labials; first pair of lower labials in contact behind the symphysial; two pairs of chin-shields, in contact with each other. 13 rows of scales. Ventrals 150; anal entire; subcaudals 16. Brown above, with six darker longitudinal lines; dirty white inferiorly, with a black line along the hinder half of the tail.

Total length 183 millim.; tail 14.

Sarawak, Borneo.

20. Calamaria rebentischii. (PLATE XVIII. fig. 3.)

Calamaria rebentischi, *Bleek. Nat. Tijdschr. Nederl. Ind.* xxi. 1860, p. 293.

Rostral broader than deep, well visible from above; frontal a little longer than broad, shorter than the parietals, about twice and a half as broad as the supraocular; no præocular; one postocular; diameter of the eye much less than its distance from the mouth; five upper labials, third and fourth entering the eye; first pair of lower labials forming a suture behind the symphysial; two pairs of chin-shields, in contact with each other. Scales in 13 rows. Ventrals 142; anal entire; subcaudals 29. Tail ending in a rather obtuse point. Purplish brown above, each scale of the outer row with a whitish spot; uniform yellowish inferiorly.

Total length 270 millim.; tail 37.

Borneo.

a. ♂ (V. 142; C. 29).   Sinkawang.   Dr. Bleeker. (Type.)

21. Calamaria agamensis.

Calamaria agamensis, *Bleek. Nat. Tijdschr. Nederl. Ind.* xxi. 1860, p. 292; *Günth. Ann. & Mag. N. H.* (3) xv. 1865, p. 89.

—— dumerili, *Bleek. l. c.*

Calamaria sinkawangensis, *Bleek. l. c.*
—— roelandti, *Bleek. l. c.* p. 294.
—— cuvieri, *Jan, Arch. Zool. Anat. Phys.* ii. 1862, p. 9, *and Icon. Gén.* 10, pl. iii. fig. 2 (1865).

Rostral broader than deep; frontal at least as broad as long, at least four times as broad as the supraocular, shorter than the parietals; no præocular; one postocular; diameter of the eye much less than its distance from the mouth; five upper labials, third and fourth entering the eye; anterior lower labials in contact behind the symphysial; two pairs of chin-shields, the posterior usually separated from each other. 13 rows of scales. Ventrals 135–174; anal entire; subcaudals 21–38. Tail ending in a point. Black above, yellowish beneath; usually a brown subcaudal line.

Total length 290 millim.; tail 43.

Sumatra, Borneo, Java.

a.	♀ (V. 153; C. 21).	Sumatra.	Dr. Bleeker. (Type.)
b.	♂ (V. 139; C. 28).	Sinkawang, Borneo.	Dr. Bleeker.
			(Type of *C. dumerili*.)
c.	♂ (V. 135; C. 27).	Sinkawang, Borneo.	Dr. Bleeker.
			(Type of *C. sinkawangensis*.)
d.	♂ (V. 167; C. 27).	Sinkawang, Borneo.	Dr. Bleeker.
			(Type of *C. roelandti*.)
e.	♂ (V. 168; C. 30).	Java.	
f–h.	♂ (V. 149, 146, 143; C. 36, 36, 38).	Salak, Java.	R. Kirkpatrick, Esq. [C.].

22. Calamaria leucocephala.

Calamaria lumbricoidea, var., *Cantor, Cat. Mal. Rept.* p. 61 (1847).
—— leucocephala, *Dum. & Bibr.* vii. p. 83 (1854); *Günth. Rept. Brit. Ind.* p. 198 (1864).
—— macrurus, *Bleek. Nat. Tijdschr. Nederl. Ind.* xxi. 1860, p. 294.
—— nigro-alba, *Günth. l. c.* pl. xviii. fig. C.
—— martapurensis, *Edeling, Nederl. Tijdschr. Dierk.* ii. 1865, p. 203.
—— iris, *Boettg. Ber. Offenb. Ver. Nat.* 1873, p. 38, pl. i.

Rostral much broader than deep; frontal at least as broad as long, four times as broad as the supraocular, shorter than the parietals; one præ- and one postocular; diameter of the eye much less than its distance from the mouth; five upper labials, third and fourth entering the eye; anterior lower labials in contact behind the symphysial; two pairs of chin-shields, the posterior usually in contact anteriorly. 13 rows of scales. Ventrals 136–171; anal entire; subcaudals 19–45. Tail ending in a point. Dark brown or black above, uniform yellowish beneath; a faint brown subcaudal line may be present.

Total length 450 millim.; tail 48.

Malay Peninsula, Java, Sumatra, Borneo.

A. Head yellowish.

a. ♂ (V. 136; C. 42). Singapore. Dr. Dennys [C.].
b. ♂ (V. 154; C. 31). Java.

B. Head coloured like the back, or somewhat lighter.

c-e. Hgr. (V. 167, 147, 145; C. 25, 30, 31). Pinang. Dr. Cantor. (Types of *C. nigro-alba.*)
f-g. ♂ (V. 154; C. 30) & ♀ (V. 167; C. 27). Singapore.
h. ♂ (V. 142; C. 45). Borneo. L. L. Dillwyn, Esq. [P.].
i. ♀ (V. 153; C. 19). Matang.
k. ♂ (V. 165; C. 28). Java. Dr. Bleeker. (Type of *C. macrurus.*)
l-n. ♂ (V. 145; C. 36) & ♀ (V.171,167; C.28,?). Salak, Java. R. Kirkpatrick, Esq. [C.].

23. Calamaria schlegelii.

Calamaria schlegelii, *Dum. & Bibr.* vii. p. 81 (1854); *Jan, Arch. Zool. Anat. Phys.* ii. 1862, p. 8, *and Icon. Gén.* 10, pl. ii. fig. 4 (1865).

Rostral broader than deep, visible from above; frontal nearly as broad as long, a little shorter than the parietals, thrice as broad as the supraocular; no præocular; one postocular; diameter of the eye less than its distance from the mouth; five upper labials, third and fourth entering the eye; symphysial in contact with the anterior chin-shields; posterior chin-shields separated from each other. Scales in 13 rows. Ventrals 138; anal entire; subcaudals 33. Tail ending in a point. Blackish above; temples and lower surfaces whitish; a brown line along the lower surface of the tail.

Total length 259 millim.; tail 37.

Borneo.

24. Calamaria linnæi.

Coluber calamarius, *Linn. Mus. Ad. Frid.* p. 23, pl. vi. fig. 3 (1754), *and S. N.* i. p. 375 (1766).
Anguis calamaria, *Laur. Syn. Rept.* p. 68 (1768).
Calamaria linnæi, *Boie, Isis,* 1827, p. 539; *Schleg. Phys. Serp.* ii. p. 28, pl. i. figs. 17 & 18 (1837), *and Abbild.* p. 15, pl. iv. (1837); *Dum. & Bibr.* vii. p. 63, pl. lxiv. (1854); *Günth. Cat.* p. 3 (1858); *Jan, Arch. Zool. Anat. Phys.* ii. 1862, p. 6, pl. v., *and Icon. Gén.* 10, pl. i. figs. 1, 2, 3, 6 (1865).
—— multipunctata, *Boie, l. c.* p. 540.
—— maculosa, *Boie, l. c.*
—— reticulata, *Boie, l. c.*
—— versicolor, *Ranzani, Mem. Soc. Ital. Modena,* xxi. 1837, p. 101, pl. v. fig. 3; *Dum. & Bibr. t. c.* p. 69; *Jan, ll. cc., Icon.* figs. 7 & 8.

Rostral as deep as broad, well visible from above; frontal longer than broad, as long as the parietals, not quite twice as broad as the supraocular; one præ- and one postocular; diameter of the eye equal to its distance from the mouth; four upper labials; second and fourth largest, second and third entering the eye; symphysial in contact with the anterior chin-shields; the two pairs of chin-shields usually in contact with each other. Scales in 13 rows. Ventrals 135–163; anal entire; subcaudals 9–23. Tail obtuse. Colour very variable.

Total length 320 millim.; tail 15.

Java, Celebes.

A. Reddish above, with black transverse bars; red inferiorly, with quadrangular black spots. (*Coluber calamarius*, L.; var. *transversalis*, Jan.)

a. ♀ (V. 144; C. 9). Java.
b. ♀ (V. 151; C. 9). Java. Dr. Ploem [C.].

B. Reddish or purplish brown above, with small black spots; red inferiorly, with quadrangular black spots. (*C. reticulata*, Boie; vars. *tessellata, bilineata, contaminata*, Jan.)

a–c, d, e, f. ♂ (V. 141, 145, 136, 138; C. 20, 18, 19, 17) & ♀ (V. 151, 159; C. 12, 11). Java.
g–h, i, k. ♂ (V. 141; C. 17) & ♀ (V. 147, 159, 159; C. 10, 10, 11). Java. Dr. Ploem [C.].
l. ♀ (V. 156; C. 10). Salak, Java. R. Kirkpatrick, Esq. [C.].
m. ♀ (V. 154; C. 10). Manado, Celebes. Dr. A. B. Meyer [C.].
n. Ad., skel. Java.

C. Like B, but belly uniform red, without spots.

a–d. ♂ (V. 139, 135, 145; C. 21, 21, 19). Java.

D. Like B, but belly uniform black.

a. ♀ (V. 163; C. 14). Willis Mts., Kediri, Java, 5000 feet. Baron v. Huegel [C.].

E. Brown above, with a few small black spots and a black vertebral line; belly red with black spots. (*C. versicolor*, Ranz.)

a. ♂ (V. 139; C. 23). Java.

F. Like E, but the vertebral line interrupted by black, light-edged rhomboidal markings. (Var. *rhomboidea*, Jan.)

a, b, c. ♂ (V. 144; C. 21) & ♀ (V. 156, 160; C. 12, 15). Java.

25. Calamaria borneensis. (PLATE XIX. fig. 1.)

Calamaria borneensis, *Bleek. Nat. Tijdschr. Nederl. Ind.* xxi. 1860, p. 296.

Snout prominent; rostral rather deeper than broad, the portion visible from above at least as long as the suture between the præfrontals; frontal longer than broad, shorter than the parietals, about twice and a half as broad as the supraocular; one præ- and one postocular; diameter of the eye rather less than its distance from the mouth; four upper labials, second much larger than first or third, second and third entering the eye; symphysial in contact with the anterior chin-shields; both pairs of chin-shields in contact with each other. 13 rows of scales. Ventrals 163–192; anal entire; subcaudals 19–21. Tail very obtusely pointed. Dark brown above, uniform or with six black longitudinal lines; some yellowish spots on the sides of the neck and tail; end of tail blackish, or spotted with black.

Total length 330 millim.; tail 24.

Borneo.

A. A white lateral streak between two black ones, the lower of which extends on the outer ends of the ventrals; belly uniform yellowish in the middle, or with small blackish dots.

a. Hgr. (V. 185; C. 21).   Sintang.   Dr. Bleeker. (Type.)
b–c. ♀ (V. 192; C. 19) & hgr. (V. 178; C. 19).   Matang.

B. No lateral streaks; belly checkered black and white.

a. ♂ (V. 163; C. 19).   Claudetown, Baram R.   C. Hose, Esq. [C.].

26. Calamaria benjaminsii.

Calamaria benjaminsii, *Edeling, Nederl. Tijdschr. Dierk.* ii. 1864, p. 202; *Boettg. Mitth. Geogr. Ges. u. Nat. Mus. Lübeck,* (2) v. 1893, p. —.

Apparently very closely allied to *C. linnæi*, but supraocular smaller and tail pointed. 145–154 ventrals; 17–20 subcaudals. Uniform blackish purple above; the scales of the outer series with a white spot; upper lip and belly reddish; a black streak along the side of the ventrals, and another along the middle of the tail.

Total length 260 millim.; tail 23.

Borneo.

27. Calamaria javanica. (PLATE XIX. fig. 2.)

Calamaria javanica, *Bouleng. Ann. & Mag. N. H.* (6) vii. 1891, p. 279.

Rostral nearly as deep as broad, visible from above; frontal a little longer than broad, shorter than the parietals, thrice as broad as the supraocular; no præocular; one postocular; diameter of the eye nearly equal to its distance from the mouth; four upper labials, second and fourth largest, second and third entering the eye; symphysial in contact with the anterior chin-shields; posterior chin-shields separated from each other. Scales in 13 rows. Ventrals 181; anal entire; subcaudals 17. End of tail obtuse. Dark brown above, each scale with a lighter dot; a yellowish collar, interrupted in the middle, some distance behind the head; upper lip and lower parts uniform yellowish.

Total length 185 millim.; tail 13.

Java.

a. ♂ (V. 181; C. 17). Java. Dr. Ploem [C.]. (Type.)

28. Calamaria brevis. (PLATE XIX. fig. 3.)

Calamaria quadrimaculata, part., *Günth. Cat.* p. 4 (1858).

Rostral as deep as broad, well visible from above; frontal longer than broad, nearly as long as the parietals, once and a half as broad as the supraocular; one præ- and one postocular; diameter of the eye exceeding its distance from the mouth; four upper labials, second and fourth largest, second and third entering the eye; first lower labial in contact with its fellow behind the symphysial; two pairs of chin-shields, in contact with each other. Scales in 13 rows. Ventrals 134; anal entire; subcaudals 19. End of tail obtuse. Pale brown above, with darker dots; yellowish inferiorly, with a few black spots.

Total length 125 millim.; tail 10.

Hab. ———?

a. ♂ (V. 134; C. 19). ———?

29. Calamaria pavimentata.

Calamaria pavimentata, *Dum. & Bibr.* vii. p. 71 (1854); *Jan, Arch. Zool. Anat. Phys.* ii. 1862, p. 7, *and Icon. Gén.* 10, pl. i. fig. 9 (1865); *Bouleng. Faun. Ind., Rept.* p. 282 (1890).
——— quadrimaculata, *Dum. & Bibr. t. c.* p. 73; *Jan, ll. cc.* fig. 10; *Günth. Rept. Brit. Ind.* p. 197 (1864).
——— siamensis, *Günth. l. c.* p. 196; *Theob. Cat. Rept. Brit. Ind.* p. 140 (1876); *Boettg. Ber. Offenb. Ver. Nat.* 1885, p. 121.

Rostral as deep as broad, well visible from above; frontal longer than broad, shorter than the parietals, twice and a half as broad as the supraocular; one præ- and one postocular; the diameter of the eye nearly equals its distance from the mouth; four upper labials, second and third entering the eye; first pair of lower labials forming a suture behind the symphysial; two pairs of chin-shields, in contact with each other. Scales in 13 rows. Ventrals 140–182; anal entire; subcaudals 13–27. Tail pointed. Reddish brown above, with five dark longitudinal lines or series of spots; nape dark brown, separated from the back by a yellow collar; a pair of yellow spots on the base and another at the end of the tail.

Total length 320 millim.; tail 15.

Java, Burma, Siam, Cochinchina, Canton.

A. Lower parts uniform yellowish, with a dark line along the tail.

a. Hgr. (V. 141; C. 13).	——?	Gen. Hardwicke [C.].
b–c. ♀ (V. 182; C. 14) & yg. (V. 176; C. 25).	Pegu.	W. Theobald, Esq. [P.].

B. Ventrals and subcaudals obscured with brown mottlings, or brown with light borders.

a. ♂ (V. 152; C. 20).	Siam.	M. Mouhot [C.].
b, c. ♂ (V. 155; C. 21) & ♀ (V. 180; C. 13).	Lao Mountains.	M. Mouhot [C.]. (Types of *C. siamensis*.)
d. ♂ (V. 160; C. 22).	North Chin hills, Upper Burma.	E. Y. Watson, Esq. [P.].

30. Calamaria septentrionalis. (Plate XX. fig. 1.)

Calamaria quadrimaculata (*non D. & B.*), *Günth. Ann. & Mag. N. H.* (6) i. 1888, p. 165.

—— septentrionalis, *Bouleng. Proc. Zool. Soc.* 1890, p. 34.

Rostral broader than deep, not or hardly visible from above; frontal as broad as long, shorter than the parietals, about twice and a half to three times as broad as the supraocular; one præ- and one postocular; the diameter of the eye nearly equal to its distance from the mouth; four upper labials, second and third entering the eye; first pair of lower labials forming a suture behind the symphysial; two pairs of chin-shields, in contact with each other. Scales in 13 rows. Ventrals 148–177; anal entire; subcaudals 9–17. Tail rounded at the end. Blackish brown above, with three longitudinal series of small black spots; each scale of the outer row with a whitish spot; a yellow nuchal collar, interrupted in the middle, and a pair of yellow spots at the base of the tail; lower parts uniform coral-red, with a black line along the middle of the tail.

Total length 300 millim.; tail 13–22.

China.

a–d, e, f. ♂ (V. 161, 159, 148; C. 17, 16, 15) & ♀ (V. 171, 172, 176; C. 9, 10, 9).	Mountains North of Kiukiang.	A. E. Pratt, Esq. [C.].	(Types.)
g. Yg. (V. 176; C. 10).	Hong Kong.	H.M.S. 'Challenger.'	
h. ♀ (V. 174; C. 9).	Mainland opposite Chusan Archipelago.	J. J. Walker, Esq. [P.].	
i. Hgr. (V. 177; C. 10).	Chusan Archipelago.	J. J. Walker, Esq. [P.].	

31. Calamaria melanota.

Calamaria linnæi, vars. melanota *et* gastrogramma, *Jan, Arch. Zool. Anat. Phys.* ii. 1862, p. 5, pl. v., *and Icon. Gén.* 10, pl. i. figs. 4 & 5 (1865).

Rostral as deep as broad, well visible from above; frontal as broad as long, three or four times as broad as the supraocular, shorter than the parietals; one præ- and one postocular; diameter of the eye less than its distance from the mouth; four upper labials, second and fourth largest, second and third entering the eye; symphysial in contact with the anterior chin-shields; two pairs of chin-shields, in contact with each other. Scales in 13 rows. Ventrals 133-142; anal entire; subcaudals 16–24. Tail ending in a point. Black above, each scale lighter in the centre; lateral scales, ventrals, and subcaudals white, with a black edge.

Total length 240 millim.; tail 30.

Borneo and Java.

a–d. ♂ (V. 134, 133; C. 24, 24) & ♀ (V. 142, 135; C. 16, 17).   S.E. Borneo.

32. Calamaria lovii. (Plate XX. fig. 2.)

Calamaria lovii, *Bouleng. Ann. & Mag. N. H.* (5) xix. 1887, p. 196, fig.

Rostral nearly as deep as broad, visible from above; frontal as broad as long, shorter than the parietals, thrice as broad as the supraocular; no præocular; one postocular; diameter of the eye about half its distance from the mouth; four upper labials, third entering the eye; symphysial in contact with the anterior chin-shields; both pairs of chin-shields in contact with each other. Scales in 13 rows. Ventrals 211–225; anal entire; subcaudals 15–22. Tail ending obtusely. Plumbeous or dark brown above, with longitudinal lines of light dots; upper lip yellowish; anterior fourth of the body with a lateral series of large yellowish spots; a few other such spots on the tail.

Total length 265 millim.; tail 22.

Borneo.

A. Belly uniform blackish, plumbeous; a yellow cross bar above the vent.

a. ♂ (V. 211; C. 22).   Rejang R., Sarawak.   Brooke Low, Esq. [P.]. (Type.)

B. Belly blotched, dark brown and yellow.

b. ♀ (V. 225; C. 15).   Niah R., Sarawak.   C. Hose, Esq. [C.].

33. Calamaria gracillima.

Calamaria gracillima, *Günth. Proc. Zool. Soc.* 1872, p. 594, pl. xxxix. fig. A.
Typhlocalamus gracillimus, *Günth. l. c.* p. 589.

Rostral as deep as broad, visible from above; frontal a little broader than long, shorter than the parietals; eye extremely minute; supraocular very small, confluent with the postocular; no præocular; four upper labials, second and fourth largest, second and third entering the eye; first lower labial in contact with its fellow behind the symphysial; two pairs of chin-shields, in contact with each other. Body extremely slender. Scales in 13 rows. Ventrals 300–320; anal entire; subcaudals 13–14. Tail obtuse, rounded at the end. Blackish, with a few yellowish (red?) spots along each side of the body and a pair at the base of the tail.

Total length 285 millim.; tail 10.

Borneo.

a. ♀ (V. 320; C. 13). Borneo. (Type.)
b. ♀ (V. 300; C. 14). Matang.

The following species is doubtful:—

CALAMARIA CATENATA, Blyth, Journ. As. Soc. Beng. xxiii. 1854, p. 287.

Frontal almost as large as the parietals. Scales in 13 rows. Ventrals 187; subcaudals 41. Predominant colour dusky above, formed by minute black specks upon a pale ground-tint; below pale buff and marked with lateral series of square black spots; four black lines throughout above, the upper bordering a pale medial streak, which is simple upon the tail, but along the body forms a concatenation of elongate oval spots; an imperfect whitish-buff collar, and similar marks before and behind the eye.

Assam.

122. TYPHLOGEOPHIS.

Typhlogeophis, *Günth. Proc. Zool. Soc.* 1879, p. 77.

Maxillary teeth 8, subequal; mandibular teeth subequal. Head not distinct from neck; eye concealed under the ocular shield; no supraocular; nostril pierced in a minute nasal; internasals small; no loreal or præocular; no temporals, the parietals in contact with the labials. Body cylindrical; scales smooth, without apical pits, in 15 rows. Tail short; subcaudals in two rows.

Philippine Islands.

1. Typhlogeophis brevis. (PLATE XX. fig. 3.)

Typhlogeophis brevis, *Günth. l. c.*

Snout rather pointed; rostral very small, nearly as deep as broad, just visible from above; suture between the internasals about half the length of that between the præfrontals; frontal small, as long as broad, shorter than its distance from the end of the snout, half as long as the parietals; five upper labials, fourth in contact with the ocular, fifth very large; two pairs of chin-shields, anterior largest.

Scales in 15 rows. Ventrals 153; anal entire; subcaudals about 15. Uniform blackish, scales and shields with whitish edge.
Total length 330 millim.
Philippine Islands.

a. ♀ (V. 153; C. ?)   Philippines.   A. Everett, Esq. [C.].
                                                (Type.)

The position of the following genera is doubtful:—

AMASTRIDIUM.

Cope, Proc. Ac. Philad. 1860, p. 370.

"Body cylindrical, elongate; tail moderate, slender. Head distinct, broad, short, tapering rather abruptly. Superior maxillary teeth in a continuous series, the last abruptly the longest, not grooved. Pupil round. Top of head flat, separated on the muzzle from the sides by an angle. Superciliaries prominent. One anterior, two postoculars. Loreal none. Nasals large, one or two, the nostril situated in the centre of the anterior. Scales on the posterior part of the body slightly keeled. Anal and subcaudal scutellæ divided."

A. VELIFERUM, Cope, *l. c.*

"Scales in 17 longitudinal rows, smooth on the anterior half of the body; posteriorly a few dorsal rows with faint keels, becoming stronger towards the tail, and extending on all the scales near the anal region, where they are tuberculous as in *Aspidura trachyprocta*. Tail nearly one third the total length. Occipital [=parietal] plates large, almost reaching the labials in front, posteriorly acuminate; vertical [=frontal] long, acute behind; superciliaries large, prominent, broad behind. Postfrontals [=præfrontals] small, their anterior outline regularly curved; præfrontals [= internasals] small, quadrangular. Rostral nearly rectangular, not appearing on the surface of the head. Postnasal high, its apex visible from above, opposite the suture between the præ- and postfrontals. Superior labials seven, eye resting on third and fourth. Inferior labials nine. Geneials two pair, the anterior shorter. Gastrosteges 127, urosteges 85. Above and below reddish brown, paler in the centres of the gastrosteges. Every fourth scale of the fifth row on each side pale, the adjacent scales on the fourth and sixth rows generally darker. Top of the head much lighter, varied anteriorly; palest behind the eye and above the labials. The latter are dark with a few light spots."

Total length 355 millim.; tail 120.
Cocuyas de Veraguas, Colombia.

ANOPLOPHALLUS.

Megalops (non *Lacép.*), *Hallow. Proc. Ac. Philad.* 1860, p. 488.
Anoplophallus, *Cope, Am. Nat.* 1893, p. 480.

"Mandibular teeth increasing in length posteriorly, recurved, nearly straight; two internasals, much smaller than the præfrontals; frontal a little longer than broad, pentangular; a frenal [loreal]; two antoculars; two postoculars; eye resting on the fourth supralabial; pupil ovoid; eyes very prominent; body slender, much compressed; abdomen angular; tail rather short." (*Hallowell*). "A long loreal and no preocular plate." (*Cope.*)

A. MACULATUS.

Megalops maculatus, Hallow. *l. c.*

"21 rows of smooth scales; body presenting numerous subquadrangular and oblique blotches above, of a brown colour; intermediate spaces white with a tinge of yellow; under surface white. Abdominal scuta 170; a bifid præanal; 61 subcaudal scutellæ."
Total length 375 millim.; tail 95.
Tahiti.

Subfam. 3. RHACHIODONTINÆ.

Leptognathiens, part., *Duméril, Mém. Ac. Sc.* xxiii. p. 464, 1853; *Duméril & Bibron, Erp. Gén.* vii. p. 25, 1854.
Rachiodontidæ, *Günther, Cat. Col. Sn.* p. 141, 1858.
Rachiodontidæ, *Jan, Elenco sist. Ofid.* p. 106, 1863.
Dasypeltinæ, part., *Cope, Proc. Am. Philos. Soc.* xxiii. 1886, p. 494.

Only a few teeth, on the posterior part of the maxillary and dentary bones and on the palatines. Some of the anterior thoracic vertebræ with the hypapophysis much developed, directed forwards, and capped with enamel.
A single genus.

123. DASYPELTIS.

Anodon (non *Lam.*), *Smith, Zool. Journ.* iv. 1829, p. 443.
Dasypeltis, *Wagler, Syst. Amph.* p. 178 (1830); *Günth. Cat. Col. Sn.* p. 141 (1858).
Rachiodon, *Jourdan, Le Temps*, 13. 6. 1833 (?); *Dum. & Bibr. Erp. Gén.* vii. p. 487 (1854); *Jan, Elenco sist. Ofid.* p. 106 (1863).
Tropidonotus, part., *Schleg. Phys. Serp.* ii. p. 297 (1837).
Deirodon, *Owen, Odontogr.* p. 220 (1845).

Teeth very small, few, 3 to 7 in each maxillary, and as many in the mandible; palatine but no pterygoid teeth. Head small, not or scarcely distinct from neck; no mental groove; eye moderate, with vertically elliptic pupil; nostril in a semidivided nasal, which is in contact with the præocular; no loreal. Body cylindrical or slightly compressed, elongate; scales strongly keeled, with apical pits, in 23 to 27 rows; scales on three or four lateral rows oblique and with more or less distinctly serrated keels; ventrals rounded. Tail moderate; subcaudals in two rows.

Tropical and South Africa.

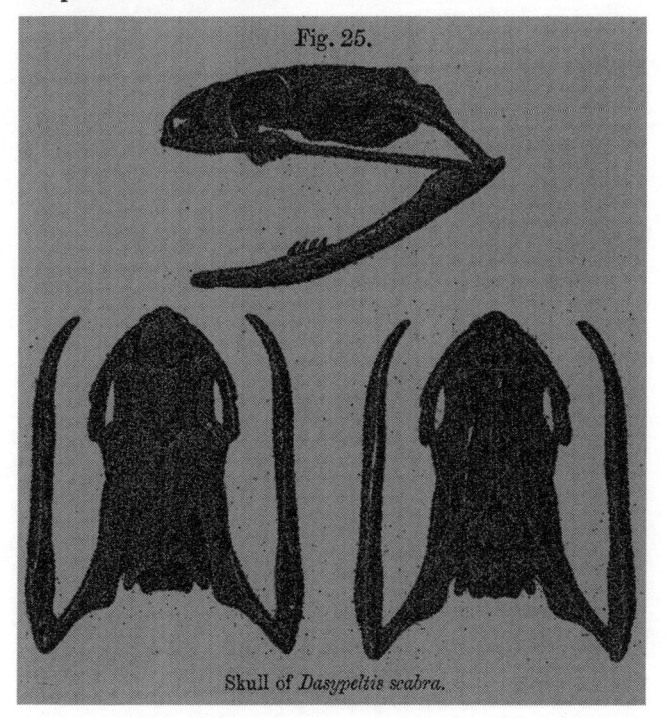

Fig. 25.

Skull of *Dasypeltis scabra*.

1. Dasypeltis scabra.

Coluber scaber, *Linn. Mus. Ad. Frid.* p. 36, pl. x. fig. 1 (1754), *and S. N.* i. p. 384 (1766); *Merr. Beitr. Nat. Amph.* i. p. 34, pl. ix. (1790); *Daud. Rept.* vi. p. 263 (1803).
—— palmarum, *Leach, Tuckey's Explor. R. Zaire*, App. p. 408 (1818).
Anodon typus, *Smith, Zool. Journ.* iv. 1829, p. 443.
Rachiodon scaber, *Jourdan, Le Temps*, 13. 6. 1833 (?); *Dum. & Bibr.* vii. p. 491, pl. lxxxi. fig. 3 (1854); *Jan, Elenco*, p. 106 (1863), *and Icon. Gén.* 39, pl. ii. fig. 4 (1872).
Tropidonotus scaber, *Schleg. Phys. Serp.* ii. p. 328, pl. xii. figs. 12 & 13 (1837); *Reinh. Vidensk. Selsk. Skrift.* x. 1843, p. 265, pl. i. fig. 24.
Dasypeltis inornatus, *Smith, Ill. Zool. S. Afr., Rept.* pl. lxxiii. (1842).
—— scaber, *Smith, l. c.*
—— fasciatus, *Smith, l. c.*; *Peters, Mon. Berl. Ac.* 1864, p. 644.
Deirodon scaber, *Owen, Odontogr.* p. 220 (1845).
Dipsas carinatus, *Hallow. Proc. Ac. Philad.* ii. 1845, p. 119, & 1857, p. 69.

Rachiodon abyssinicus, *Dum. & Bibr. t. c.* p. 496.
—— inornatus, *Dum. & Bibr. t. c.* p. 498.
Dasypeltis scabra, *Günth. Cat.* p. 142 (1858); *Peters, Reise n. Mossamb.* iii. p. 120 (1882); *Boettg. Ber. Senck. Ges.* 1887, p. 163, & 1888, p. 75.
—— palmarum, *Günth. l. c.*; *Stejneger, Proc. U. S. Nat. Mus.* xvi. 1894, p. 733.
Dipsas medici, *Bianconi, Mem. Acc. Bologna*, x. 1859, p. 501, pl. xxvi.
Rachiodon scaber, vars. abissinicus, unicolor, subfasciatus, *Jan, Elenco,* p. 106.
Dasypeltis scaber, var. mossambica, *Peters, Mon. Berl. Ac.* 1864, p. 644, *and Reise n. Mossamb.* iii. p. 120.
—— scaber, vars. breviceps, abyssinicus, *Peters, Mon. Berl. Ac.* 1864, p. 645.
Rachiodon scaber, var. inornatum, *Bocage, Jorn. Sc. Lisb.* i. 1867, p. 227.
Dasypeltis scaber, var. fasciolata, *Peters, Mon. Berl. Ac.* 1868, p. 451.
—— scabra, var. medici, *Peters, Reise n. Mossamb.* iii. p. 121; *Mocquard, Mém. Cent. Soc. Philom.* 1888, p. 131.
—— scabra, var. fasciata, *Mocquard, Bull. Soc. Philom.* (7) xi. 1887, p. 81; *Boettg. Ber. Senck. Ges.* 1888, p. 76.
—— scabra, var. palmarum, *Bouleng. Zoologist,* 1887, p. 179; *Boettg. l. c.* p. 77.
—— elongata, *Mocquard, Mém. Cent. Soc. Philom.* 1888, p. 131, pl. xii. fig. 2.
—— abyssina ?, *Stejneger, l. c.*

Snout short, rounded, very convex; rostral broader than deep, just visible from above; internasals a little shorter than the præfrontals; frontal often more or less distinctly grooved along the middle, a little longer than broad, longer than its distance from the end of the snout, as long as the parietals or shorter; one (rarely two) præ- and two (rarely one or three) postoculars; temporals $2+3$ or $3+4$ (rarely $1+3$); seven (rarely six or eight) upper labials, third and fourth (or second and third) entering the eye; a pair of large chin-shields, followed by a pair of smaller ones. Scales in 23 to 27 rows. Ventrals 185–263; anal entire; subcaudals 41–94. Pale olive or pale brown above, uniform or with dark brown spots, usually disposed in three longitudinal series; a $\wedge$-shaped dark marking on the nape, preceded by one or two on the head; the latter may be broken up into spots; upper labials with brown vertical bars; belly yellowish, uniform or dotted or spotted with brown or blackish.

Total length 760 millim.; tail 105.

Africa, as far north as Sennar and Sierra Leone.

A. Vertebral spots elongate and more or less confluent into a zigzag vertebral band; a black stripe along upper surface of tail; belly spotted with blackish.

a–c. Hgr. ♂ (Sc. 25; V. 189; C. 66), ♀ (Sc. 27; V. 204; C. 59), & yg. (Sc. 26; V 207; C. 52). Port Elizabeth. H. A. Spencer, Esq. [P.].

356   COLUBRIDÆ.

d. Yg. (Sc. 25; V. 185; C. 59). Matabele-land. C. Beddington, Esq. [P.].
e–f. ♂ (Sc. 25; V. 193, 188; C. 67, 64). S. Africa.

B. A dorsal series of large squarish or rhomboidal dark spots, separated by light intervals, alternating with a lateral series of spots or cross bars; belly spotted or dotted only at the sides. (*C. scaber*, L.)

a. ♀ (Sc. 27; V. 221; C. 48). Cape of Good Hope. H.M.S. 'Herald.'
b. ♀ (Sc. 27; V. 214; C. 41). Cape of Good Hope.
c. Yg. (Sc. 23; V. 208; C. 60). S. Africa. Leyden Museum.
d. ♀ (Sc. 27; V. 237; C. 54). Zomba, Brit. C. Africa. H. H. Johnston, Esq. [P.].
e–g. ♂ (Sc. 25; V. 210; C. 73) & ♀ (Sc. 25, 27; V. 225, 229; C. 60, 58). Zanzibar. Sir J. Kirk [C.].
h. Yg. (Sc. 25; V. 233; C. 59). Kilimanjaro. F. J. Jackson, Esq. [P.].
i. Yg. (Sc. 23; V. 214; C. 68). Wadelai. Dr. Emin Pasha [P.].
k. Hgr. ♂ (Sc. 25; V. 201; C. 53). Kithu-Uri, E. Kikuyu. Dr. J. W. Gregory [P.].
l. Yg. (Sc. 23; V. 209; C. 67). W. Africa. Mr. Rich [C.].
m. ♂ (Sc. 23; V. 226; C. 89). Old Calabar. R. Logan, Esq. [P.].
n. ♀ skeleton. E. Kikuyu. Dr. J. W. Gregory [P.].

C. Pale reddish brown above, with the markings very much effaced. Intermediate between B and F.

a, b. ♀ (Sc. 25, 27; V. 237, 245; C. 80, 79). Zomba, Brit. C. Africa. H. H. Johnston, Esq. [P.].

D. Dorsal markings as in B, but ventrals edged with blackish.

a. Yg. (Sc. 23; V. 242; C. 63). Abyssinia. W. T. Blanford, Esq. [P.].

E. Dorsal spots confluent with lateral ones, forming cross bands; belly unspotted. (*D. medici*, Bianc.; *D. fascioluta*, Peters.)

a. Hgr. (Sc. 25; V 238; C. 60). Mouth of the Zambezi. Capt. Bedingfield [P.].

F. No spots or markings of any kind. (*C. palmarum*, Leach; *D. inornata*, Smith.)

a. Hgr. ♂ (Sc. 23; V. 224; C. 74). Boma, Congo. Dr. Leach. (Type of *C. palmarum*.)
b. ♀ (Sc. 25; V. 229; C. ?). Cacuaca, Angola. Dr. Welwitsch [C.].
c. Hgr. ♂ (Sc. 25; V. 233; C. 85). Natal. E. Howlett, Esq. [P.].

ADDENDA.

d. ♀ (Sc. 25; V. 224; C. 68). Port Natal. Mr. T. Ayres [C.].
e. ♂ (Sc. 24; V. 214; C. 85). S. Africa. Sir A. Smith [P.]. (One of the types of *D. inornata*.)
f. ♂ (Sc. 25; V. 231; C. 76). S. Africa. Sir A. Smith [P.].
g. ♀ (Sc. 24; V. 231; C. 79). Kilimanjaro. F. J. Jackson, Esq. [P.].
h. ♀ (Sc. 25; V. 243; C. 74). Uganda. Scott Elliot, Esq. [P.].

ADDENDA.

Add:— Page 13. **Drymobius boddaertii.**

A. (*C. boddaertii*).
e. ♀ (V. 187; C. 91). Chontalez Mines, Nicaragua. R. A. Rix, Esq. [C.]; W. M. Crowfoot, Esq. [P.].

Lower parts bright orange.

Add:— Page 15. **Drymobius rhombifer.**

f. ♂ (V. 151; C. 93). Chontalez Mines, Nicaragua. R. A. Rix, Esq. [C.]; W. M. Crowfoot, Esq. [P.].

Add:— Page 16. **Drymobius dendrophis.**

r–t. ♀ (V. 157; C. ?), hgr. (V. 165; C. 123), and yg. (V. 162; C. 111). Chontalez Mines, Nicaragua. R. A. Rix, Esq. [C.]; W. M. Crowfoot, Esq. [P.].

Add:— Page 37. **Coluber helena.**

s–w. ♀ (Sc. 27, 26; V. 238, 239; C. 81, 87) & yg. (Sc. 27; V. 237, 223, 220; C. 86, 91, 100). Ceylon. Miss Layard [P.].

Add:— Page 52. **Coluber longissimus.**

Coluber æsculapii, *Lütken, Vidensk. Meddel.* 1894, p. 72.

Add:— Page 57. **Coluber oxycephalus.**

n. Ad., head and tail. Balabac. A. Everett, Esq. [C.].

Add:— Page 58. **Coluber janseni.**

Coluber (Gonyosoma) jansenii, *F. Müll. Verh. nat. Ges. Basel*, x. 1894, p. 829.

ADDENDA.

Add:— Page 63. **Coluber erythrurus.**

Coluber erythrurus, *F. Müll. Verh. nat. Ges. Basel*, x. 1894, p. 828.

A. (*C. erythrurus*).
g. ♂ (V. 235; C. 112).   Palawan.   A. Everett, Esq. [C.].

Add:— Page 79. **Dendrophis pictus.**
η. ♀ (V. 189; C. 154).   Balabac.   A. Everett, Esq. [C.].

Add:— Page 80. **Dendrophis bifrenalis.**
d. ♀ (V. 164; C. 155).   Ceylon.   Miss Layard [P.].

Add:— Page 89. **Dendrelaphis tristis.**
q-r. ♂ (V. 182; C. 128) & yg.   Ceylon.   Miss Layard [P.].
(V. 182; C. 134).

Add:— Page 90. **Dendrelaphis caudolineatus.**
r-t. ♀ (V. 185; C. 109) & yg.   Palawan.   A. Everett, Esq. [C.].
(V. 182, 182; C. 107, 106).
u-v. Hgr. (V. 177; C. 101) &   Balabac.   A. Everett, Esq. [C.].
yg. (V. 178; C. 108).
w-x. ♂ (V. 183; C. 118) & ♀   Sibutu.   A. Everett, Esq. [C.].
(V. 189; C. 108).

Add:— Page 96. **Chlorophis heterolepidotus.**
g. ♂ (V. 183; C. 132).   Mouths of the Niger.   Alvan Millson, Esq. [P.].

Add:— Page 98. **Chlorophis heterodermus.**
h. ♂ (V. 147; C. ?).   Mouths of the Niger.   Alvan Millson, Esq. [P.].
i-k. ♂ (V. 151, 156; C.   Mouth of the Loango.   Mr. H. J. Duggan [C.].
93, 95).

Add:— Page 105. **Thrasops flavigularis.**
e-f. Yg. (V. 207, 202;   Mouth of the Loango.   Mr. H. J. Duggan [C.].
C. 146, 143).
Olive in front, with black edges to the scales; variegated yellowish brown and black behind; ventrals and subcaudals black, with large round or oval yellowish spots. One of the specimens has nine upper labials, fifth and sixth entering the eye.

Add:— Page 111. **Leptophis bilineatus.**
b-e. ♂ (V. 145; C.   Chontalez Mines, Nicara-   R. A. Rix, Esq. [C.];
125) & ♀ (V. 147,   gua.   W. M. Crowfoot,
148, 146; C. ?, 126,   Esq. [P.].
131).
Dorsal keels very feeble in the females.

Add:— Page 147. **Xenodon colubrinus.**

r-s. ♀ (V. 137; C. 46) & yg. (V. 150; C. 45).   Chontalez Mines, Nicaragua.   R. A. Rix, Esq. [C.]; W. M. Crowfoot, Esq. [P.].

Add:— Page 177. **Rhadinæa decorata.**

f-g. ♂ (V. 119; C. 95) & yg. (V. 115; C. 111).   Chontalez Mines, Nicaragua.   R. A. Rix, Esq. [C.]; W. M. Crowfoot, [P.].

One of the specimens has nine labials on one side, fifth and sixth entering the eye.

Add:— Page 183. **Urotheca elapoides.**

g-i. Hgr. (V. 139; C. ?) & yg. (V. 133, 133; C. 105, 114).   Chontalez Mines, Nicaragua.   R. A. Rix, Esq. [C.]; W. M. Crowfoot, Esq. [P.].

20 to 24 black annuli on the body, broader than the red interspaces.

Add:— Page 187. **Hydrops triangularis.**

Pseuderyx inagnitus, *Bocourt, Le Natur.* 1894, p. 155.

Add:— Page 194. **Coronella amaliæ.**

c-m. ♂ (V. 180; C. 64), ♀ (V. 189, 179, 186, 191; C. 52, 52, 63, 69), & yg. (V. 186, 174, 184, 181, 177; C. 65,65,58,65,60).   Near Tangier.   M. H. Vaucher [C.].

The largest specimen measures 500 millim.; tail 85.

Add:— Page 196. **Coronella semiornata.**

d. ♀ (V. 176; C. 82).   Lake Nyassa.

Add:— Page 210. **Hypsiglena torquata.**

Liophis janii, *Dugès, Mém. Ac. Montp.* vi. 1866, *Proc.-Verb.* p. 32.

Add:— Page 222. **Simotes formosanus.**

Simotes hainanensis, *Boettg. Ber. Senckenb. Ges.* 1894, p. 133, pl. iii. fig. 2.

Add:— Page 230. **Simotes arnensis.**

q-r. ♀ (V. 181; C. 46) & hgr. (V. 171; C. 49).   Ceylon.   Miss Layard [P.].

Add:— Page 242. **Oligodon templetonii.**

d. ♂ (V. 139; C. 29).   Ceylon.   Miss Layard [P.].

Page 245. Add :—

16 a. Oligodon tæniurus.

Oligodon tæniurus, *F. Müll. Verh. nat. Ges. Basel*, x. 1894, p. 826.

Nasal entire. Portion of rostral visible from above as long as its distance from the frontal; suture between the internasals a little longer than that between the præfrontals; frontal once and a half as long as broad, longer than its distance from the end of the snout, a little shorter than the parietals; loreal as long as deep; one præ- and two postoculars; temporals 1+2; seven upper labials, third and fourth entering the eye; four lower labials in contact with the anterior chin-shields, which are longer than the posterior. Scales in 15 rows. Ventrals 145–154; anal divided; subcaudals 23–28. Dark brown above, anteriorly with small orange dots, posteriorly with a reddish vertebral stripe; flanks pale grey, separated from the belly, on the anterior third of the body, by a black stripe occupying the outer ends of the ventrals; lower parts yellowish white, clouded with fine greyish dots.

Total length 275 millim. ; tail 35.

Celebes.

a. ♂ (V. 145; C. 28).　Pinogo, Bona Valley.　Drs. P. & F. Sarasin [P.].
(One of the types.)

Page 324. Add :—

113 a. AGROPHIS.

Agrophis, *F. Müll. Verh. nat. Ges. Basel*, x. 1894, p. 827.

Maxillary teeth 14, subequal; mandibular teeth slightly decreasing in size posteriorly. Head small, not distinct from neck; eye very small, with round pupil; nostril between two nasals, the anterior of which is very small; no præocular, loreal and præfrontal entering the eye; internasals and supraoculars very small; parietals in contact with labials. Body cylindrical; scales smooth, without apical pits, in 15 rows; ventrals rounded. Tail short; subcaudals in two rows.

Celebes.

1. Agrophis sarasinorum.

Agrophis sarasinorum, *F. Müll. l. c.* fig.

Snout long, obtusely pointed. Rostral large, the portion visible from above a little shorter than its distance from the frontal; præfrontals twice as long as nasals; frontal large, rhomboidal, as broad as long, as long as its distance from the end of the snout, shorter than the parietals; supraocular very small; a minute postocular; five upper labials, third and fourth entering the eye, fifth largest and forming a long suture with the parietal; symphysial not quite touching the anterior chin-shields, which are a little longer than the posterior. Scales in 15 rows. Ventrals 139–141; anal entire; subcaudals 37–40. Tail pointed. Blackish brown above, strongly iridescent, whitish beneath; ventrals and subcaudals darker in front.

Total length 190 millim.; tail 30.

Celebes.

a. ♀ (V. 141; C. 37).　Gunung Masarang.　Drs. P. & F. Sarasin [P.].

ALPHABETICAL INDEX.

abacura (Farancia), 291.
abacurum (Calopisma), 291.
abacurus (Coluber), 291.
abacurus (Farancia), 291.
abacurus (Helicops), 291.
abacurus (Hydrops), 291.
Abastor, 289.
Ablabes, 9, 25, 160, 188, 255, 277.
abnorma (Coronella), 203.
abyssina (Dasypeltis), 355.
abyssinicum (Homalosoma), 276.
abyssinicus (Rachiodon), 355.
Acanthophallus, 144.
Achirhina, 255.
Adelphicos, 300.
adspersus (Dromicus). 120.
ægyptiaca (Coronella), 192.
æneus (Xenodon), 150.
æqualis (Elapochrus), 183.
æqualis (Liophis), 183.
æqualis (Pleiocercus), 182.
æruginosus (Leptophis), 107.
æruginosus (Philothamnus), 107.
æsculapii (Callopeltis), 53.
æsculapii (Coluber), 52, 357.
æsculapii (Elaphis), 52.
æsculapii (Zamenis), 52.
æstiva (Contia), 258.
æstivus (Coluber), 258.
æstivus (Cyclophis), 258.
æstivus (Herpetodryas), 15, 258.
æstivus (Leptophis), 258.
æstivus (Liopeltis), 258.
æstivus (Opheodrys), 258.
æstivus (Phyllophilophis), 258.

affinis (Dromicus), 172, 173.
affinis (Drymobius), 14.
affinis (Herpetodryas), 14.
affinis (Hypsiglena), 210.
affinis (Oligodon), 236.
affinis (Pituophis), 69.
affinis (Rhadinæa), 172.
affinis (Simotes), 218.
agamensis (Calamaria), 343.
agassizii (Ablabes), 259.
agassizii (Coluber), 65.
agassizii (Contia), 259.
agassizii (Eirenis), 259.
agassizii (Rhinechis), 65.
Agrophis, 360.
Ahætulla, 77, 91, 98, 102, 105, 111.
ahætulla (Ahætulla), 113.
ahætulla (Coluber), 78, 113.
ahætulla (Leptophis), 78, 113.
ahætulla (Natrix), 78.
ahætulla (Thrasops), 113.
albiventer (Calamaria), 336.
albiventer (Changulia), 336.
albiventer (Simotes), 229.
albiventris (Coluber), 99.
albiventris (Liophis), 130.
albocincta (Coronella), 220.
albocinctus (Leptognathus), 295.
albocinctus (Simotes), 218, 220.
albopunctatus (Scaphiophis), 254.
albovariata (Dendrophis), 96.
albovariata (Philothamnus), 96.

alkeni (Calamaria). 333.
alleghaniensis (Coluber), 50.
alleghaniensis (Elaphis), 40, 50.
alleghaniensis (Pantherophis), 50.
alleghaniensis (Scotophis), 50.
Allophis, 25.
almadensis (Liophis), 134.
almadensis (Natrix), 134.
Alopecophis, 25.
alpestris (Coluber), 46.
alpinus (Coluber), 192.
Alsophis, 118.
alternans (Coluber), 131.
alternatus (Coryphodon), 11.
alternatus (Drymobius), 11.
alticolus (Liophis), 130.
alticolus (Opheomorphus), 130.
amabilis (Coronella), 207.
amabilis (Diadophis), 207.
amabilis (Dromicus), 158.
amabilis (Simotes), 221.
amaliæ (Coronella), 193, 359.
amaliæ (Rhinechis), 193.
Amastridium, 352.
amaura (Lampropeltis), 203.
ambigua (Prosymna), 248.
Amblymetopon, 270.
amboinense (Rabdosoma), 237.
amœna (Calamaria), 324.
amœna (Carphophiops), 324.
amœna (Celuta), 324.
amœnus (Aporophis), 160.
amœnus (Brachyorrhos), 324.

amœnus (Carphophis), 324.
amœnus (Coluber), 324.
amœnus (Enicognathus), 160.
ancoralis (Simotes), 225.
ancorus (Xenodon), 224.
andamanensis (Dendrophis), 78.
andreæ (Liophis), 140.
angolensis (Chlorophis), 95.
angolensis (Philothamnus), 95.
angulifer (Alsophis), 120.
angulifer (Dromicus),120.
angusticeps (Contia), 262.
angustirostris (Xenodon), 146.
annectens (Pituophis), 67.
annularis (Galedon), 296.
annulata (Coronella), 203.
annulata (Lampropeltis), 201, 203.
annulatum (Tropidoclonium), 296.
annulatus (Geophis), 296.
annulatus (Heterodon), 155.
annulatus (Ophibolus), 201, 203.
annulifer (Simotes), 226.
annulifera (Tropidodipsas), 297.
Anodon, 353.
anomala (Coronella), 165, 203.
anomala (Rhadinæa), 165.
anomalolepis (Spilotes), 23.
anomalus (Alsophis), 125.
anomalus (Aporophis), 165.
anomalus (Dromicus), 125.
anomalus (Zamenis), 125.
Anoplophallus, 353.
anthracops (Leptognathus), 297.
anthracops (Tropidodipsas), 297.
antillensis (Alsophis), 123.
antillensis (Dromicus), 122, 123.
antillensis (Psammophis), 122, 123.
antonii (Rhinochilus), 213.
aphanospilus (Simotes), 225.
Aporophis, 157, 160.

arcticeps(Calamaria), 341.
arctiventris (Calamaria), 274.
arctiventris (Coluber), 274.
arctiventris (Duberria), 274.
arctiventris (Homalosoma), 274.
argonauta (Tyria), 260.
argus (Spilotes), 20.
Arizona, 25.
arizonæ (Coluber), 66.
arnensis (Coluber), 229.
arnensis (Simotes), 229, 359.
arnyi (Diadophis), 208.
arnyi (Liophis), 208.
Arrhyton, 251.
aruensis (Dendrophis), 80.
asclepiadeus (Coluber), 52.
ater (Alsophis), 121.
ater (Dromicus), 121.
ater (Ocyophis), 121.
ater (Zamenis), 139.
atmodes (Heterodon),155.
atra (Natrix), 121.
Atractus, 300.
atrostriata (Dendrophis), 85.
auribundus (Spilotes), 33.
australis (Coronella), 168.
austriaca (Coronella, 53, 191.
austriacus (Coluber),191.
austriacus(Zacholus),192.

badia (Calamaria), 308, 309.
badium (Rabdosoma), 307, 308.
badius (Atractus), 308.
badius (Brachyorrhos), 308.
badius (Geophis), 308.
bairdi (Coluber), 40.
baliodeira (Coronella), 283.
baliodeirus (Diadophis), 283.
baliodirus (Ablabes), 283.
baliolum (Homalosoma), 282.
balteata (Coronella), 197.
baluensis(Calamaria),335.
beccarii (Calamaria), 343.
beddomii (Simotes), 229.
bellii (Ahætulla), 78.
bellii (Herpetæthiops),97.
bellona (Churchillia), 69.
bellona (Pituophis), 69.

Bellophis, 188.
benjaminsii (Calamaria), 347.
bernouillii (Leptognathus), 296.
bernouillii (Tropidodipsas), 296.
bertholdi (Xenodon),146.
bicarinata (Natrix), 73.
bicarinatus (Coluber), 73.
bicatenatus (Simotes), 219.
bicincta (Urotheca), 184.
bicinctus (Coluber), 184.
bicinctus (Leiosophis), 184.
bicinctus (Liophis), 184.
bicinctus (Xenodon), 184.
bicolor (Calamaria), 342.
bicolor (Coluber),131,168.
bicolor (Dirosema), 298.
bicolor (Geophis), 298.
bicolor (Liophis), 168.
bifossatus (Coluber), 10.
bifossatus (Drymobius), 10.
bifrenalis (Dendrophis), 80, 358.
bilineata (Diplotropis), 111.
bilineatus (Drymobius), 11.
bilineatus (Elaphis), 59.
bilineatus (Enicognathus), 173.
bilineatus (Herpetodryas), 11.
bilineatus (Leptophis), 111, 358.
binotatus (Simotes), 235, 243.
bipræocularis (Xenodon), 146.
bipunctatus (Diadophis), 285.
biserialis (Herpetodryas), 119.
biserialis (Orophis), 120.
bistrigatus(Cynophis),36.
bitorquata (Coronella), 196.
bitorquatum (Meizodon), 196.
bitorquatus (Oligodon), 237.
bitorques (Calamaria), 338.
bivittatum (Arrhyton), 252.
bivittatus (Drymobius), 15.

ALPHABETICAL INDEX. 363

bivittatus (Leptophis), 15.
bivittatus (Thamnosophis), 15.
bocagii (Ahætulla), 99.
bocourti (Atractus), 306.
boddaertii (Coluber), 11.
boddaertii (Drymobius), 11, 357.
boddaertii (Eudryas), 12.
boddaertii (Herpetodryas), 11.
bogorensis (Calamaria), 340.
boiga (Coluber), 113.
boii (Dendrophis), 88.
borneensis (Calamaria), 347.
boursieri (Dromicus), 174.
boylii (Lampropeltis), 197.
boylii (Ophibolus), 197.
brachycephalum (Catostoma), 299.
brachycephalum (Dirosema), 299.
brachycephalus (Colobognathus), 299.
brachycephalus (Elapoïdis), 299.
Brachyorrhos, 300.
brachyorrhos (Calamaria), 218.
brachyura (Coronella), 206.
brachyurus (Opheomorphus), 136.
brachyurus (Zamenis), 206.
brevicauda (Liopeltis), 259.
brevicauda (Oligodon), 240.
breviceps (Coluber), 149.
breviceps (Dasypeltis), 355.
breviceps (Dendrophis), 86.
breviceps (Liophis), 164.
breviceps (Ophiomorphus), 164.
breviceps (Rhadinæa), 164.
brevifrenum (Rabdosoma), 302.
brevirostris (Dromicus), 174.
brevis (Calamaria), 348.
brevis (Typhlogeophis), 351.
brunneus (Drymobius), 16.

brunneus (Herpetodryas), 15.
brunneus (Masticophis), 15.
cacodæmon (Coluber), 154.
cærulescens (Coluber), 96.
cæruleus (Dromicus), 11.
cæruleus (Drymobius), 12.
cæruleus (Hapsidophrys), 103.
cæsar (Xenurophis), 288.
Calamaria, 188, 233, 255, 273, 300, 324, 330.
calamaria (Ablabes), 282.
calamaria (Anguis), 345.
calamaria (Cyclophis), 282.
calamarius (Coluber), 345.
californiæ (Coluber), 197.
californiæ (Coronella), 197.
californiæ (Ophibolus), 197.
californiæ (Ophis), 197.
callicephalus (Coluber), 34.
callicephalus (Coronella), 34.
calligaster (Ablabes), 198.
calligaster (Coluber), 198.
calligaster (Contia), 164.
calligaster (Coronella), 198.
calligaster (Dendrophis), 80.
calligaster (Lampropeltis), 199.
calligaster (Ophibolus), 199.
calligaster (Rhadinæa), 164.
calligaster (Scotophis), 40.
callilæma (Dromicus), 142.
callilæma (Natrix), 142.
callilæmus (Liophis), 142.
Callopeltis, 24.
callostictus (Hydrops), 187.
Calonotus, 160.
Calopisma, 185, 186, 289, 290.
cana (Ficimia), 272.
caninana (Natrix), 23.
cantoris (Coluber), 35.
canum (Gyalopion), 272.

capistrata (Natrix), 121.
capistratus (Coluber), 10.
carinata (Cynophis), 36.
carinata (Phyllophis), 55.
carinatus (Chironius), 73.
carinatus (Coluber), 73.
carinatus (Dipsas), 354.
carinatus (Herpetodryas), 72, 73, 75, 76.
carolinianus (Coluber), 39.
Carphophiops, 324.
Carphophis, 288, 324.
catenata (Calamaria), 351.
catenatus (Cyclophis), 274.
catenifer (Coluber), 67.
catenifer (Pituophis), 67, 69.
catenifer (Simotes), 218.
catesbyi (Ahætulla), 115.
catesbyi (Dendrophis), 115.
catesbyi (Heterodon), 156.
catesbyi (Leptophis), 115.
catesbyi (Uromacer), 115.
Catostoma, 314.
caucasica (Coronella), 192.
caucasicus (Coluber), 192.
caudalineatus (Leptophis), 89.
caudolineata (Ahætula), 89.
caudolineata (Dendrophis), 89.
caudolineatus (Dendrelaphis), 89, 358.
caudolineolatus (Dendrophis), 85.
caymanus (Alsophis), 120.
celebensis (Elaphis), 62.
Celuta, 324.
Cemophora, 213.
cenchrus (Coluber), 166.
cervone (Elaphis), 46.
chairecacos (Dendrophis), 88.
chalybæa (Geophis), 317, 318, 319, 320, 321, 322.
chalybæum (Catostoma), 318.
chalybæus (Elapoides), 319.
chalybeus (Alopecophis), 57.
chalybeus (Geophis), 318, 320.
chamissonis (Coronella), 119.

ALPHABETICAL INDEX.

chamissonis (Dromicus). 119.
chamissonis (Opheomorphus), 120.
championi (Geophis), 321.
Cheilorhina, 188.
chenonii (Dendrophis), 96.
chenonii (Leptophis), 96.
chiametla (Natrix), 168.
Chilomeniscus, 272.
chinensis (Simotes), 228.
Chionactis, 255.
Chironius, 24, 71, 118.
chloris (Herpetodryas), 279.
chlorophæa (Hypsiglena), 209.
Chlorophis, 91.
Chlorosoma, 255
chlorosoma (Coluber), 38.
ch oroticum (Dendrophidium), 16.
chloroticus (Drymobius), 16.
chrysobronchus (Phrynonax), 22.
chrysobronchus (Spilotes), 22.
chrysostoma (Rhadinæa), 167.
Churchillia, 25.
cincta (Carphophis), 273.
cinctus (Ablabes), 283.
cinctus (Chilomeniscus), 273.
cinereus (Alsophis), 124.
cinereus (Simotes), 222.
cinnamomea (Natrix), 72.
clavata (Rhadinæa), 177.
clavatus (Dromicus), 177.
clericus (Ophibolus), 200.
climacophorus (Coluber), 54.
cobella (Coluber), 166.
cobella (Coronella), 166.
cobella (Elaps), 166.
cobella (Liophis), 166, 167.
cobella (Ophiomorphus), 166.
cobella (Rhadinæa), 166.
coccinea (Cemophora), 214.
coccinea (Coronella), 205.
coccinea (Lampropeltis), 205.
coccineus (Coluber), 214.
coccineus (Elaps), 214.

coccineus (Heterodon), 214.
coccineus (Ophibolus), 201.
coccineus (Rhinostoma), 214.
coccineus (Simotes), 214.
cochinchinensis (Simotes), 219.
Cœlognathus, 25.
cognatus (Heterodon), 155.
collaris (Ablabes), 261.
collaris (Coluber), 260.
collaris (Contia), 260.
collaris (Coronella), 261.
collaris (Cyclophis), 261.
collaris (Eirenis), 260, 261.
collaris (Geoptyas), 31.
collaris (Homalosoma), 262.
collaris (Idiopholis), 327.
collaris (Liophis), 167.
collaris (Ophibolus), 200.
Colobognathus, 314.
Colophrys, 314.
Colorhagia, 251.
Coluber, 24.
colubrinus (Xenodon), 146, 359.
Comastes, 208.
compressus (Coluber), 39.
Compsosoma, 25.
concolor (Hydromorphus), 185.
confinis (Scotophis), 49.
conirostris (Aporophis), 135.
conirostris (Liophis), 134.
conirostris (Lygophis), 134.
conjuncta (Coronella), 203.
conjuncta (Lampropeltis), 197.
Conopsis, 255.
conspicillata (Coronella), 51.
conspicillatus (Callopeltis), 51.
conspicillatus (Coluber), 51.
conspicillatus (Elaphis), 51.
constrictor (Coryphodon), 31.
Contia, 255.
copii (Cemophora), 214.
corais (Coluber), 31.
corais (Spilotes), 31.

corallioides (Synchalinus), 70.
coralliventris (Aporophis), 159.
coronata (Calamaria), 196.
coronata (Coronella), 196.
coronatus (Cerastes), 23.
coronatus (Coluber), 23.
coronatus (Mizodon), 196.
Coronella, 25, 126, 160, 188 214. 277.
coronella (Ablabes), 264.
coronella (Calamaria), 264.
coronella (Coluber), 191.
coronella (Contia), 264.
coronella (Homalosoma), 264.
coronelloides (Homalosoma), 264.
coronilla (Natrix), 191.
Coryphodon, 7. 9. 188.
Cosmiosophis, 180.
couperi (Coluber), 31.
couperi (Georgia), 31.
couperi (Spilotes), 31.
crassicaudatum (Rhabdosoma), 304, 310.
crassicaudatus (Atractus), 310.
crassus (Simotes), 219.
Crossanthera, 9.
cruciatum (Rabdion), 245.
cruentatus (Coluber), 41.
cruentatus (Simotes), 231.
Cryptodacus, 251.
cubensis (Dromicus), 140.
cuculliceps (Leptognathus), 296.
cuculliceps (Tropidodipsas), 296.
cupreum (Rhinostoma), 247.
cupreus (Coluber), 191.
cupreus (Leptophis), 109.
cupreus (Thrasops), 109.
cursor (Coluber), 139.
cursor (Dromicus), 139, 140.
cursor (Herpetodryas), 139, 140.
cursor (Liophis), 139.
cuvieri (Calamaria), 344.
cyanopleurus (Aporophis), 142.
Cyclagras, 144, 180.
Cyclophiops, 277.

ALPHABETICAL INDEX.

Cyclophis, 255, 277.
cyclura (Coronella), 219.
cyclurus (Simotes), 219.
Cynophis, 24.

darnleyensis (Dendrophis), 80.
Dasypeltinæ, 353.
Dasypeltis, 353.
daudini (Pseudoeryx), 186.
davidi (Coluber), 56.
davidi (Tropidonotus), 56.
decalepis (Herpetodryas), 75.
decemlineata (Contia), 260.
decemlineatus (Ablabes), 260.
decemlineatus (Eirenis), 260.
decipiens (Ablabes), 181.
decorata (Coronella), 174, 176.
decorata (Rhadinæa), 176, 359.
decoratus (Diadophis), 176, 178.
decorus (Ahætulla), 78.
decorus (Coluber), 78.
defilippii (Oxyrhina), 269.
Deirodon, 353.
Dendrelaphis, 87.
Dendrophidium, 9.
Dendrophis, 71, 77, 87, 98, 102, 105, 115.
dendrophis (Dendrophidium), 15.
dendrophis (Drymobius), 15, 357.
dendrophis (Herpetodryas), 14, 15.
dennysi (Simotes), 218.
deppei (Elaphis), 66.
deppei (Pituophis), 66.
deppei (Spilotes), 67.
deppii (Coluber), 66.
deppii (Elapochrous), 182.
depressirostris (Leptophis), 107.
depressirostris (Philothamnus), 107.
deserticola (Pituophis), 68.
Diadophis, 160, 188, 277.
diasii (Chionactis), 268.
diastema (Liophis), 183.
dichroa (Herpetodryas), 30.

dichrous (Coluber), 30.
Dicraulax, 215.
dilepis (Aporophis), 158
dilepis (Lygophis), 158.
Dimades, 185.
dimidiata (Calamaria), 333.
dimidiatus (Elapochrus), 183.
dimidiatus (Pliocercus), 183.
dione (Chironius), 44.
dione (Cœlopeltis), 44.
dione (Coluber), 44.
dione (Elaphis), 44, 46.
Diplotropis, 105.
diplotropis (Ahætulla), 110.
diplotropis (Hapsidophrys), 110.
diplotropis (Leptophis), 110.
Dipsas, 292.
Dirosema, 298.
discolor (Hypsiglena), 211.
discolor (Leptodira), 211.
docilis (Diadophis), 207, 208.
doliata (Coronella), 200, 201, 203, 205.
doliata (Lampropeltis), 201.
doliatus (Coluber), 131, 205.
doliatus (Ophibolus), 200, 201, 203, 205.
doliatus (Ophiomorphus), 132.
dolichocephalum (Catostoma), 320.
dolichocephalus (Colobognathus), 320.
dolichocephalus (Elapoides), 320.
dolichocephalus (Geophis), 320.
dorbignyi (Heterodon), 151.
dorbignyi (Lystrophis), 151.
doriæ (Ablabes), 279.
doriæ (Cyclophiops), 279.
dorsale (Oligodon), 234.
dorsalis (Ahætulla), 101.
dorsalis (Dromicus), 119.
dorsalis (Drymobius), 12.
dorsalis (Elaps), 241.
dorsalis (Leptophis), 101.
dorsalis (Liophis), 170.

dorsalis (Oligodon), 241.
dorsalis (Ophiomorphus), 170.
dorsalis (Philothamnus), 101.
dougesii (Diadophis), 208.
Dromicus, 9, 118, 126, 157, 160.
drummondi (Farancia), 291.
Drymobius, 8, 25.
Duberria, 273.
duberria (Coluber), 274.
duberria (Elaps), 274.
dubium (Geophidium), 322.
dubium (Rabdosoma), 308.
dubium (Xenodon), 243.
dubius (Geophis), 322, 323.
duboisi (Atractus), 310.
duboisi (Rabdosoma), 310.
dugesi (Elapoidis), 317.
dugesii (Geophis), 317.
dumerili (Calamaria), 343.
dumerilii (Dendrophis), 8.
dumerilii (Dromicus), 181.
dumerilii (Leptognathus), 296.
dumerilii (Urotheca), 181.
dumfrisiensis (Coluber), 205.
dysopes (Diadophis), 206.

Eirenis, 255.
eiseni (Ophibolus), 197.
elaiocroma (Ficimia), 271.
Elaphis, 24.
elaphis (Coluber), 45.
elaphis (Natrix), 46.
Elapochrous, 180.
Elapoides, 314.
elapoides (Liophis), 183.
elapoides (Pliocercus), 182.
elapoides (Urotheca), 182, 359.
elaps (Atractus), 302.
elaps (Geophis), 302.
elaps (Rhabdosoma), 302.
elapsoidea (Calamaria), 205.

elapsoidea (Osceola), 205.
elapsoideus (Lampropeltis), 205.
elegans (Arizona), 66.
elegans (Coronella), 175, 196.
elegans (Dendrophis), 86.
elegans (Enicognathus), 173.
elegans (Pityophis), 66.
elegans (Rhinechis), 66.
elegans (Virginia), 289.
ellioti (Oligodon), 242.
elongata (Dasypeltis), 355.
emini (Abætulla), 92.
emini (Chlorophis), 92.
emmeli (Atractus), 311.
emmeli (Geophis), 311.
emoryi (Coluber), 40.
emoryi (Natrix), 40.
emoryi (Scotophis), 40.
enganensis (Coluber), 63.
Enicognathus, 160.
Enulius, 249.
ephippicus (Chilomeniscus), 273.
epinephelus (Liophis), 137.
episcopa (Contia), 265, 266.
episcopum (Homalosoma), 265.
episcopum (Lamprosoma), 265.
erebennus (Spilotes), 31.
eremita (Coluber), 44.
Erpetodryas, 71.
erythrogrammum (Calopisma), 290.
erythrogrammus (Abastor), 290.
erythrogrammus (Coluber), 290.
erythrogrammus (Helicops), 290.
erythrogrammus (Homalopsis), 290.
erythrogrammus (Hydrops), 290.
erythrurum (Compsosoma), 62.
erythrurus (Coluber), 62, 358.
erythrurus (Plagiodon), 62.
Eudryas, 9.
Eurypholis, 277.
euryzona (Elapochrus), 182.
euryzona (Urotheca), 182.

euryzonus (Pliocercus), 182.
eutropis (Phrynonax), 22.
evansii (Coronella), 199.
evansii (Ophibolus), 199.
everetti (Calamaria), 340.
everetti (Oligodon), 239.
exiguus (Dromicus), 126.
eximia (Coronella), 200.
eximius (Coluber), 197, 200.
eximius (Ophibolus), 200.
Exorhina, 255.
extenuatum (Stilosoma), 325.

Farancia, 290.
fasciata (Contia), 262.
fasciata (Dendrophis), 85.
fasciata (Farancia), 291.
fasciata (Geophis), 296.
fasciata (Higina), 187.
fasciata (Leptognathus), 294.
fasciata (Tropidodipsas), 294, 295, 296.
fasciata (Virginia). 296.
fasciatus (Ablabes), 262.
fasciatus (Chilomeniscus), 273.
fasciatus (Cyclophis), 262.
fasciatus (Dasypeltis), 354.
fasciatus (Eirenis), 262.
fasciatus (Elapoides), 296.
fasciatus (Oligodon), 243.
fasciatus (Phrynonax), 21.
fasciatus (Spilotes), 21, 22.
fasciolatus (Simotes), 219.
favæ (Atractus), 313.
favæ (Calamaria), 313.
favæ (Rabdosoma), 313.
ferox (Hypsirhynchus), 117.
ferruginosus (Coluber), 191.
Ficimia, 270.
fischeri (Tropidodipsas), 296.
fischeri (Zamenis), 195.
fitzingeri (Zacholus), 192.
flammigerus (Brachyorrhos), 308.
flavescens (Callopeltis), 52.
flavescens (Coluber), 52.
flavescens (Elaphis), 53.
flavescens (Phyllosira), 75.

flaviceps (Calamaria), 333.
flavifrenatus (Aporophis), 158.
flavifrenatus (Dromicus), 158.
flavifrenatus (Lygophis), 158.
flavigularis (Dendrophis), 105.
flavigularis (Thrasops), 105, 358.
flavilatus (Dromicus), 143.
flavilatus (Liophis), 143.
flavirufus (Coluber), 39.
flavirufus (Natrix), 39.
flaviventris (Geoptyas), 31.
flaviventris (Liophis), 167.
flavolineatus (Herpetodryas), 73.
floridanus (Coluber), 40.
forbesii (Simotes). 225.
formosa (Coronella), 203.
formosa (Dendrophis), 84.
formosanus (Simotes), 222, 359.
formosus (Dendrophis), 84.
formosus (Leptophis), 78.
forsteni (Rabdion), 328.
forsteni (Rhabdophidium), 328.
forsteri (Natrix), 136.
fraseri (Liophis), 131.
frenata (Abætulla), 116.
frenatum (Gonyosoma), 58.
frenatus (Ablabes), 280.
frenatus (Coluber), 58.
frenatus (Cyclophis), 280.
frenatus (Dromicus), 181.
frenatus (Herpetodryas), 58.
frenatus (Leptophis), 141.
frenatus (Uromacer), 116.
frontalis (Contia), 270.
frontalis (Ficimia), 270.
frontalis (Geagras), 270.
frontalis (Prosymna), 248.
frontalis (Pseudoficimia), 270.
frontalis (Temnorhynchus), 247, 248.
frontalis (Toluca), 270.
fugax (Coluber), 52.
fugitivus (Coluber), 139.

ALPHABETICAL INDEX. 367

fugitivus (Dromicus), 139, 140, 141.
fugitivus (Liophis), 139.
fulviceps (Rhadinæa), 179.
fulvivittis (Diadophis), 178.
fulvivittis (Dromicus), 178.
fulvivittis (Rhadinæa), 178.
fulvum (Arrhyton), 252.
funereus (Alsophis), 142.
furcata (Grayia), 287.
fusca (Ahætula), 82.
fusca (Dendrophis), 82.
fusca (Rhadinæa), 169.
fuscus (Coluber), 11, 75.
fuscus (Dendrophis), 82.
fuscus (Geophis), 322.
fuscus (Herpetodryas), 73, 75.
fuscus (Liophis), 169.
fuscus (Opheomorphus), 169.

Galedon, 294.
gallicus (Coluber), 192.
gastrogramma (Calamaria), 349.
Gastropyxis, 102.
gastrostictus (Dendrophis), 86.
Geagras, 326.
genimaculata (Liophis), 170.
genimaculata (Lygophis), 170.
genimaculata(Rhadinæa), 170.
gentilis (Coronella), 201.
gentilis (Ophibolus), 201.
Geophidium, 314.
Geophis, 314.
Geoptyas, 25.
Georgia, 25.
gervaisii (Calamaria), 338.
getula (Coronella), 197.
getula (Lampropeltis), 197.
getulus (Coluber), 197.
getulus (Coronella), 197.
getulus (Herpetodryas), 197.
getulus (Ophibolus), 197.
giardi (Grayia), 288.
gigas (Cyclagras), 144.
gigas (Xenodon), 144.
gigas (Leiosophis), 144.
girardi (Philothamnus), 102.

girondica (Coronella), 194.
girondicus (Coluber), 194.
glabra (Herpetodryas), 75.
Glaniolestes, 286.
godmani (Coronella), 180.
godmani (Geophis), 322.
godmani (Rhadinæa), 179.
godmanii (Henicognathus), 179.
godmanni (Dromicus), 179.
Gonyophis, 70.
Gonyosoma, 24.
grabowskii (Calamaria), 336.
grabowskyi (Elaphis), 47.
gracile (Trimetopon), 184.
gracilis (Ablabes), 184.
gracilis (Dendrophis), 82.
gracilis (Leptophis), 103.
gracillima (Ahætulla), 95.
gracillima (Calamaria), 350.
gracillimus (Typhlocalamus), 350.
græca (Elaphis), 46.
gramineum (Gonyosoma), 59.
grandisquamis (Herpetodryas), 76.
grandisquamis (Spilotes), 76.
grandoculis (Dendrophis), 84.
graphicus (Coluber), 137.
grayi (Calamaria), 338.
Grayia, 286.
gregorii (Dendrophis), 85.
guentheri (Atractus), 305.
guentheri (Geophis), 305.
guentheri (Philothamnus), 96.
guentheri (Phrynonax), 20.
guentheri (Xenodon), 147.
guttatus (Coluber), 39, 198.
guttatus (Elaphis), 40, 49.
guttatus (Scotophis), 40.
guttulatum (Rhabdosoma), 298.
Gyalopion, 270.

habeli (Dromicus), 119.
hæmatois (Pityophis), 67.

hainanensis (Simotes), 359.
Hapsidophrys, 102, 103.
harperti (Virginia), 289.
harpertii (Carphophis), 289.
heathii (Drymobius), 11.
heathii (Herpetodryas), 11.
helena (Coluber), 36, 357.
helena (Cynophis), 36.
helena (Dendrophis), 88.
helena (Herpetodryas), 36.
helena (Plagiodon), 36.
helenæ (Carphophiops), 325.
helenæ (Carphophis), 325.
helenæ (Celuta), 324.
Helicops, 185, 289.
hermanni (Coluber), 65.
Herpetæthiops, 91.
Herpetodryas. 8, 24, 71, 126, 157, 188, 255, 277.
herzi (Simotes), 43.
heteroderma (Ahætulla), 97.
heterodermus (Chlorophis), 97, 358.
heterodermus (Philothamnus), 97.
Heterodon, 153, 213, 253.
heterodon (Coluber), 154, 156.
heterolepidota (Ahætulla), 95.
heterolepidota (Leptophis), 95.
heterolepidotus (Chlorophis), 95, 358.
heterolepidotus (Philothamnus), 95.
Heteronotus, 286.
hexagonotus (Ptyas), 8.
hexagonotus (Xenelaphis), 8.
hexahonotus (Coluber), 8.
hexahonotus (Xenelaphis), 8.
hexanotus (Coryphodon), 8.
Higina, 186.
hispanica (Coronella), 194.
histricus (Heterodon), 152.
histricus (Lystrophis), 152.
hodgsonii (Coluber), 35.

ALPHABETICAL INDEX.

hodgsonii (Compsosoma), 35.
hodgsonii (Spilotes), 35.
hoevenii (Calamaria), 337.
hoffmanni (Colobognathus), 319.
hoffmanni (Elapoides), 319.
hoffmanni (Geophis), 319.
hohenackeri(Coluber),42.
Holarchus, 215.
holbrookii (Elaphis), 50.
holochlorus (Herpetodryas), 75.
Homalopsis, 185, 186, 289, 290.
Homalosoma, 233, 255, 273, 277.
hoplogaster (Ahætulla), 93.
hoplogaster (Chlorophis), 93.
hoplogaster (Philothamnus), 93.
Hydromorphus, 185.
Hydrops, 186.
Hypsiglena, 208.
Hypsirhynchus, 117.

Idiopholis, 327.
ignita (Coronella), 176.
ignita (Rhadinæa), 176.
ignitus (Dromicus), 176.
iheringii (Coronella), 172.
inagnitus (Pseuderyx), 359.
inconstans (Dromicus), 121.
inornata (Coronella), 195.
inornatus (Dasypeltis), 354.
inornatus (Eirenis), 260.
inornatus (Rachiodon), 355.
inornatus (Uromacer), 116.
intermedius (Pityophis), 37.
iris (Calamaria), 344.
irregularis (Ahætulla), 92, 94, 96.
irregularis (Chlorophis), 96.
irregularis (Coluber), 96.
irregularis (Philothamnus), 96, 99.
irregularis (Xenodon), 150.
isolepis (Xenodon), 136.
Isoscelis, 300.

isozona (Contia), 266.
isthmicus (Atractus), 307.
italica (Coronella), 192.

jægeri (Coronella), 170.
jægeri (Rhadinæa), 170.
jani (Arizona), 66.
jani (Prosymna), 249.
janii (Liophis), 359.
janseni (Coluber), 57, 357.
jansenii (Gonyosoma), 57, 357.
javanica (Calamaria),347.
joberti (Enicognathus), 174.
juliæ (Aporophis), 139.
juliæ (Dromicus), 139.
juliæ (Liophis), 139.

katowens's (Dendrophis), 80.
kennerlyi (Heterodon), 156.
kirkii (Ahætulla), 99.

labuanensis (Simotes), 218.
lachrymans (Dromicus), 174.
lachrymans (Lygophis), 174.
lachrymans (Rhadinæa), 174.
lætus (Coluber), 49.
lætus (Diadophis), 208.
lætus (Liophis), 208.
lætus (Scotophis), 49.
lævicollis (Coluber), 73.
lævis (Coluber), 191, 286.
lævis (Coronella), 192, 194.
lævis (Drymobius), 11.
lævis (Herpetodryas), 11.
lævis (Zacholus), 192.
lagoensis (Ahætulla), 100.
lagoensis (Philothamnus), 100.
Lampropeltis, 188.
Lamprosoma, 255.
lateralis (Calamaria), 342.
lateralis (Macrocalamus), 327.
lateristriga (Dromicus), 181.
lateristriga (Liophis), 181.
lateristriga (Urotheca), 181.
latifasciata (Hypsiglena), 211.
latifrons (Atractus), 303.

latifrons (Geophis), 303.
latifrontalis (Atractus), 304.
latifrontalis (Geophis), 304.
laureata (Rhadinæa), 179.
laurentus (Dromicus), 179.
lebaris (Coluber), 154.
lecontii (Rhinochilus), 212.
Leionotus, 286.
Leiosophis, 144, 180.
leonis (Coronella), 199.
leopardina (Coronella), 192.
leopardina (Natrix), 41.
leopardinus (Callopeltis), 41.
leopardinus (Coluber), 41.
leprosus (Coluber), 52.
Leptocalamus, 249.
Leptognathiens, 353.
Leptognathus, 292, 294.
Leptophis, 9, 77, 87, 91, 102, 105, 255.
leucocephala (Calamaria), 344.
leucogaster (Calamaria), 341.
leucogaster (Liophis), 163.
leucogaster (Rhadinæa), 163.
leucomelas (Alsophis), 123.
leucomelas (Dromicus), 123.
leucostomus (Leptognathus), 296.
lichtensteinii (Coluber), 10.
Ligonirostra, 246.
lindheimeri (Coluber), 50.
lindheimeri (Pantherophis), 50.
lindheimeri (Scotophis), 50.
lineata (Hapsidophrys), 104.
lineata (Toluca), 269.
lineaticollis (Arizona), 64.
lineaticollis (Coluber), 64.
lineaticollis (Pituophis), 64.
lineatum (Rabdosoma), 312.

lineatus (Aporophis), 158.
lineatus (Coluber), 158.
lineatus (Conopsis), 268.
lineatus (Dromicus), 141, 158, 170.
lineatus (Geophis), 312.
lineatus (Herpetodryas), 158.
lineatus (Lygophis), 158.
lineolata (Dendrophis), 80, 85.
lineolatus (Dendrophis), 80, 85.
lineolatus (Sphenocalamus), 326.
linnæi (Calamaria), 336, 345, 349.
liocercus (Ahætulla), 111, 113.
liocercus (Coluber), 113.
liocercus (Dendrophis), 111, 113.
liocercus (Leptophis), 111, 113.
Liopeltis, 255, 277.
Liophis, 118, 126, 160, 180.
lippiens (Geophis), 188.
lippiens (Sympholis), 188.
longicauda (Ablabes), 284.
longicaudata (Grayia), 288.
longicaudatum (Rabdosoma), 313.
longicaudatus (Geagras), 250.
longiceps (Atractus), 305.
longiceps (Calamaria), 329.
longiceps (Oxycalamus), 329.
longiceps (Pseudorhabdium), 329.
longiceps (Rhabdosoma), 305.
longifrenatus (Philothamnus), 96.
longissima (Natrix), 52.
longissimus (Callopeltis), 53.
longissimus (Coluber), 52, 357.
loreata (Rhadinæa), 179.
lovii (Calamaria), 350.
lumbricoidea (Calamaria), 333, 344.
lunulata (Tropidodipsas), 21.
VOL. II.

lunulatus (Phrynonax), 21.
lunulatus (Spilotes), 21.
lutrix (Coluber), 274.
lutrix (Homalosoma), 274, 275.
Lystrophis, 151,

macclellanii (Pituophis), 69.
Macrocalamus, 327.
Macrophis, 286.
macrophthalma (Herpetodryas), 73.
Macrops, 71.
macrops (Dendrophis), 85.
macrurus (Calamaria), 344.
maculata (Ficimia), 268.
maculata (Isoscelis), 306.
maculata (Oxyrhina), 268.
maculatum (Rhabdosoma), 305, 306, 307, 308.
maculatus (Anoplophallus), 353.
maculatus (Atractus), 306.
maculatus (Coluber), 39.
maculatus (Conopsis), 268.
maculatus (Megalops), 353.
maculivittis (Alsophis), 11.
maculivittis (Dromicus), 11.
maculolineata (Calamaria), 336.
maculosa (Calamaria), 345.
mæoticus (Coluber), 44.
majalis (Leptophis), 258.
majalis(Philophyllophis), 258.
major (Ablabes), 279.
major (Atractus), 307.
major (Cyclophis), 279.
malabaricus (Cynophis), 36.
malabaricus (Herpetodryas), 36.
mancas (Leptophis), 88.
mandarinus (Coluber), 42.
maniar (Dendrophis), 88.
manillensis (Elaphis), 62.
margaritatum (Gonyosoma), 71.

margaritatus (Gonyophis), 71.
margaritifera (Calamaria), 336.
margaritiferus (Coryphodon), 196.
margaritiferus (Dromicus), 17.
margaritiferus (Drymobius), 17.
margaritiferus (Herpetodryas), 17.
margaritiferus (Leptophis), 17.
margaritiferus (Thamnosophis), 17.
margaritophora (Calamaria), 336.
marginatus (Leptophis), 112.
marginatus (Thrasops), 113.
martapurensis (Calamaria), 344.
martii (Calopisma), 187.
martii (Homalopsis), 187.
martii (Hydrops), 187.
medici (Dipsas), 355.
megalolepis (Spilotes), 24.
Megalops, 353.
Meizodon, 188.
melanauchen (Enicognathus), 175.
melanauchen (Rhadinæa), 175.
melanichnus (Alsophis), 122.
melanocephala (Coronella), 246.
melanocephalum (Homalosoma), 246.
melanocephalus (Dromicus), 173.
melanocephalus (Enicognathus), 172, 174.
melanocephalus (Oligodon), 246.
melanocephalus (Rhynchocalamus), 246.
melanoleucus (Coluber), 68.
melanoleucus (Pituophis), 68.
melanolomus (Masticophis), 11.
melanolomus (Zamenis), 12.
melanonotus (Liophis), 134.
melanorhynchus (Calamaria), 333.

2 B

melanostigma (Dendrophis), 99.
melanostigma (Dromicus), 142.
melanostigma (Liophis), 142.
melanostigma (Natrix), 142.
melanota (Calamaria), 349.
melanotropis (Coluber), 33.
melanotropis (Dendrophidium), 33.
melanotropis (Elaphis), 33.
melanotus (Coluber), 134.
melanotus (Dromicus), 134.
melanotus (Liophis), 134.
melanurum (Compsosoma), 60, 62.
melanurus (Coluber), 60.
melanurus (Elaphis), 60, 62.
melanurus (Spilotes), 31, 35, 60, 62.
melas (Herpetodryas), 76.
meleagris (Calamaria), 249.
meleagris (Opheomorphus), 132.
meleagris (Prosymna), 249.
meleagris (Temnorhynchus), 249.
meridionalis (Coluber), 194.
meridionalis (Coronella), 194.
merremii (Coluber), 168.
merremii (Coronella), 131, 168.
merremii (Liophis), 131, 138, 168, 169.
merremii (Ophiomorphus), 168, 169.
merremii (Ophis), 150.
merremii (Rhadinæa), 168.
merremii (Xenodon), 150.
mexicana (Ahætulla), 108.
mexicana (Coronella), 201.
mexicanus (Cerastes), 33.
mexicanus (Hapsidophrys), 108.
mexicanus (Leptophis), 108.

mexicanus (Ophibolus), 201.
mexicanus (Pituophis), 66, 69.
mexicanus (Thrasops), 108.
meyerinkii (Simotes), 224.
michahelles (Xenodon), 65.
micropholis (Coronella), 203.
micropholis (Lampropeltis), 203.
microrhynchum (Rhabdosoma), 308.
miliaris (Coluber), 168.
mimus (Opheomorphus), 164.
mimus (Rhadinæa), 164.
miolepis (Dromicus), 175.
mite (Homalosoma), 267.
mitis (Contia), 267.
M-nigrum (Coluber), 131.
modesta (Ahætulla), 107.
modesta (Calamaria), 340.
modesta (Contia), 261.
modesta (Coronella), 260.
modestus (Ablabes), 260, 261, 262.
modestus (Atractus), 304.
modestus (Cyclophis), 261.
modestus (Dendrelaphis), 91.
modestus (Diadophis), 208.
modestus (Homalosoma), 260.
modestus (Leptophis), 107.
modestus (Oligodon), 238.
modestus (Philothamnus), 107.
moellendorffii (Coluber), 56.
moellendorffii (Cynophis), 56.
mœsta (Geophis), 319.
mœstum (Rhabdosoma), 319.
mœstus (Geophis), 319.
molossus (Coluber), 39.
moniliger (Psammophis), 261.
monochrous (Calamaria), 340.
monticola (Dipsas), 280.
monticolus (Coluber), 229.

mossambica (Dasypeltis), 355.
multicinctum (Rabdosoma), 308
multicinctus (Ophibolus), 197.
multifasciata (Coronella), 202.
multifasciatus (Simotes), 222.
multilineatus (Dromicus), 181.
multipunctata (Calamaria), 345.
multistratus (Ophibolus), 201.
multistriata (Lampropeltis), 201.
murinus (Enulius), 250.
mutabilis (Coluber), 37.
mutabilis (Natrix), 37.
mutabilis (Scotophis), 37.
mutitorques (Geophis), 304.
mutitorques (Rhabdosoma), 304.

nasale (Catostoma), 318.
nasale (Rhabdosoma), 318.
nasalis (Cyclophis), 282.
nasicus (Heterodon), 156.
nasus (Conopsis), 268.
nasus (Contia), 268.
nasus (Ficimia), 268.
natalensis (Ahætulla), 94.
natalensis (Chlorophis), 94.
natalensis (Dendrophis), 94.
natalensis (Philothamnus), 94.
Natrix, 24, 25.
nattereri (Heterodon), 152.
nauii (Coluber), 45.
nebulatus (Cerastes), 293.
nebulatus (Coluber), 293.
nebulatus (Dipsas), 293.
nebulatus (Leptognathus), 293.
nebulatus (Petalognathus), 293.
nebulosus (Coluber), 192.
nebulosus (Cyclophis), 278.
neglecta (Ahætulla), 94.
neglectus (Chlorophis), 94.
neglectus (Philothamnus), 94.
neovidii (Xenodon), 148.

neuwiedii (Xenodon), 148.
nicaga (Rhadinæa), 174.
nicagus (Lygophis), 174.
nicobariensis (Ablabes), 285.
niger (Hapsidophrys), 105.
niger (Heterodon), 155.
niger (Ophibolus), 197.
niger (Scytale), 154.
nigricaudus (Allophis), 58.
nigricaudus (Elaphis), 58.
nigricollis (Coluber), 260.
nigro-alba (Calamaria), 344.
nigrofasciatus (Psammophis), 34.
nigrofasciatus (Philothamnus), 99.
nigromarginata (Ahætulla), 112.
nigromarginatus (Leptophis), 112.
nitida (Ahætulla), 100, 101.
nitidus (Philothamnus), 100.
notospilus (Oligodon), 239.
novæ-hispaniæ (Coluber), 33.
nuchalis (Herpetodryas), 15.
nuntius (Dromicus), 181.
nuthalli (Coluber), 47.
nyctenurus (Elaphis), 64.

obscurus (Simotes), 219.
obsoleta (Georgia), 31.
obsoletus (Coluber), 31, 50.
obsoletus (Elaphis), 50.
obsoletus (Scotophis), 50.
obsoletus (Spilotes), 31.
obtusa (Coronella), 171.
obtusa (Rhadinæa), 171.
occidentalis (Ahætulla), 111.
occidentalis (Drymobius), 17.
occidentalis (Leptophis), 111.
occidentalis (Thrasops), 111.
occipitale (Lamprosoma), 266.
occipitale (Rhinostoma), 266.
occipitalis (Ablabes), 206.
occipitalis (Calamaria), 342.
occipitalis (Chionactis), 266.
occipitalis (Contia), 266.
occipitalis (Diadophis), 206.
occipitalis (Enicognathus), 175.
occipitalis (Herpetodryas), 30.
occipitalis (Ophibolus), 203.
occipitalis (Rhadinæa), 175.
occipito-album (Rhabdosoma), 310.
occipito-albus (Atractus), 310.
ocellatus (Xenodon), 150.
ochrorhynchus (Hypsiglena), 209.
octolineata (Coronella), 224.
octolineata (Dendrophis), 89.
octolineatus (Coluber), 224.
octolineatus (Elaps), 224.
octolineatus (Simotes), 224.
Ocyophis, 118.
oldhami (Chlorophis), 93.
oldhami (Cyclophis), 93.
Oligodon, 233.
oligodon (Calamaria), 237.
oligozona (Coronella), 203.
olivacea (Ahætulla), 82.
olivacea (Dendrophis), 82.
olivacea (Ficimia), 270, 271.
omiltemana (Geophis), 299.
omiltemanum (Dirosema), 299.
omiltemanus (Dromicus), 178.
Ophibolus, 188.
Ophiomorus, 160.
Ophis, 144.
orientalis (Coronella), 167.
ornata (Ficimia), 271.
ornata (Hypsiglena), 211.
ornata (Liophis), 138.
ornatus (Chlorophis), 93.
ornatus (Coluber), 139.
ornatus (Comastes), 211.
ornatus (Dromicus), 139.
ornatus (Liophis), 139.
ornatus (Macrophis), 287.
ornatus (Philothamnus), 93.
Orophis, 118.
ortonii (Leptophis), 114.
Osceola, 188.
ovivorus (Coluber), 291.
owenii (Ablabes), 282.
Oxycalamus, 328.
oxycephalum (Gonyosoma), 57.
oxycephalum (Pseudorhabdium), 329.
oxycephalum (Rhabdosoma), 329.
oxycephalus (Coluber), 56, 357.
oxycephalus (Herpetodryas), 57.
oxycephalus (Oxycalamus), 329.
Oxyrhina, 255.
oxyrhyncha (Ahætulla), 117.
oxyrhyncha (Leptophis), 117.
oxyrhynchus (Uromacer), 117.

pachyura (Contia), 267.
pallidus (Diadophis), 206.
palmarum (Coluber), 354.
palmarum (Dasypeltis), 355.
pannonicus (Coluber), 52.
pantherinus (Coluber), 10, 40.
pantherinus (Coryphodon), 10.
pantherinus (Drymobius), 10.
pantherinus (Ptyas), 10.
Pantherophis, 25.
papuæ (Dendrophis), 86.
Parageophis, 314.
parallelus (Ophibolus), 200.
pardalinus (Elaphis), 39.
parreysii (Elaphis), 46.
parvifrons (Dromicus), 141.
parvifrons (Liophis), 141.

pavimentata (Calamaria), 348.
percarinatus (Drymobius), 16.
perfuscus (Dromicus), 133.
perfuscus (Liophis), 133.
persica (Contia), 263.
persicus (Cyclophis), 263.
persicus (Pseudocyclophis), 263.
persimilis (Liophis), 173.
peruvianum (Rabdosoma), 305.
peruvianus (Atractus), 305.
Petalognathus, 292.
petersii (Geophis), 321.
phænochalinus (Simotes), 224.
philippii (Leptognathus), 295.
philippii (Tropidodipsas), 295.
philippinensis (Dendrophis), 90.
philippinica (Calamaria), 338.
Philothamnus, 91, 98.
Phrynonax, 18.
Phyllophilophis, 255.
Phyllophis, 25.
phyllophis (Coluber), 55.
Phyllosira, 71.
piceus (Spilotes), 30.
pickeringii (Coluber), 75.
picta (Ahætulla), 78.
picta (Dendrophis), 78, 80, 89, 90.
pictus (Coluber), 46, 78.
pictus (Dendrophis), 78, 88, 358.
pictus (Leptophis), 78, 88.
Pituophis, 24.
Plagiodon, 25.
planiceps (Simotes), 232.
platyrhinus (Heterodon), 154, 156.
pleii (Dromicus), 11, 142.
pleii (Drymobius), 12.
pleurostictus (Elaphis), 66.
pleurostictus (Pituophis), 67.
plicatile (Calopisma), 186.
plicatilis (Cerastes), 186.
plicatilis (Coluber), 186.
plicatilis (Dimades), 186.
plicatilis (Elaps), 186.
plicatilis (Homalopsis), 186, 290.

Pliocercus, 180.
plumbiceps (Coronella), 195.
plutonius (Coluber), 23.
pœcilocephalus (Coluber), 46.
pœcilogyrus (Coluber), 131.
pœcilogyrus (Liophis), 131.
pœcilolæmus (Coronella), 168.
pœcilonotus (Phrynonax), 20.
pœcilonotus (Spilotes), 20, 21.
pœcilopogon (Coronella), 173.
pœcilopogon (Rhadinæa), 173.
pœcilostoma (Coluber), 19.
pœcilostoma (Spilotes), 19.
poeppigii (Geophis), 316.
poeppigii (Rabdosoma), 316.
poitei (Herpetodryas), 15.
polychroa (Dendrophis), 78.
polyhemizona (Ablabes), 8.
polylepis (Ahætulla), 21.
polyzona (Coronella), 203.
polyzona (Lampropeltis), 203.
polyzonus (Ophibolus), 203.
ponticus (Coluber), 192.
porphyraceus (Ablabes), 34.
porphyraceus (Coluber), 34.
portoricensis (Alsophis), 122.
præstans (Leptophis), 111.
præstans (Thrasops), 111.
prakkii (Calamaria), 337.
prasina (Liophis), 135.
prasinus (Coluber), 59.
proboscideus (Rhinaspis), 253.
propinquus (Oligodon), 240.
Prosymna, 246.
protenus (Dromicus), 141.
Psammophis, 118.
psephota (Geophis), 299.
psephotum (Catostoma), 299.

psephotum (Dirosema), 299.
psephotus (Elapoidis), 299.
Pseudocyclophis, 255.
Pseudodipsas, 208.
Pseudoeryx, 185, 289.
Pseudoficimia, 255.
pseudogetulus (Coronella), 197.
Pseudorhabdium, 328.
Psilosoma, 255.
publia (Ficimia), 271.
pulchella (Coronella), 165.
pulchellus (Diadophis), 207.
pulcher (Alsophis), 11.
pulcher (Heterodon), 153.
pulcher (Liophis), 165.
pulchra (Pseudoficimia), 270.
pulchriceps (Masticophis), 11.
pullata (Tyria), 23.
pullatus (Coluber), 23, 33.
pullatus (Spilotes), 23.
punctata (Calamaria), 206.
punctata (Coronella), 206.
punctatolineata (Cyclophis), 261.
punctatovittatum (Rabdosoma), 312.
punctatus (Ablabes), 206.
punctatus (Coluber), 206.
punctatus (Diadophis), 206, 207, 208.
punctatus (Leptophis), 99.
punctatus (Philothamnus), 99.
punctulata (Dendrophis), 82.
punctulatus (Coronella), 220.
punctulatus (Dendrophis), 80, 82, 85.
punctulatus (Leptophis), 82.
punctulatus (Simotes), 220.
purpurans (Ablabes), 167.
purpurans (Diadophis), 167.
purpurans (Liophis), 168.
purpurans (Rhadinæa), 167.
purpurascens (Simotes), 218, 220, 225, 226.
purpurascens (Xenodon), 218, 220.

purpureocauda (Ablabes), 267.
pustulatus (Thrasops), 105.
putnamii (Liophis), 139.
pygmæus (Liophis), 129.
pyromelanus(Ophibolus), 202.
pyrrhomelas (Ophibolus), 202.
pyrrhopogon (Coluber), 73.

quadrangularis(Ficimia), 272.
quadrangularis(Geophis), 320.
quadricarinatus (Erpetodryas), 72.
quadrifasciatus(Coluber), 61.
quadrilineata(Coronella), 41.
quadrilineata (Eirenis), 260.
quadrilineata (Liophis), 130.
quadrilineatus (Ablabes), 41.
quadrilineatus (Callopeltis), 41.
quadrilineatus (Coluber), 41, 45.
quadrilineatus (Elaphis), 46.
quadrilineatus (Simotes), 227.
quadrimaculata (Calamaria), 348, 349.
quadristriatus (Coluber), 45.
quadrivirgatum (Adelphicos), 312.
quadrivirgatum (Compsosoma), 59.
quadrivirgatus(Atractus), 312.
quadrivirgatus (Coluber), 59.
quadrivirgatus (Elaphis), 59.
quadrivittatus (Coluber), 50.
quadrivittatus (Elaphis), 50.
quaterradiatus (Coluber), 45.
quaterradiatus (Elaphis), 46.
quatuorlineatus (Coluber), 45, 53.

quincunciatus(Comastes), 210.
quinque (Tropidonotus), 61.
quinquelineata (Coronella), 178.
quinquelineata (Rhadinæa), 178.
quinquelineatus(Herpetodryas), 11.
quinquestriatus(Ablabes), 284.

Rabdion, 328.
rabdocephalus (Coluber), 146, 150.
Rabdosoma, 300, 314.
Rachiodon, 353.
Rachiodontidæ, 353.
radiatum (Compsosoma), 61.
radiatus (Coluber), 61.
radiatus (Elaphis), 62.
radiatus (Spilotes), 62.
raffreyi (Scaphiophis), 254.
raninus (Coluber), 134.
rappii (Ablabes), 282.
rappii (Drymobius), 11.
rappii (Herpetodryas),11.
rebentischii (Calamaria), 343.
redimita (Colorhogia), 252.
redimita (Geophis), 326.
redinitum (Arrhyton), 252.
redimitus (Cryptodacus), 252.
redimitus (Geagras), 326.
regalis (Coronella), 208.
regalis (Diadophis), 208.
regalis (Liophis), 208.
reginæ (Coluber), 137.
reginæ (Coronella), 138.
reginæ (Liophis), 130, 132, 137, 142, 175.
regularis (Coronella),196.
regularis(Meizodon), 196.
reinwardti (Calopisma), 291.
reinwardtii(Homalopsis), 291.
reinwardtii (Hydrops), 291.
reissii (Drymobius), 11.
reissii (Herpetodryas),11.
reticulare (Compsosoma), 35.
reticularis (Coluber), 31, 35.

reticularis (Spilotes), 35.
reticulata (Calamaria), 345.
reticulata(Herpetodryas), 11.
reticulatus (Atractus), 311.
reticulatus (Coluber),260.
reticulatus (Elaphis), 67.
reticulatus (Geophis), 311.
reticulatus(Pituophis),68.
rhabdocephalus (Xenodon), 146, 148, 150.
Rhabdophidium, 328.
Rhachiodontinæ, 353.
Rhadinæa, 160.
Rhegnops, 300.
Rhinaspis, 253.
Rhinechis, 24.
Rhinochilus, 212.
rhinomegas (Coluber), 40.
Rhinostoma, 246, 253.
rhinostoma (Heterodon), 253.
rhinostoma (Simophis), 253.
rhodogaster (Colophrys), 317.
rhodogaster (Geophis), 317.
rhombifer (Coryphodon), 14.
rhombifer (Drymobius), 14, 357.
rhombifer (Spilotes), 14.
rhombifer (Zamenis), 14.
rhombomaculata (Coronella), 198.
rhombomaculata (Lampropeltis), 199.
rhombomaculatus (Ophibolus), 198.
Rhynchocalamus, 233.
riccioli (Coluber), 194.
riccioli (Coronella), 194.
riccioli (Zamenis), 194.
richardi (Ahætula), 113.
richardii (Coluber), 113.
rijersmæi (Alsophis), 124.
rodriguezii (Elaphis), 39.
roelandti (Calamaria), 344.
rohdii (Rhinaspis), 254.
rohdii (Simophis), 254.
romanus (Coluber), 52.
rosaceus (Coluber), 49.
rostrale (Rhabdosoma), 323.
rostralis (Elapoides), 323.
rostralis (Geophis), 323.

rothi (Ablabes), 262.
rothi (Contia), 262.
rothi (Eirenis), 262.
rubens (Coluber), 194.
rubescens (Diadophis), 120.
rubinianum(Rabdosoma), 308.
rubriceps (Elaphis), 40.
rubriventris (Coluber), 42.
rufiventris (Alsophis), 124.
rufiventris (Dromicus), 124.
rufodorsatus (Ablabes), 43.
rufodorsatus (Coluber), 43.
rufodorsatus (Tropidonotus), 43.
rufus (Liophis), 129.
russelii (Coronella), 229.
russelius (Coluber), 229.
russellii (Simotes), 229.
rutilus (Lygophis), 165.

sagittifer (Chlorosoma), 165.
sagittifer (Liopeltis), 165.
sagittifera (Rhadinæa), 165.
sallæi (Geophis), 318.
salomonis (Dendrophis), 80.
salvinii (Spilotes), 33.
sanctæ-crucis(Dromicus), 122.
sancticrucis (Alsophis), 122.
sansibaricus (Philothamnus), 99.
sarasinorum (Agrophis), 360.
sargii (Adelphicus), 313.
sargii (Ahætulla), 111.
sargii (Leptophis), 111.
sargii (Pliocercus), 183.
sargii (Rhegnops), 313.
sargii (Thrasops), 111.
sartorii (Leptognathus), 296.
sartorii (Tropidodipsas), 296.
saturatus(Hapsidophrys), 110.
saturatus (Leptophis), 110.
saturninus (Coluber), 75.
saturocephalus (Coluber), 149.

sauromates (Coluber), 45, 52.
sauromates (Elaphis), 46, 55.
sauromates (Tropidonotus), 46.
sayi (Coluber), 68.
sayi (Coronella), 197.
sayi (Lampropeltis), 197.
sayi (Ophibolus), 197.
sayi (Pituophis), 69.
scaber (Coluber), 354.
scaber (Dasypeltis), 354.
scaber (Deirodon), 354.
scaber (Rachiodon), 354.
scaber (Tropidonotus), 354.
scabra (Dasypeltis), 354.
scalaris (Coluber), 65.
scalaris (Hypsirhynchus), 117.
scalaris (Rhinechis), 65.
Scaphiophis, 254.
schlegeli (Leionotus),286.
schlegelii (Ablabes), 282.
schlegelii (Calamaria), 333, 345.
schlegelii (Rhinostoma), 253.
schokari (Dipsas), 78, 88.
schranckii (Elaps), 184.
schrenckii (Coluber), 48.
schrenckii (Elaphis), 48.
sclateri (Leptocalamus), 251.
scopolii (Coluber), 52.
Scotophis, 25.
scriptus (Ablabes), 284.
scurrula (Herpetodryas), 75.
scurrula (Natrix), 75.
sebastus (Herpetodryas), 75.
sellatus (Coluber), 40.
sellmanni (Coluber), 52.
semianulata (Contia), 268.
semiannulata (Sonora), 268.
semiaureus (Ophiomorphus), 169.
semicarinatus (Ablabes), 278.
semicarinatus (Eurypholis), 278.
semicinctus (Heterodon), 153.
semicinctus (Lystrophis), 153.
semidoliatum (Catostoma), 316.

semidoliatum (Rabdosoma), 316.
semidoliatus (Elapoides), 316.
semidoliatus (Geophis), 316.
semifasciatus (Simotes), 222.
semilineata (Natrix), 138.
semimaculata (Ablabes), 261.
semiornata (Coronella), 195, 359.
semivariegata (Ahætulla), 99.
semivariegata (Dendrophis), 99.
semivariegatus (Philothamnus), 99.
septentrionalis (Calamaria), 349.
seriatus (Coluber), 290.
serpentinus (Coluber), 166.
serperastra (Ablabes),172.
serperastra (Rhadinæa), 172.
severus (Cerastes), 149.
severus (Coluber), 149.
severus (Xenodon), 146, 148, 149, 150.
sexcarinata (Natrix), 72.
sexcarinatus (Herpetodryas), 72.
sexlineata (Coronella),43.
sexlineatus (Ablabes), 43.
sexscutatus (Leptognathus), 296.
shirana (Ahætulla), 96.
shiranum (Homalosoma), 276.
siamensis (Calamaria), 348.
sibon (Coluber), 293.
sibonius (Alsophis), 123.
sieboldi (Elapoides), 318.
sieboldii (Ninia), 318.
signatus (Simotes), 226.
silurophaga (Grayia),286.
Simophis, 253.
Simotes, 213, 214.
simus (Coluber), 156.
simus (Heterodon), 156.
sinkawangensis (Calamaria), 343.
smaragdina (Ahætulla), 103.
smaragdina (Dendrophis), 103.
smaragdina (Gastropyxis), 103.

ALPHABETICAL INDEX. 375

smaragdina (Hapsidophrys), 103.
smaragdinus (Leptophis), 103.
smithii (Philothamnus), 99.
smythii (Coluber), 286.
smythii (Grayia), 286.
Sonora, 255.
Sphenocalamus, 326.
spiloides (Coluber), 50.
spiloides (Elaphis), 50.
spilonotus (Oligodon), 243.
Spilotes, 18, 23, 25.
spixii (Coluber), 75.
splendens (Liophis), 182.
splendida (Lampropeltis), 197.
splendidus (Ophibolus), 197.
splendidus (Simotes), 217.
stahlknechtii (Calamaria), 335.
Stasiotes, 213.
stictogenys (Diadophis), 207.
Stilosoma, 325.
stoliczkæ (Ablabes), 281.
straminea (Carphophis), 273.
stramineus (Chilomeniscus), 273.
striolatus (Dendrophis), 85.
stuhlmanni (Ligonirostra), 248.
subannulatus (Leptognathus), 294.
subannulatus (Tropidodipsas), 295.
subcarinatus (Dendrophis), 91.
subcarinatus (Simotes), 226.
subfasciatus (Liophis), 132.
subfasciatus (Rachiodon), 355.
subgriseus (Oligodon), 243.
sublineatus (Liophis),132.
sublineatus (Oligodon), 242.
sublutescens (Coryphodon), 8.
subocularis (Dendrelaphis), 89.
subocularis (Dendrophis), 89.

subquadratus (Oligodon), 237.
subradiatum (Compsosoma), 64.
subradiatus (Coluber), 64.
subradiatus (Elaphis), 62, 64.
subradiatus (Tropidonotus), 15.
sulphurea (Natrix), 19.
sulphureus (Phrynonax), 19.
sumatrana (Calamaria), 333, 339.
sumichrasti (Geagras), 250.
sumichrasti (Leptocalamus), 250.
sundevalli (Temnorhynchus), 247.
sundevallii (Prosymna), 247.
suspectus (Xenodon),147.
swinhonis (Simotes), 222.
Sympholis, 188.
Synchalinus, 70.
syspylus (Ophibolus),201.

tæniata (Rhadinæa), 178.
tæniatum (Arrhyton),252.
tæniatus (Dromicus),178.
tæniatus (Simotes), 227.
tæniogaster (Liophis), 166.
tæniolata (Coronella), 174.
tæniolata (Rhadinæa), 174.
tæniolatus (Enicognathus), 174.
Tæniophis, 118.
tæniurus (Aporophis), 130.
tæniurus (Coluber), 47.
tæniurus (Elaphis), 47.
tæniurus (Liophis), 130, 138.
tæniurus (Oligodon), 360.
tantillus (Tæniophis), 119.
taylori (Contia), 265.
temminckii (Calamaria), 333.
temminckii (Dromicus), 119.
temminckii (Liophis), 119.
temminckii (Psammophis), 119.
Temnorhynchus, 246.

templetonii (Oligodon), 241, 359.
temporalis (Dromicus), 143.
temporalis (Liophis), 143.
temporalis (Ophibolus), 200.
terrificus (Dendrelaphis), 90.
terrificus (Dendrophis), 90.
tessellatus (Ablabes), 11.
tessellatus (Rhinochilus), 213.
tetragonus (Coluber), 274.
tetratænia (Herpetodryas), 15.
texana (Hypsiglena), 209.
texensis (Diadophis), 207.
Thamnosophis, 9.
theobaldi (Simotes), 230.
thomensis (Philothamnus), 101.
thominoti (Rhinochilus), 213.
thraso (Coluber), 155.
Thrasops, 104, 105.
tigrina (Coronella), 199.
tigrinus (Heterodon), 155.
Toluca, 255.
torquata (Contia), 266.
torquata (Hypsiglena), 209, 210, 359.
torquata (Leptodira), 210.
torquatum (Pseudorabdion), 329.
torquatum (Rabdion), 329.
torquatum (Rabdosoma), 309.
torquatus (Atractus), 309.
torquatus (Coluber), 206.
torquatus (Leptocalamus), 250.
torquatus (Simotes), 232.
trabalis (Coluber), 44.
travancoricus (Oligodon), 236.
triangula (Lampropeltis), 200.
triangularis (Coronella), 286.
triangularis (Elaps), 187.
triangularis (Grayia), 287.

ALPHABETICAL INDEX.

triangularis (Heteronotus), 286.
triangularis (Hydrops), 187, 359.
triangulum (Ablabes), 198, 200.
triangulum (Coluber), 200.
triangulum (Coronella), 200.
triangulus (Ophibolus), 199, 200, 201.
triaspis (Coluber), 37.
triaspis (Natrix), 37.
tricinctus (Liophis), 183.
tricolor (Ablabes), 281.
tricolor (Cylophis), 281.
tricolor (Herpetodryas), 281.
tricolor (Liopeltis), 281.
trilineatus (Atractus), 312.
trilineatus (Oligodon), 238.
trilineatus (Simotes), 238.
Trimetopon, 184.
trinotatus (Simotes), 218.
Tripeltis, 233.
triscalis (Coluber), 129.
triscalis (Dromicus), 129.
triscalis (Liophis), 129.
tristis (Coluber), 88.
tristis (Dendrelaphis), 88, 358.
trivirgatum (Rhabdosoma), 312.
Tropidoclonium, 294.
Tropidodipsas, 294.
Tropidogeophis, 294.
Tropidonotus, 9, 353.
Typhlocalamus, 330.
Typhlogeophis, 351.
typhlus (Coluber), 136.
typhlus (Liophis), 135, 136.
typhlus (Opheomorphus), 136.
typhlus (Xenodon), 136.
typus (Anodon), 354.
Tyria, 23, 24.

undulata (Rhadinæa), 174.
undulatus (Coluber), 174.
undulatus (Dromicus), 174.
unicolor (Dromicus), 120.
unicolor (Geophis), 250.
unicolor (Leptocalamus), 250.
unicolor (Rachiodon), 355.
univittatum (Rabdosoma), 305.
Uromacer, 115.
urosticta (Ahætulla), 115.
urostictus (Leptophis), 115.
Urotheca, 180.

vaillanti (Simotes), 228.
valeriæ (Virginia), 289.
variabilis (Calamaria), 333.
variabilis (Coluber), 23.
variabilis (Spilotes), 23, 33.
varians (Conopsis), 268.
varians (Oxyrhina), 268.
variegata (Ficimia), 271.
variegatum (Amblymetopon), 271.
variegatum (Homalosoma), 276.
varium (Rabdosoma), 309.
veliferum (Amastridium), 352.
venustus (Oligodon), 235.
venustus (Simotes), 235.
verecundus (Liophis), 134.
vermiculaticeps (Coronella), 177.
vermiculaticeps (Rhadinæa), 177.
vermiculaticeps (Tæniophis), 177.
vermiformis (Calamaria), 333.
vermiformis (Carphophiops), 324.
vermis (Celuta), 324.
vermis (Carphophiops), 325.
vernalis (Chlorosoma), 258.
vernalis (Coluber), 258.
vernalis (Contia), 258.
vernalis (Cyclophis), 258.
vernalis (Herpetodryas), 258.
vernalis (Liopeltis), 259.
versicolor (Calamaria), 345.
versicolor (Coluber), 149, 191.
vertebralis (Coluber), 67.
vertebralis (Oligodon), 245.
vertebralis (Pituophis), 66, 67, 68.
vertebralis (Simotes), 245.
villarsii (Cheilorhina), 188.
vincenti (Herpetodryas), 73.
violacea (Coronella), 222.
violaceus (Coluber), 137.
violaceus (Simotes), 222.
viperinus (Dromicus), 174.
virgatus (Coluber), 54.
virgatus (Elaphis), 47, 54.
Virginia, 288.
virgulata (Calamaria), 338, 340.
viride (Gonyosoma), 57.
viridicyanea (Liophis), 132.
viridis (Dendrophis), 75.
viridis (Liophis), 135.
viridis (Opheomorphus), 135.
visoninus (Adelphicus), 313.
visoninus (Rhegnops), 313.
vittata (Rhadinæa), 178.
vittatum (Arrhyton), 252.
vittatus (Atractus), 304.
vittatus (Coluber), 134.
vittatus (Cryptodacus), 252.
vittatus (Enicognathus), 176, 178.
vittatus (Liophis), 134.
vudii (Alsophis), 120.
vulneratus (Coluber), 59.
vulpinus (Coluber), 49.
vulpinus (Elaphis), 49.
vulpinus (Scotophis), 49.

waandersii (Oligodon), 245.
waandersii (Rabdion), 245.
wagleri (Liophis), 134.
walteri (Contia), 263.
walteri (Pseudocyclophis), 263.
whymperi (Coronella), 174.
wilkesii (Pituophis), 67.
woodmasoni (Simotes), 223.

wuchereri (Dromicus), 175.
wuchereri (Lygophis), 175.

xanthogaster (Coluber), 46.
xanthurus (Spilotes), 31.
Xenelaphis, 7.
Xenodon, 24, 126, 144, 151, 180, 214.

Xenurophis, 288.

Y (Pseudoelaps), 200.
y-græcum (Liophis), 135.
y-græcum (Lygophis), 135.
yunnanensis (Elaphis), 47.

Zacholus, 188.
Zamenis, 24.

zebrina (Geophis), 308.
zebrinum (Rabdosoma), 306, 307.
zebrinus (Rhegnops), 308.
zonata (Coronella), 202.
zonatus (Bellophis), 202.
zonatus (Coluber), 202.
zonatus (Ophibolus), 203.
zonatus (Zacholus), 202.

LIST OF PLATES.

Plate I.

Fig. 1. *Phrynonax eutropis*, Blgr., p. 22.
    2. *Coluber janseni*, Blkr., p. 57.
        Upper and side views of head and neck.

Plate II.

Spilotes megalolepis, Gthr., p. 24.
    Upper, lower, and side views of head and neck.

Plate III.

Fig. 1. *Coluber frenatus*, Gray, p. 58.
    2. *Leptophis occidentalis*, Gthr., p. 111.
    3. —— *nigromarginatus*, Gthr., p. 112.
        Upper view of head and anterior part of body, and side view of head.

Plate IV.

Fig. 1. *Dendrophis bifrenalis*, Blgr., p. 80. Upper view of head and neck.
    1 a. —— ——. Side view of head.
    1 b. —— ——. Side view of middle of body.
    2. —— *grandoculis*, Blgr., p. 84. Upper view of head and neck.
    2 a. —— ——. Side view of head.

LIST OF PLATES.

Fig. 3. *Dendrophis gastrostictus*, Blgr., p. 86. Upper view of head and neck.

    3 a. —— ——. Side view of head.

    3 b. —— ——. Lower view of middle of body.

    4. *Dendrelaphis modestus*, Blgr., p. 91. Upper view of head and neck.

    4 a. —— —— Side view of head.

Plate V.

Fig. 1. *Chlorophis emini*, Gthr., p. 92.

    2. —— *hoplogaster*, Gthr., p. 93.

    3. —— *heterolepidotus*, Gthr., p. 95.

    4. *Philothamnus nitidus*, Gthr., p. 100.

    Upper, lower, and side views of head and neck.

Plate VI.

Fig. 1. *Hypsirhynchus ferox*, Gthr., p. 117.

    2. *Liophis fraseri*, Blgr., p. 131.

    3. —— *callilæmus*, Gosse, p. 142.

    Upper view of head and anterior part of body, and side view of head.

Plate VII.

Fig. 1. *Xenodon guentheri*, Blgr., p. 147.

    2. *Aporophis coralliventris*, Blgr., p. 159.

    3. *Rhadinæa jægeri*, Gthr., p. 170.

    Upper and side views of head and anterior part of body.

Plate VIII.

Fig. 1. *Hypsiglena affinis*, Blgr., p. 210. Upper view of head and anterior part of body.

    1 a. —— ——. Side view of head and anterior part of body.

    2. *Simotes formosanus*, Gthr., p. 222. Upper view of head and anterior part of body.

    2 a. —— ——. Side view of head and anterior part of body.

Fig. 3. *Simotes annulifer*, Blgr., p. 226. Upper view of head and anterior part of body.

3 *a*. —— ——. Side view of head and anterior part of body.

3 *b*. —— ——. Enlarged upper view of head.

Plate IX.

Fig. 1. *Simotes chinensis*, Gthr., p. 228.

2. —— *beddomii*, Blgr., p. 229.

3. —— *theobaldi*, Gthr., p. 230.

Upper and side views of head and anterior part of body, and enlarged upper and side views of head.

Plate X.

Fig. 1. *Simotes cruentatus*, Gthr., p. 231.

2. *Oligodon travancoricus*, Bedd., p. 236.

3. —— *modestus*, Gthr., p. 238.

Upper, lower, and side views of head and anterior part of body.

Plate XI.

Fig. 1. *Oligodon everetti*, Blgr., p. 239.

2. —— *vertebralis*, Gthr., p. 245.

3. —— *waandersii*, Blkr., p. 245.

Upper and side views of head and anterior part of body, and enlarged upper and side views of head.

Plate XII.

Fig. 1. *Leptocalamus sclateri*, Blgr., p. 251. Upper and side views of head and anterior part of body, and enlarged upper and side views of head.

2. *Arrhyton tæniatum*, Gthr., p. 252. Enlarged upper, lower, and side views of head.

3. *Contia taylori*, Blgr., p. 265. Upper and side views of head and anterior part of body, and enlarged upper and side views of head.

Plate XIII.

Fig. 1. *Homalosoma shiranum*, Blgr., p. 276. Upper view of head and anterior part of body, and enlarged upper and side views of head.
2. —— *abyssinicum*, Blgr., p. 276. Upper view of head and anterior part of body, and enlarged upper and side views of head.
3. *Grayia smythii*, Leach, p. 286. Upper and side views of head and anterior part of body.

Plate XIV.

Fig. 1. *Tropidodipsas annulifera*, Blgr., p. 297.
2. *Dirosema bicolor*, Gthr., p. 298.
3. *Atractus maculatus*, Gthr., p. 306.
Upper and side views of head and anterior part of body.

Plate XV.

Fig. 1. *Atractus modestus*, Blgr., p. 304. Upper and side views of head and anterior part of body.
2. —— *vittatus*, Blgr., p. 304. Upper, lower, and side views of head and anterior part of body.
3. —— *reticulatus*, Blgr., p. 311. Upper and side views of head and anterior part of body.

Plate XVI.

Fig. 1. *Geophis sallæi*, Blgr., p. 318. ×3.
2. —— *petersii*, Blgr., p. 321. ×3.
3. —— *championi*, Blgr., p. 321. ×3.
4. —— *godmani*, Blgr., p. 322. ×2½.
Upper, lower, and side views of head.

Plate XVII.

Fig. 1. *Calamaria baluensis*, Blgr., p. 335.
    2. —— *margaritifera*, Blkr., p. 336.
        Upper, lower, and side views of head and anterior part of body, and lower view of tail.

Plate XVIII.

Fig. 1. *Calamaria everetti*, Blgr., p. 340.
    2. —— ——, young.
    3. —— *rebentischii*, Blkr., p. 343.
        Upper and side views of head and anterior part of body, and enlarged side views of head.

Plate XIX.

Fig. 1. *Calamaria borneensis*, Blkr., p. 347.
    2. —— *javanica*, Blgr., p. 347.
    3. —— *brevis*, Blgr., p. 348.
    4. —— *lovii*, Blgr., p. 350.
        Upper and side views of head and anterior part of body, and enlarged upper and side views of head.

Plate XX.

Fig. 1. *Calamaria septentrionalis*, Blgr., p. 349.
    2. *Typhlogeophis brevis*, Gthr., p. 351.
        Upper, lower, and side views of head and anterior part of body, lower and side views of tail, and enlarged upper view of head.

PRINTED BY TAYLOR AND FRANCIS, RED LION COURT, FLEET STREET.

BRIT. MUS. N.H.   Pl. I.

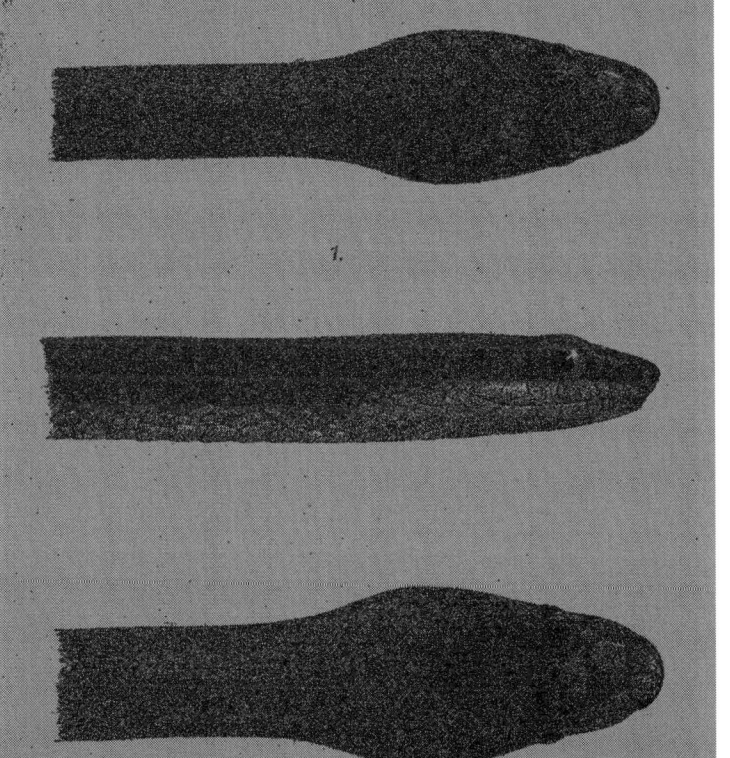

1.

2.

P. Mintern & J.G. del. et lith.   Mintern Bros. imp.
1. *Phrynonax eutropis*. 2. *Coluber jansenii*.

Spilotes megalolepis.

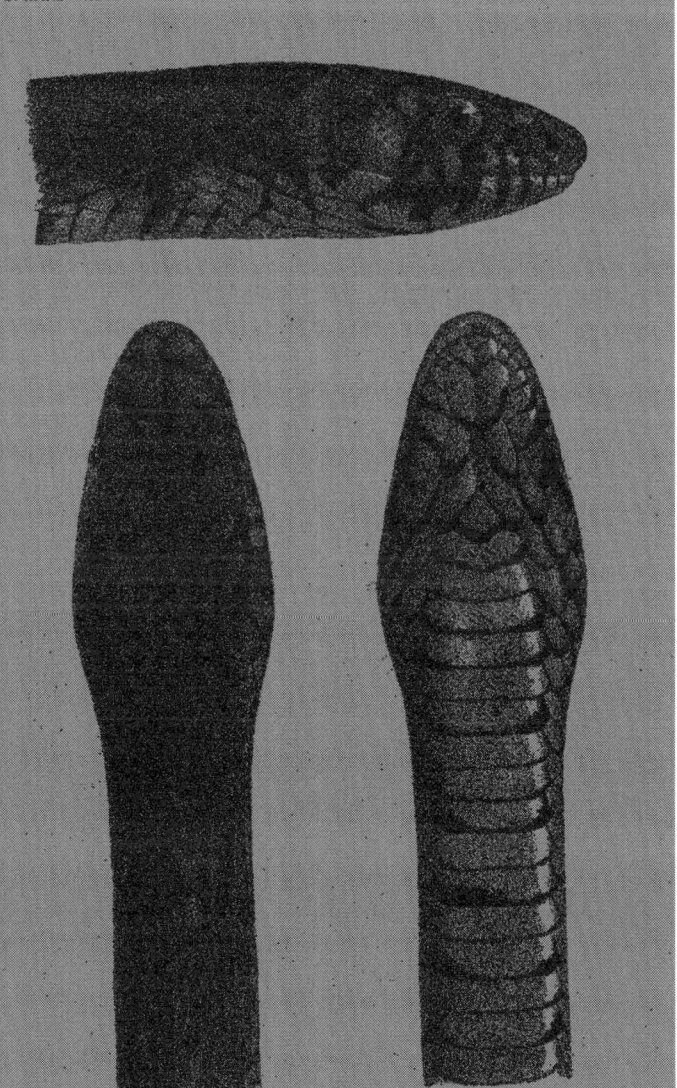

BRIT.MUS.N.H.
Pl. III.

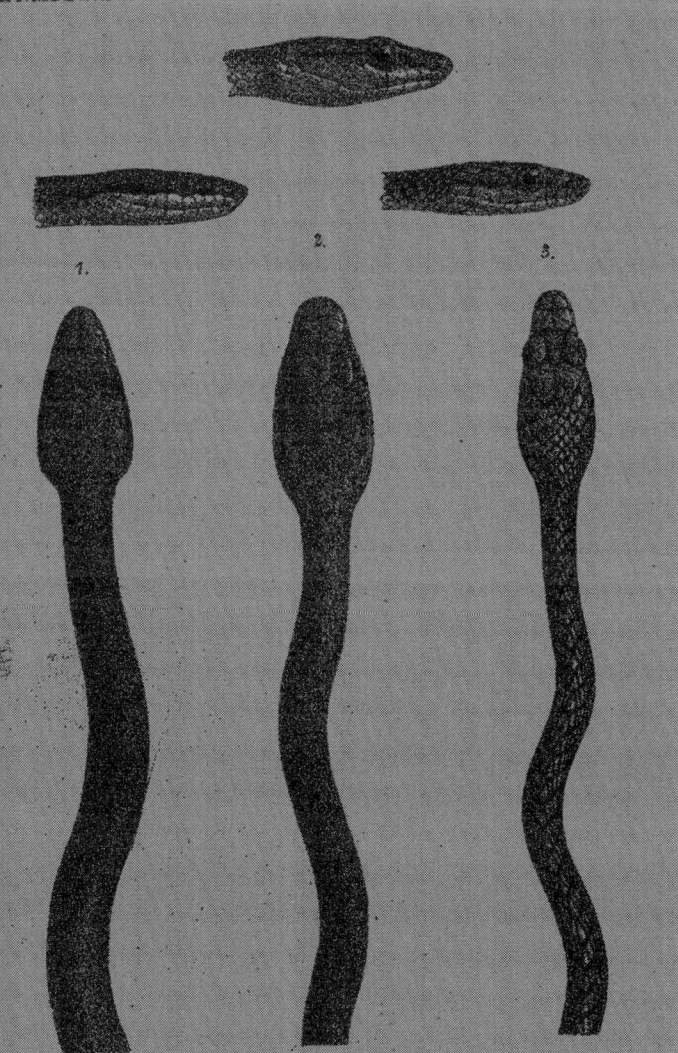

1. 2. 3.

R. Mintern & J.Green del. et lith. Mintern Bros. imp.

1. *Coluber frenatus.*   2. *Leptophis occidentalis.*
3. *Leptophis nigromarginatus.*

BRIT. MUS. N.H.   Pl. IV.

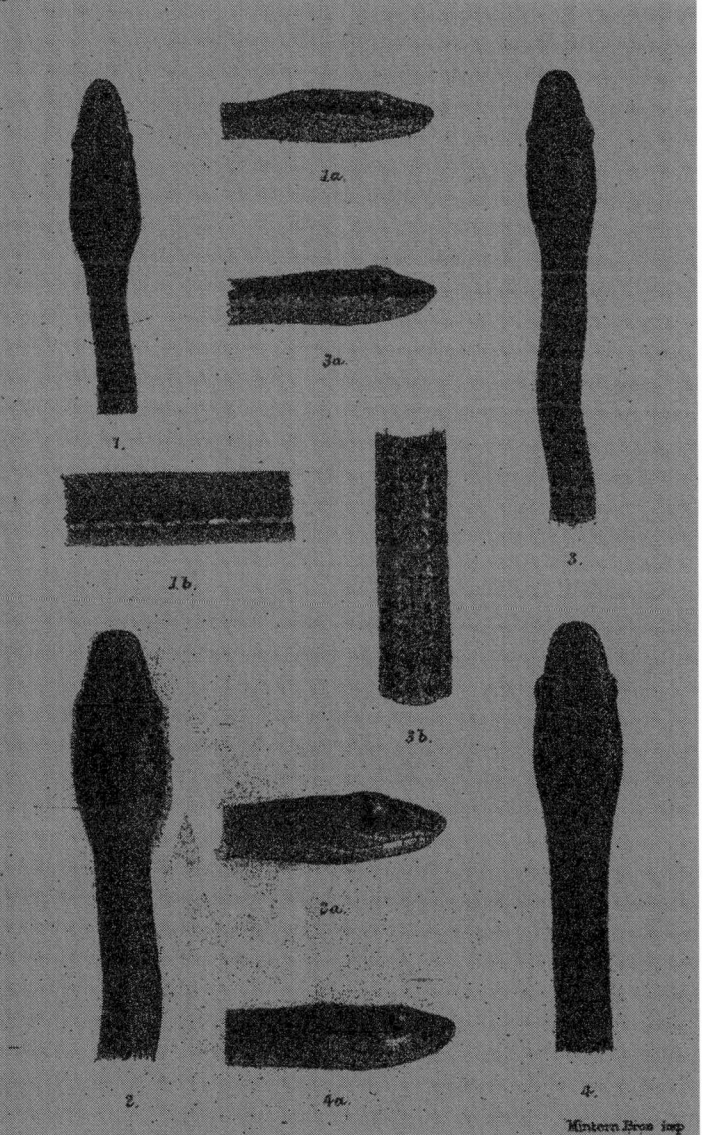

1. Dendrophis bifrenatus. 2. Dendrophis grandoculis.
3. Dendrophis gastrostictus. 4. Dendrelaphis modestus.

Pl. V.

BRIT. MUS. N.H.

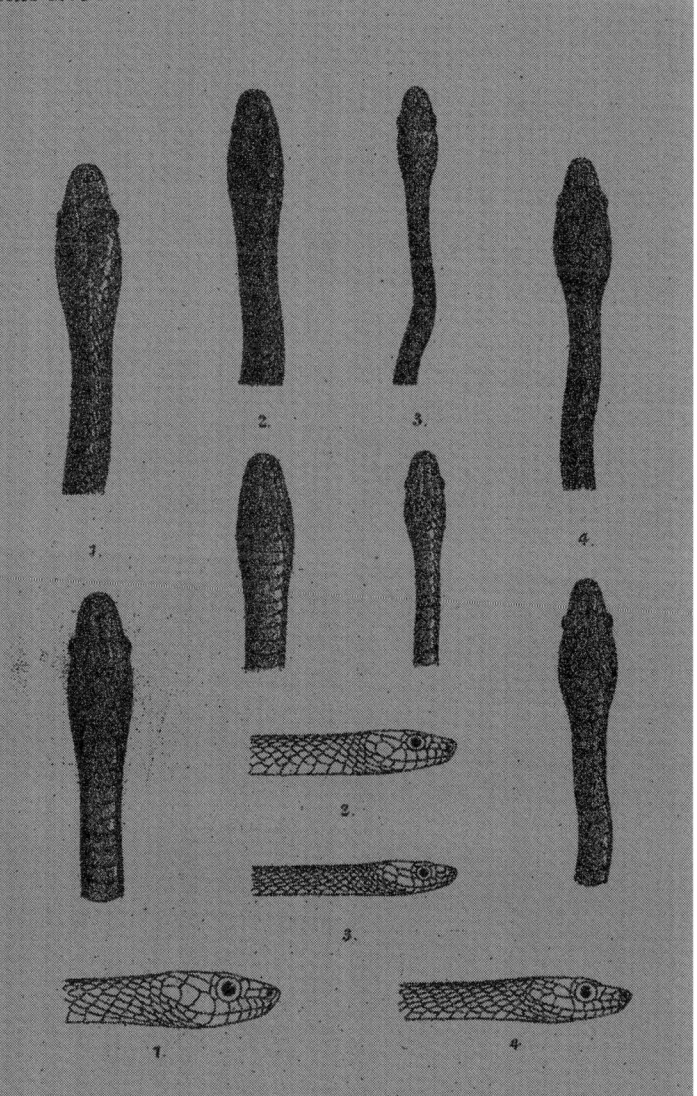

P. Mintern & J.S. del et lith.   Mintern Bros. imp.
1. *Chlorophis emini.* 2. *Chlorophis hoplogaster.* 3. *Chlorophis heterolepidotus.*
4. *Philothamnus nitidus.*

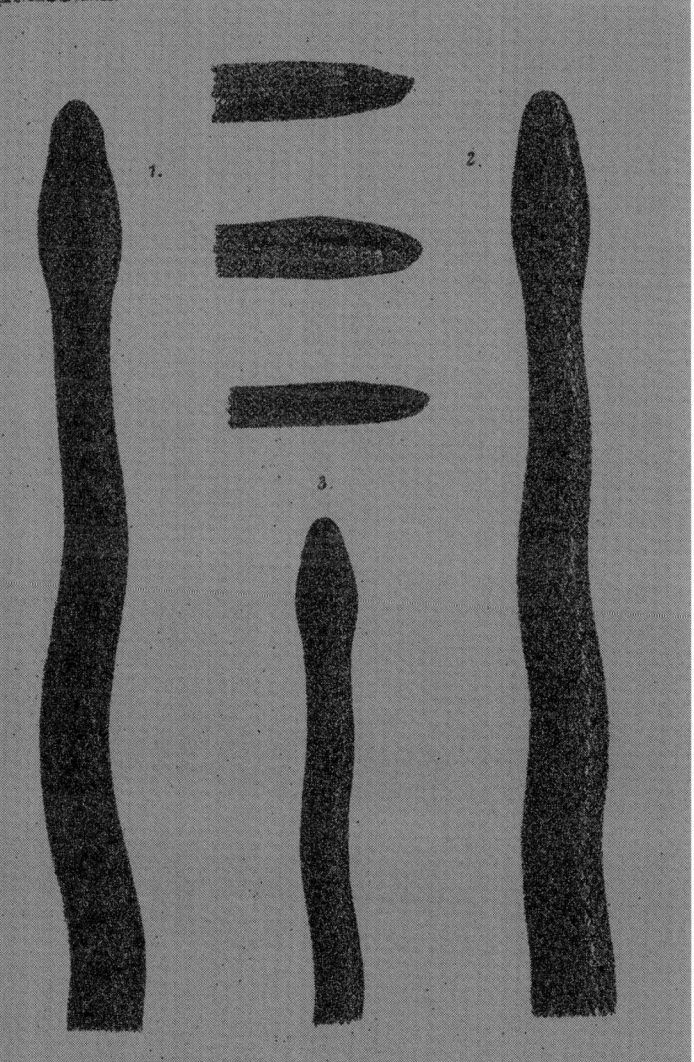

1. Hypsirhynchus ferox. 2. Liophis fraseri.
3. Liophis callilaemus.

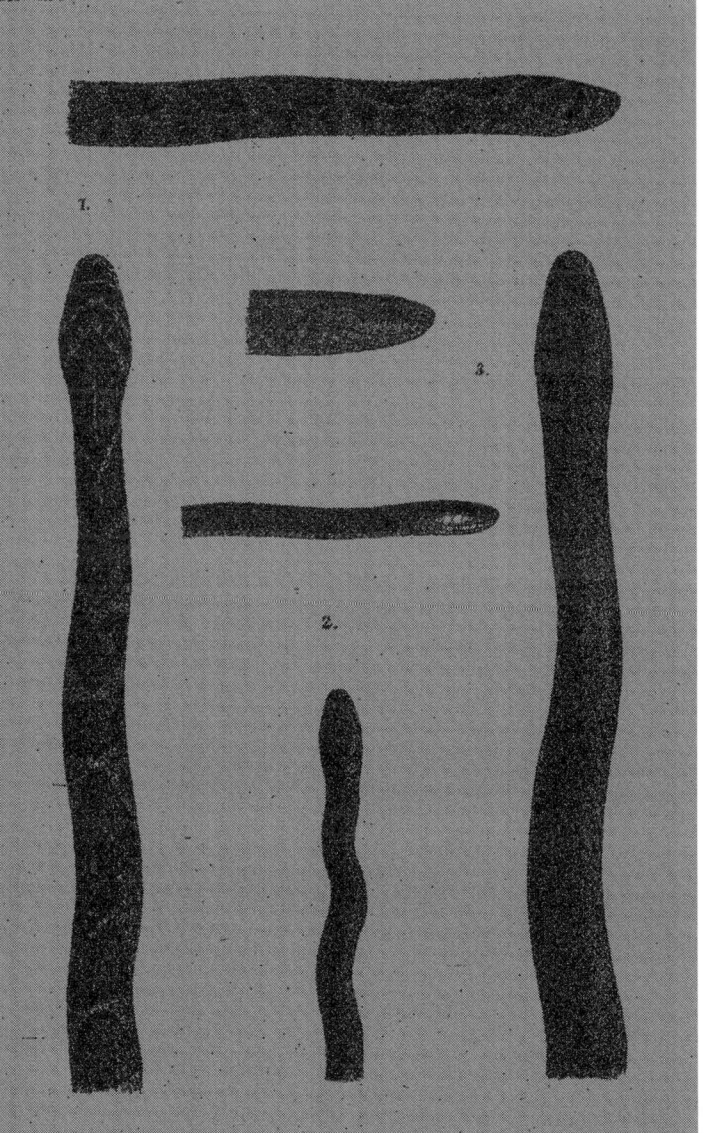

1. *Xenodon guentheri*.   2. *Aporophis coralliventris*.
3. *Rhadinaea jaegeri*.

Pl. VIII.

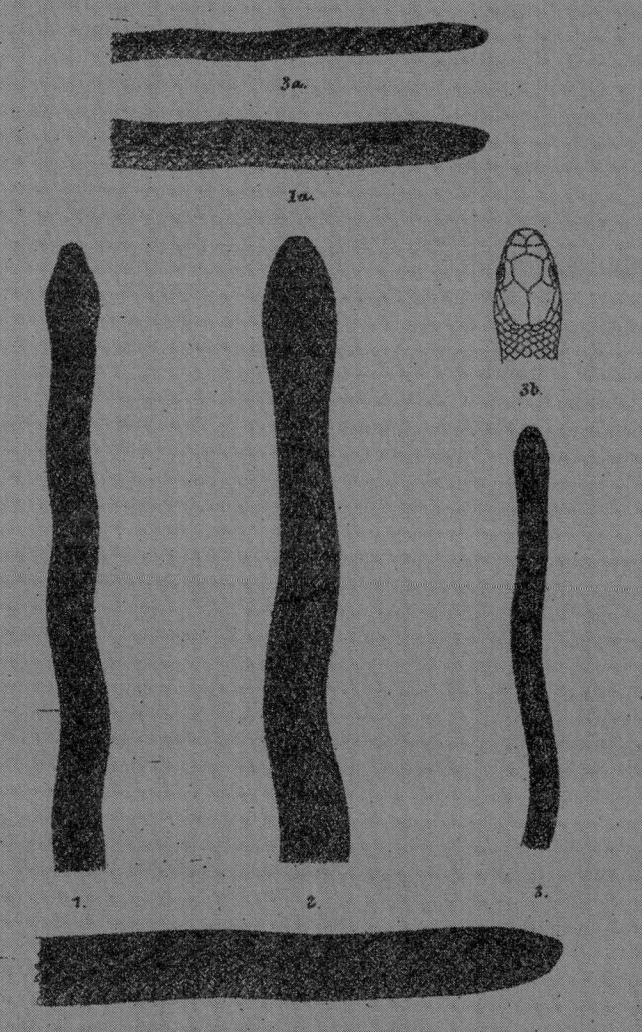

1. *Hypsiglena affinis.* 2. *Simotes formosanus.*
3. *Simotes annulifer.*

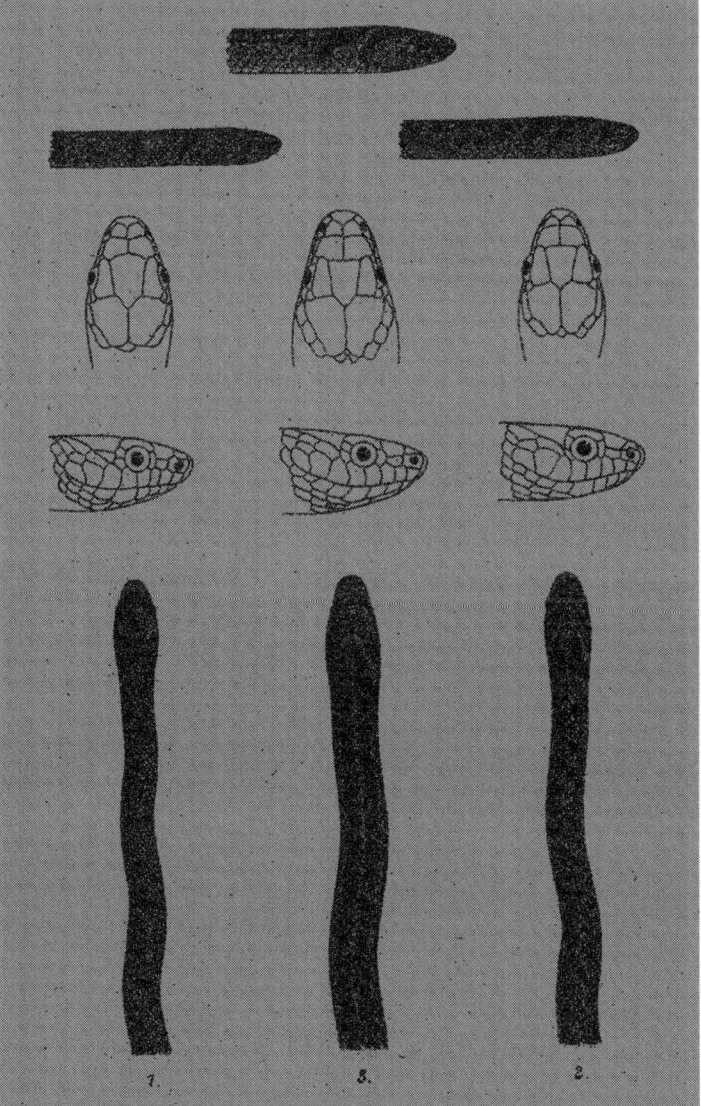

BRIT.MUS.N.H.   Pl. IX.

1. *Simotes chinensis*.   2. *Simotes beddomii*.
3. *Simotes theobaldi*.

BRIT. MUS. N.H.

Pl. X.

1. Simotes cruentatus. 2. Oligodon travancoricus.
3. Oligodon modestus.

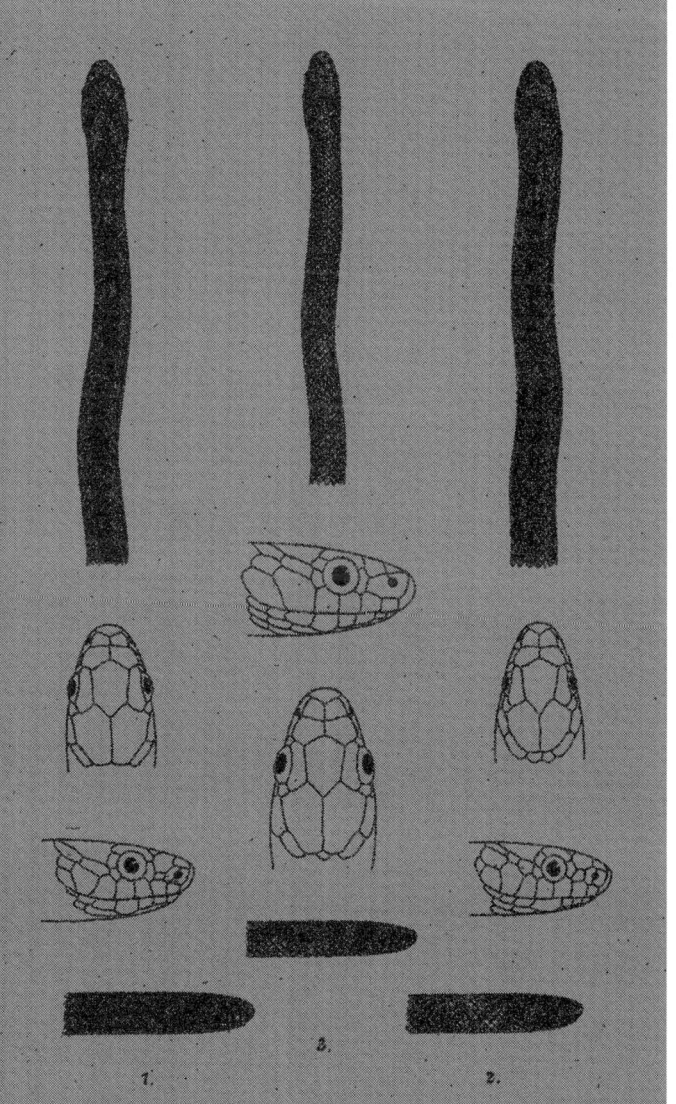

1. *Oligodon everetti.* 2. *Oligodon vertebralis.*
3. *Oligodon waandersii.*

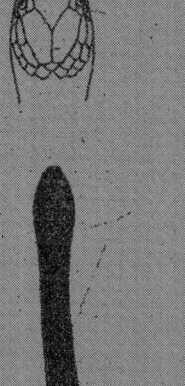

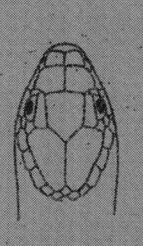

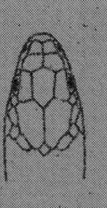

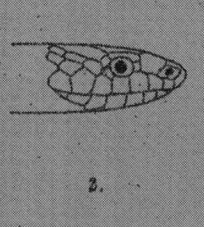

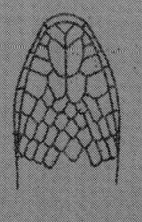

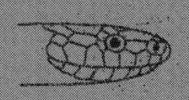

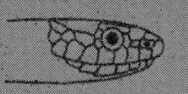

BRIT.MUS.N.H. Pl. XII.

1. Leptocalamus sclateri. 2. Arrhyton taeniatum.
3. Contia taylori.

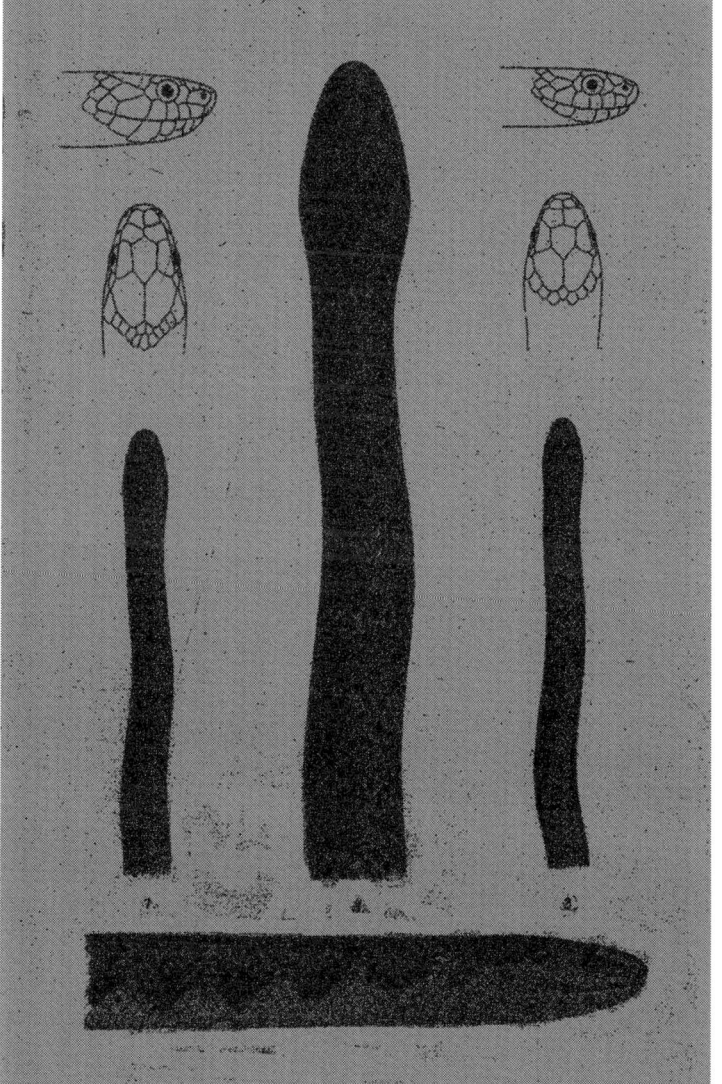

BRIT. MUS. N.H. PL. XIII.

1. *Homalosoma chinense.* 2. *Homalosoma abyssinicum*
3. *Grayia smythii.*

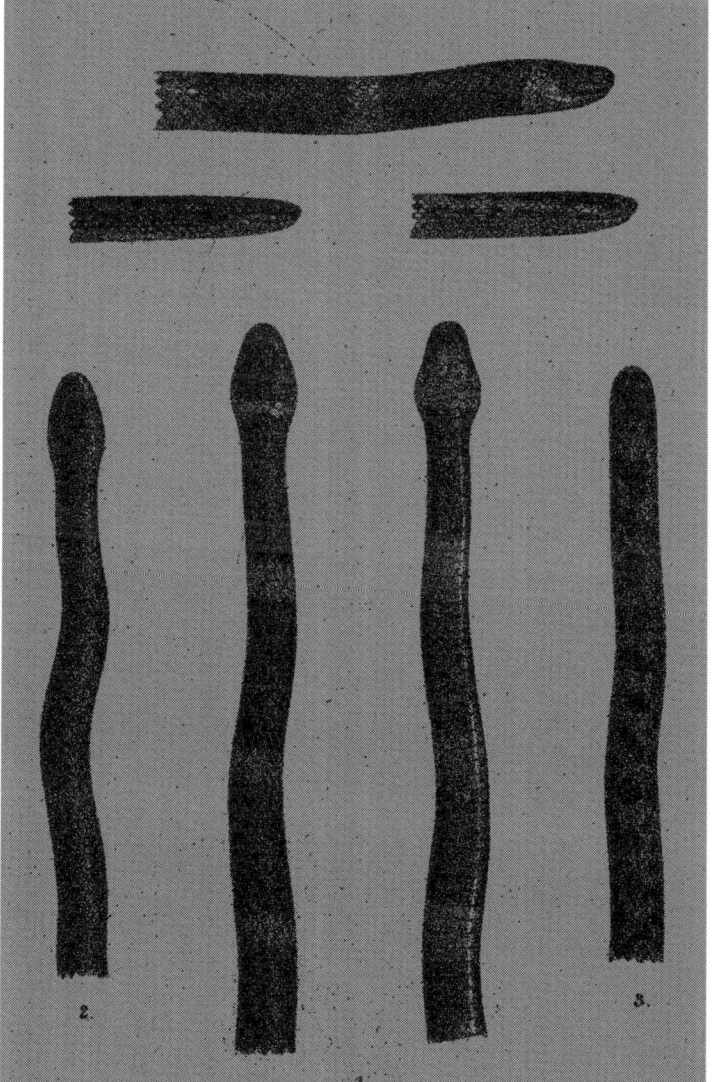

1. *Tropidodipsas annulifera.* 2. *Dirosema bicolor.*
3. *Atractus maculatus.*

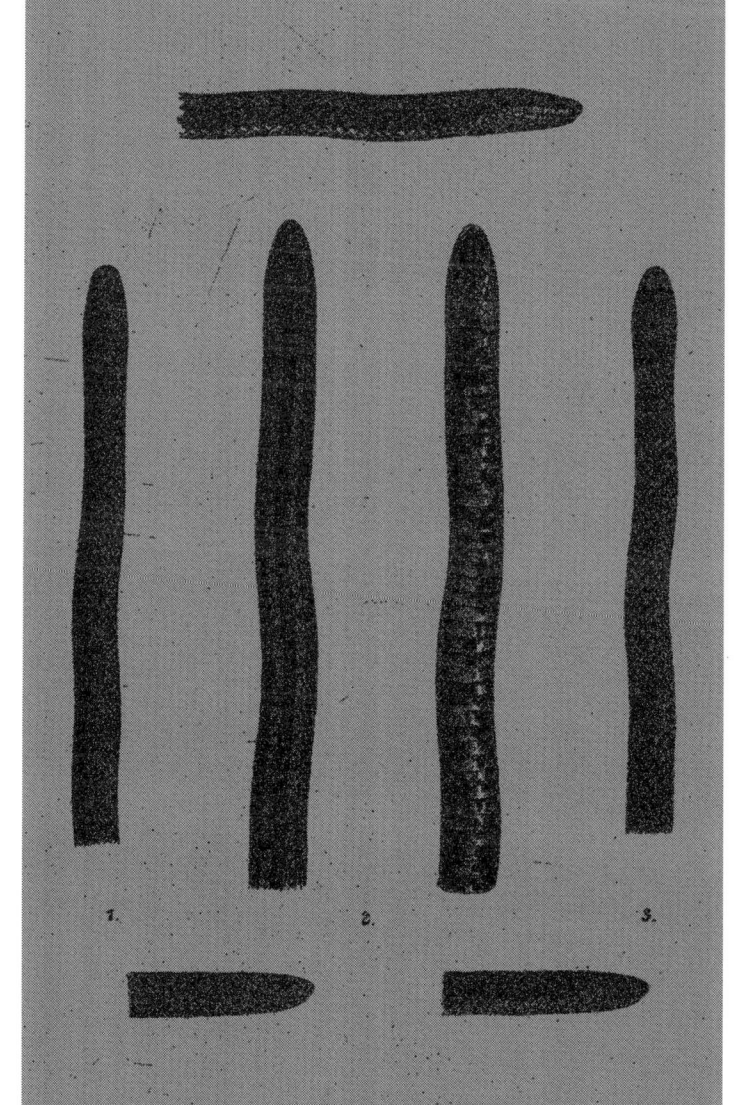

BRIT. MUS. N. H.   Pl. XV.

R. Mintern & J. G. del. et lith.   Mintern Bros. imp.

1. Atractus modestus.   2. Atractus vittatus.
3. Atractus reticulatus.

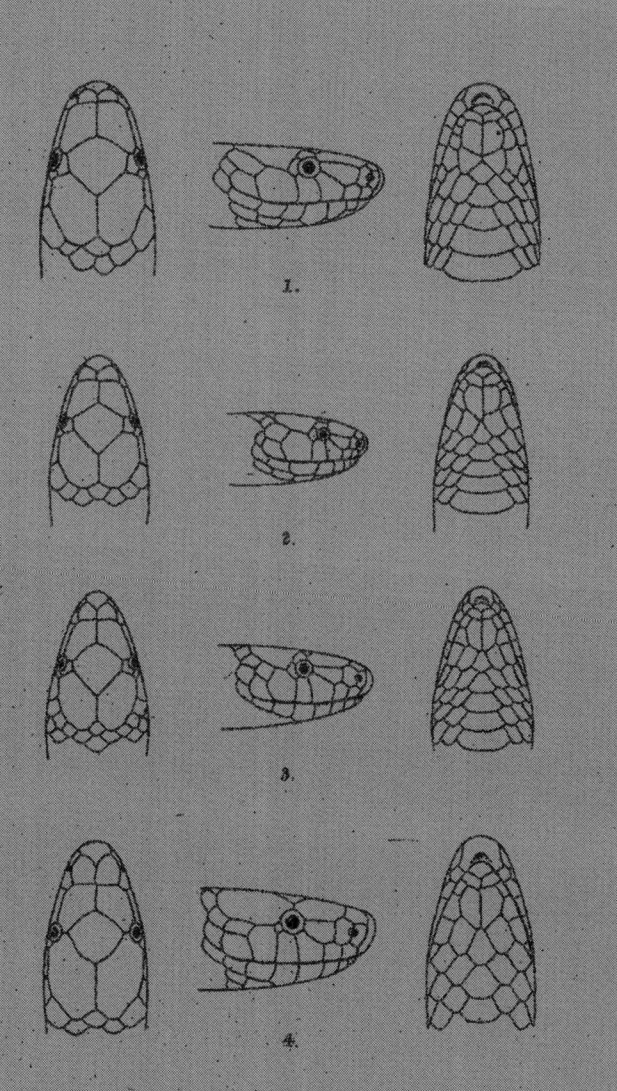

BRIT. MUS. N. H.

Pl. XVI.

J. Green del. et lith.

Mintern Bros., imp.

Geophis.

1. *G. sallæi.* 2. *G. petersii.* 3. *G. championi.* 4. *G. godmani.*

Pl. XVII

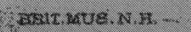

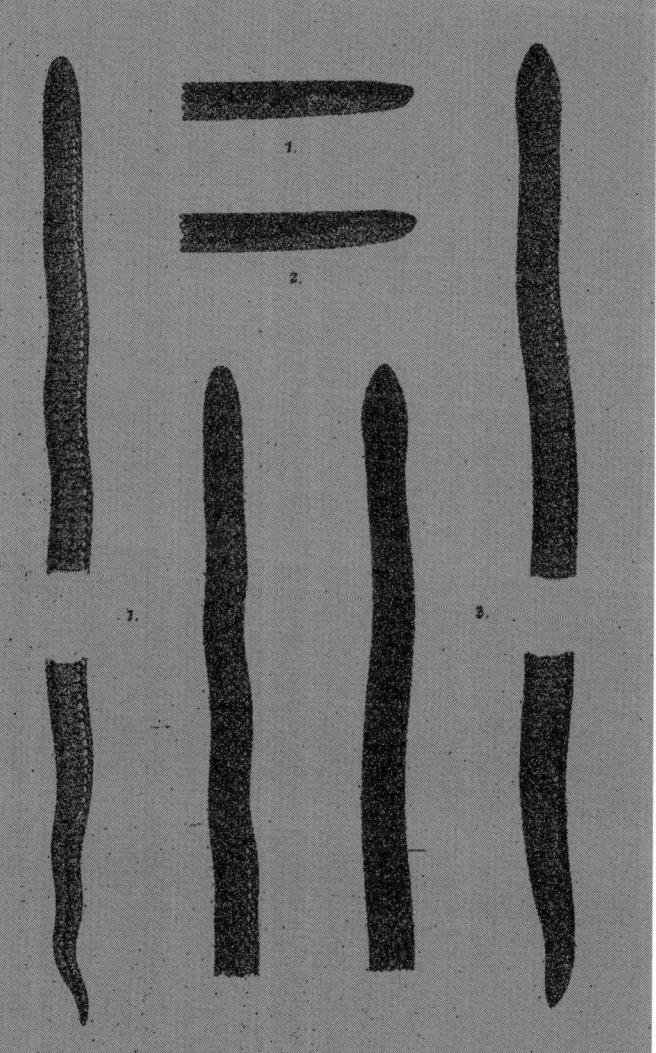

1. *Calamaria baluensis.*   2. *Calamaria margaritifera.*

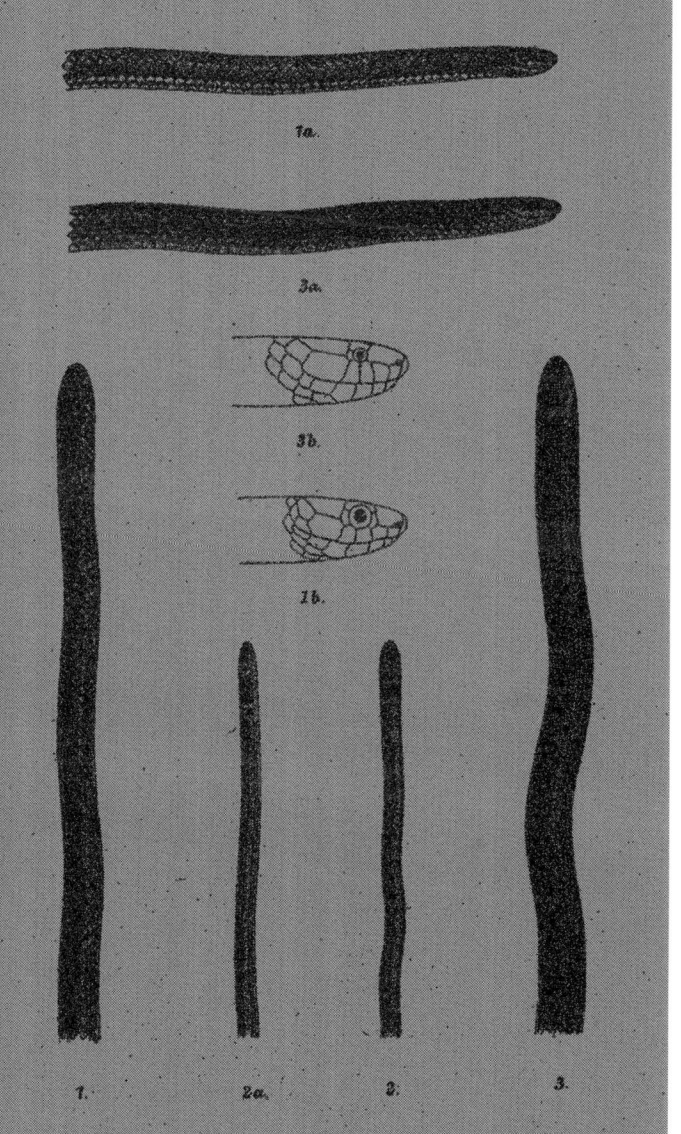

BRIT. MUS. N.H.   Pl. XVIII.

1, 2. Calamaria everetti.   3. Calamaria rebentischii.

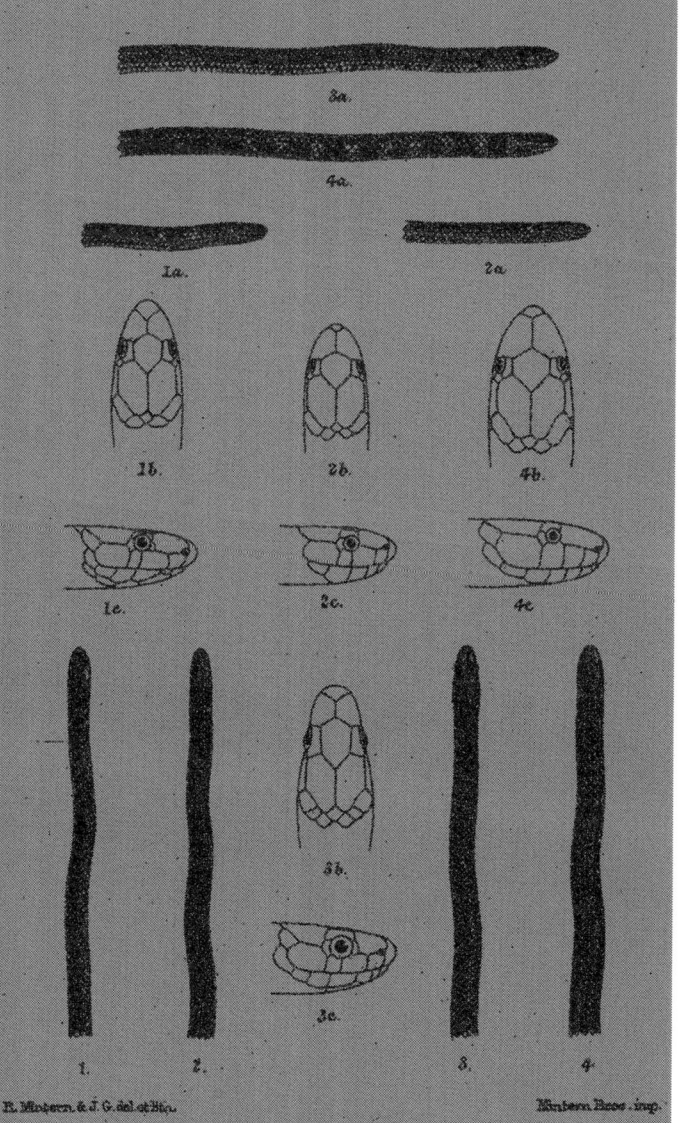

Calamaria.
1. C. borneensis. 2. C. javanica. 3. C. brevis. 4. C. lovii.

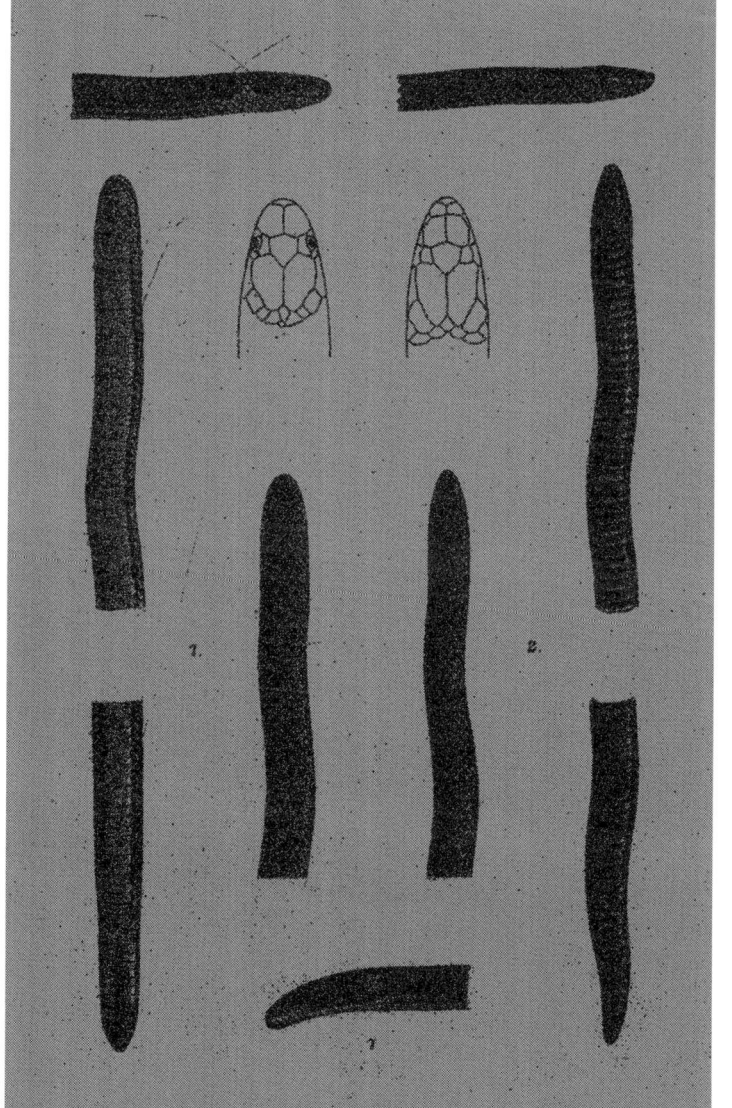

BRIT. MUS. N. H.
Pl. XX.

1. Calamaria septentrionalis. 2. Typhlogeophis brevis.

LIST OF THE CURRENT

NATURAL HISTORY PUBLICATIONS OF THE TRUSTEES OF THE BRITISH MUSEUM.

The following publications can be purchased through the Agency of Messrs. LONGMANS & CO., 39, *Paternoster Row;* Mr. QUARITCH, 15, *Piccadilly;* Messrs. KEGAN PAUL, TRENCH, TRÜBNER & Co., *Paternoster House, Charing Cross Road;* and *Messrs.* DULAU & CO., 37, *Soho Square;* or at the NATURAL HISTORY MUSEUM, *Cromwell Road, London, S.W.*

Catalogue of the Specimens and Drawings of Mammals, Birds, Reptiles, and Fishes of Nepal and Tibet. Presented by B. H. Hodgson, Esq., to the British Museum. 2nd edition. By John Edward Gray. Pp. xii., 90. [With an account of the Collection by Mr. Hodgson.] 1863, 12mo. 2s. 3d.

Report on the Zoological Collections made in the Indo-Pacific Ocean during the voyage of H.M.S. "Alert," 1881–2. Pp. xxv., 684. 54 Plates. 1884, 8vo.

Summary of the Voyage	By Dr. R. W. Coppinger.
Mammalia	,, O. Thomas.
Aves	,, R. B. Sharpe.
Reptilia, Batrachia, Pisces	,, A. Günther.
Mollusca	,, E. A. Smith.
Echinodermata	,, F. J. Bell.
Crustacea	,, E. J. Miers.
Coleoptera	,, C. O. Waterhouse.
Lepidoptera	,, A. G. Butler.
Alcyonaria and Spongiida	,, S. O. Ridley.

1*l.* 10s.

MAMMALS.

Catalogue of the Bones of Mammalia in the Collection of the British Museum. By Edward Gerrard. Pp. iv., 296. 1862, 8vo. 5s.

Catalogue of Monkeys, Lemurs, and Fruit-eating Bats in the Collection of the British Museum. By Dr. J. E. Gray, F.R.S., &c. Pp. viii., 137. 21 Woodcuts. 1870, 8vo. 4s.

Catalogue of Carnivorous, Pachydermatous, and Edentate Mammalia in the British Museum. By John Edward Gray, F.R.S., &c. Pp. vii., 398. 47 Woodcuts. 1869, 8vo. 6s. 6d.

Hand-List of Seals, Morses, Sea-Lions, and Sea-Bears in the British Museum. By Dr. J. E. Gray, F.R.S., &c. Pp. 43. 30 Plates of Skulls. 1874, 8vo. 12s. 6d.

Catalogue of Seals and Whales in the British Museum. By John Edward Gray, F.R.S., &c. 2nd edition. Pp. vii., 402. 101 Woodcuts. 1866, 8vo. 8s.

—————— Supplement. By John Edward Gray, F.R.S., &c. Pp. vi., 103. 11 Woodcuts. 1871, 8vo. 2s. 6d.

List of the Specimens of Cetacea in the Zoological Department of the British Museum. By William Henry Flower, LL.D., F.R.S., &c. [With Systematic and Alphabetical Indexes.] Pp. iv., 36. 1885, 8vo. 1s. 6d.

Catalogue of Ruminant Mammalia (*Pecora*, Linnæus) in the British Museum. By John Edward Gray, F.R.S., &c. Pp. viii., 102. 4 Plates. 1872, 8vo. 3s. 6d.

Hand-List of the Edentate, Thick-skinned, and Ruminant Mammals in the British Museum. By Dr. J. E. Gray, F.R.S., &c. Pp. vii., 176. 42 Plates of Skulls, &c. 1873, 8vo. 12s.

Catalogue of the Marsupialia and Monotremata in the Collection of the British Museum. By Oldfield Thomas. Pp. xiii., 401. 4 coloured and 24 plain Plates. [With Systematic and Alphabetical Indexes.] 1888, 8vo. 1l. 8s.

BIRDS.

Catalogue of the Birds in the British Museum:—

> Vol. III. Catalogue of the Passeriformes, **or Perching** Birds, in the Collection of the British Museum. *Coliomorphæ*, containing the families Corvidæ, Paradiseidæ, Oriolidæ, Dicruridæ, and Prionopidæ. By R. Bowdler Sharpe. Pp. xiii., 343. Woodcuts and 14 coloured Plates. [With Systematic and Alphabetical Indexes.] 1877, 8vo. 17s.
>
> Vol. IV. Catalogue of the Passeriformes, or Perching Birds, in the Collection of the British Museum. *Cichlomorphæ*: Part I., containing the families Campophagidæ and Muscicapidæ. By R. Bowdler Sharpe. Pp. xvi., 494. Woodcuts and 14 coloured Plates. [With Systematic and Alphabetical Indexes.] 1879, 8vo. 1l.
>
> Vol. V. Catalogue of the Passeriformes, or Perching Birds, in the Collection of the British Museum. *Cichlomorphæ*: Part II., containing the family Turdidæ (Warblers and Thrushes). By Henry Seebohm. Pp. xvi., 426. Woodcuts and 18 coloured Plates. [With Systematic and Alphabetical Indexes.] 1881, 8vo. 1l.

Catalogue of the Birds in the British Museum—*continued.*
Vol. VI. Catalogue of the Passeriformes, or Perching Birds, in the Collection of the British Museum. *Cichlomorphæ:* Part III., containing the first portion of the family Timeliidæ (Babbling Thrushes). By R. Bowdler Sharpe. Pp. xiii., 420. Woodcuts and 18 coloured Plates. [With Systematic and Alphabetical Indexes.] 1881, 8vo. 1*l.*

Vol. VII. Catalogue of the Passeriformes, or Perching Birds, in the Collection of the British Museum. *Cichlomorphæ:* Part IV., containing the concluding portion of the family Timeliidæ (Babbling Thrushes). By R. Bowdler Sharpe. Pp. xvi., 698. Woodcuts and 15 coloured Plates. [With Systematic and Alphabetical Indexes.] 1883, 8vo. 1*l.* 6*s.*

Vol. VIII. Catalogue of the Passeriformes, or Perching Birds, in the Collection of the British Museum. *Cichlomorphæ:* Part V., containing the families Paridæ and Laniidæ (Titmice and Shrikes); and *Certhiomorphæ* (Creepers and Nuthatches). By Hans Gadow, M.A., Ph.D. Pp. xiii., 386. Woodcuts and 9 coloured Plates. [With Systematic and Alphabetical Indexes.] 1883, 8vo. 17*s.*

Vol. IX. Catalogue of the Passeriformes, or Perching Birds, in the Collection of the British Museum. *Cinnyrimorphæ*, containing the families Nectariniidæ and Meliphagidæ (Sun Birds and Honey-eaters). By Hans Gadow, M.A., Ph.D. Pp. xii., 310. Woodcuts and 7 coloured Plates. [With Systematic and Alphabetical Indexes.] 1884, 8vo. 14*s.*

Vol. X. Catalogue of the Passeriformes, or Perching Birds, in the Collection of the British Museum. *Fringilliformes:* Part I., containing the families Dicæidæ, Hirundinidæ, Ampelidæ, Mniotiltidæ, and Motacillidæ. By R. Bowdler Sharpe. Pp. xiii., 682. Woodcuts and 12 coloured Plates, [With Systematic and Alphabetical Indexes.] 1885. 8vo. 1*l.* 2*s.*

Vol. XI. Catalogue of the Passeriformes, or Perching Birds, in the Collection of the British Museum. *Fringilliformes:* Part II., containing the families Cœrebidæ, Tanagridæ, and Icteridæ. By Philip Lutley Sclater, M.A., F.R.S. Pp. xvii., 431. [With Systematic and Alphabetical Indexes.] Woodcuts and 18 coloured Plates. 1886, 8vo.1*l.*

Vol. XII. Catalogue of the Passeriformes, or Perching Birds, in the Collection of the British Museum. *Fringilliformes:* Part III., containing the family Fringillidæ. By R. Bowdler Sharpe. Pp. xv., 871. Woodcuts and 16 coloured Plates. [With Systematic and Alphabetical Indexes.] 1888, 8vo. 1*l.* 8*s.*

Vol. XIII. Catalogue of the Passeriformes, or Perching Birds, in the Collection of the British Museum. *Sturni-*

Catalogue of the Birds in the British Museum—*continued*.
formes, containing the families Artamidæ, Sturnidæ, Ploceidæ, and Alaudidæ. Also the families Atrichiidæ and Menuridæ. By R. Bowdler Sharpe. Pp. xvi., 701. Woodcuts and 15 coloured Plates. [With Systematic and Alphabetical Indexes.] 1890, 8vo., 1*l.* 8*s.*

Vol. XIV. Catalogue of the Passeriformes, or Perching Birds, in the Collection of the British Museum. *Oligomyodæ*, or the families Tyrannidæ, Oxyrhamphidæ, Pipridæ, Cotingidæ, Phytotomidæ, Philepittidæ, Pittidæ, Xenicidæ, and Eurylæmidæ. By Philip Lutley Sclater, M.A., F.R.S. Pp. xix., 494. Woodcuts and 26 coloured Plates. [With Systematic and Alphabetical Indexes.] 1888, 8vo. 1*l.* 4*s.*

Vol. XV. Catalogue of the Passeriformes, or Perching Birds, in the Collection of the British Museum. *Tracheophonæ*, or the families Dendrocolaptidæ, Formicariidæ, Conopophagidæ, and Pteroptochidæ. By Philip Lutley Sclater, M.A., F.R.S. Pp. xvii., 371. Woodcuts and 20 coloured Plates. [With Systematic and Alphabetical Indexes.] 1890, 8vo. 1*l.*

Vol. XVI. Catalogue of the Picariæ in the Collection of the British Museum. *Upupæ* and *Trochili*, by Osbert Salvin. *Coraciæ*, of the families Cypselidæ, Caprimulgidæ, Podargidæ, and Steatornithidæ, by Ernst Hartert Pp. xvi. 703. Woodcuts and 14 coloured Plates. [With Systematic and Alphabetical Indexes.] 1892, 8vo. 1*l.* 16*s.*

Vol. XVII. Catalogue of the Picariæ in the Collection of the British Museum. *Coraciæ* (contin.) and *Halcyones*, with the families Leptosomatidæ, Coraciidæ, Meropidæ, Alcedinidæ, Momotidæ, Totidæ, and Coliidæ, by R. Bowdler Sharpe. *Bucerotes* and *Trogones*, by W. R. Ogilvie Grant. Pp. xi., 522. Woodcuts and 17 coloured Plates. [With Systematic and Alphabetical Indexes.] 1892, 8vo. 1*l.* 10*s.*

Vol. XVIII. Catalogue of the Picariæ in the Collection of the British Museum. *Scansores*, containing the family Picidæ. By Edward Hargitt. Pp. xv., 597. Woodcuts and 15 coloured Plates. [With Systematic and Alphabetical Indexes.] 1890, 8vo. 1*l.* 6*s.*

Vol. XIX. Catalogue of the Picariæ in the Collection of the British Museum. *Scansores* and *Coccyges*: containing the families Rhamphastidæ, Galbulidæ, and Bucconidæ, by P. L. Sclater; and the families Indicatoridæ, Capitonidæ, Cuculidæ, and Musophagidæ, by G. E. Shelley. Pp. xii., 484: 13 coloured Plates. [With Systematic and Alphabetical Indexes.] 1891, 8vo. 1*l.* 5*s.*

Vol. XX. Catalogue of the Psittaci, or Parrots, in the Collection of the British Museum. By T. Salvadori.

Pp. xvii., 658 : woodcuts and 18 coloured Plates. [With Systematic and Alphabetical Indexes.] 1891, 8vo. 1*l*. 10*s*.

Vol. XXI. Catalogue of the Columbæ, or Pigeons, in the Collection of the British Museum. By T. Salvadori. Pp. xvii., 676: 15 coloured plates. [With Systematic and Alphabetical Indexes.] 1893, 8vo. 1*l*. 10*s*.

Vol. XXII. Catalogue of the Game Birds (*Pterocletes, Gallinæ, Opisthocomi, Hemipodii*), in the Collection of the British Museum. By W. R. Ogilvie Grant. Pp. xvi., 585 : 8 coloured plates. [With Systematic and Alphabetical Indexes.] 1893, 8vo. 1*l*. 6*s*.

Hand-List of Genera and Species of Birds, distinguishing those contained in the British Museum. By G. R. Gray, F.R.S., &c. :—

Part II. Conirostres, Scansores, Columbæ, and Gallinæ. Pp. xv., 278. [Table of Genera and Subgenera: Part II.] 1870, 8vo. 6*s*.

Part III. Struthiones, Grallæ, and Anseres, with Indices of Generic and Specific Names. Pp. xi., 350. [Table of Genera and Subgenera: Part III.] 1871, 8vo. 8*s*.

List of the Specimens of Birds in the Collection of the British Museum. By George Robert Gray :—

Part III., Sections III. and IV. Capitonidæ and Picidæ. [With Index.] Pp. 137. 1868, 12mo. 1*s*. 6*d*.

Part IV. Columbæ. [With Index.] Pp. 73. 1856, 12mo. 1*s*. 9*d*.

Part V. Gallinæ. Pp. iv., 120. [With an Alphabetical Index.] 1867, 12mo. 1*s*. 6*d*.

Catalogue of the Birds of the Tropical Islands of the Pacific Ocean in the Collection of the British Museum. By George Robert Gray, F.L.S., &c. Pp. 72. [With an Alphabetical Index.] 1859, 8vo. 1*s*. 6*d*.

REPTILES.

Catalogue of the Tortoises, Crocodiles, and Amphisbænians in the Collection of the British Museum. By Dr. J. E. Gray, F.R.S., &c. Pp. viii., 80. [With an Alphabetical Index.] 1844, 12mo., 1*s*.

Catalogue of Shield Reptiles in the Collection of the British Museum. By John Edward Gray, F.R.S., &c. :—

Part I. Testudinata (Tortoises). Pp. 79. 50 plates. 1855, 4to. 2*l*. 10*s*.

Supplement. With Figures of the Skulls of 36 Genera. Pp. ix., 120. 40 Woodcuts. 1870, 4to. 10*s*.

Appendix. Pp. 28. 1872, 4to. 2*s*. 6*d*.

Part II. Emydosaurians, Rhynchocephalia, and Amphisbænians. Pp. vi., 41. 25 Woodcuts. 1872, 4to. 3*s*. 6*d*.

Hand-List of the Specimens of Shield Reptiles in the British Museum. By Dr. J. E. Gray, F.R.S., F.L.S., &c. Pp. iv., 124. [With an Alphabetical Index.] 1873, 8vo. 4s.

Catalogue of the Chelonians, Rhynchocephalians, and Crocodiles in the British Museum (Natural History). New Edition. By George Albert Boulenger. Pp. x., 311. 73 Woodcuts and 6 Plates. [With Systematic and Alphabetical Indexes.] 1889, 8vo. 15s.

Gigantic Land Tortoises (living and extinct) in the Collection of the British Museum. By Albert C. L. G. Günther, M.A., M.D., Ph.D., F.R.S. Pp. iv., 96. 55 Plates, and two Charts of the Aldabra group of Islands, north-west of Madagascar. [With a Systematic Synopsis of the Extinct and Living Gigantic Land Tortoises.] 1877, 4to. 1l. 10s.

Catalogue of the Lizards in the British Museum (Natural History). Second edition. By George Albert Boulenger:—

Vol. I. Geckonidæ, Eublepharidæ, Uroplatidæ, Pygopodidæ, Agamidæ. Pp. xii., 436. 32 Plates. [With Systematic and Alphabetical Indexes.] 1885, 8vo. 20s.

Vol. II. Iguanidæ, Xenosauridæ, Zonuridæ, Anguidæ, Anniellidæ, Helodermatidæ, Varanidæ, Xantusiidæ, Teiidæ, Amphisbænidæ. Pp. xiii., 497. 24 Plates. [With Systematic and Alphabetical Indexes.] 1885, 8vo. 20s.

Vol. III. Lacertidæ, Gerrhosauridæ, Scincidæ, Anelytropidæ, Dibamidæ, Chamæleontidæ. Pp. xii., 575. 40 Plates. [With a Systematic Index and an Alphabetical Index to the three volumes.] 1887, 8vo. 1l. 6s.

Catalogue of the Snakes in the British Museum (Natural History). Vol. I., containing the families Typhlopidæ, Glauconiidæ, Boidæ, Ilysiidæ, Uropeltidæ, Xenopeltidæ, and Colubridæ aglyphæ, part. By George Albert Boulenger. Pp. xiii., 448. 26 Woodcuts and 28 plates. [With Systematic and Alphabetical Indexes.] 1893, 8vo. 1l. 1s.

Catalogue of Colubrine Snakes in the Collection of the British Museum. By Dr. Albert Günther. Pp. xvi., 281. [With Geographical, Systematic, and Alphabetical Indexes.] 1858, 12mo. 4s.

BATRACHIANS.

Catalogue of the Batrachia Salientia in the Collection of the British Museum. By Dr. Albert Günther. Pp. xvi., 160. 12 Plates. [With Systematic, Geographical, and Alphabetical Indexes.] 1858, 8vo. 6s.

Catalogue of the Batrachia Salientia, s. Ecaudata, in the Collection of the British Museum. Second Edition. By George Albert Boulenger. Pp. xvi., 503. Woodcuts and 30 Plates. [With Systematic and Alphabetical Indexes.] 1882, 8vo. 1*l*. 10*s*.

Catalogue of the Batrachia Gradientia, s. Caudata, and Batrachia Apoda in the Collection of the British Museum. Second edition. By George Albert Boulenger. Pp. viii., 127. 9 Plates. [With Systematic and Alphabetical Indexes.] 1882, 8vo. 9*s*.

FISHES.

Catalogue of the Fishes in the Collection of the British Museum. By Dr. Albert Günther, F.R.S., &c. :—

Vol. III. Acanthopterygii (Gobiidæ, Discoboli, Oxudercidæ, Batrachidæ, Pediculati, Blenniidæ, Acanthoclinidæ, Comephoridæ, Trachypteridæ, Lophotidæ, Teuthididæ, Acronuridæ, Hoplognathidæ, Malacanthidæ, Nandidæ, Polycentridæ, Labyrinthici, Luciocephalidæ, Atherinidæ, Mugilidæ, Ophiocephalidæ, Trichonotidæ, Cepolidæ, Gobiesocidæ, Psychrolutidæ, Centriscidæ, Fistularidæ, Mastacembelidæ, Notacanthi). Pp. xxv., 586. Woodcuts. [With Systematic and Alphabetical Indexes, and a Systematic Synopsis of the families of the Acanthopterygian Fishes.] 1861, 8vo., 10*s*. 6*d*.

Vol. IV. Acanthopterygii pharyngognathi and Anacanthini. Pp. xxi., 534. [With Systematic and Alphabetical Indexes.] 1862, 8vo. 8*s*. 6*d*.

Vol. V. Physostomi (Siluridæ, Characinidæ, Haplochitonidæ, Sternoptychidæ, Scopelidæ, Stomiatidæ). Pp. xxii., 455. Woodcuts. [With Systematic and Alphabetical Indexes.] 1864, 8vo. 8*s*.

Vol. VII. Physostomi (Heterophygii; Cyprinidæ, Gonorhynchidæ, Hyodontidæ, Osteoglossidæ, Clupeidæ, Chirocentridæ, Alepocephalidæ, Notopteridæ, Halosauridæ). Pp. xx., 512. Woodcuts. [With Systematic and Alphabetical Indexes.] 1868, 8vo. 8*s*.

Vol. VIII. Physostomi (Gymnotidæ, Symbranchidæ, Muræmidæ, Pegasidæ), Lophobranchii, Plectognathi, Dipnoi, Ganoidei, Chondropterygii, Cyclostomata, Leptocardii. Pp. xxv., 549. [With Systematic and Alphabetical Indexes.] 1870, 8vo. 8*s*. 6*d*.

List of the Specimens of Fish in the Collection of the British Museum. Part I. Chondropterygii. By J. E. Gray. Pp. x., 160. 2 Plates. [With Systematic and Alphabetical Indexes.] 1851, 12mo. 3*s*.

Catalogue of Fish collected and described by Laurence Theodore Gronow, now in the British Museum. Pp. vii., 196. [With a Systematic Index.] 1854, 12mo. 3*s*. 6*d*.

Catalogue of Lophobranchiate Fish in the Collection of the British Museum. By J. J. Kaup, Ph.D., &c. Pp. iv., 80. 4 Plates. [With an Alphabetical Index.] 1856, 12mo. 2s.

MOLLUSCA.

Guide to the Systematic Distribution of Mollusca in the British Museum. Part I. By John Edward Gray, Ph.D., F.R.S., &c. Pp. xii., 230. 121 Woodcuts. 1857, 8vo. 5s.

Catalogue of the Collection of Mazatlan Shells in the British Museum, collected by Frederick Reigen. Described by Philip P. Carpenter. Pp. xvi., 552. 1857, 12mo. 8s.

List of Mollusca and Shells in the Collection of the British Museum, collected and described by MM. Eydoux and Souleyet in the "Voyage autour du Monde, exécuté pendant les années " 1836 et 1837, sur la Corvette 'La Bonite,'" and in the " Histoire naturelle des Mollusques Ptéropodes," Par MM. P. C. A. L. Rang et Souleyet. Pp. iv., 27. 1855, 12mo. 8d.

Catalogue of Pulmonata, or Air Breathing Mollusça, in the Collection of the British Museum. Part I. By Dr. Louis Pfeiffer. Pp. iv., 192. Woodcuts. 1855, 12mo. 2s. 6d.

Catalogue of the Auriculidæ, Proserpinidæ, and Truncatellidæ in the Collection of the British Museum. By Dr. Louis Pfeiffer. Pp. iv., 150. Woodcuts. 1857, 12mo. 1s. 9d.

List of the Mollusca in the Collection of the British Museum. By John Edward Gray, Ph.D., F.R.S., &c.
    Part I. Volutidæ. Pp. 23. 1855, 12mo. 6d.
    Part II. Olividæ. Pp. 41. 1865, 12mo. 1s.

Catalogue of the Conchifera, or Bivalve Shells, in the Collection of the British Museum. By M. Deshayes :—
    Part 1. Veneridæ, Cyprinidæ, Glauconomidæ, and Petricoladæ. Pp. iv., 216. 1853, 12mo. 3s.
    Part II. Petricoladæ (concluded); Corbiculadæ. Pp. 217-292. [With an Alphabetical Index to the two parts.] 1854, 12mo. 6d.

BRACHIOPODA.

Catalogue of Brachiopoda Ancylopoda or Lamp Shells in the Collection of the British Museum. [*Issued as* "Catalogue of the Mollusca, Part IV."] Pp. iv., 128. 25 Woodcuts. [With an Alphabetical Index.] 1853, 12mo. 3s.

POLYZOA.

Catalogue of Marine Polyzoa in the Collection of the British Museum. Part III. Cyclostomata. By George Busk, F.R.S. Pp. viii., 39, 38 Plates. [With a Systematic Index.] 1875, 8vo, 5s.

CRUSTACEA.

Catalogue of Crustacea in the Collection of the British Museum. Part I. Leucosiadæ. By Thomas Bell, V.P.R.S., Pres. L.S., &c. Pp. iv., 24. 1855, 8vo. 6d.

Catalogue of the Specimens of Amphipodous Crustacea in the Collection of the British Museum. By C. Spence Bate, F.R.S., &c. Pp. iv., 399. 58 Plates. [With an Alphabetical Index.] 1862, 8vo. 1l. 5s.

INSECTS

Coleopterous Insects.

Nomenclature of Coleopterous Insects in the Collection of the British Museum:—

> Part V. Cucujidæ, &c. By Frederick Smith. [*Also issued as* "List of the Coleopterous Insects. Part I."] Pp. 25. 1851, 12mo. 6d.
>
> Part VI. Passalidæ. By Frederick Smith. Pp. iv., 23. 1 Plate. [With Index.] 1852, 12mo. 8d.
>
> Part VII. Longicornia, I. By Adam White. Pp. iv., 174. 4 Plates. 1853, 12mo. 2s. 6d.
>
> Part VIII. Longicornia, II. By Adam White. Pp. 237. 6 Plates. 1855, 12mo. 3s. 6d.
>
> Part IX. Cassididæ. By Charles H. Boheman, Professor of Natural History, Stockholm. Pp. 225. [With Index.] 1856, 12mo. 3s.

Illustrations of Typical Specimens of Coleoptera in the Collection of the British Museum. Part I. Lycidæ. By Charles Owen Waterhouse. Pp. x., 83. 18 coloured Plates. [With Systematic and Alphabetical Indexes.] 1879, 8vo. 16s.

Catalogue of the Coleopterous Insects of Madeira in the Collection of the British Museum. By T. Vernon Wollaston, M.A., F.L.S. Pp. xvi., 234 : 1 plate. [With a Topographical Catalogue and an Alphabetical Index.] 1857, 8vo. 3s.

Catalogue of the Coleopterous Insects of the Canaries in the Collection of the British Museum. By T. Vernon Wollaston, M.A., F.L.S. Pp. xiii., 648. [With Topographical and Alphabetical Indexes.] 1864, 8vo. 10s. 6d.

Catalogue of Halticidæ in the Collection of the British Museum. By the Rev. Hamlet Clark, M.A., F.L.S. Physapodes and Œdipodes. Part I. Pp. xii., 301. Frontispiece and 9 Plates. 1860, 8vo. 7s.

Catalogue of Hispidæ in the Collection of the British Museum. By Joseph S. Baly, M.E.S., &c. Part I. Pp. x., 172. 9 Plates. [With an Alphabetical Index.] 1858, 8vo. 6s.

Hymenopterous Insects.

Catalogue of Hymenopterous Insects in the Collection of the British Museum. By Frederick Smith. 12mo.:—
  Part I. Andrenidæ and Apidæ. Pp. 197. 6 Plates. 1853, 2s. 6d.
  Part II. Apidæ. Pp. 199–465. 6 Plates. [With an Alphabetical Index.] 1854, 6s.
  Part III. Mutillidæ and Pompilidæ. Pp. 206. 6 Plates. 1855, 6s.
  Part IV. Sphegidæ, Larridæ, and Crabronidæ. Pp. 207–497. 6 Plates. [With an Alphabetical Index.] 1856, 6s.
  Part V. Vespidæ. Pp. 147. 6 Plates. [With an Alphabetical Index.] 1857, 6s.
  Part VI. Formicidæ. Pp. 216. 14 Plates. [With an Alphabetical Index.] 1858, 6s.
  Part VII. Dorylidæ and Thynnidæ. Pp. 76. 3 Plates. [With an Alphabetical Index.] 1859, 2s.

Descriptions of New Species of Hymenoptera in the Collection of the British Museum. By Frederick Smith. Pp. xxi., 240. [With Systematic and Alphabetical Indexes.] 1879, 8vo. 10s.

List of Hymenoptera, with descriptions and figures of the Typical Specimens in the British Museum. Vol. I., Tenthredinidæ and Siricidæ. By W. F. Kirby. Pp. xxviii., 450. 16 coloured Plates. [With Systematic and Alphabetical Indexes.] 1882, 8vo. 1l. 18s.

Dipterous Insects.

List of the Specimens of Dipterous Insects in the Collection of the British Museum. By Francis Walker, F.L.S. 12mo.:—
  Part II. Pp. 231–484. 1849. 3s. 6d.
  Part IV. Pp. 689–1172. [With an index to the four parts, and an Index of Donors.] 1849. 6s.
  Part V. Supplement I. Stratiomidæ, Xylophagidæ, and Tabanidæ. Pp. iv., 330. 2 Cuts. 1854. 4s. 6d.
  Part VI. Supplement II. Acroceridæ and part of the family Asilidæ. Pp. ii., 331–506. 8 Cuts. 1854. 3s.
  Part VII. Supplement III. Asilidæ. Pp. ii., 507–775. 1855. 3s. 6d.

Lepidopterous Insects.

Illustrations of Typical Specimens of Lepidoptera Heterocera in the Collection of the British Museum:—
  Part I. By Arthur Gardiner Butler. Pp. xiii., 62. 20 Coloured Plates. [With a Systematic Index.] 1877, 4to. 2l.

Illustrations of Typical Specimens of Lepidoptera Heterocera, &c.—*continued.*
> Part III. By Arthur Gardiner Butler. Pp. xviii., 82. 41–60 Coloured Plates. [With a Systematic Index.] 1879, 4to. 2*l.* 10*s.*
> Part V. By Arthur Gardiner Butler. Pp. xii., 74. 78–100 Coloured Plates. [With a Systematic Index.] 1881, 4to. 2*l.* 10*s.*
> Part VI. By Arthur Gardiner Butler. Pp. xv., 89. 101–120 Coloured Plates. [With a Systematic Index.] 1886, 4to. 2*l.* 4*s.*
> Part VII. By Arthur Gardiner Butler. Pp. iv., 124. 121–138 Coloured Plates. [With a Systematic List.] 1889, 4to. 2*l.*
> Part VIII. The Lepidoptera Heterocera of the Nilgiri District. By George Francis Hampson. Pp. iv., 144. 139–156 Coloured Plates. [With a Systematic List.] 1891, 4to. 2*l.*
> Part IX. The Macrolepidoptera Heterocera of Ceylon. By George Francis Hampson. Pp. v., 182. 157–176 Coloured Plates. [With a General Systematic List of Species collected in, or recorded from, Ceylon.] 1893, 4to. 2*l.* 2*s.*

Catalogue of Diurnal Lepidoptera of the family Satyridæ in the Collection of the British Museum. By Arthur Gardiner Butler, F.L.S., &c. Pp. vi., 211. 5 Plates. [With an Alphabetical Index.] 1868, 8vo. 5*s.* 6*d.*

Catalogue of Diurnal Lepidoptera described by Fabricius in the Collection of the British Museum. By Arthur Gardiner Butler, F.L.S., &c. Pp. iv., 303. 3 Plates. 1869, 8vo. 7*s.* 6*d.*

Specimen of a Catalogue of Lycænidæ in the British Museum. By W. C. Hewitson. Pp. 15. 8 Coloured Plates. 1862, 4to. 1*l.* 1*s.*

List of Lepidopterous Insects in the Collection of the British Museum. Part I. Papilionidæ. By G. R. Gray, F.L.S. Pp. 106. [With an Alphabetical Index.] 1856, 12mo. 2*s.*

List of the Specimens of Lepidopterous Insects in the Collection of the British Museum. By Francis Walker. 12mo.:—
> Part VI. Lepidoptera Heterocera. Pp. 1258–1507. 1855, 3*s.* 6*d.*
> Part XIX. Pyralides. Pp. 799–1036. [With an Alphabetical Index to Parts XVI.–XIX.] 1859, 3*s.* 6*d.*
> Part XX. Geometrites. Pp. 1–276. 1860, 4*s.*
> Part XXI. ——————— Pp. 277–498. 1860, 3*s.*
> Part XXII. ——————— Pp. 499–755. 1861, 3*s.* 6*d.*
> Part XXIII. ——————— Pp. 756–1020. 1861, 3*s.* 6*d.*
> Part XXIV. ——————— Pp. 1021–1280. 1862, 3*s.* 6*d.*
> Part XXV. ——————— Pp. 1281–1477. 1862, 3*s.*
> Part XXVI. ——————— Pp. 1478–1796. [With an Alphabetical Index to Parts XX.–XXVI.] 1862, 4*s.* 6*d.*

List of the Specimens of Lepidopterous Insects, &c.—*continued.*
  Part XXVII.   Crambites and Tortricites.   Pp. 1–286. 1863, 4s.
  Part XXVIII. Tortricites and Tineites.   Pp. 287–561. 1863, 4s.
  Part XXIX.   Tineites.   Pp. 562–835.   1864, 4s.
  Part XXX.   ———   Pp. 836–1096.   [With an Alphabetical Index to Parts XXVII.–XXX.]   1864, 4s.
  Part XXXI.   Supplement.   Pp. 1–321.   1864, 5s.
  Part XXXII.   ———   Part 2.   Pp. 322–706.   1865, 5s.
  Part XXXIII.   ———   Part 3.   Pp. 707–1120.   1865, 6s.
  Part XXXIV.   ———   Part 4.   Pp. 1121–1533.   1865, 5s. 6d.
  Part XXXV.   ———   Part 5.   Pp. 1534–2040.   [With an Alphabetical Index to Parts XXXI.–XXXV.]   1866, 7s.

Neuropterous Insects.

Catalogue of the Specimens of Neuropterous Insects in the Collection of the British Museum. By Francis Walker. 12mo. :—
  Part I.   (Phryganides—Perlides.)   Pp. iv., 192.   1852, 2s. 6d.
  Part II.   Sialidæ—Nemopterides.   Pp. ii., 193–476.   1853, 3s. 6d.
  Part III.   Termitidæ—Ephemeridæ.   Pp. ii., 477–585.   1853, 1s. 6d.
  Part IV.   Odonata.   Pp. ii., 587–658.   1853, 12mo.   1s.

Catalogue of the Specimens of Neuropterous Insects in the Collection of the British Museum. By Dr. H. Hagen. Part I. Termitina. Pp. 34. 1858, 12mo. 6d.

Orthopterous Insects.

Catalogue of Orthopterous Insects in the Collection of the British Museum. Part I. Phasmidæ. By John Obadiah Westwood, F.L.S., &c. Pp. 195. 48 Plates. [With an Alphabetical Index.] 1859, 4to. 3l.

Catalogue of the Specimens of Blattariæ in the Collection of the British Museum. By Francis Walker, F.L.S., &c. Pp. 239. [With an Alphabetical Index.] 1868, 8vo. 5s. 6d.

Catalogue of the Specimens of Dermaptera Saltatoria [Part I.] and Supplement to the Blattariæ in the Collection of the British Museum. Gryllidæ. Blattariæ. Locustidæ. By Francis Walker, F.L.S., &c. Pp. 224. [With an Alphabetical Index.] 1869, 8vo. 5s.

Catalogue of the Specimens of Dermaptera Saltatoria in the Collection of the British Museum. By Francis Walker, F.L.S., &c.—
    Part II. Locustidæ (continued). Pp. 225–423. [With an Alphabetical Index.] 1869, 8vo. 4s. 6d.
    Part III. Locustidæ (continued).—Acrididæ. Pp. 425–604. [With an Alphabetical Index.] 1870, 8vo. 4s.
    Part IV. Acrididæ (continued). Pp. 605–809. [With an Alphabetical Index.] 1870, 8vo. 6s.
    Part V. Tettigidæ.—Supplement to the Catalogue of Blattariæ.— Supplement to the Catalogue of Dermaptera Saltatoria (with remarks on the Geographical Distribution of Dermaptera). Pp. 811–850; 43; 116. [With Alphabetical Indexes.] 1870, 8vo. 6s.

Hemipterous Insects.

List of the Specimens of Hemipterous Insects in the Collection of the British Museum. By W. S. Dallas, F.L.S. :—
    Part I. Pp. 368. 11 Plates. 1851, 12mo. 7s.
    Part II. Pp. 369–590. Plates 12–15. 1852, 12mo. 4s.

Catalogue of the Specimens of Heteropterous Hemiptera in the Collection of the British Museum. By Francis Walker, F.L.S., &c. 8vo.:—
    Part I. Scutata. Pp. 240. 1867. 5s.
    Part II. Scutata (continued). Pp. 241–417. 1867. 4s.
    Part III. Pp. 418–599. [With an Alphabetical Index to Parts I., II., III., and a Summary of Geographical Distribution of the Species mentioned.] 1868. 4s. 6d.
    Part IV. Pp. 211. [Alphabetical Index.] 1871. 6s.
    Part V. Pp. 202. ——————— 1872. 5s.
    Part VI. Pp. 210. ——————— 1873. 5s.
    Part VII. Pp. 213. ——————— 1873. 6s.
    Part VIII. Pp. 220. ——————— 1873. 6s. 6d.

Homopterous Insects.

List of the Specimens of Homopterous Insects in the Collection of the British Museum. By Francis Walker. Supplement. Pp. ii., 369. [With an Alphabetical Index.] 1858, 12mo. 4s. 6d.

VERMES.

Catalogue of the Species of Entozoa, or Intestinal Worms, contained in the Collection of the British Museum. By Dr. Baird. Pp. iv., 132. 2 Plates. [With an Index of the Animals in which the Entozoa mentioned in the Catalogue are found; and an Index of Genera and Species.] 1853, 12mo. 2s.

ANTHOZOA.

Catalogue of Sea-pens or Pennatulariidæ in the Collection of the British Museum. By J. E. Gray, F.R.S., &c. Pp. iv., 40. 2 Woodcuts. 1870, 8vo. 1s. 6d.

Catalogue of Lithophytes or Stony Corals in the Collection of the British Museum. By J. E. Gray, F.R.S., &c. Pp. iv., 51. 14 Woodcuts. 1870, 8vo. 3s.

Catalogue of the Madreporarian Corals in the British Museum (Natural History). Vol. I. The Genus Madrepora. By George Brook. Pp. xi., 212. 35 Collotype Plates. [With Systematic and Alphabetical Indexes, Explanation of Plates, and a Preface by Dr. Günther.] 1893, 4to. 1l. 4s.

BRITISH ANIMALS.

Catalogue of British Birds in the Collection of the British Museum. By George Robert Gray, F.L.S., F.Z.S., &c. Pp. xii., 248. [With a List of Species.] 1863, 8vo. 3s. 6d.

Catalogue of British Hymenoptera in the Collection of the British Museum. Second edition. Part I. Andrenidæ and Apidæ. By Frederick Smith, M.E.S. New Issue. Pp. xi. 236. 11 Plates. [With Systematic and Alphabetical Indexes.] 1891, 8vo. 6s.

Catalogue of British Fossorial Hymenoptera, Formicidæ, and Vespidæ in the Collection of the British Museum. By Frederick Smith, V.P.E.S. Pp. 236. 6 Plates. [With an Alphabetical Index.] 1858, 12mo. 6s.

A Catalogue of the British Non-parasitical Worms in the Collection of the British Museum. By George Johnston, M.D., Edin., F.R.C.L. Ed., LL.D. Marischal Coll. Aberdeen, &c. Pp. 365. Woodcuts and 24 Plates. [With an Alphabetical Index.] 1865, 8vo. 7s.

Catalogue of the British Echinoderms in the British Museum (Natural History). By F. Jeffrey Bell, M.A. Pp. xvii. 202. Woodcuts and 16 Plates (2 coloured). [With Table of Contents, Tables of Distribution, Alphabetical Index, Description of the Plates, &c.] 1892, 8vo. 12s. 6d.

List of the Specimens of British Animals in the Collection of the British Museum; with Synonyma and References to figures. 12mo. :—

> Part V. Lepidoptera. By J. F. Stephens. 1850. 2nd Edition. By H. T. Stainton and E. Shepherd. Pp. iv. 224. 1856, 12mo. 1s. 9d.
>
> Part VII. Mollusca, Acephala, and Brachiopoda. By Dr. J. E. Gray. Pp. iv., 167. 1851, 12mo. 3s. 6d.
>
> Part XI. Anoplura or Parasitic Insects. By H. Denny. Pp. iv., 51. 1852, 1s.

List of the Specimens of British Animals, &c.—*continued.*
Part XIII. Nomenclature of Hymenoptera. By Frederick Smith. Pp. iv., 74. 1853, 12mo. 1s. 4d.
Part XIV. Nomenclature of Neuroptera. By Adam White. Pp. iv., 16. 1853, 12mo. 6d.
Part XV. Nomenclature of Diptera, I. By Adam White. Pp. iv., 42. 1853, 12mo. 1s.

PLANTS.

List of British Diatomaceæ in the Collection of the British Museum. By the Rev. W. Smith, F.L.S., &c. Pp. iv., 55. 1859, 12mo. 1s.

FOSSILS.

Catalogue of the Fossil Mammalia in the British Museum (Natural History). By Richard Lydekker, B.A., F.G.S.:—
Part I. Containing the Orders Primates, Chiroptera, Insectivora, Carnivora, and Rodentia. Pp. xxx., 268. 33 Woodcuts. [With Systematic and Alphabetical Indexes.] 1885, 8vo. 5s.
Part II. Containing the Order Ungulata, Suborder Artiodactyla. Pp. xxii., 324. 39 Woodcuts. [With Systematic and Alphabetical Indexes.] 1885, 8vo. 6s.
Part III. Containing the Order Ungulata, Suborders Perissodactyla, Toxodontia, Condylarthra, and Amblypoda. Pp. xvi., 186. 30 Woodcuts. [With Systematic Index, and Alphabetical Index of Genera and Species, including Synonyms.] 1886, 8vo. 4s.
Part IV. Containing the Order Ungulata, Suborder Proboscidea. Pp. xxiv., 235. 32 Woodcuts. [With Systematic Index, and Alphabetical Index of Genera and Species, including Synonyms.] 1886, 8vo. 5s.
Part V. Containing the Group Tillodontia, the Orders Sirenia, Cetacea, Edentata, Marsupialia, Monotremata, and Supplement. Pp. xxxv., 345. 55 Woodcuts. [With Systematic Index, and Alphabetical Index of Genera and Species, including Synonyms.] 1887, 8vo. 6s.

Catalogue of the Fossil Birds in the British Museum (Natural History). By Richard Lydekker, B.A. Pp. xxvii., 368. 75 Woodcuts. [With Systematic Index, and Alphabetical Index of Genera and Species, including Synonyms.] 1891, 8vo. 10s. 6d.

Catalogue of the Fossil Reptilia and Amphibia in the British Museum (Natural History). By Richard Lydekker, B.A., F.G.S.:—
Part I. Containing the Orders Ornithosauria, Crocodilia, Dinosauria, Squamata, Rhynchocephalia, and Proterosauria. Pp. xxviii., 309. 69 Woodcuts. [With Systematic Index, and Alphabetical Index of Genera and Species, including Synonyms.] 1888, 8vo. 7s. 6d.

Catalogue of the Fossil Reptilia and Amphibia—*continued.*
   Part II. Containing the Orders Ichthyopterygia and Sauropterygia. Pp. xxi., 307. 85 Woodcuts. [With Systematic Index, and Alphabetical Index of Genera and Species, including Synonyms.] 1889, 8vo. 7s. 6d.
   Part III. Containing the Order Chelonia. Pp. xviii., 239. 53 Woodcuts. [With Systematic Index, and Alphabetical Index of Genera and Species, including Synonyms.] 1889, 8vo. 7s. 6d.
   Part IV. Containing the Orders Anomodontia, Ecaudata, Caudata, and Labyrinthodontia; and Supplement. Pp. xxiii., 295. 66 Woodcuts. [With Systematic Index, Alphabetical Index of Genera and Species, including Synonyms, and Alphabetical Index of Genera and Species to the entire work.] 1890, 8vo. 7s. 6d.

Catalogue of the Fossil Fishes in the British Museum (Natural History). By Arthur Smith Woodward, F.G.S., F.Z.S.:—
   Part I. Containing the Elasmobranchii. Pp. xlvii., 474. 13 Woodcuts and 17 Plates. [With Alphabetical Index, and Systematic Index of Genera and Species.] 1889, 8vo. 21s.
   Part II. Containing the Elasmobranchii (Acanthodii), Holocephali, Ichthyodorulites, Ostracodermi, Dipnoi, and Teleostomi (Crossopterygii and Chondrostean Actinopterygii). Pp. xliv., 567. 58 Woodcuts and 16 Plates. [With Alphabetical Index, and Systematic Index of Genera and Species.] 1891, 8vo. 21s.

Systematic List of the Edwards Collection of British Oligocene and Eocene Mullusca in the British Museum (Natural History), with references to the type-specimens from similar horizons contained in other collections belonging to the Geological Department of the Museum. By Richard Bullen Newton, F.G.S. Pp. xxviii., 365. [With table of Families and Genera, Bibliography, Correlation-table, Appendix, and Alphabetical Index.] 1891, 8vo. 6s.

Catalogue of the Fossil Cephalopoda in the British Museum (Natural History). By Arthur H. Foord, F.G.S.:—
   Part I. Containing part of the Suborder Nautiloidea, consisting of the families Orthoceratidæ, Endoceratidæ, Actinoceratidæ, Gomphoceratidæ, Ascoceratidæ, Poterioceratidæ, Cyrtoceratidæ, and Supplement. Pp. xxxi., 344. 51 Woodcuts. [With Systematic Index, and Alphabetical Index of Genera and Species, including Synonyms.] 1888, 8vo. 10s. 6d.
   Part II. Containing the remainder of the Suborder Nautiloidea, consisting of the families Lituitidæ, Trochoceratidæ, Nautilidæ, and Supplement. Pp. xxviii., 407. 86 Woodcuts. [With Systematic Index, and Alphabetical Index of Genera and Species, including Synonyms.] 1891, 8vo. 15s.

A Catalogue of British Fossil Crustacea, with their Synonyms and the Range in Time of each Genus and Order. By Henry Woodward, F.R.S. Pp. xii., 155. [With an Alphabetical Index.] 1877, 8vo. 5s.

Catalogue of the Blastoidea in the Geological Department of the British Museum (Natural History), with an account of the morphology and systematic position of the group, and a revision of the genera and species. By Robert Etheridge, jun., of the Department of Geology, British Museum (Natural History), and P. Herbert Carpenter, D.Sc., F.R.S., F.L.S. (of Eton College). [With Preface by Dr. H. Woodward, Table of Contents, General Index, Explanations of the Plates, &c.] Pp. xv., 322. 20 Plates. 1886, 4to. 25s.

Catalogue of the Fossil Sponges in the Geological Department of the British Museum (Natural History). With descriptions of new and little known species. By George Jennings Hinde, Ph.D., F.G.S. Pp. viii., 248. 38 Plates. [With a Tabular List of Species, arranged in Zoological and Stratigraphical sequence, and an Alphabetical Index.] 1883, 4to. 1l. 10s.

Catalogue of the Fossil Foraminifera in the British Museum (Natural History). By Professor T. Rupert Jones, F.R.S., &c. Pp. xxiv., 100. [With Geographical and Alphabetical Indexes.] 1882, 8vo. 5s.

Catalogue of the Palæozoic Plants in the Department of Geology and Palæontology, British Museum (Natural History). By Robert Kidston, F.G.S. Pp. viii., 288. [With a list of works quoted, and an Index.] 1886, 8vo. 5s.

GUIDE-BOOKS.

(*To be obtained only at the Museum.*)

A General Guide to the British Museum (Natural History), Cromwell Road, London, S.W. [By W. H. Flower.] With 2 Plans, 2 views of the building, and an illustrated cover. Pp. 78. 1893, 8vo. 3d.

Guide to the Galleries of Mammalia (Mammalian, Osteological, Cetacean) in the Department of Zoology of the British Museum (Natural History). [By A. Günther.] 4th Edition. Pp. 126. 57 Woodcuts and 2 Plans. Index. 1892, 8vo. 6d.

Guide to the Galleries of Reptiles and Fishes in the Department of Zoology of the British Museum (Natural History). [By A. Günther.] 3rd Edition. Pp. iv. 119. 101 Woodcuts and 1 Plan. Index. 1893, 8vo. 6d.

Guide to the Shell and Starfish Galleries (Mollusca, Echinodermata, Vermes), in the Department of Zoology of the British Museum (Natural History). [By A. Günther.] 2nd Edition. Pp. iv., 74. 51 Woodcuts and 1 Plan. 1888, 8vo. 4d.

A Guide to the Exhibition Galleries of the Department of Geology and Palæontology in the British Museum (Natural History), Cromwell Road, London, S.W. [New Edition. By Henry Woodward.]—
    Part I. Fossil Mammals and Birds. Pp. xii., 103. 119 Woodcuts and 1 Plan. 1890, 8vo. 6d.
    Part II. Fossil Reptiles, Fishes, and Invertebrates. Pp. xii., 109. 94 Woodcuts and 1 Plan. 1890, 8vo. 6d.

Guide to the Collection of Fossil Fishes in the Department of Geology and Palæontology, British Museum (Natural History), Cromwell Road, South Kensington. [By Henry Woodward.] 2nd Edition. Pp. 51. 81 Woodcuts. Index. 1888, 8vo. 4d.

Guide to Sowerby's Models of British Fungi in the Department of Botany, British Museum (Natural History). By Worthington G. Smith, F.L.S. Pp. 82. 93 Woodcuts. With Table of Diagnostic Characters and Index. 1893, 8vo. 4d.

A Guide to the Mineral Gallery of the British Museum (Natural History). [By L. Fletcher.] Pp. 32. Plan. 1893, 8vo. 1d.

An Introduction to the Study of Minerals, with a Guide to the Mineral Gallery of the British Museum (Natural History), Cromwell Road, S.W. [By L. Fletcher.] Pp. 120. With numerous Diagrams, a Plan of the Mineral Gallery, and an Index. 1894, 8vo. 6d.

The Student's Index to the Collection of Minerals, British Museum (Natural History). New Edition. Pp. 32. With a Plan of the Mineral Gallery. 1893, 8vo. 2d.

An Introduction to the Study of Meteorites, with a List of the Meteorites represented in the Collection. [By L. Fletcher.] Pp. 91. [With a Plan of the Mineral Gallery, and an Index to the Meteorites represented in the Collection.] 1893, 8vo. 3d.

                                          W. H. FLOWER,
                                                  Director.

British Museum
    (Natural History),
        Cromwell Road,
            London, S.W.

    February 15th, 1894.

Printed in Great Britain
by Amazon.co.uk, Ltd.,
Marston Gate.

BRIT. MUS. N.H. Pl. XXXII.

P. Smit del. et lith. Mintern Bros. imp.

Uromastix loricatus.

Printed in Great Britain
by Amazon.co.uk, Ltd.,
Marston Gate.